Physical Chemistry

Developing a Dynamic Curriculum

Richard W. Schwenz, Editor
University of Northern Colorado

Robert J. Moore, Editor
Haverford College

Developed from a symposium sponsored
by the Divisions of Chemical Education
and Physical Chemistry
of the American Chemical Society
at the Fourth Chemical Congress of North America
(202nd National Meeting of the American Chemical Society)
New York, New York
August 25-30, 1991

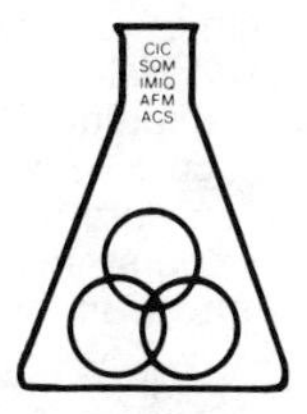

American Chemical Society, Washington, DC 1993

Library of Congress Cataloging-in-Publication Data

Physical chemistry: developing a dynamic curriculum / edited by Robert J. Moore, Richard W. Schwenz.

p. cm.

Includes bibliographical references and index.

ISBN 0–8412–2503–6

1. Chemistry, Physical and theoretical—Study and teaching (Higher)

I. Moore, Robert J., 1956– . II. Schwenz, Richard W. (Richard William), 1955–

QD455.5.P49 1992
541.3′071′1—dc20 92–35619
CIP

The paper used in this publication meets the minimum requirements of American National Standard for Information Sciences—Permanence of Paper for Printed Library Materials, ANSI Z39.48–1984. ∞

The citation of trade names and/or names of manufacturers in this publication is not to be construed as an endorsement or as approval by ACS of the commercial products or services referenced herein; nor should the mere reference herein to any drawing, specification, chemical process, or other data be regarded as a license or as a conveyance of any right or permission to the holder, reader, or any other person or corporation, to manufacture, reproduce, use, or sell any patented invention or copyrighted work that may in any way be related thereto. Registered names, trademarks, etc., used in this publication, even without specific indication thereof, are not to be considered unprotected by law.

PRINTED IN THE UNITED STATES OF AMERICA

1993 Advisory Board

Contents

Contributors

Marianne H. Begemann
Department of Chemistry
Vassar College
Poughkeepsie, NY 12601

Joseph J. BelBruno
Department of Chemistry
Dartmouth College
Hanover, NH 03755

S. Bluestone
Department of Chemistry
California State University, Fresno
Fresno, CA 93740

Franklin B. Brown
Science and Mathematics Division
Tallahassee Community College
Tallahassee, FL 32304–4052

Kristen Brubaker
Department of Chemistry
Lafayette College
Easton, PA 18042

Richard A. Butera
Department of Chemistry
University of Pittsburgh
Pittsburgh, PA 15260

Robert N. Compton
Department of Chemistry
University of Tennessee
Knoxville, TN 37996-1600

Andrew R. Cook
Department of Chemistry
Massachusetts Institute
of Technology
Cambridge, MA 02139

A. W. Czanderna
Measurements and
Characterization Branch
National Renewable Energy
Laboratory
Golden, CO 80401

Paul J. Dagdigian
Department of Chemistry
The Johns Hopkins University
Baltimore, MD 21218

Julio C. de Paula
Department of Chemistry
Haverford College
Haverford, PA 19041

M.A. Duncan
Department of Chemistry
University of Georgia
Athens, GA 30602

Charles S. Feigerle
Department of Chemistry
University of Tennessee
Knoxville, TN 37996-1600

Lee K. Fraiji
Department of Chemistry
Union College
Schenectady, NY 12308

Betty J. Gaffney
Department of Chemistry
The Johns Hopkins University
Baltimore, MD 21218

Matthew Gardner
Department of Chemistry
Haverford College
Haverford, PA 19041

George L. Hardgrove
Department of Chemistry
St. Olaf College
Northfield, MN 55057

David M. Hayes
Department of Chemistry
Union College
Schenectady, NY 12308

Patrick L. Holt
Department of Chemistry
Trinity University
715 Stadium Drive
San Antonio, TX 78212

Dennis Huang
Chemistry Department
Knox College
Galesburg, IL 61401

Virginia M. Indivero
Department of Chemistry
Swarthmore College
Swarthmore, PA 19081

Alpa J. Khandhar
Department of Chemistry
University of Pennsylvania
Philadelphia, PA 19104-6323

Roy W. King
Department of Chemistry
University of Florida
Gainesville, FL 32611

Robert G. Kooser
Chemistry Department
Knox College
Galesburg, IL 61401

Mark Ledeboer
Department of Chemistry
Lafayette College
Easton, PA 18042

Marsha I. Lester
Department of Chemistry
University of Pennsylvania
Philadelphia, PA 19104-6323

Jeffrey Lind
Department of Chemistry
Haverford College
Haverford, PA 19041

Andrew Loomis
Department of Physics
University of Northern Colorado
Greeley, CO 80639

Colin F. MacKay
Department of Chemistry
Haverford College
Haverford, PA 19041

Susan Mathison
Chemistry Department
Knox College
Galesburg, IL 61401

Gary L. Miessler
Department of Chemistry
St. Olaf College
Northfield, MN 55057

Richard S. Moog
Department of Chemistry
Franklin and Marshall College
Lancaster, PA 17604

Robert J. Moore
Department of Chemistry
Haverford College
Haverford, PA 19041

Ken Morton
Department of Chemistry
Carson-Newman College
Jefferson City, TN 37760

Keith A. Nelson
Department of Chemistry
Massachusetts Institute
of Technology
Cambridge, MA 02139

J.S. Pilgrim
Department of Chemistry
University of Georgia
Athens, GA 30602

William F. Polik
Department of Chemistry
Hope College
Holland, MI 49423

Arvind Rajpal
Chemistry Department
Knox College
Galesburg, IL 61401

J.E. Salcido
Department of Chemistry
University of Georgia
Athens, GA 30602

Richard W. Schwenz
Department of Chemistry and
Biochemistry
University of Northern Colorado
Greeley, CO 80639

Don Stedman
Department of Chemistry
University of Denver
Denver, CO 80208

Jack K. Steehler
Department of Chemistry
Roanoke College
Salem, VA 24153

J.I. Steinfeld
Department of Chemistry
Massachusetts Institute
of Technology
Cambridge, MA 02139

Thomas A. Stephenson
Department of Chemistry
Swarthmore College
Swarthmore, PA 19081

Donald A. Tarr
Department of Chemistry
St. Olaf College
Northfield, MN 55057

Thomas M. Ticich
Department of Chemistry
Gustavus Adolphus College
St. Peter, MN 56082

Jane F. Trinkle
Department of Chemistry
University of Pennsylvania
Philadelphia, PA 19104-6323

David A. Van Dyke
Department of Chemistry
University of Pennsylvania
Philidelphia, PA 19104-6323

Gerald R. Van Hecke
Department of Chemistry
Harvey Mudd College
Claremont, CA 91711

Valerie A. Walters
Department of Chemistry
Lafayette College
Easton, PA 18042

Weining Wang
Department of Chemistry
Massachusetts Institute
of Technology
Cambridge, MA 02139

T.C. Werner
Department of Chemistry
Union College
Schenectady, NY 12308

Kathryn R. Williams
Department of Chemistry
University of Florida
Gainesville, FL 32611

Preface

THE FIELD OF PHYSICAL CHEMISTRY has become increasingly instrument-oriented, involved in the details of atomic and molecular structure and dynamics. But too often today's physical chemistry lecture and laboratory courses are relics of the way physical chemistry used to be performed. Most of them lag behind the changes that have occurred in the field in the past 20 or more years. The slow rate of change may be due, at least in part, to the expensive equipment presently needed in the laboratory and to the relatively small number of students electing the course.

Students choose physical chemistry either because they are interested in understanding molecules and their interactions or because the course is required. The typical class includes future physical chemistry and chemistry graduate students, chemistry majors heading for industry, chemical engineers, pre-health majors, and (increasingly) teachers. All are taking the year-long physical chemistry program for their own reasons, and they come with widely varying expectations. This growing diversity among physical chemistry students generates a challenge for those presenting the lecture and laboratory courses. Each group is largely unsatisfied with the lecture material and the laboratory exercises. Many consider the course irrelevant to their present and future professional development. Both students and teachers need to feel the excitement of modern physical chemistry.

Two independent efforts to modernize the undergraduate physical chemistry curriculum generated this book. The chemistry departments of the 10 institutions that form the Pew Charitable Trust Mid-Atlantic Cluster (Bryn Mawr, Bucknell, Franklin and Marshall, Haverford, Lafayette, Muhlenberg, Pennsylvania, Princeton, Swarthmore, and Vassar) began a project in April 1988 that focused on modernization of the undergraduate physical chemistry laboratory. In the summers of 1990 and 1991, the National Science Foundation funded the Undergraduate Faculty Enhancement Workshops on Physical Chemistry Curriculum Development at the University of Northern Colorado. More than 35 scientists from institutions around the country participated in these workshops. Both of these projects dealt with the growing gap between undergraduate physical chemistry courses and the actual practice of physical chemistry. The materials they produced reflect the practice of contemporary physical chemistry.

This book continues their discussion of what materials should be stressed in the physical chemistry curriculum. Similar discussions are taking place about the "best" methods and materials for teaching introductory chemistry, but the movement does not yet seem to include the other chemical subdisciplines. The American Chemical Society has published two related volumes within the past 10 years: *Content of Undergraduate Physical Chemistry Courses* and *Essays in Physical Chemistry: A Sourcebook for Physical Chemistry Teachers*. Both of these publications, representing the best judgment of the committees that prepared them, recommend that physical chemistry be revised and make suggestions as to what material should be included. Many traditional topics such as electrochemistry, analytical applications of spectroscopy, and complex phase diagrams are missing. The suggestion that today's educators may find most startling is the proposed reduction of the mathematically rigorous treatment of quantum mechanics. Concrete examples illustrate how these ideas can be incorporated into lecture and laboratory courses, demonstrations, and homework exercises.

The first several chapters of this book present broadly organized essays on the content and presentation of various topics. Through the content, background, and references of these chapters an instructor can become acquainted with the relevant literature. Each chapter also contains a section on specific lecture materials (examples and demonstrations), specific problems that could be done outside of class, and general laboratory experiments appropriate for undergraduates who do not have access to research-level equipment.

This material is followed by chapters containing descriptions of tested modern experiments that are now in use at a variety of institutions across the country. Still other chapters describe a consortium approach to laboratory development and an innovative writing program for physical chemistry students.

Our hope is that all of these materials will help physical chemistry educators to improve and modernize both their lectures and their laboratory courses.

Acknowledgments

We thank the authors, who took the lead in the development of new laboratory experiments and lecture materials, and the reviewers, who gave selflessly of their time to evaluate and offer suggestions for improvement of the manuscripts. We also thank Regina Cody and L. Jewell Nicholls, the safety reviewers, and Maureen Matkovich, who compiled the safety information. We also acknowledge the Pew Charitable Trust and the National Science Foundation, Division of Undergraduate Education, Undergraduate Faculty Enhancement Program, for providing financial support for much of the work presented in this book. We thank our co-workers, who gave generously of their time and effort during this project. We especially thank Robert Coombe of the University of Denver for his hospitality during R. W. Schwenz's sabbatical, which made the completion of this book possible. Finally, we thank our wives and children for their support and patience while we were putting this project together. They are Mary Ellen Moore and Andy, Patti, Tom, and Brad Moore, and Lynn Geiger and Caroline and Robert Schwenz.

RICHARD W. SCHWENZ
University of Northern Colorado
Department of Chemistry and Biochemistry
Greeley, CO 80639

ROBERT J. MOORE
Haverford College
Haverford, PA 19041

July 1992

Laboratory Safety

DISCLAIMER

Safety information is included in each chapter as a precaution to the readers. Although the materials, safety information, and procedures contained in this book are believed to be reliable, they should serve only as a starting point for laboratory practices. They do not purport to specify minimal legal standards or to represent the policy of the American Chemical Society. No warranty, guarantee, or representation is made by the American Chemical Society, the authors, or the editors as to the accuracy or specificity of the information contained herein, and the American Chemical Society, the authors, and the editors assume no responsibility in connection therewith. The added safety information is intended to provide basic guidelines for safe practices. Therefore, it cannot be assumed that all necessary warnings or additional information and measures may not be required. Users of this book and the procedures contained herein should consult the primary literature and other sources of safe laboratory practices for more exhaustive information. As a starting point, a partial list of sources follows.

SOURCES OF INFORMATION

General Information on Chemicals

- *A Comprehensive Guide to the Hazardous Properties of Chemical Substances* by Pradyot Patniak. Van Nostrand Reinhold, 1992. This guide classifies chemicals by functional groups, structures, or general use, and the introductions to these chapters discuss the general hazards. Cross-indexed by name–CAS number and CAS number–name.
- *Catalog of Teratogenic Agents* by Thomas H. Shepard. 6th Ed. Johns Hopkins, 1989. Among books on this subject, this is perhaps the most thoughtful source list.
- *Handbook of Reactive Chemical Hazards* by Leslie Bretherick. 4th Ed. Butterworths, 1990. This book is a "must" for chemistry researchers or for anyone "experimenting" in the laboratory. It has a very useful format for determining the possible explosive consequences of mixing chemicals.

- *Fire Protection Guide to Hazardous Materials*. 10th Ed. National Fire Protection Association, 1991. Laboratory personnel are not the primary audience of NFPA publications, but this document has clear, concise, and easily accessed information on most commercial chemicals and a number of laboratory chemicals.
- *Safe Storage of Laboratory Chemicals* edited by David A. Pipitone. 2nd Ed. John Wiley & Sons, 1991. The scope of this book is far broader than the title suggests. Techniques for the proper dispensing of flammable solvents and spill procedures are discussed.

High-Pressure Systems

- *Handbook of Compressed Gases* by the Compressed Gas Association. 3rd Ed. Van Nostrand Reinhold, 1990. The primary coverage is bulk handling and storage of compressed gases; this book is not absolutely necessary for the average laboratory user.
- *Improving Safety in the Chemical Laboratory: A Practical Guide* edited by Jay A. Young. 2nd Ed. John Wiley & Sons, 1991. Section 11.8, "High Pressures and Compressed Gases", pp 184–192.
- *Chemical Safety Manual for Small Businesses* by the ACS Committee on Chemical Safety. 2nd Ed. American Chemical Society, 1992. "Compressed Gases", pp 42–44, contains almost everything one needs to know about handling compressed gas cylinders in the laboratory. Also see p 39, "Reduced Pressure Operations".
- *The Laboratory Handbook of Materials, Equipment, and Technique* by Gary S. Coyne. Prentice Hall, 1992. Chapter 5, "Compressed Gases", pp 217–244.
- *Standard on Fire Protection for Laboratories Using Chemicals* (NFPA 45). National Fire Protection Association, 1991. Especially relevant is Chapter 8, Section 8-2, "Compressed and Liquefied Gases in Cylinders".

Cryogens

- *Standard on Fire Protection for Laboratories Using Chemicals* (NFPA 45). National Fire Protection Association, 1991. Especially relevant is Chapter 8, Section 8-3, "Cryogenic Fluids".
- *Handbook of Compressed Gases* by the Compressed Gas Association. 3rd Ed. Van Nostrand Reinhold, 1990.
- "Cryogenic Fluids in the Laboratory" Data Sheet I–688–Rev. 86. National Safety Council, 1986.

- *Improving Safety in the Chemical Laboratory: A Practical Guide* edited by Jay A. Young. 2nd Ed. John Wiley & Sons, 1991. Section 11.11, "Low Temperature and Cryogenic Systems", pp 192–195.

Lasers

- *American National Standard for the Safe Use of Lasers*. ANSI Z136.1–1986. American National Standards Institute, 1986. Provides guidance for the safe use of lasers and laser systems.
- *Light, Lasers, and Synchrotron Radiation* edited by M. Grandolfo et al. Plenum Press, 1990. The chapter titled "Laser Safety Standards: Historical Development and Rationale" puts ANSI Z136.1–1986 in perspective.
- "Lasers—The Nonbeam Hazards" by R. James Rockwell, Jr. In *Lasers and Optronics*, August 1989, p 25.
- *Improving Safety in the Chemical Laboratory: A Practical Guide* edited by Jay A. Young. 2nd Ed. John Wiley & Sons, 1991. Section 11.3.5.1, "Laser Radiation", pp 177–178.

High Voltage

- "Electrical Hazards in the High Energy Laboratory" by Lloyd B. Gordon. In *IEEE Transactions on Education*, Vol. 34, No. 3, August 1991, pp 231–242. This article contains primarily physics laboratory safety; it gives excellent information on electrical hazards. Highly recommended.
- *Improving Safety in the Chemical Laboratory: A Practical Guide* edited by Jay A. Young. 2nd Ed. John Wiley & Sons, 1991. Section 11.5, "Electric Currents and Magnetic Fields", pp 181–182.

Vacuum Techniques

- *The Laboratory Handbook of Materials, Equipment, and Technique* by Gary S. Coyne. Prentice Hall, 1992. Chapter 7, "Vacuum Systems", pp 275–408.
- *Improving Safety in the Chemical Laboratory: A Practical Guide* edited by Jay A. Young. 2nd Ed. John Wiley & Sons, 1991. Section 11.7, "Vacuum and Dewar Flasks", p 183.

Compiled by Maureen Matkovich, American Chemical Society

The Laboratory Notebook

NOTEKEEPING IS AN ACQUIRED SKILL that can be of tremendous benefit in almost any career. Whether notes are written by hand, typed on a lap-top computer, or spoken into a recording device is immaterial: the same principles apply. Proper notekeeping should be a matter of habit, not a chore and, therefore, should be developed early in a student's academic experience, certainly before the student begins a scientific career. Good communication skills are important in every job, and notewriting is a fundamental personal communication skill.

The ACS Committee on Professional Training stated,[1] "Employers of chemists report to the Committee that a large fraction of baccalaureate chemists write and speak poorly." The Committee recommended, "Laboratory work should give students hands-on knowledge of chemistry and the self-confidence and competence to keep neat, complete, experimental records. Frequent exercises in writing and speaking, critically evaluated by the chemistry faculty, are an essential part of a sound program in chemistry."

Teachers often express concern that companies have specific rules for notekeeping. Specific notekeeping duties will of course be different at different companies; notekeeping duties within a company will vary at different jobs. However, the principles of good notekeeping for working efficiently and the legal requirements for protecting inventions are the same at any job. An understanding of these principles and a knowledge of how to implement them are precisely the sort of skills that students should have before they begin job-hunting.

To make notekeeping habitual, the student should be taking notes each time work is done in the lab or field. Students should feel comfortable with a notebook in hand at the start of every experiment, and they should feel just a bit "naked" without a notebook. In the same way in which they get used to taking notes in a lecture class, students should expect to write notes while doing an experiment. Too often, teaching assistants "reward" students who quickly finish the lab sessions by allowing them to leave as soon as the experiment is done; notemaking and writing lab reports are postponed until some later time. Such a practice reinforces several wrong attitudes including, "Finishing the work quickly is more important than doing the job completely" and "Writing down the work is a necessary evil, better left for later." Although lab grades are frequently based on lab reports, many industrial supervisors would argue that it is more important to teach respect for primary data.

A common myth is that scientific notekeeping ought to be done in the third person and in the passive voice: "It was observed that the mixture changed color after a few minutes had elapsed." According to this myth, such a tone lends an air of professionalism or objectivity to the writing. In fact, most major guidelines for authors of scientific publications stress that authors should use the active voice in the first person, when appropriate, to express an opinion or to state a fact. The same is true for notekeeping: "I saw the mixture turn blue after 10 minutes" leaves no doubt as to who did the work, what happened, and when it happened. This tone is vigorous, concise, and to the point. It is the tone of choice for basic descriptive writing, and it is the tone of choice for laboratory notekeeping.

Student notekeeping was the subject of several papers published in the *Journal of Chemical Education*[2-7] between 1925 and 1930. Precious little on the same subject has appeared[8,9] since that time. Even though some of these papers were written more than 50 years ago, they addressed some of the basic concerns that teachers face today—how to assess student performance, how to motivate better performance, and how to do these things efficiently. Today's teachers would be well advised to read these short, eloquent papers. Most of the papers conclude that the best system for assessing student performance is to look over the notebooks during the laboratory period while the students are working. This practice serves several purposes:

1. Students are aware that they must write notes during the laboratory investigation, not later. They begin to make notekeeping habitual.

2. Students get assistance and critical review of their work (immediate feedback), leading to improvement in their skill level throughout the semester. The teacher can get a much better impression of the student's attitude and level of achievement than if the notebooks are quickly skimmed once or twice a semester.

3. The teacher does not have a thick stack of notebooks to grade at the middle and end of the semester, and each student's work gets the attention it deserves at the critical time.

The single most important criterion for assessing student writing is simply, Does the notebook clearly tell what was done? Allowances can be made for the conditions under which the student's notekeeping is done: time constraints to finish the experiment or cooperation or the lack of it among lab partners, for example. Here, the teacher is grading the notes against an arbitrary, but achievable, standard. John T. Stock, formerly professor of analytical chemistry at the University of Connecticut, took one of his student's finest examples of a notebook and actually chained it to a lab bench for all to see! It became the standard by which all other student notebooks were judged.

Students can be given examples of how much detail to record, such as the following:

- Record all information necessary to unambiguously identify chemical reagents and other research materials, including the source (manufacturer), lot number, purity, and age or expiration date.
- Note whether water was distilled (singly or multiply?) or deionized (how many megaohms?). Was it stored before use? What is your check on its purity?
- When using an instrument, note the last calibration date. How do you know it is performing properly? When was the last time that oven or furnace temperature sensors were checked? What is the spectrometer's wavelength and resolution supposed to be?
- Use proper names for labware and vessels. Was the sample weighed in a boat, dish, crucible, beaker, or flask? What kind of flask? Record the materials that vessels are made of: Are your crucibles platinum, gold, ceramic, graphite, or something else? What procedures were used to clean and prepare glassware, mixers, reactors, or other vessels?
- In what sequence were the reagents mixed? Was A added to B or vice versa? How precisely were reagents measured? Did you use a pipet, buret, graduated cylinder, or automated dispenser? Was the balance readable to 0.01 gram or 0.00001 gram? How were materials heated? What was the heating rate? How were materials stirred? Gently or vigorously? Mechanically or by hand?
- If you are recording the color of a material, are you observing it under fluorescent, incandescent, or natural light? How long did it take to go from step A to step B? At what point did you stop working? Did the experiment sit unattended since last week?

Following are guidelines for evaluating laboratory notes. All of these questions should be answered in the affirmative if the notebooks are being kept properly.

- Are the notes in black, ball-point pen? If using a computer, are notes backed up? What provision is made to keep from overwriting notes once they are written?
- Is the handwriting (or font) clearly legible? Are numbers and symbols unambiguous? (Does a 7 look like a 2, a 1 like an "el", or an "oh" like a zero?)
- Is the table of contents up-to-date?
- Does each section have a clear, grammatical heading?

- Is each entry signed and dated unambiguously? (Tues. 9 October 1984, not 10-9-84).
- Is the work described completely so that it can be understood without additional explanation by the writer?
- Are ideas and observations entered immediately and directly into the book and not on scraps of paper that are transcribed later?

Experimental science, the subject of most laboratory notekeeping, is descriptive science. Communicating the results of scientific investigations, whether theoretical or experimental, requires strong skills of description. Writing notes strengthens the skills of reasoning and exposition. As a matter of philosophy, students should understand that chemists perform experiments and record and interpret the results of these experiments to understand chemistry. Merely filling in the blanks of an answer sheet is unsatisfying and, in the long run, unfair to the student.

Howard M. Kanare
author of *Writing the Laboratory Notebook*
Construction Technology Laboratories
5420 Old Orchard Road
Skokie, IL 60077

References

[1]"Undergraduate Professional Education in Chemistry: Guidelines and Evaluation Procedures;" American Chemical Society Committee on Professional Training; fall 1983.
[2]Gould, H. W. *J. Chem. Ed.* **1927,** *4(7),* 890.
[3]Mortensen, J. C. *J. Chem. Ed.* **1927,** *4(7),* 892.
[4]Bowers, W. G. *J. Chem. Ed.* **1926,** *3(4),* 419.
[5]Stubbs, M. F. *J. Chem. Ed.* **1926,** *3(3),* 296.
[6]Walker, W. O. *J. Chem. Ed.* **1925,** *2(6),* 489.
[7]Graham, H. C. *J. Chem. Ed.* **1930,** *7(5),* 1122.
[8]Hancock, C. K. *J. Chem. Ed.* **1954,** *31(8),* 433.
[9]Wilson, L. R. *J. Chem. Ed.* **1969,** *46(7),* 477.

Today's Physical Chemistry Classroom

CHAPTER 1

Computational Chemistry in the Physical Chemistry Laboratory: Ab Initio Molecular Orbital Calculations

Franklin B. Brown

Historically, most fundamental research in many of the physical sciences could be categorized as either theoretical or experimental. Often the theoretician attempted to predict or explain experimental observations by constructing models and solving the resulting mathematical equations. However, in many instances the systems that the theoreticians could investigate and solve fully were much smaller or more "idealized" than the "real" chemical systems that the experimentalist could observe. In 1929, P. A. M. Dirac noted that:

> "The underlying physical laws necessary for the mathematical theory of a large part of physics and the whole of chemistry are thus completely known, and the difficulty is only that the exact application of these laws leads to equations much too complicated to be soluble."*(1)*

In the past decade, scientists have had an unparalleled access to computers ranging from large supercomputers with very fast central processing units (CPU's) and large memories to desktop workstations with powerful CPU's and graphics capabilities to laptop microcomputers. With this access, a third area of fundamental research has developed and been recognized: computational science. In contrast to using experimental equipment or pencil and paper, the computational scientist uses one or more computers and a variety of computational algorithms to solve complex sets of equations that result from modelling physical systems. Today, computational chemists can tackle problems of interest to experimental chemists. For example many pharmaceutical and chemical companies own large computers and employ computational chemists. D. A. Dixon of Du Pont has stated that:

> "Numerical simulation is now becoming a trusted partner with experiment. Simulations ... can replace experiments [that] cannot be done because of cost or experimental difficulty."*(2)*

Computational chemistry can be very useful especially with respect to making predictions and investigating chemical systems that are difficult to study experimentally. Specifically, reasonably reliable calculations can be done on chemical species that are highly toxic, short-lived, highly reactive, or in an excited state almost as easily as calculations can be performed on stable molecules in their groundstates. (However, it should be noted that often more sophisticated techniques are required to achieve highly accurate results for molecules in excited states than in their groundstate.) As computers become both faster and less expensive, and are found increasingly in chemists' laboratories, various types of computational

2503–6/93/0002$06.00/0

investigations of chemical systems will be performed by chemists trained in subdisciplines other than computational and theoretical chemistry.

In light of these observations, it is imperative that the undergraduate chemistry curriculum introduce students to computational chemistry. In general, today's college students are computer literate. Thus, chemistry instructors can focus their instruction on how to use computational chemistry software and analyze the data resulting from calculations. Additionally, with computer graphics and visualization, these exercises can complement lecture material concerning a variety of concepts including, for example, molecular bonding, molecular structure, vibrational modes, reaction mechanisms, reaction rates, oscillating chemical reactions, and protein or macromolecular structure and dynamics. Finally, in many instances, undergraduate students can utilize these computer codes in various research projects and obtain data which is suitable for publication.

Recently, Handy and Colwell *(3-4)* developed the MICROMOL codes which run on IBM and IBM-compatible microcomputers. These codes, which have been derived from the CADPAC codes *(5)* allow the user to perform ab initio electronic structure calculations on a variety of small molecular systems. Specifically, using MICROMOL, one can calculate molecular orbital wavefunctions and energies based upon the Hartree-Fock self-consistent-field (HFSCF) method. Additionally, these codes can be used to calculate a variety of molecular properties including, for example, dipole moments, energy-optimized molecular structures, and vibrational frequencies. Two additional features of the complete MICROMOL package include a TUTOR module which allows the user to interactively set up the data for the ab initio calculation and a DRAW program which can be used to graphically display molecular orbitals, wavefunctions, electron densities, and normal mode vibrations.

During the last two summers, this author has introduced MICROMOL to physical chemistry instructors for incorporation in their physical chemistry laboratory curriculum. The response to this program has been very positive. In this paper, the MICROMOL codes are described and a brief synopsis of the theory, necessary to intelligently use the codes, is presented. Finally, a discussion of ideas concerning the use of MICROMOL in the physical chemistry laboratory curriculum in presented.

Theory

In this section, a brief synopsis of the background theory that is necessary to understand and interpret the output of the MICROMOL codes is presented. Detailed derivations of the theory and computational algorithms can be found in a number of monographs and papers which have been listed in Appendix A. Since the MICROMOL codes can only calculate HFSCF wavefunctions and their associated properties, the following theoretical discussion will be largely limited in scope to this method, although a brief description of correlated methods will be mentioned later in this section.

For purposes of computationally investigating the stationary states of atomic and molecular systems, one can construct wavefunctions, Ψ, which are solutions of the time-independent Schroedinger equation *(6)*:

$$\hat{H} \Psi = E \Psi \qquad (1)$$

where $\hat{H}$ is the quantum mechanical Hamiltonian operator for the system and E is the total energy of the system. Various theoretical models and methods differ in both the approximations and computational algorithms used in solving equation 1, as well as in the assumed mathematical form of the wavefunction Ψ.

Consider a molecular system consisting of N nuclei and n electrons. The nonrelativistic Hamiltonian operator for this system can be expressed as:

$$\hat{H} = \sum_{A=1}^{N} \hat{T}_A + \sum_{A<B}^{N} \hat{V}_{AB} + \sum_{i}^{n} \hat{T}_i + \sum_{A}^{N} \sum_{i}^{n} \hat{V}_{Ai} + \sum_{i<j}^{n} \hat{V}_{ij} \tag{2}$$

where the first and third sums are the kinetic energy operators, $\hat{T}$, for the nuclei and electrons, respectively. The second sum determines the nuclear-nuclear repulsion energy and the fourth sum gives the nuclear-electronic attraction energy. Finally, the last sum is the electron-electron repulsion term. The wavefunction Ψ (**R**, **r**) which is an eigenfunction of the $\hat{H}$ operator, is a function of both the nuclear coordinates, **R**, and the electronic coordinates, **r**.

For most molecular systems, one separates the motion of the nuclei from that of the electrons using the Born-Oppenheimer Approximation *(7)*. In this approximation, the electrons, which are very light and move rapidly compared to the heavy nuclei, are assumed to adjust essentially instantaneously to any changes in the positions of the nuclei. This allows one to rewrite, Ψ (**R**, **r**) as a product of two functions:

$$\Psi(\mathbf{R}, \mathbf{r}) = \chi(\mathbf{R})\, \psi_{el}(\mathbf{r}; \mathbf{R}) \tag{3}$$

where χ (**R**) describes the nuclear motion and ψ_{el} (**r**; **R**) describes the electronic motion. Note that the nuclear coordinates are parameters in ψ_{el}. In the "clamped-nuclei" approximation (no nuclear motion) the Schroedinger equation reduces to

$$\hat{H}_{el}(\mathbf{r};\mathbf{R})\, \psi_{el}(\mathbf{r};\mathbf{R}) = E_{el}(\mathbf{R})\, \psi_{el}(\mathbf{r};\mathbf{R}) \tag{4}$$

where $\hat{H}_{el}$ only includes the last three terms in equation 2 and E_{el} is the portion of the energy involved electrons. Furthermore the total energy of the molecular system in the "clamped-nuclei" approximation is given by

$$E(\mathbf{R}) = E_{el}(\mathbf{R}) + V_{nuc}(\mathbf{R}) \tag{5}$$

where V_{nuc} is the nuclear repulsion energy - the second sum in equation 2.

In the HFSCF method, the electronic wavefunction ψ_{el} is approximated by ψ_{HFSCF} which is a spin-adapted, antisymmetrized (by $\hat{A}$) product of one electron functions called spin orbitals:

$$\psi_{HFSCF}(\mathbf{r};\mathbf{R}) = \hat{A} \prod_{i} \phi_i(\mathbf{r}_i;\mathbf{R}) \tag{6}$$

The wavefunction is spin-adapted so that it is an eigenfunction of the square of the total spin operator for the system. It is also antisymmetrized so that it satisfies the Pauli exclusion principle for fermion systems *(8)*. Generally, this type of wavefunction is referred to as a Slater Determinant (SD) *(9)*. Each spin orbital ϕ is a product of a spatial function and spin function. (Note that in this equation, the vector $\mathbf{r}_i$ is used to denote the spatial coordinates of the ith electron as well as its spin: ± 1/2.) In most HFSCF calculations, the spatial functions corresponding to a pair of spin orbitals - one with α and one with β spin, are constrained to be identical. As a result, for example, both electrons in the $1\sigma_g$ orbital in H_2 have the same spatial functions. HFSCF wavefunctions with this constraint are called restricted HFSCF wavefunctions. For a closed shell system containing n electrons, one needs only to determine $n/2$ unique spatial orbitals.

Using the variational principle, the exact groundstate energy, E, of any system satisfies the equation:

$$E \leq \langle \Phi_{el} | \hat{H}_{el} | \Phi_{el} \rangle \, / \, \langle \Phi_{el} \, | \, \Phi_{el} \rangle \tag{7}$$

where Φ_{el} is a well-behaved trial wavefunction. The equality holds in equation 7 when the trial wavefunction Φ_{el} is identical to the exact wavefunction ψ_{el}. In theory, one approximates E and ψ_{el} by minimizing the functional on the right-hand-side of equation 7 with respect to varying the trial wavefunction Φ_{el}.

In most algorithms, the variation of the trial wavefunction is achieved by incorporating parameters in Φ_{el} and minimizing the corresponding functional in equation 7 with respect to these parameters. Specifically, in the linear-combination of atomic orbital (LCAO) approach, each of the spin orbitals χ_i is taken to be a linear combination of functions that are usually centered at the nuclei of the atoms in molecular system *(10)*. That is

$$\Phi_i\,(\mathbf{r}) \quad = \quad \sum_{p=1}^{m} c_{ip}\, \chi_p\,(\mathbf{r}) \tag{8}$$

where c_{ip} is a variational parameter known as the molecular orbital expansion coefficient for the ith orbital and the pth basis function χ_p. $\mathbf{r}$ denotes the coordinates of a single electron. The complete set of expansion functions $\{\chi\}$ used in a calculation is called the basis set and ultimately defines the mathematical function space that is available for describing the electronic wavefunction.

Consider a closed shell system containing n electrons and utilizing a basis set containing m, real (not complex) basis functions. When one substitutes equation 8 into equation 6 and evaluates $\langle \Phi_{el} | \hat{H} | \Phi_{el} \rangle$, the corresponding energy is given by

$$E = \sum_{i}^{n/2} 2h_{ii} + \sum_{i}^{n/2} \sum_{j}^{n/2} (2J_{ij} - K_{ij}) \tag{9}$$

where h_{ii} is the one-electron operator and is given by

$$h_{ii} = \int \phi_i\,(\mathbf{r}_1)\,(\hat{T}_1 + \hat{V}_{N1})\,\phi_i\,(\mathbf{r}_1)\,d\tau_1 \tag{10}$$

$\hat{T}_1$ is the kinetic energy operator for a single electron and $\hat{V}_{N1}$ is the nuclear-electron attraction potential between electron 1 and all of the nuclei. The coulomb integral J_{ij} is given by

$$J_{ij} = \int\int \phi_i\,(\mathbf{r}_1)\,\phi_j\,(\mathbf{r}_2)\,(1/\mathbf{r}_{12})\,\phi_i\,(\mathbf{r}_1)\,\phi_j\,(\mathbf{r}_2)\,d\tau_1\,d\tau_2 \tag{11}$$

and corresponds classically to the coulombic repulsion of the two electrons in orbitals i and j. Finally, the exchange integral K_{ij}, which does not have a classical analogue, is given by

$$K_{ij} = \int\int \phi_i(\mathbf{r}_1)\, \phi_j(\mathbf{r}_2)\, (1/r_{12})\, \phi_i(\mathbf{r}_2)\, \phi_j(\mathbf{r}_1)\, d\tau_1\, d\tau_2 \qquad (12)$$

When applying the minimization condition of the variational principle, the additional constraint that the orbitals remain mutually orthonormal, namely:

$$\int \phi_i(\mathbf{r})\, \phi_j(\mathbf{r})\, d\tau = \delta_{ij} \qquad (13)$$

where the Kronecker delta, $\delta_{ij} = 1$ if i equals j and 0 if i does not equal j, is incorporated. This results in a set of pseudoeigenvalue equations, which are known as the Hartree-Fock equations *(10)*. Each equation has the general form:

$$\hat{F}_i(\mathbf{r})\, \phi_i(\mathbf{r}) = \epsilon_i\, \phi_i(\mathbf{r}) \qquad (14)$$

where ϵ_i is the orbital energy of the ith orbital and $\hat{F}_i$ is the corresponding Fock operator. In practice, the Fock operator is projected into the mathematical function space spanned by the basis set and the Fock operator is represented as a matrix whose individual elements are linear combinations of one- and two-electron integrals: h_{ii}, J_{ij} and K_{ij}.

The Hartree-Fock equations are a set of integro-differential equations and therefore must be solved iteratively. This can easily be observed when one considers that equation 14 is used to determine the ith orbital; however the Fock operator, $\hat{F}_i$ requires knowledge of the other orbitals through the coulomb and exchange integrals. Thus one assumes an initial set of orbitals $\{\phi\}^0$. The $\hat{F}_i$'s are constructed and a new set of orbitals $\{\phi\}^1$ are calculated. This allows one to construct new $\hat{F}_i$'s and the process is iterated to convergence. The final energy of the converged wavefunction is the lowest energy that can be produced from a wavefunction consisting of a single antisymmetrized product of spinorbitals within the mathematical function space spanned by the basis set. If the basis is increased to be mathematically complete over all of the function space, the resulting HFSCF wavefunction and energy corresponds to the Hartree-Fock limit.

The HFSCF method is essentially an independent particle method inasmuch as the orbital for a given electron is determined independently of the instantaneous motion of the remaining $n - 1$ electrons. The electron–electron repulsion enters the calculation through the coulomb and exchange integrals and contributes to a field constructed from the averaged motions of the $n - 1$ electrons. Furthermore, the electrons are constrained to occupy a single molecular orbital. In practice the HFSCF wavefunction cannot describe electronic rearrangements within degenerate or nearly degenerate orbitals. For example, in describing the molecular dissociation of H_2 to two H atoms, the $1\sigma_g$ and $1\sigma_u$ orbitals become degenerate at large H–H internuclear distances as they correspond to the atomic 1s orbitals of the isolated H atoms. The groundstate HFSCF wavefunction for H_2 restricts the two electrons to be in the $1\sigma_g$ orbital. This is usually an adequate description of molecules that have geometries near their equilibrium geometries; but this is often a poor description at large molecular displacements from equilibrium. As a result, dissociation energies, D_e, calculated at the HFSCF level are not reliable. In conclusion, the

HFSCF gives a very reasonable wavefunction and energy for systems which can be described by a single product of orbitals, and usually, such a wavefunction is reliable near the equilibrium geometries of closed shell molecules.

At this point, it is appropriate to digress briefly and mention ab initio computational methods that go beyond the HFSCF method and are suitable for describing molecular systems that have geometries significantly removed from equilibrium. The contribution to the energy that is missing from the HFSCF calculation is the correlation energy which is defined as the difference between the HFSCF energy and the exact solution to the nonrelativistic Schroedinger equation*(11)*. This correlation energy contains two contributions. The first is from the rearrangement of electrons within nearly degenerate orbitals and is required for adequately describing the dissociation of chemical bonds. The second contribution to the correlation energy comes from the instantaneous correlated motion of the electrons. These effects can be included in the wavefunction by including additional orbital occupancies (configurations) in the wavefunction. Thus, the wavefunction becomes a linear combination of SD's and the SD expansion coefficients become additional variational parameters:

$$\psi_{el} = \sum_j C_j \Phi_j \tag{15}$$

In the multiconfiguration self-consistent-field method, both the molecular orbital coefficients and the SD expansion coefficients are varied. In the configuration interaction method, the molecular orbital coefficients are held constant as determined in a previous calculation, and the expansion coefficients are optimized. In short, various methods for determining correlated energies and wavefunctions differ in how the expansion coefficients are optimized. Though the total correlation energy determined from these calculations is a small fraction of the total energy of most systems, it is of the same order of magnitude as bond energies and thus is significant when trying to describe bond dissociation. These calculations require significantly more computational effort than HFSCF calculations and thus are not included in most microcomputer packages.

As mentioned above, the choice of the set of basis functions $\{\chi\}$ ultimately limits the mathematical space that is available to describe the electrons in the molecular system. Early computational chemistry codes employed Slater-type-orbitals (STO) *(12)* which are atom-centered, hydrogenic functions of the form:

$$\chi_{nlm}(r, \theta, \phi) = N r^{n-1} \exp(-\zeta r) Y_{lm}(\theta, \phi) \tag{16}$$

where n, l, and m are the principal, azimuthal, and magnetic quantum numbers, respectively. N is a normalization constant; ζ is a parameter which is called the orbital exponent and $Y_{lm}(\theta, \phi)$ is a spherical harmonic. While the STO basis functions can yield very good wavefunctions and properties, the evaluation of the three- and four-center coulomb and exchange integrals is computationally quite demanding. (These two types of integrals contain four basis functions and if each of the basis functions is on a different center, a four-center integral arises.) Thus, most programs utilize the cartesian Gaussian-type orbital (GTO) *(13-14)* which has the general form:

$$\chi_{lmn}(x, y, z) = N x^l y^m z^n \exp(-\alpha r^2) \tag{17}$$

where N is again a normalization constant and α is the orbital exponent. The sum of the positive integers, l, m, and n is the azimuthal quantum number. Thus, for a s function (s orbital), these integers are all zero, while for a p_x function, l will be one and n and m will be zero. The major advantage of using GTO's is that they have the property that the product of two GTO functions on different centers is equal to a third GTO function on another center. Thus, all four-center integrals can easily be reduced to two-center integrals which can be quickly evaluated. The disadvantage of using GTO's compared with STO's is that GTO's do not behave correctly as $r \rightarrow 0$ nor asymptotically as $r \rightarrow \infty$. Thus larger GTO basis sets are required to achieve the same accuracy as a given STO basis set. Nonetheless, with current computers and software, it is more efficient to use a larger GTO basis set than use a smaller STO basis set.

With respect to the computational cost of a HFSCF calculation, the majority of the CPU time is spent evaluating the two-electron integrals. The number of these integrals is approximately $m^4/8$ where m is again the number of basis functions. In solving the Hartree-Fock equations, the computational time is proportional to m^3. From a chemical viewpoint, the inner shell electrons of the atoms are not involved in bonding to any great extent and, for example, the carbon 1s orbital looks about the same regardless of whether carbon is in methane or benzene. In light of this, contracted basis sets are often employed. A contracted basis function is itself a linear combination of atom centered basis functions called primitive functions. The expansion coefficients of the primitive functions have usually been determined by HFSCF calculations on the atom. These coefficients are held constant during HFSCF calculations on molecules containing the atom. By using contracted basis sets, the number of variational parameters in the molecular calculation is reduced without significantly reducing the function space spanned by the GTO's.

The selection of a reliable basis set for HFSCF calculations on small systems is not trivial. However, there is a large literature which gives many different basis sets and analyzes the reliability of various basis sets *(15)*. There are two major sources of basis sets that are used extensively in the literature. The first and possibly the most widely used are Pople's basis sets*(16)*. Additionally, Dunning *(17)* has developed basis sets that have often been employed. One final note concerning basis sets. They are often classified as being single zeta or double zeta and/or containing polarization and/or diffuse functions. Single zeta refers to a single basis function for each occupied atomic orbital while a double zeta basis set contains two basis functions for each occupied orbital. Furthermore, to describe the distortion of atomic orbitals in the molecular environment, functions with a higher azimuthal quantum number are added and are referred to as polarization functions. These include p orbitals for H and He and d orbitals for second row atoms. Finally, to describe anions or the electron density far from the nucleus, diffuse functions which have very small orbital exponents are required. A typical basis set that is suitable for many small molecules is double zeta with polarization. If there is an art to performing HFSCF calculations, much of it is associated with the choice of a basis set. One wants a basis set that is capable of describing the molecular system, but does not want it to be any larger than necessary since the computational time is dependent on its size to the fourth power.

One important question which ab initio calculations can often address is the lowest energy geometry of a molecule. All stable geometries of small molecules occur where the gradient of the energy, with respect to the nuclear coordinates, is zero. The lowest energy minimum corresponds to the global minimum of the energy whereas the other minima are local. Many ab initio programs have subroutines that analytically determine the energy gradient. This gradient can also be used to determine optimized or stable geometries using various optimization algorithms*(18)*. The simplest of these methods is the Method of Steepest Descent in which the next geometry, in an optimization, is chosen along the direction of the negative of the gradient at the current geometry. While this method is relatively simple, it has a relatively small radius of convergence.

Finally, the second derivative matrix of the energy with respect to nuclear coordinates, which is the Hessian matrix, can be used, along with the masses of the atoms in the molecule, to determine vibrational frequencies and normal modes. While some programs calculate the Hessian matrix analytically, many codes including MICROMOL, determine it by numerical differencing of the gradient vectors.

In conclusion, the steps necessary to initiate an ab initio HFSCF calculation include the following: (1) Choose the molecular system that is to be studied. (2) Choose a geometry of the molecule. (3) Choose the basis set for each of the atoms in the molecule. (4) Choose the type of calculation (i.e. single geometry, geometry optimization, vibrational frequency, etc.). (5) Run the HFSCF code.

The MICROMOL Package

One of the currently available ab initio HFSCF computer codes that runs on a microcomputer is MICROMOL. As mentioned in the introduction, Colwell and Handy *(3-4)* developed this code by adapting CADPAC *(5)* to run on a microcomputer. (This code is available from S. Colwell for a nominal fee.) The complete MICROMOL package consists of three modules which have been named TUTOR, MICROMOL, and DRAW. The major module is MICROMOL which calculates HFSCF energies and wavefunctions using Gaussian basis sets containing s, p, and d GTO's. Additionally, MICROMOL can be used to calculate gradients analytically as well as to construct the Hessian matrix which is used to determine the vibrational frequencies and normal modes. Currently, MICROMOL is limited to 12 atoms, 63 basis functions, and 110 Gaussian primitive functions. In general, these limitations are quite reasonable considering the CPU time needed for larger calculations.

One of the major strengths of the entire MICROMOL package is that undergraduate students can learn to utilize the code very quickly and calculate meaningful numbers before understanding all of the theoretical underpinnings of the code. This is due in large part to the TUTOR module. This module, which is written in PASCAL, leads the beginning user through the steps needed to set up the input for MICROMOL. This is done interactively with appropriate optional online documentation which explains most of the choices that one needs to make. (There is also a manual that explains in detail all of the options available for MICROMOL.) Some of the options through which TUTOR leads the user, include specifying the geometry of the molecular system, choosing an appropriate basis set, and choosing the type of calculation to be performed. This module has a library containing several of the more popular small basis sets and thus the user does not have to type in the details. The output from TUTOR is an ASCII file which contains the input to MICROMOL. As the user becomes more familiar with the entire package, it becomes faster to simply use an editor to create new input files or modify existing input files rather than use TUTOR.

After TUTOR has created an appropriate input file which MICROMOL expects to be named SOURCE.DAT, MICROMOL can be run. Depending on the molecular system, the details of the calculation, and the computer hardware characteristics, this may require from a few seconds to several days. Reasonable calculations on H_2 take a few seconds while a complete force constant calculation on benzene may take in excess of 24 hours on an IBM AT compatible.

After MICROMOL has run, its output files can be utilized by the third module DRAW to graphically display various quantities derived from the calculation. Specifically, DRAW can be used to view wavefunctions and electron densities of the individual orbitals or the entire system from a variety of 3-D perspectives as well as 2-D contour plots. Also, this module can provide animation of the vibrational modes of a molecular system. From an educational perspective, this is the highlight of the entire package since many concepts introduced in the standard freshman general chemistry course as well as in the undergraduate physical chemistry course can be investigated by the students and observed graphically. For example, one can easily distinguish bonding and antibonding molecular orbitals as well as nonbonding atomic orbitals. Additionally, one can observe the difference between a plot of an orbital or wavefunction and an electron density. Finally, and perhaps most importantly, the students can proceed to investigate other interesting molecular systems on their own. In fact, Colwell and Handy have published a paper presenting data on a van der Waals system*(3)*. Thus one can use this program in an undergraduate research project and produce publishable results. However, it would probably be more efficient to introduce the student to ab initio computational chemistry research using this code and then perform the calculations on a larger computer either in the department, college or at one of the national supercomputer centers.

Utilizing MICROMOL in the Physical Chemistry Laboratory

As described in the last section, MICROMOL is an ideal package for use in the physical chemistry laboratory. What are the hardware requirements? Very simply one needs an IBM or IBM compatible microcomputer (XT, AT, 80386, 80386sx, 80486, PS/2) with at least 640 K of memory and a hard disk. To run DRAW it is currently necessary to have a math coprocessor and VGA monitor. However, neither are necessary to run MICROMOL. Calculations on molecular systems which illustrate important concepts can be performed with simple hardware.

One of the most important challenges is to get the students on the computers doing calculations and looking at the results without having to give them several hours of background material. The students can be directed to either their textbook or a written handout containing the essential theory necessary to have some grasp of the calculations prior to the laboratory session. Then, at the beginning of lab, only the highlights need to be discussed. After this, the instructor can lead the students through a simple calculation on the hydrogen molecule starting with TUTOR and ending with DRAW. There are many points that can be nicely illustrated in such a simple calculation which will only take a few seconds of computer time. Next, the students can determine the optimized geometry of H_2 and its vibrational frequency. After that, it is useful for students to study a potential energy curve of this system. Note that even though the dissociation energy will be too large, the general shape of the curve will be quite reasonable, especially near the equilibrium bond length.

After the students have examined the simple hydrogen molecule, perhaps they can look at a triatomic molecule, linear or nonlinear, and study the orbitals, electron densities, and vibrational frequencies and modes. For the linear triatomics, the degeneracies among modes can be nicely illustrated. Finally, the students might be encouraged to investigate a system in which they have an interest. These exercises will easily consume an entire 3-4 hour laboratory session.

Finally, it should be noted that MICROMOL can nicely complement the classical physical chemistry laboratory experiment in which the IR spectrum of HCl is measured. Usually, the spectrum of another isotopic species is observed. One can use MICROMOL to predict the appropriate shifts in frequencies since it is possible to perform calculations on various isotopic species*(19)*.

In conclusion, MICROMOL is user friendly and students can very quickly get "up and running" with the program without a sophisticated understanding of the formal theory. Furthermore, MICROMOL allows for open-ended laboratory exercises where students may use their creativity in choosing additional systems to study.

Safety

In the laboratory exercises described within this paper, the only "instrument" utilized is a "desktop-type" microcomputer. Thus there are no hazards other than the electrical hazards and fields associated with a CRT and computer.

Conclusion

Today's chemistry students need to be exposed to computational chemistry as it is making a significant contribution in both academic and industrial environments. Having described the ab initio MICROMOL codes, the underlying theoretical concepts, and how to utilize these codes in the physical chemistry laboratory, this author hopes to encourage and facilitate that student exposure.

Acknowledgments

This author is indebted to Professor Richard W. Schwenz for his encouragement, hospitality and many helpful discussions during the course of this project. Furthermore, the participants of the NSF-funded, physical chemistry workshops held at the University of Northern Colorado during the summers of 1990 and 1991 provided helpful insights. The author acknowledges helpful discussions with Dr. David Moncrieff. Finally, the support of The Florida State University Supercomputer Computations Research Institute and the IBM Education Development Program is greatly appreciated. This work was partially supported by the U. S. Department of Energy through Contract No. DE-FC05-85ER250000.

Literature Cited

1. Dirac, P. A. M. *Proc. Roy. Soc. (London)* **1929**, *123*, 714.
2. Dixon, D. A. in "Improved Access to Supercomputers Boosts Chemical Applications" by Forman, S. *Chem. & Eng. News* **1989**, *67*, 29.
3. Colwell, S. M.; Handy, N. C. *J. of Molec. Struct. (Theochem)* **1988**, *70*, 197.
4. Colwell, S. M.; Handy, N. C. *J. Chem. Educ.* **1988**, *65*, 21.
5. Amos, R. D. *The Cambridge Analytic Derivatives Package*; SERC: Daresbury, 1984, CCP1/84/4.
6. Schroedinger, E. *Ann. Phys.* **1926**, *79*, 361.
7. Born, M.; Oppenheimer, J. R. *Ann. Phys.* **1927**, *84*, 457.
8. Pauli, W. *Z. Phys.* **1925**, *31*, 765.
9. Slater, J. C. *Phys. Rev.* **1929**, *34*, 1293.
10. Roothaan, C. C. J. *Rev. Mod. Phys.* **1951**, *23*, 69.
11. Shavitt, I. In *Methods of Electronic Structure Theory*; Schaefer, H. F., Ed.; Plenum, New York, New York, 1977, Vol. 3; 189.
12. Slater, J. C. *Phys. Rev.* **1930**, *36*, 57.
13. Boys, S. F. *Proc. Roy. Soc. (London)*, **1950**, *A200*, 542.
14. Shavitt, I. in *Methods of Computational Physics*; Wiley: New York, NY, 1962, Vol. 2.
15. Davidson, R. R.; Feller, D. *Chem. Rev.* **1986**, *86*, 681.
16. Ditchfield, R.; Hehre, W. J.; Pople, J. A. *J. Chem. Phys.* **1971**, *54*, 724.
17. Dunning, T. H. *J. Chem. Phys.* **1970**, *53*, 2823.
18. Schlegel, H. B. In *Ab Initio Methods in Quantum Chemistry, Part II.*; Lawley, K. P.,Ed. Wiley, Chichester, 1987; 249.
19. Schwenz, R. W.; Brown, F. B. (to be published.)

Appendix A

References for Background Material for Computational Chemistry

1. Carsky, P.; Urban, M. *Ab Initio Calculations. Methods and Applications in Chemistry*; Springer-Verlag: Berlin, 1980.
2. Clark, T. *A Handbook of Computational Chemistry: A Practical Guide to Chemical Structure and Energy Calculations*; Wiley: New York, 1985.
3. *Methods in Computational Molecular Physics*; Diercksen, G. H. F.; Wilson, S., Eds.; Reidel: Dordrecht, 1984.
4. Hehre, W. J.; Radom, L.; Schleyer, P. v. R.; Pople, J. A. *Ab Initio Molecular Orbital Theory*; Wiley: New York, NY, 1986.
5. Hinchliffe, A. *Computational Quantum Chemistry*; Wiley: Chichester, 1988.
6. Hirst, D. M. *A Computational Approach to Chemistry*; Blackwell, London, 1990.
7. *Ab Initio Methods in Quantum Chemistry, Parts I and II*; Lawley, K. P. Ed.; Wiley: Chichester, 1987.
8. Richards, W. G., Cooper, D. L. *Ab Initio Molecular Orbital Calculations for Chemistry, 2nd ed.*; Clarendon Press: Oxford, 1983.
9. Rogers, D. W. *Computational Chemistry Using the PC*; VCH Publishers: New York, NY, 1990.

10. Roothaan, C. C. J. *Rev. Mod. Phys.* **1951**, *23*, 69.
11. Schaefer, H. F. *The Electronic Structure of Atoms and Molecules: A Survey of Rigorous Quantum Mechanical Results*; Addison-Wesley: Reading, MA, 1972.
12. Schaefer, H. F. *Methods of Electronic Structure Theory*; Plenum Press: New York, NY, 1977.
13. Schaefer, H. F. *Applications of Electronic Structure Theory*; Plenum Press: New York, NY, 1977.
14. Schaefer, H. F. *Quantum Chemistry: The Development of Ab Initio Methods in Molecular Electronic Structure Theory*; Oxford University Press: Oxford, 1984.
15. Szabo, A., Ostlund, N. S. *Modern Quantum Chemistry: Introduction to Advanced Electronic Structure Theory*; Macmillan: New York, NY, 1982.
16. Wilson, S. *Chemistry by Computer*; Plenum Press: New York, NY, 1986.

RECEIVED October 1, 1992

CHAPTER 2

State-to-State Dynamics

Richard W. Schwenz

One of the principal subjects which interests the physical chemist is understanding the detailed mechanism of reactions. This interest manifests itself in many forms, several of which are the examination of temperature dependence of reaction rates, the effects of isotopic substitution, and a changing of substitutent groups. Recent efforts are detailed regularly in Chemical and Engineering News, and have resulted in the awarding of the 1986 Nobel prizes for detailed investigations in reaction dynamics. Most recently, this has taken the form of following the detailed movements of atoms during photodissociation. This essay will concern itself with the manifestation of this interest in following how particular pre-collision reactant quantum states evolve into post collision quantum states during bimolecular collision events using both theoretical and experimental methods *(1-2)*. This essay will be organized as follows, first, a general introduction on the relationship between quantum state resolved collision dynamics and the temperature dependent rate coefficients. Second, a section describing the theoretical basis for studying collisional dynamics on potential energy surfaces and a discussion of the advantages and disadvantages of various methods of performing the dynamics calculations. Third, a section discussing experimental methods for examining the results of single collisions. Lastly, a short section will deal with methods for incorporating molecular dynamics into portions of the physical chemistry curriculum.

Several immediate questions are raised, namely why would one want to study a reaction process in such excruciating detail, and then how does this detail relate to the previously observed temperature averaged rate coefficients. The study of quantum state resolved reactions provides the most direct evidence for particular features on a potential energy surface that governs the collision, and allows a precise comparison of theoretical and experimental results for collisions. In state to state dynamics the results that are obtained by an experimentalist can often be directly compared with those obtained by a theoretician to the benefit of both. An increased understanding of the system under study results, as evidenced by the activity on the H + H_2 *(3-5)*, F + H_2 *(6-7)*, and now several other reactions.

The inherent averaging involved in the determination of the reaction's temperature dependance is eliminated in most studies of state-resolved reactions. The sources of averaging are immediately apparent when one considers that total reaction probabilities are functions of the initial quantum states, total energies and the initial orientations of all of the reactants, and of the product quantum states and orientations. To be more explicit, the relationship between the cross section σ and the overall temperature-dependent state averaged rate coefficient, k, is given by

$$k(T) = \Sigma P_i(T)\, k_i(T).$$

The state specific rate coefficients, $k_i(T)$, are given by

2503–6/93/0014$06.00/0

$$k_i(T) = (\pi\mu)^{-1/2}(2/k_BT)^{3/2} \int_0^\infty E\, \sigma_i \exp(-E/k_BT)\, dE$$

and the $P_i(T)$ are the populations of quantum state i *(8)*, in general given by the Boltzmann distribution. Here μ is the reduced mass of the collision pair, k_B is the Boltzmann constant, T is the temperature, and E is the energy of the translational and internal energy states of the reactant species. The integral is over the energy of the reactants, classically this would be for energies sufficient to cross the transition state to infinity. A reaction dynamicist will then typically speak of the cross-section, σ, as the important quantity at constant energy, while recognizing that σ_i is a function of many orientational and quantum state (i). The cross section becomes the important quantity at constant energy because it removes the velocity dependence of the rate coefficient. The classical interpretation of σ is the molecular size which can lead to a given process. Thus a larger cross section means that a greater proportion of collisions will lead to products.

The remaining parts of the chapter will focus on the theoretical and experimental means of determining the σ_i's, and on methods of incorporating these ideas into the physical chemistry lecture and laboratory courses.

Theoretical Background to State-to-State Dynamics

The basis for understanding chemical dynamics lies in the potential energy surface for the reaction. In most cases, the dynamics have been studied for processes which occur on the adiabatic Born–Oppenheimer potential energy surface, or for processes which occur on only a select few of the potential surfaces. Within the Born–Oppenheimer approximation, the previous chapter has shown how to calculate a single potential energy surface at a single geometry of the atoms *(9)*. The difficulties in performing these calculations at a single geometry are not immense. In order to obtain reasonable results for an entire potential energy surface an enormous amount of effort is expended. The amount of effort is due to the large number of geometries necessary to describe the entire system for anything larger than an atom-atom system, and the requirement for "chemical" accuracy (of order 10 kJ mol^{-1}) of the surface in order to correctly model the collisional dynamics. The large number of geometries is a function of the number of internuclear coordinates and the range over which one must study each of the coordinates. For the reaction of O with HD for example, one should look at 10–15 points of varying O–H distance for constant H–D distance at a given approach angle, and both the H–D distance and approach angles must also be varied. Quickly it becomes obvious that hundreds of ab initio points are required at each approach angle to describe the potential energy surface. A representation of the O – H – H(D) potential energy surface is given as figure 1. This graphical representation shows the potential energy as a function of O – H and H – H distances, here in a collinear geometry (O – H – H). The saddle point (transition state) for the reaction is clearly evident at R_{OH} of 1.1 Å, R_{HH} = 0.9 Å, with the reactant and product channels (valleys) leading away from there. Note that there is not a deep well corresponding to the water molecule because this is the lowest triplet surface, correlating to ground state oxygen atoms and hydrogen molecules, while the water molecule is on the singlet surface.

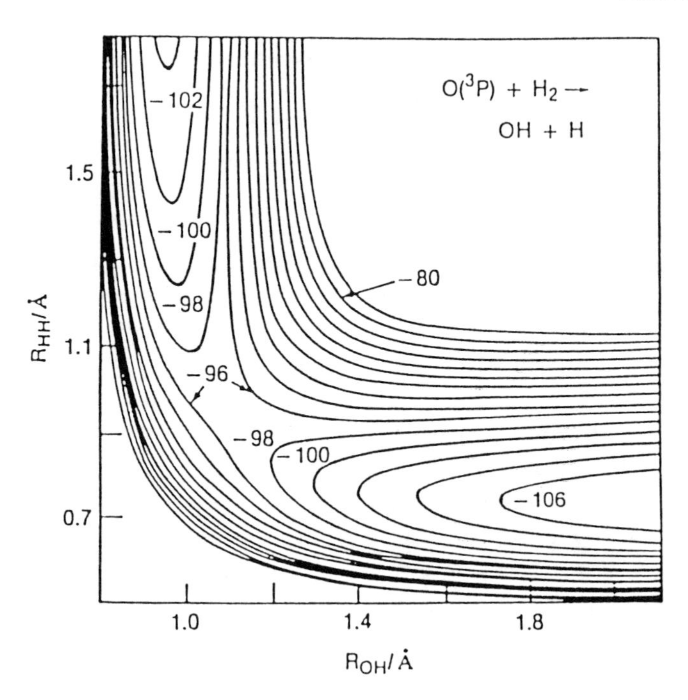

Figure 1. Potential energy surface for O + H_2. Reproduced with permission from ref. 1. Copyright 1987 Oxford University Press.

The requirement of chemical accuracy of the resulting potential energy surface means the each point is required to be within several kJ mol^{-1}. A calculation at this level of accuracy requires more than the HF–SCF level of accuracy. Typically configuration interaction or perturbation theory must be used with fairly large basis sets to obtain this level of accuracy. The large basis sets are necessary in order to describe the interaction of atoms at longer distances. These requirements make the computation of points for a potential energy surface a computer time intensive task, with the result that very few potential energy surfaces have been completely described in a ab initio manner.

In addition to the chemical requirements of reasonable shape of the potential energy surface and of chemical accuracy of the points on the surface, running the dynamics imposes the additional requirement of continuous first and second derivatives, or smooth surfaces . This requirement arises from the forms of Schrödinger's equation for quantum mechanics and of the classical equations of motion. This does not imply that a single functional form applies for all possible combinations of atoms and molecules. Each potential energy surface must be fit individually, and the important regions fit must be more exactly than in less important regions, typically extemely high energy regimes. A recent review *(10)* provides a reasonably complete summary of the art of generating potential energy surfaces.

The use of an accurate ab initio surface is then to be preferred over semi-empirical surfaces because, in principle, accurate ab initio methods provide a representation of the surface without the use of adjustable parameters for the fitting of the surface. In practice, the production of an ab initio surface is much more difficult and time consuming. Various features of the surface are often modelled in unimportant regions by simpler functional forms taken from a number of model surfaces. Some of these simplifications are necessary because the use of an ab initio surface becomes much more time consuming than using a semi-empirical surface.

Where the use of an ab initio surface becomes impractical a number of types of semi-empirical surfaces have arisen. Generally speaking, semi-empirical surfaces consist of a mathematically complex functional form containing a number of adjustable parameters used to fit the form to a combination of theoretical and experimental data. Semi-empirical surfaces have a number of advantages over an ab initio surface. Typically it is possible to evaluate, determine, the value of the potential energy for a semi-empirical surface in less time than an ab initio surface for a given system. The functional form used for a semi-empirical surface, while complex, can be made to be smooth, and in fact, it may be possible to analytically solve for the derivatives of the potential energy surface with respect to the coordinates, saving even more computational hours in the solution of the dynamics.

Among the possible disadvantages of semi-empirical surfaces include the ad hoc nature of the functional forms used. Some people will argue that the use of adjustable parameters, modified to fit experimental data constitutes an unacceptable forcing of the surface to fit the data. Others will argue that such a modification gives the surface a more correct physical description of the potential energy than could be obtained without the cost of high accuracy ab initio calculations, which may not even be possible. There is no doubt that a semi-empirical surface can effectively model the behavior of a quantum mechanical surface in some cases. Some semi-empirical potential energy surfaces are of simple enough mathematical form that it is possible to identify the particular forces associated with each term.

In order to determine the cross section for various processes as a function of quantum state, it is necessary to solve the dynamics, using either quantum or classical mechanics on the potential energy surface.

Quantum dynamics Most often quantum dynamics is performed on simpler systems because of the number of quantum states involved where reducing the number of atoms greatly reduces the number of states, and aids in increasing the symmetry of the system. The essence of a quantum mechanical solution to the dynamics of a system is finding the overlap of the initial state wavefunction with that of the final state where the two wavefunctions depend on each other in a non-trival manner. This process requires that the wavefunction, and its derivatives, be evaluated at a large number of points over the entire potential energy surface. There are a number of approximations which can be made for the simplification of the problem, as well as a number of approaches to the solution of the problem. Reviews of the topics are available which will inform the reader of the difficulties involved *(3,6)*. The current state of the theory is such that three atom systems are the most that can be solved exactly, although some progress is being made on the formal theory for four atom systems, and various approximations can be made which allow the solution of the scattering problem for many atom systems. The level of difficulty of the calculations is such that only a few systems have had the complete quantum dynamics solved on a realistic potential energy surface.

Classical dynamics Where in quantum dynamics it is necessary to solve for the dynamics of all quantum states simultaneously, classical (or quasi-classical) dynamics allows for the solution of the dynamics in a Monte Carlo sense. The Monte Carlo method allows for the determination of a cross section (or rate coefficient) via the statistics of random paths on the potential energy surface, i.e., what percentage of trajectories reach the product channel starting from reactants.

Although simple, two atom systems tend to be uninteresting since only inelastic energy transfer can occur on a single potential energy surface. Two atoms on multiple potential energy surfaces, for example, Ca ($^1P_1$) + He can prove to be quite interesting because the electronic surface can change during the course of the reaction.

As a more complex example consider a reaction system, A + BC. When applying the dynamics of macroscopic bodies to the microscopic interactions of atoms it is necessary to make several approximations of varying quality. First, the atoms themselves are point particles. Second, quantum mechanical effects are negligible, i.e., no tunneling, no quantized energy levels, and no uncertainty principle. Third, the motion of the atoms can be solved exactly using an appropriate set of the equations of motion (Newton's, Lagrange's, or Hamilton's). Probably the easiest of these to solve are Hamilton's equations of motion, written in a generalized coordinate system as:

$$\partial q_k / \partial t = \partial H / \partial p_k \qquad \partial p_k / \partial t = - \partial H / \partial q_k$$

where the q_ks and p_ks are the positions and conjugate momenta of each atom in each of the coordinates (k), t is the time, and H is the classical Hamiltonian, a constant of the motion *(11)*. Thus for a simple three atom system, there are nine Cartesian coordinates to be worried about, of which three can be eliminiated as the motion of the center of mass. In order to solve for the motion of this simple system, it is necessary to solve the six coupled first order partial differential equations during the time of the collision. Thus the problem is reduced to a case where simple numerical integration of the coupled equations to any desired accuracy allows for the solution of the dynamics given a set of initial conditions, i.e., that the classical dynamics of a collision is deterministic. This simple minded view doesn't account for the random nature of some of the relevant variables within the atomic collision partners. For example, in most experimental arrangements it is impossible for the impact parameter of the reaction, the distance between the center of masses of the collision partners along the velocity vector to be held constant. A calculation of the reaction probability must then average over appropriate variables. The Monte Carlo technique does this averaging by randomly selecting values from the range of possible values and thus determining the reaction probabilities from how many trajectories reach the desired products *(12)*.

A review of many trajectory studies is available *(13)*. The results of such studies include the size of the cross section, its dependence on translational energy, and the motions of the products into which energy is dispersed. One advantage to trajectory studies is the ease with which the effects of different potential energy surface features can be identified. In examining the results of classical trajectories it is necessary to remember that the quantum mechanical effects mentioned earlier are not included. There are a number of modifications of classical trajectory methods which attempt to account for the quantized energy levels (quasiclassical), tunneling (semiclassical), or zero point energy effects at the transition state (vibrationally adiabatic effects).

This article specifically does not include transition state theory methods, statistical methods or unimolecular events. A general review is availible *(2)* with indications of further review articles in appropriate chapters.

Experimental Measurement of State-Resolved Collision Processes

The basic concept behind the measurement of state-resolved reaction rates involves the observation of gaseous collision products after a single collision but prior to succeeding collisions. The kinetic molecular theory of gases provides an expression for the collision rate per atom A as

$$Z_{AB} = \sigma_{AB} <v>_{AB} N_B$$

where Z_{AB} is the collision rate per atom A, σ_{AB} is the cross section, $<v>_{AB}$ is the mean speed of the A–B pair, and N_B is the number density of molecules B. The reciprocal of this collision rate gives an estimate of the mean time between collisions. Thus to observe the results of a single collision, one must look (observe) the products quickly, i.e., before the next collision. One alternative is to use a somewhat reduced pressure (of order 1 torr) and to look in the nanoseconds between collisions; easily done with nanosecond or faster laser pulses. A second alternative is to lengthen the time between collisions, by reducing the pressure to order 10^{-6} torr, and then using a slower observational technique. A combination of these two techniques is commonly used.

The molecular beam technique was the first technique which evolved to examine the results of a single collision *(1)*. It does this by lengthening the time between collisions to the order of the flight time across the apparatus. A block diagram of the essentials of the apparatus is included as figure 2.

Basically two molecular beams (one of each reactant A and B) cross at an angle, usually 90°, in a small region, *V*, defining the collision zone. The number of collision products is measured at one or more scattering angles and velocities by detector C, and the data transformed to the center of mass frame. In the center of mass frame, one can easily tell what direction the products are scattered. Knowledge of the directionality of scattering can give indications about the type of mechanism for reaction, and about the lifetime of the reaction complex. Molecular beam techniques using angular resolved measurements tend not to give information on the quantum states of products. The resolution of cross sections, or probabilities, into the internal quantum states of collision products is left to spectroscopic methods, such as luminescence and induced fluorescence. In principle, the observation of the internal state distribution, $P(v,J)$, as a function of vibrational state (v) and rotational state (J) will give more information than the angular distribution, and information which is complementary to the angular distribution *(1)*.

A second technique for the observation of the results of single collision evolved from the flash photolysis techniques pioneered by Norrish and Porter *(13)*. In this general method, a short pulse (typically ps–ns) of light will prepare one of the reactants from a stable mixture of gases containing all of the reactants less one, and a precursor to that last reactant. For example, to examine collisions of H with CO_2, a mixture of CO_2 and H_2S might be photolyzed at 193 nm to dissociate the H_2S. One of the nice features of this type of experiment is that the

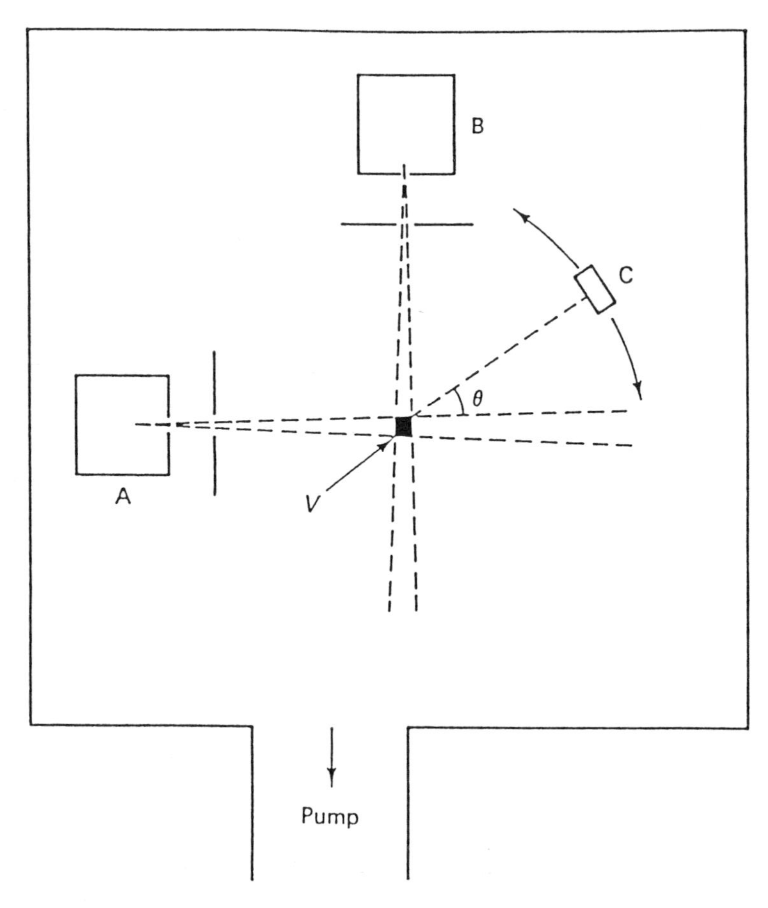

Figure 2. Schematic for a crossed molecular beam apparatus. Reproduced from ref. 2. Reprinted with permission from Pergamon, copyright 1970.

H atoms which result from the photodissociation event are very nearly monoenergetic, eliminating the need for energy averaging. The single collision products of OH and CO might each be examined by laser induced fluorescence 20 ns later to determine the reaction probability into each product quantum state. Typically this type of experiment involves the use of two laser pulses accurately timed with respect to each other in a reasonably low pressure (approximately 1 mm Hg) reactor. These experiments are often called "pulse-probe" experiments because the first pulse creates the reactants, and the second probes the resulting products. There exist several good review articles discussing the types of information which these experiments provide *(14)*.

It is possible to combine these two types of experiments and do a quantum state resolved detection in a molecular beams experiment. The principle here is the preparation of two intersecting molecular beams with the products detected using optical methods rather than using a detector located at a given laboratory scattering angle. Typical detection schemes might examine the proportion of various electronically excited states using their luminescence properties, ionic products by collecting the resulting product ions in a Faraday cup, or ground state products using laser induced fluorescence, multiphoton ionization, or coherent anti-Stokes Raman spectroscopy *(1)*. Each of these detection schemes would typically be carried out at the collision center rather than at some scattering angle.

Given the preceding general introduction, let us now examine several examples which provide experimental evidence for some of the particulars in collisional dynamics. The intent of the following sections is to show how the modern experimentalist can eliminate some of the averaging inherent in many of the experiments that were done in the past. These examples will be presented in chronological order so that the increasing complexity of the experiments will become evident.

It should be obvious from much of the preceding discussion that the reaction probability should depend on the exact quantum states of reactants which are colliding. Thus if it is possible to alter the relative proportion of the various quantum states, then the relative reactivity should be observable. Imagine a molecular beam apparatus in which a laser beam is used to excite a vibrational, rotational or electronic transition in one of the reactants. This laser would be capable of altering the populations within the reactant molecular beam. With sensitive detection techniques, possibly a second laser, the changes in product quantum states become measurable. These types of experiments show that the quantum states do have dramatically differing reactivities.

One such experiment probed the increased reactivity of the combination of Ba with HF to produce BaF ($X\ ^1\Sigma$) *(15)*. When the HF is in vibrational state 0, the reaction is very slow, in part the reaction is endothermic. When HF is promoted to vibrational state 1 by pumping with an HF laser the reaction rate increases by four orders of magnitude at least in part because the additional energy makes the reaction exothermic. In this system, the reaction rate is observed to have a maximum in the reactivity with rotational state, *J*, which is non-zero in the first vibrational state. This maximum is explained by the requirement of the metal atom striking the fluorine end of the HF molecule in the gas phase. At higher *J*'s the fluorine atom is essentially shielded by the hydrogen because of the high rotational frequency of the hydrogen rotating about the fluorine. At lower *J*'s, this hydrogen is less likely to shield the entire fluorine atom, but can only shield less than half of the fluorine from the barium.

This type of experiment has also been extended to the reactions of electronically excited states pumped. A number of groups from different universities all began studying the influence of different electronically excited states on chemical reactivities because they access vastly different potential energy surfaces. In many cases the effects of electronic excitation are quite dramatic, the $O(^3P) + H_2$ reaction has a rate coefficient of 3.5×10^{-18} cm^3 $molecule^{-1}$ s^{-1} at room temperature, while that for the excited state $O(^1D) + H_2$ is 1.0×10^{-12} cm^3 $molecule^{-1}$ s^{-1}, a difference of many orders of magnitude. These effects have been seen in state-to-state experiments more directly, by laser pumping of electronically excited states of one of the collision partners. Another example, the 4s4p $^1P_1$ state of calcium which can be reached in a single photon excitation from the ground state. This prepared state can then be used as a collision partner for reaction. Laser pumping does not provide the only method for the excitation of electronically excited states, electron bombardment, resonance lamp excitation, and thermal excitation have all been proposed as methods for the production of electronically excited species of differing energies.

Laser pumping of electronically excited species imparts another property on the excited atoms. The polarization property of the exciting laser beam allows specific orientations of the atom to be prepared *(16)*. If one starts with an isotropic distribution of Ca $4s^2\ ^1S_0$ such as is found in a typical molecular beam, and excites with a polarized laser resonant with the transition to $4s5p\ ^1P_1$, an anisotropic distribution of the upper state obtains. This distribution is described by the degree of polarization of the laser with the maximum of the distribution in the plane of laser polarization. This maximum can be pointed in the direction of motion of the molecular beam, or any angle away from the beam by rotating the plane of laser polarization using a double Fresnel rhomb polarization rotator. Thus one can "point" the 5p orbital at the collision partner, or perpendicular to the collision partner. When the collision partner is an atom, the dynamics is much simplified, and one can see the variation of the efficiency of energy transfer from $4s5p\ ^1P_1$ to $4s5p\ ^3P$ with the angle of approach of the 5p aligned orbital relative to the rare gas atom. The energy transfer is most probable for He when the 5p orbital is aligned with the direction of motion, while for Xe the energy transfer is most probable when the 5p orbital is aligned perpendicular to the direction of approach. These observations can be interpreted in terms of the potential energy curves for the approach of a rare gas atom to a p orbital, in more exactness, with the depth and tightness of the potential wells, and the surface crossings for the potential energy surfaces.

The examples just given have presented cases where the quantum states of one or more of the reactants have been specified through the specific preparation of the quantum states. The relative positioning of the collision partners has not been specified although that of orbitals has. It is possible to prepare collision partners through at least two techniques using the condensation properties of a supersonic molecular beam. For many years, spectroscopists have studied molecules in supersonic molecular beams because of the cooling properties of the beam; rotational temperatures of 4 K are commonly recorded dramatically reducing the number of spectral lines. Within the last fifteen years, the so called van der Waal's molecules have come under study using a supersonic molecular beam for use in dynamics. Dynamicists are now starting to use these van der Waal's molecules as precursors to the collision event *(17)*. If one coexpands HBr with CO_2, the van der Waal's molecule $HBrCO_2$ results. This molecule is known from microwave spectroscopy to be quasilinear (Br - H - O - CO), although rather floppy. If one then exposes this van der Waal's molecule to 193 nm light from an ArF laser, the H - Br bond will absorb the photon and dissociate. The H atom will collect the bulk of the energy of the photon, and carry this translational energy along the direction of the H - Br bond. The H atom will then be colliding with the central oxygen at a given energy, with an impact parameter of 0, given the collinear geometry of the van der Waal's complex. The resulting OH product state distributions are observable using laser induced fluorescence.

A second method for the preparation of collision partners with known geometries uses photoelectron spectroscopy to examine the structure of the transition state for a collision *(18)*. In this method the collision partners are prepared in negative ion form, i.e., for the Cl + HCl collision pair one would prepare $ClHCl^-$, which is a stable negative ion in a molecular beam. This species has been well characterized by the normal types of spectroscopy so that the structure of the ion is known. When the photoelectron spectrum is taken, the Franck–Condon principle applies, so that the energies of the electron which result are the quantized energies corresponding to the energies of the Cl – HCl neutral species at the geometry of the anion. The anion geometry corresponds to the geometry of the transition state for the collision of a Cl atom with an HCl molecule, and thus the observed quantized energies to the energies of the transition state. A well defined set of energies at the transition state allows the back calculation of the potential energy surface for the neutral species near the transistion state. This experiment is one of the most powerful tools yet for the study of the transition state region of the potential energy surface, and thus for the predictions of theory, and its comparison with experimental results.

An exciting technique in the study of reaction dynamics involves the use of femtosecond pulse length lasers to watch the motions of atoms within molecules. Recalling that a typical vibrational period is 10^{-13} s, we readily note that the femtosecond pulse give stoboscopic pictures of atoms. In one example of this technique, the iodine atom is followed as it leaves the CN fragment in the photodissociation of ICN. Observations of this sort allow the colloboration between the theoretician and the experimentalist to examine the results of interactions between two potential energy surfaces in exceedingly small regions.

State-to-State Dynamics in the Classroom

State-to-state dynamics provides a plethora of examples for use in the classrooom throughout the physical chemistry curriculum. For example, the formation of a supersonic molecular beam is essentially the adiabatic expansion of a gas into a vacuum. The uses of such an example, and the low temperatures achieved might prove interesting to some students. In the discussion of kinetic molecular theory we often discuss topics related to the development of state to state experimental methods, for example the mean free path being larger than the apparatus, and what that implies. We also discuss the hard sphere collision at great length, why not illustrate the concept of what happens if the molecules interact with each other?

In a discussion of chemical kinetics, how many of us stop to ask whether the reaction cross sections for given quantum states are changing, or is it simply the change in populations of the quantum states which cause the rate to change as the temperature increases? Lastly, of course, infrequently, we mention that there is a potential energy surface for a stable molecule; but fail to extend the analogy to the unstable molecule, the reaction complex, which is as important as the stable species.

Several people have developed exercises for physical chemistry students to use which examine the classical trajectories involved in reactions more closely *(19)*. The programs which result typically display the positions of the various atoms at various times during the evolution of the reaction, and show the bond formation and breaking during the course of the chemical reaction. Experimentally, it is more difficult to develop meaningful laboratory exercies because of the cost of the equipment *(20)*. The cost of the equipment should not be a determining factor because the same, typically laser-based, items can be used for many experiments. For examples of some of these experiments see the materials later in this book.

Literature Cited

1. Levine, R.D. and Bernstein, R.B. *Molecular Reaction Dynamics and Chemical Reactivity*; Oxford: New York, NY, 1987.
2. Steinfeld, J.I., Francisco, J.S., and Hase, W.L. *Chemical Kinetics and Dynamics*; Prentice Hall: Englewood Cliffs, NJ, 1989.
3. Miller, W. H. *Ann. Rev. Phys. Chem.* **1990**, *41*, 245.
4. Nieh, J-C. and Valentini, J. J. *J. Chem. Phys.*, **1990**, *92*, 1083 and references therein.
5. Kliner, D. A. V., Adelmann, D. E., and Zare, R. N. *J. Chem. Phys.*, **1991**, *94*, 1069.
6. Schatz, G. C. *Ann. Rev. Phys. Chem.* **1988**, *39*, 317.
7. Faubel, M., Schlemmer, S., Sondermann, F., and Toennies, J. P. *J. Chem. Phys.*, **1991**, *94*, 4676.
8. Bernstein, R.B. in *State to State Dynamics*; Brooks, P.R. and Hayes, E.F., Eds.; ACS Symposium Series, American Chemical Society: Washington, D.C., vol. 56, pp. 3-21.
9. Brown, F.B. in this volume
10. Schatz, G. C. *Rev. Mod. Phys.*, **1989**, *61*, 669.
11. Symon, K. R. *Mechanics, third edition*; Addison Wesley: Reading, MA, 1971.
12. Bunker, D. L. in *Methods in Computational Physics*, **1971**, *10*, 287.
13. Christie, M. I., Norrish, R. G. N., and Porter, G. *Proc. Roy. Soc. (London)*, **1952**, *A216*, 152.
14. Flynn, G. W. *Science*, **1989**, *246*, 1009.
15. Altkorn, R., Bartoszek, F. E., Dehaven, J., Hancock, G., Perry, D. S., and Zare, R. N. *Chem. Phys. Lett.*, **1983**, *98*, 212.
16. Bussert, W., Neuschaefer, D., and Leone, S. R. *J. Chem. Phys.*, **1987**, *87*, 3833.
17. Rice, J., Hoffmann, G., and Wittig, C. *J. Chem. Phys.*, **1988**, *88*, 2841.
18. Hertz, R. B., Kitsopoulos, T., Weaver, A., and Neumark, D. M. *J. Chem. Phys.*, **1988**, *88*, 1463.
19. Eaker, C. W. and Jacobs, E. L. *J. Chem. Educ.*, **1982**, *59*, 939.
20. Kovalenko, L. J. and Leone, S. R. *J. Chem. Educ.*, **1988**, *65*, 681.

RECEIVED October 1, 1992

CHAPTER 3

Atmospheric Chemistry

Don Stedman

POTENTIAL HAZARDS: Vacuums

Atmospheric chemistry provides a wonderful example of physical chemistry at work. A wealth of detailed information is available in books on atmospheres as a whole (*1*), or on the troposphere (*2-3*) and stratosphere (*4*). There are several reasons to use atmospheric examples to illustrate principles of physical chemistry. The examples are both rich, and plentiful. Students have some intuitive understanding of the examples as they come from the world around them. Most students not only breathe, but are aware of the fact. Students have enormous interest in atmospheric pollution, what causes it, how it effects them, and what can be done about it. Lastly, part of our responsibility as scientists is to educate the public, including students, about the world, its interactions and its problems.

Atmospheric science is a combination of meteorology, physics, chemistry, biology, and geology. Physical and analytical chemistry provide the chemical information involved in studies of the atmosphere. Physical chemists are an absolute necessity in the studies of many atmospheric processes. Examples from the earth's atmosphere can be given for most of the principal topics in physical chemistry including thermodynamics, transport properties, chemical kinetics, quantum chemistry, and photochemistry.

The organization of this essay will present the structure of the atmosphere, some cycles which involve physical chemistry, chemistry in the troposphere and stratosphere, atmospheric pollution, the detection of atmospheric species, and suggestions for lecture examples, lecture demonstrations, and laboratory exercises for students.

Structure of the Atmosphere

Figure 1 shows a schematic diagram of the atmosphere as cross sections of a series of spherical shells. On the left side of the diagram is the approximate total density in molecules per cm^3. Next to the left side is a list of the molecules which are the major species at a given altitude. The troposphere is the lowest shell extending from the ground up to 10 to 20 km in height depending on latitude. All major and all permanent gases in the troposphere are well mixed by wind and turbulence regardless of molecular weight. The ratio of helium to argon is constant up to 100 km in altitude. By 100 km the pressure is reduced to 10^{-5} of an atmosphere. Since the molecular diffusion coefficient is inversely proportional to pressure, molecular diffusion dominates over wind and turbulence at altitudes above 100 km. The turbopause labelled on the right indicates the start of the heterosphere, the region in which each gas diffuses according to its own molecular weight.

The dominant molecule at ground level is N_2, with approximately 19% impurity of O_2. As noted in Figure 1 this situation persists as altitude increases up

2503–6/93/0025$06.00/0

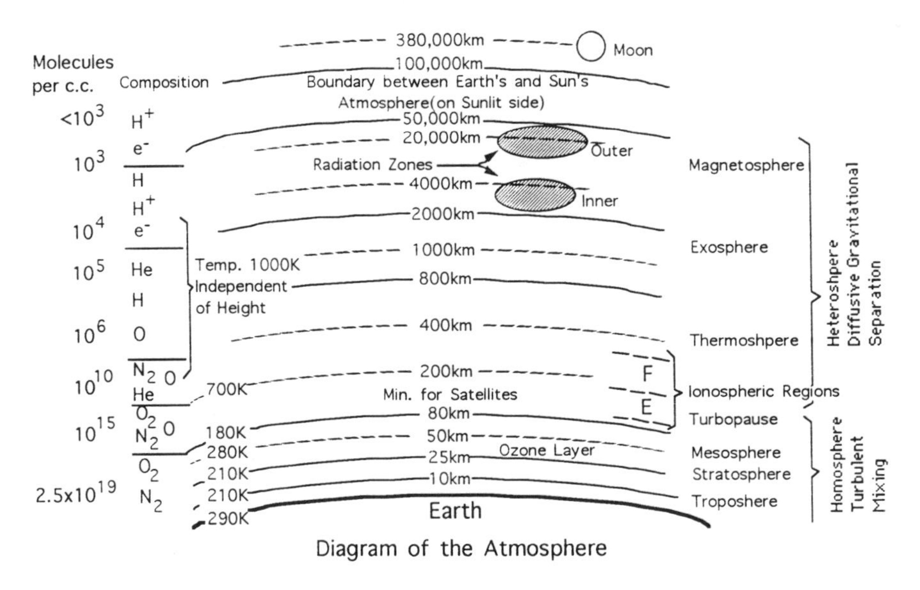

Figure 1. Schematic diagram of the atmospheric cross section.

to 100 km, above which oxygen atoms become dominant, followed by helium, then hydrogen. Thus, at high altitudes the atmosphere of the earth merges chemically with the atmosphere of the sun which is also dominated by the presence of hydrogen.

The atmosphere of Venus has a surface pressure of about 10^2 Atm of CO_2 with a little nitrogen. The atmosphere of Mars has a surface pressure of about 10^{-2} Atm mostly of CO_2 with a little nitrogen. It would be quite safe for a traveller from outer space to interpolate (logarithmically) the surface pressures to come up with one Atm for the earth, however the atmosphere is not currently dominated by CO_2. There is every reason to believe that the primordial atmosphere for the first billion years was a CO_2 atmosphere with a little nitrogen. The geological evidence unequivocally prohibits the presence of significant amounts of oxygen, since ferrous iron was soluble in surface waters worldwide. Current theories do not include other than trace amounts of methane or ammonia.

The removal of carbon dioxide, mostly as limestone, and the formation of oxygen, mostly by means of photosynthesis, are both the result of biological activity. Interestingly, the stability of nitrogen itself, once thought to be thermodynamically stable, is also in control of the biosphere. At the current ocean pH of 8.6 and at the current oxygen partial pressure the thermodynamically stable form of nitrogen is as nitrate in the oceans. Since oceanic organisms fix nitrogen at a rate rapid enough to deplete the nitrogen from the atmosphere in a few million years, apparently there are other organisms which liberate nitrogen fast enough to keep the atmospheric pressure reasonably constant. The nitrogen balance is the first geochemical cycle. Many of the cycles need to be elucidated to understand the atmosphere.

The temperature of the atmosphere is not constant with altitude. Figure 2 shows an approximate temperature structure for 40N at the equinox (the latitude of Denver, Colorado and Philadelphia, Pennsylvania). Notice that the graph is plotted

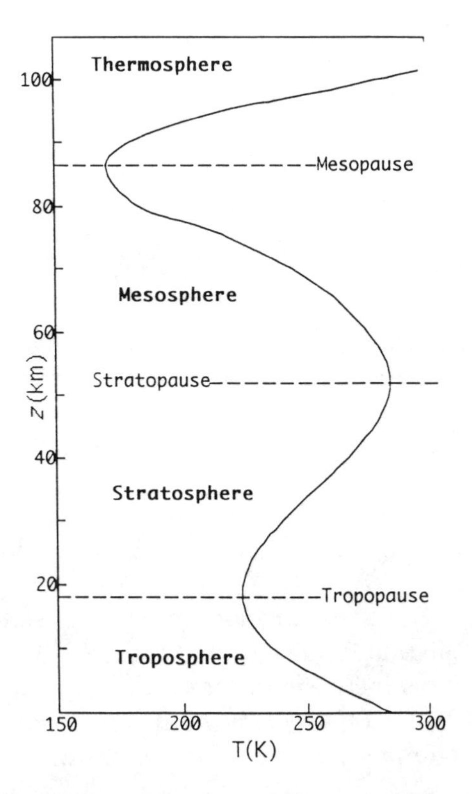

Figure 2. Temperature structure of the atmosphere at 40N at the equinox.

with altitude upwards (the *y* axis). For obvious reasons this is the convention for atmospheric scientists even though it violates the normal attribution of independent and dependent variables. Each of the "-spheres", tropo- strato- meso-, etc. are named from peak to trough of the temperature structure.

Pressure Gradient The decrease in pressure and temperature with altitude in the troposphere can be calculated from first principles. Consider the forces acting upon a stationary cylinder of air of area A, of height dz and mass m. The gravitational force *acting on m is then mg* which must be balanced by the pressure gradient, more pressure acting on the bottom than on the top. For consistency we write the force as $P\,A$ on the lower face and $(P + \mathrm{d}P)$ A on the upper. Since this parcel of air is stationary and not accelerating the forces must be in balance and:

$$-A\ \mathrm{d}P = \mathrm{mg}$$

if we define the density ($\rho = m/V$), since $m = \rho\ V = \rho\ A\ \mathrm{dz}$

$$\mathrm{d}P = -\rho\ \mathrm{dz}$$

and from the perfect gas laws $\rho = MP/RT$ where M is the molecular weight of air thus:

$$-\mathrm{d}P/P = Mg\mathrm{dz}/RT$$

from which by integration one obtains

$$P = P_0 \exp(-z/Z_0)$$

where Z_0 is called the "scale height", and is the altitude over which the pressure decreases by a factor of e. From this derivation we obtain:

$$Z_0 = RT/Mg$$

where M is the mean molecular weight of air, close to 29 atomic mass units. At 300 K the scale height is 8.3 km. Although temperature is not constant with height (Figure 2) when log (P) is plotted against z the changes in slope are subtle compared to the very rapid overall decrease in pressure. There is approximately an order of magnitude decrease in pressure every $8.3 \times 2.3 = 19$ km in height. Thus at 100 km in altitude the atmospheric pressure is down by about five orders of magnitude to 7.6 microns of Hg pressure, which is lower than can be obtained by most mechanical vacuum pumps.

It is an interesting exercise to sharpen up ones skills at differentiation to write the (Beer's law) equation for absorption of monochromatic solar radiation by a molecule whose density is increasing exponentially as the photons approach the earth. Not surprisingly, since one obtains an exponentially increasing exponential term, as soon as the photons penetrate to an altitude at which absorption is appreciable all the photons are absorbed in a layer only about one scale height thick. Absorption and degradation to heat of ultraviolet solar photons by molecules in the stratosphere and thermosphere is the major cause of the higher temperatures in those regions.

For many atmospheric chemical calculations it is appropriate to integrate the atmospheric pressure equation to derive a single, dynamically impossible but chemically convenient, spherical shell of height 8.3 km at a constant pressure of one Atm. Under this approximation the oxygen becomes a layer 1.75 km thick, thus over each m^2 of planet there is 1,750 m^3 (about 72,000 moles) of oxygen. If all the living biosphere, trees, fish, plankton, humans etc. are combined and layered over the earth in the same way they amount to only about 8 kg of carbon per m^2, i.e., 700 moles/m^2. From this comparison it is apparent that if all the living biosphere were to be instantaneously burned the immediate effect on the oxygen partial pressure would not be very large. In the long run, photosynthesis would be shut off but many geochemical processes (such as $FeS \rightarrow Fe_2O_3$ and H_2SO_4) removing oxygen would continue, and the oxygen would, over geological time, be depleted to a zero level.

Above 100 km molecular diffusion dominates over turbulent mixing thus all species decrease in pressure with further increase in altitude according to their individual molecular weights. Helium (Molecular weight, 4) decreases with a scale height an order of magnitude less than that for argon (Molecular weight, 40). In the next 70 km the argon partial pressure decreases by five orders of magnitude while the helium pressure decreases by only a factor of three. No wonder helium becomes the majority gas at 800 km, ultimately to be overtaken by hydrogen as altitude increases.

Temperature Gradient If a large mass of air is forced to rise, for instance by wind forcing the air over a range of mountains, the pressure will decrease, the volume increase and the temperature decrease. In the absence of condensation of water the process is adiabatic, the temperature decrease caused mainly by the *PV* work done on the rest of the atmosphere by the necessary volume increase. From thermodynamics for an ideal gas with initial temperature T_1 one obtains the new temperature T_2 as:

$$T_2 = T_1 (P_1/P_2)^{\gamma-1}$$

where γ is the ratio of specific heats at constant pressure and volume, C_p/C_v. For air γ is approximately 7/5. The adiabatic temperature decrease, termed the dry adiabatic lapse rate is, in log terms 2/5 the rate of decrease of pressure, thus if pressure decreases by a factor of *e* in 8.3 km absolute temperature would decrease by the same factor in 21 km. Near the ground the decrease in pressure (about 26 torr per thousand feet altitude) is accompanied by a decrease in temperature of 10 K per km or 5.6 ^{o}F per thousand feet.

The Various cycles (O_x, HO_x, NO_x)

The relative proportions of the various constituents of the atmosphere influence not only the rates of reaction at differing altitudes, but also the structure of the atmosphere itself, through the amounts of absorption which occur at different altitudes. This section will describe the basic atmospheric cycles which are operative at all altitudes; following sections will describe the relative importance of each of the cycles within the troposphere and stratosphere.

A deceptively simple looking example to illustrate atmospheric photochemistry is with the formation and disappearance of odd oxygen, O and O_3, by the Chapman mechanism *(5)*. This mechanism was first proposed to explain the increased mixing ratio of ozone in the 10–60 km region of the atmosphere. In the Chapman mechanism the first reaction is the solar photolysis of oxygen. Photolysis is a first order reaction. The frequency of photolysis is given the symbol *j* (in units of s^{-1}) since it is used like a rate constant. In fact the frequency of photolysis is not constant since it depends on the amount and availability of sunlight. The appropriate value of *j* can be calculated from the solar spectrum intensity, $I(\lambda,t)$, the absorption cross section, $\sigma(\lambda)$, and the quantum yield, $\phi(\lambda)$, thus:

$$j = \int_0^{\infty} \left(\frac{dI(\lambda,t)}{d\lambda} \sigma(\lambda) \phi(\lambda) \right) d\lambda$$

In practice the limits of integration are from the shortest wavelength (λ) available at a given altitude to the long wavelength cutoff of the chemical process in question.

	<u>*k* or *j* at 30 km</u> (midday, 40N latitude) *(6)*		
$O_2 + h\nu \rightarrow O + O$	5.2×10^{-11}	s^{-1}	1)
$O + O_2 + M \rightarrow O_3 + M$	5.6×10^{-34}	$cm^6\ mol^{-2}\ s^{-1}$	2)

$O_3 + h\nu \rightarrow O + O_2$ $\quad 9.5 \times 10^{-4}$ $\quad s^{-1}$ $\quad$ 3)

$O + O_3 \rightarrow 2O_2$ $\quad 1 \times 10^{-15}$ $\quad cm^3\ mol^{-1}\ s^{-1}$ $\quad$ 4)

M here is any third body gas in the atmosphere. Even though this mechanism only includes a single element, it does not include all the "oxygen only" reactions important at all levels in the atmosphere. If the differential equations for the four reactions above are written out it turns out that their solution is simplified by adding the terms for $d[O]/dt$ and $d[O_3]/dt$ to form a single term $d[O+O_3]/dt$, i.e., d/dt of "odd oxygen". The first and fourth reactions are very much slower than the second and third. The second and third do not make any net change in odd oxygen, they only set up a steady state ratio such that $[O]/[O_3] = j_3/k_2[O_2][M]$. Since both O_2 and M are very dependent upon altitude the ratio is very altitude dependent, with O_3 dominating over O atoms in the stratosphere while O dominates over O_3 in the mesosphere. Since both O and O_3 are formed photochemically one might expect them to disappear at night. In the stratosphere as the sun goes down the few residual O atoms are more likely to make O_3 than remove it. In the mesosphere the few residual ozone molecules are removed by the oxygen atoms, but then the atoms persist since all their removal processes are pressure dependent, and the pressure is too low for the reactions to proceed in only twelve hours. At an altitude of about 80 km the O and O_3 are at equal concentrations of about 10^{11} cm^{-3}. Since the concentrations are equal one might expect them to annihilate one another. This does not take place for two reasons, the first is that some of the oxygen atoms form O_3 rather than remove it. The second reason is that, even neglecting O_3 formation, with a rate constant of 10^{-15} and a concentration of 10^{11} the half life (perfect second order kinetics because of equal initial concentrations and a stoichiometric reaction) would be 10^4 seconds, the next half life would be 2×10^4 seconds and the next 4×10^4 seconds. Since 7×10^4 seconds is longer than the nighttime period, the concentrations would not deplete overnight by more than a factor of eight, even if O_3 reformation (the major process) is neglected.

Since the rate coefficients for all four reactions in the mechanism are known, it is possible to calculate the steady state concentrations of ozone at any altitude. Using the data given above, it is readily found that the calculated ozone molecule concentration is 8.2×10^{12} cm^{-3} at 30 km and that the oxygen atom concentration is 4.7×10^8 cm^{-3} at the same altitude.

The predicted concentrations of ozone arising from this mechanism are approximately a factor of two larger than the measured values. This implies that there must be some loss mechanism(s) for ozone in the atmosphere which are not due solely to oxygen containing species. There are three cycles involving nitrogenated (NO_x), hydrogenated (HO_x), and chlorinated (ClO_x) species which serve to catalytically transform ozone to O_2, thus reducing the amounts of ozone present in the atmosphere below that predicted by the Chapman mechanism.

Molecular nitrogen does not typically involve itself in the chemistry of the atmosphere except as a third body because of its exceptional bond strength, although many compounds containing nitrogen are of major importance in the atmosphere. Among these is the ubiquitous NO, which is formed in a number of different ways. NO can be of natural or anthropogenic origin. Naturally occurring sources range from lightning produced dissociation of N_2 followed by abstraction of an oxygen atom from O_2, to microbial processes which release NO. The man-made sources of

NO arise from high temperature combustion processes, where NO is in equilibrium with N_2 and O_2, and biological emission of NO in the process of degradation of man made nitrogenous fertilizers. Most NO in the stratosphere is produced by the reaction with N_2O of the first excited state of the oxygen atom $O(^1D)$, mainly arising from photolysis of ozone.

NO can be further oxidized to NO_2 and subsequently on to HNO_3. Stratospheric NO is one of the catalytic species responsible for the destruction of ozone through the following cycle.

	k (298 K)	
$NO + O_3 \rightarrow NO_2 + O_2$	1.4×10^{-14}	$cm^3\ molecule^{-1}\ s^{-1}$
$NO_2 + O \rightarrow NO + O_2$	1×10^{-11}	$cm^3\ molecule^{-1}\ s^{-1}$

The rates of these reactions are such that this cycle competes effectively with the Chapman mechanism for the destruction of ozone. Most oxygen atoms reform ozone. For an oxygen atom to remove ozone it must either react with ozone with a rate constant of 10^{-15} or with NO_2 with a rate constant of 10^{-11}. Thus, when the mixing ratio of ozone is 5 ppm, only 5 ppb of NO_2 would double the rate of loss, thus halving the steady state ozone level all on its own. Actual measured NO_2 concentrations are of the order of 2 ppb. NO_x contributes significantly to ozone removal, giving rise to congressional hearings and National Academy Reports as to whether stratospheric flying SST aircraft should be allowed to deposit NO in the lower stratosphere *(7)*.

Chlorine atoms seem like an unexpected species to be involved in atmospheric chemistry because there don't seem to be any naturally occurring chlorine sources. The oceans are however a major source of methyl chloride. Another source of methyl chloride is biomass burning, some from natural causes and some from man-made fires. Most of the methyl chloride is removed from the atmosphere by OH initiated oxidation and incorporation into rain. However, a small amount of the CH_3Cl gets to the stratosphere during its one year tropospheric lifetime. Once into the stratosphere approximately one third of the methyl chloride is photolyzed producing chlorine atoms. Chlorine atoms then catalytically destroy ozone through the following cycle.

	k (298 K)	
$Cl + O_3 \rightarrow ClO + O_2$	1.2×10^{-11}	$cm^3\ molecule^{-1}\ s^{-1}$
$ClO + O \rightarrow Cl + O_2$	3.8×10^{-11}	$cm^3\ molecule^{-1}\ s^{-1}$

Note that the rate constant of the second reaction is even larger than the reaction of NO_2, thus making ClO a four times more effective catalytic ozone remover than NO_2 on a molecule for molecule basis.

Various forms of hydrogen atoms are present throughout the atmosphere, from ground level to outer space. In fact as one reaches the outermost limits of the atmosphere, the major gas becomes hydrogen because of its large scale height above the turbopause. At lower levels in the atmosphere, the hydrogen is mostly tied up in the form of water, a small but significant, fraction is present in the form of OH and HO_2 radicals, the "odd hydrogen". These two radicals are involved in the catalytic destruction of ozone through the following two cycles.

Cycle 1

$$H + O_3 \rightarrow OH + O_2$$

$$OH + O \rightarrow H + O_2$$

Cycle 2

$$OH + O_3 \rightarrow HO_2 + O_2$$

$$HO_2 + O_3 \rightarrow OH + 2O_2$$

OH is also one of the important species in the combustion of hydrocarbon fuels.

If one models the concentrations that should result in the atmosphere for the measured concentrations of the NO_x and ClO_x, uses the rates of the reactions, then one would quickly find that the resulting concentrations of ozone would be too low. This prediction then means that either the rates are wrong, or that some other species must be tying up the active species in the catalytic cycles. In fact, there are several stratospheric reservoir species which serve to tie up hydrogen, chlorine and nitrogen catalytic chain carriers. Some of these species, such as HNO_3 (nitric acid or hydrogen nitrate) and $ClONO_2$ (chlorine nitrate) serve to tie up two of the catalytic species. $ClONO_2$ also serves to illustrate that all of the catalytic cycles are interwoven, with species serving various functions in various cycles.

Stratospheric Chemistry

Stratospheric chemistry is dominated by the chemistry and physics of ozone since ozone is the molecule which plays such a large part in the profile of the reactions occurring in the stratosphere. Figures 3–5 present Nicolet diagrams which symbolically represent the interactions among the major species in each of the catalytic cycles in the stratosphere. In the NO_x cycle ClO causes the conversion of NO_2 to $ClONO_2$, a step which is also present in the diagram for ClO_x as the conversion of ClO to $ClONO_2$ by NO_2.

Many of the species involved in the stratospheric chemistry are formed in the stratosphere by photochemical or kinetic processes. Upper stratospheric photochemistry is dominated by the absorption of photons with wavelengths shorter than 220 nm by the oxygen molecule. Absorption in the Schumann–Runge region results in the dissociation of the O_2 molecule via either direct dissociation or predissociation. The resulting oxygen atoms are formed in the ^1D state (one atom) and the ^3P ground state (the other atom). Quenching and third order combination with oxygen then lead directly to the production of O_3. The photochemistry of the lower stratosphere is dominated by processes which occur at photon energies between the oxygen cutoff and the visible (220–380 nm). Here a dominant photochemical event is the photodissociation of ozone, which has a broad absorption

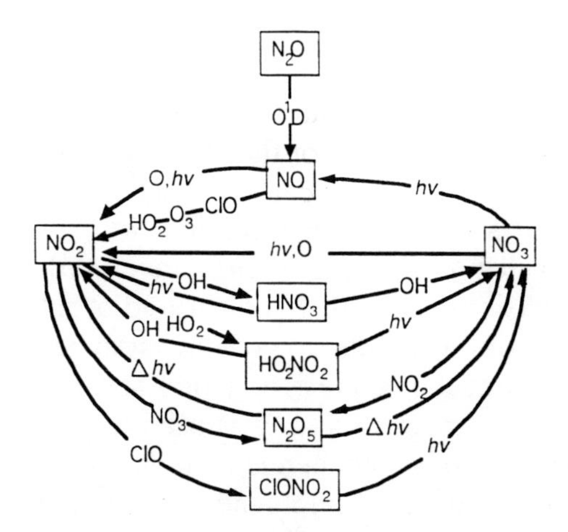

Figure 3. Chemical cycles for NO_x species.

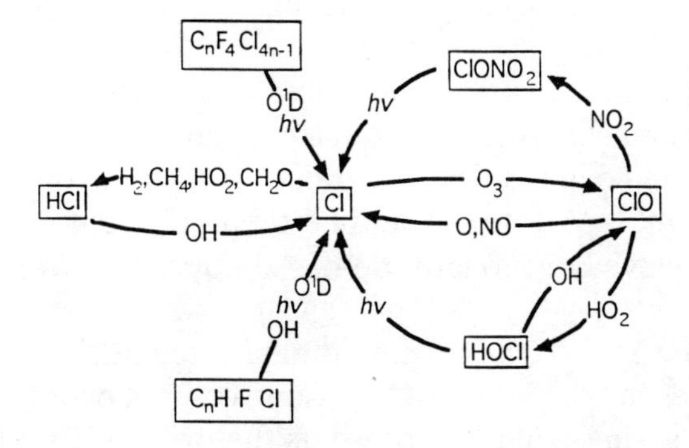

Figure 4. Chemical cycles for ClO_x species.

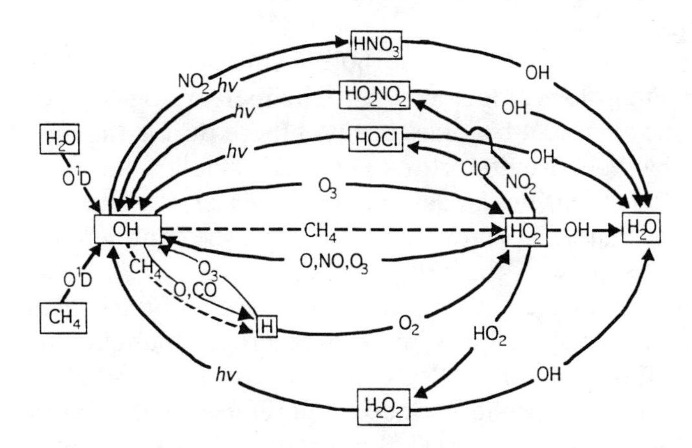

Figure 5. Chemical cycles for HO_x species.

feature from 220–320 nm. This absorption results in the dissociation of ozone to an oxygen molecule in its ground state and an oxygen atom in the ^1D state.

Another primary photochemical event of importance in the stratosphere occurs because of the increasing energy of the accessible photons as you increase the altitude. Whenever a chlorinated hydrocarbon reaches high enough altitudes in the stratosphere to intercept some 200 nm radiation, it's immediate response is the cission of a carbon-chlorine bond. For example, a man-made chlorofluorocarbon will photodissociate to a chlorofluoromethyl radical and a chlorine atom. This is an important event because of the production of the catalytically active chlorine atom.

As discussed, most of the ozone production occurs at high altitudes over the equator where there is the most hard UV sunshine. The ozone produced then participates in a combination of photochemistry and atmospheric flow. As the ozone flows polewards and downwards the photochemical time constants become longer and longer, and the ozone pattern in the lower stratosphere is dominated by flow. Because there is less heating at the poles the stratosphere comes nearer to the ground at the poles. Although the highest mixing ratios of ozone are found over the equator, the highest concentrations, and the highest column abundances are expected over the poles.

Since there is more sunlight and less ozone over the equator, equatorial peoples have developed skin pigmentation (melanin) to absorb solar UV. Scotsmen were not designed to live in tropical Australia. If man's activities were to lower the average ozone column by 5% this would be the equivalent of involuntarily moving the population of Denver to the latitude of Albuquerque *(8)*.

The Antarctic Ozone Hole One of the dominant scientific news stories of the last several years involves the depletion of the ozone column over the antarctic continent during the onset of spring. Figure 6 presents the original data from the Halley Bay station showing a depletion of the ozone column by some 70% during October (immediately following sunrise after the dark winter) in the Antarctic *(9)*. The level of interest in the popular press initially exceeded the ability of scientists to explain this phenomenon.

We now know that the ozone hole forms because of the combination of chemical and meteorological factors *(10)*. The meteorological factor is that the air over the antarctic continent becomes trapped there during the antarctic winter and into the spring by a persistent vortex. There is a similar, but less persistent vortex over the North Pole. During the dark antarctic winter, the barrier to air circulation appears at approximately 73S latitude, the air over the continent becomes very cold, and does not exchange with the rest of the atmosphere. There is very little water in the stratosphere, but the extreme cold allows ice and nitric acid/ice particles to freeze. These ice particles allow some new heterogeneous chemistry to take place, namely the transformation of $ClONO_2$ and HCl to $Cl_2(g)$ and $HNO_3(s)$. The latter compound forms the stratospheric clouds, and removes the stratospheric NO_2. The former is capable of absorbing the 350 nm radiation available at those latitudes at sunrise (spring). The solar dissociation of Cl_2 releases chlorine atoms which remove O_3 and form the ClO free radical. If there were any NO_2 left in the stratosphere it would remove the ClO as $ClNO_3$, but it has all been removed. The result is a massive buildup of ClO. Eventually the ClO radicals reach a high enough concentration that their self reaction becomes appreciable. The reaction between two ClO free radicals is not fully understood, but it has been known since 1967 that active forms of chlorine result. Thus the ice allows for the removal of NO_2 and the release of Cl_2; the Cl_2 can be photolyzed by the available near UV sunlight to liberate chlorine atoms, and in the total absence of NO_2 the ClO radicals build up to the point that their self reaction liberates further ozone destroying chlorine atoms in a catalytic cycle requiring only near UV (350 nm) radiation.

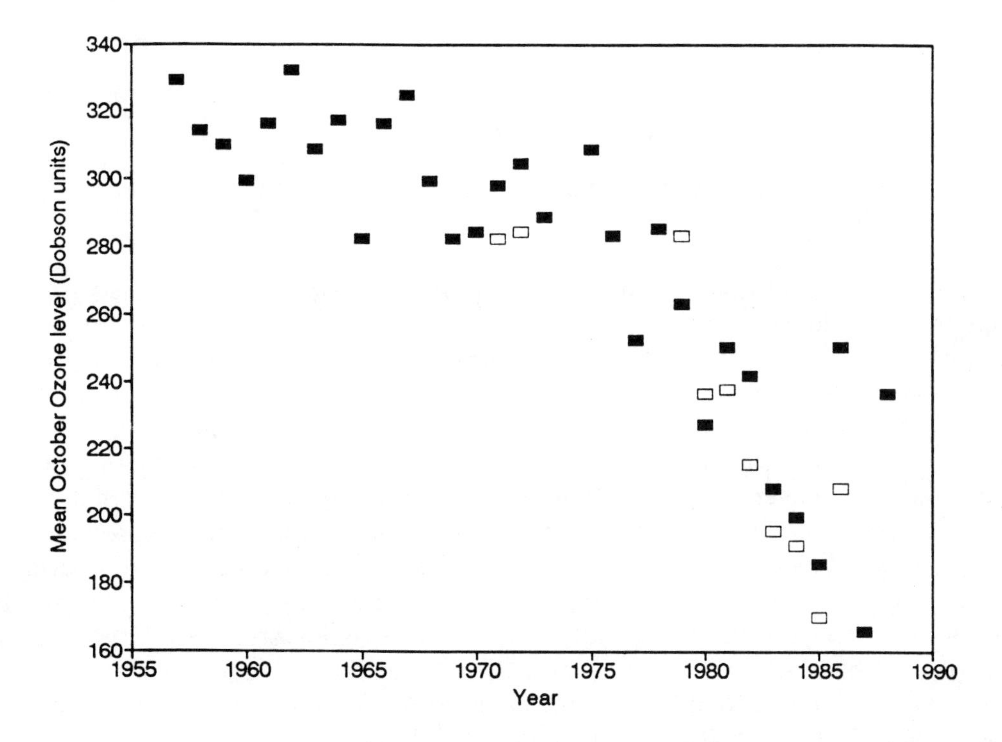

Figure 6. October Ozone concentrations over Halley bay. The closed squares are from ground based instruments, while the open squares are from space based instruments.

Tropospheric Chemistry

Tropospheric chemistry is very complex because there are many more species involved with widely divergent lifetimes. Many of the reactions which occur in the stratosphere occur in the troposphere, however in the stratosphere there are very few molecules containing more than one carbon. Tropospheric chemistry includes a full, and highly complex range of multiple carbon organic oxidation chemistry. Atmospheric oxidation of organic species involves an impressive range of oxidized and partially oxidized products. Even in the simplest case, methane oxidation, intermediates on the way to CO_2 include CH_3COO, CH_3O, CH_2O, CHO, CH_3OOH and CO. Some of the step connecting these species, and the central role of the odd hydrogen species in the various transformations are illustrated in Figure 7.

Among the species which become important in the troposphere are liquid water and olefinic hydrocarbons. Automobiles produce propene, trees produce isoprene, both species are so reactive in the troposphere that they have no chance

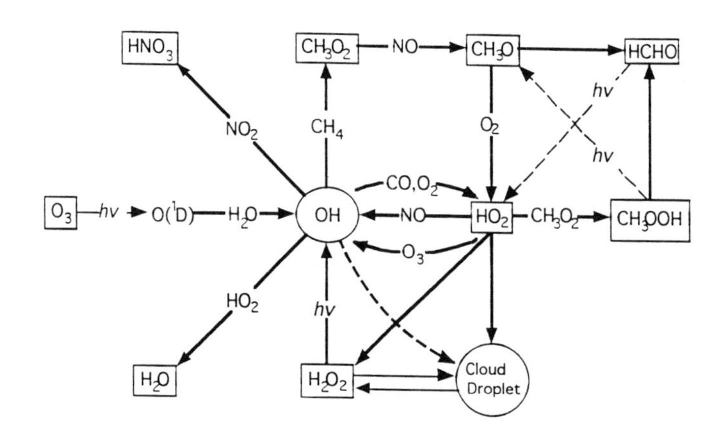

Figure 7. Schematic diagram of the chemistry of the troposphere emphasizing the OH and HO_2 radicals.

of entering the stratosphere. Water is important in vapor, liquid and ice phases. The water cycle is a complete course in a hydrology curriculum. Using our simplified homogeneous spherical assumption there is an average of 1 m of waterprecipitated globally per year. There is an average of about 1 cm of water (liquid) mostly in the form of vapor over the whole globe, thus on average a water molecule spends 1/100 years or three days in the atmosphere before being rained out. There is an average of only 1 mm of water in the form of cloud droplets, 90% of which evaporate and only 10% of which grow enough to fall to the ground.

In the vapor phase water is both the sink and the source for odd hydrogen containing species frequently abbreviated HO_x. Liquid water plays an important role in the troposphere, that of condensation. As droplet size gets larger, those molecules which are water soluble will dissolve in the liquid water droplets and eventually fall to the ground. Rain and cloud formation are important removal and conversion mechanisms for the pollutants present in the atmosphere.

Photochemistry in the troposphere There are basically three conditions which we consider for air pollution, atmospheric haze, "London" (sulfurous) smog, and "Los Angeles" (photochemical) smog. In all three cases the visible "haze" arises from particle formation. The particles in the atmosphere may arise from a number of sources, dust storms, smoke or photochemistry. The hazy appearance is caused by the scattering of light by particles of a size similar to the wavelengths of visible light, and results in a loss of resolution available to the eye. The atmospheric haze often observed on the east coast of the US can be attributed for most of its mass to the growth of water droplets in the atmosphere at relative humidities lower than 100%. This growth arises because materials are dissolved in the water which lower the vapor pressure sufficiently for the water to condense. These hygroscopic materials can be as simple as sea salt previously evaporated from ocean spray, however in the Northern Hemisphere the major hygroscopic nucleus is usually partially neutralized sulfuric acid in the approximate form ammonium bisulfate. The sulfate is mostly man made by photochemical oxidation of emitted SO_2. Details of light scattering, particle growth, etc. are particularly well presented in the text book "An Introduction to Air Pollution" by Butcher and Charlson *(11)*.

The worst "London" smogs were multiday continuous episodes of dirty foggy air. They occur in winter when there is little sunlight at 52 N latitude. Elderly or asthmatic patients showed a significantly increased death rate. These episodes are becoming less common as the households use less coal in badly controlled hearths. The first control laws required only the burning of "smokeless fuels". This law caused a reduction in the soot content of the air without significant reduction in the SO_2 concentration. The death rate seemed to go down sharply. Apparently the combination of SO_2 and soot was capable of killing susceptible individuals. Soot particles are just the same size as cigarette smoke particles (0.05 - 0.5 μm diameter), and are perfectly designed to lodge in the deeper recesses of the lung. They are small enough to travel with the air around the curved passageways of the nose, trachea, and bronchii, but diffuse to the walls of the tiny alueoli. Soot is an extremely complex and heterogeneous material. The chemistry of soot in the atmosphere has been studied by many groups, but much remains to be learned.

"Los Angeles" or photochemical smog is formed by a different mechanism entirely. It relies upon the sunlight available to initiate the chemistry involved. The breakthrough was in 1952 when Haagen-Smit et al *(12)* recognized LA smog as a new, and ozone related phenomenon. It was shown that a mixture of auto exhaust and air could produce ozone when exposed to sunlight. The important ingredients are oxygen, olefinic hydrocarbons and the oxides of nitrogen. It is curious that photochemists had not discovered this interesting chemistry. All three compounds were added to other photochemical systems to inhibit free radical reactions, but their photochemistry together at low concentrations had not previously been investigated. Most photochemical studies were carried out with pure materials, or with dilution down to percent levels with inert gases. The photochemical formation of ozone from the above ingredients does not occur at percent concentrations, and a further three orders of magnitude of dilution are required before the interesting effects are observed.

If a mixture of ppm levels of NO and propene in air is photolyzed with sunlight, or in a so called "smog chamber" during a period of a few hours all the NO is converted to NO_2. This is in itself a counterintuitive phenomenon since the only significant light absorber, namely NO_2, increases in concentration as photolysis proceeds. After this conversion ozone begins to appear, reaching a peak a few hours later as both the propene and the NO_2 disappear from the system. Figure 8 shows the results of a typical smog chamber run.

There is always a little ozone in clean air, so when an automobile emits NO some of it is oxidized to NO_2 via

$$NO + O_3 \rightarrow NO_2 + O_2$$

this NO_2 is photolyzed by sunlight thus:

$$NO_2 + h\nu \rightarrow O + NO$$

At one atmosphere pressure the oxygen atom formed becomes ozone within a few microseconds by the reaction:

$$O + O_2 + M \rightarrow O_3 + M$$

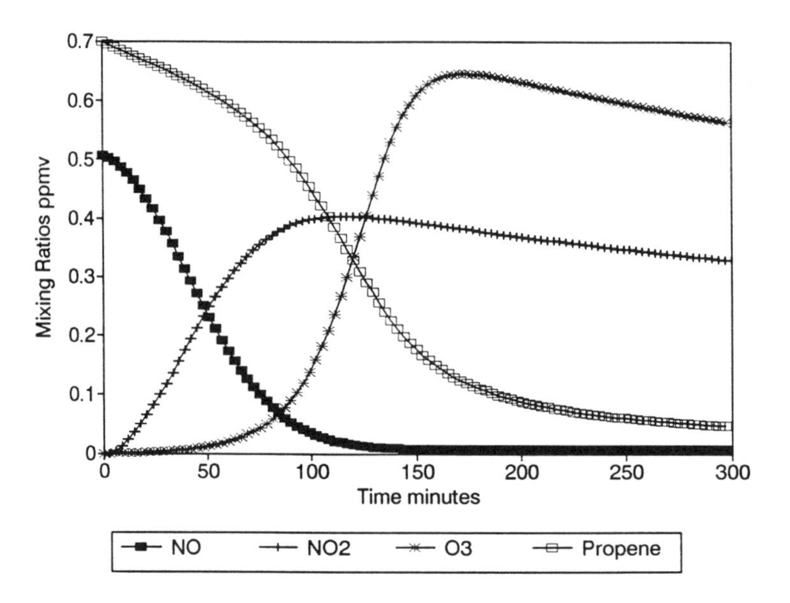

Figure 8. Results from a typical smog chamber run.

These three reactions were recognized as important early in the 1950's, however it was also recognized that these three reactions by themselves were cyclic and could neither form significant amounts of ozone, nor convert significant fractions of the ambient NO to NO_2.

The three reactions are all fast. To a first approximation one can consider them to be in steady state, thus the rate of formation and removal of NO_2 can be set equal, and the equation rearranged to show that:

$$[O_3] = j(NO_2)[NO_2]/k(NO+O_3)[NO]$$

It was only in the late 1960s that it was recognized that a few HO_x species could enter the cycle catalytically and cause the observed effects.

$HO_2 + NO \rightarrow NO_2 + OH$ is a reaction which converts NO to NO_2 without removing an ozone. The result is to bypass one side of the cycle, and to leave an extra ozone every time the HO_2 reaction occurs. If this process were to occur without limit, however slowly, then one can imagine the [NO] in the denominator being reduced to zero and the ozone concentration going to infinity. This never occurs, but since the ozone is made from oxygen, it is not unusual to observe more ozone than the reaction mixture originally contained NO or propene.

The role of the propene is to recycle the OH free radical back to HO_2. Propene oxidation which accompanies this reaction actually oxidizes many HO_2s to OH. The details of the chemistry by which this occurs are beyond the scope of this chapter, however the last step involves the formation of a carbon monoxide molecule. This oxidizes the last of the OHs by:

$CO + OH \rightarrow CO_2 + H$ which is followed immediately by

$$H + O_2 + M \rightarrow HO_2 + M$$

Oxidation of NO and olefins to ozone and NO_2 has (in itself) produced no "haze". Haze is formed when gas phase molecules are oxidized to products with negligible vapor pressure. If these products are also hygroscopic then haze formation is enhanced. The active free radical chemistry going on in photochemical smog ensures that any sulfur gases (H_2S or SO_2) rapidly become H_2SO_4, a very effective haze forming material. High molecular weight volatile olefins (including biogenic terpenes) are oxidized to the corresponding involatile carboxylic acids. See later for a lecture demonstration of this phenomenon.

Detection of Atmospheric Pollutants

Government regulations are such that a number of organizations such as electrical utilities and oil refineries were required to monitor air pollution in the neighborhood of their facilities, but no longer have to do so. The result is that there are often available more or less older, but usually well maintained and functioning air monitors often accessible as gifts. A little knowledge of the ways in which air monitors function make useful free equipment available for research or for teaching purposes. For instance the fact that all chemiluminescent NO_x detectors for ambient air monitoring contain an electrically cooled red-sensitive photomultiplier tube together with power supply and electrometer can be used to turn a monochromator into a high sensitivity spectrophotometer.

Non-Methane Hydrocarbon Detector There have been several devices sold, none of which was ever entirely satisfactory. All are based on the concept that a flame ionization detector (FID) measures an approximation of the total HC including methane. The most successful of these devices consisted of a short gc column. A sample of air was injected and left on the column until the methane peak eluted. The column was then backflushed to the detector providing a single non-methane HC (NMHC) peak. These devices contain lots of plumbing, solenoid valves, an FID and an electrometer.

Non-dispersive IR devices Carbon monoxide, carbon dioxide and hydrocarbons in automobile exhaust are routinely determined by means of nondispersive infrared (NDIR). These devices usually contain IR detectors, usually but not always with built-in interference filters for the wavelengths of interest. They always contain a useful sampling pump and a flow-meter or two.

Ozone detectors Almost all modern ozone detectors are based on UV absorption at 257 nm. These devices contain a mercury lamp and very stable power supply, as well as a sampling pump and associated electronics. Older ozone monitoring equipment is based on the chemiluminescent reaction of ozone with ethylene. These instruments are very often available since they have been superceded. They contain a lot of useful stuff, not the least of which is a high sensitivity blue sensitive photomultiplier tube and all associated electronics. In most versions the PM tube was electrically cooled.

Other Chemilluminescence based detectors Nitric oxide is detected based on its chemiluminescence with ozone. Since the chemiluminescence is both weak and in the red, a cooled red-sensitive photomultiplier tube is always a component. Because ozone must be made in-situ, the instruments are also equipped with a high voltage (corona discharge) ozone generator.

Sulfur containing compounds are frequently determined by means of the blue and near-UV S_2 emission resulting when the compounds are present in a hydrogen flame. Again the emission is weak so high quality photon detecting devices are a part of the apparatus.

SO_2 fluorescence detectors A second method of detecting SO_2 in air is based on its fluorescence when exposed to UV radiation. The UV source is either a D_2 lamp, or in some units a pulsed xenon arc lamp. In both cases a sampling pump and high quality photon detecting devices are a part of the system. In every case the instrument manuals explain the details of the instrument operation and components.

The Classroom

Automobile exhaust to brown cloud to acid rain all in twenty minutes. It is possible to synthesize NO gas but easier if a cylinder is available. Insert about 4 mL of water and about 150 torr of NO into an evacuated 1-L glass flask with vacuum stopcock. During lecture first ask the color of nitric oxide. Someone is sure to say "brown". Show the flask and tell the class that there is a lot of NO in it, and that NO is a common component of auto exhaust. At high concentrations the rate of $2NO + O_2 \rightarrow 2NO_2$ is very fast so when you open the cock air rushes in and the contents turn russet brown. Set the flask on one side with the stopcock closed and return to it every few minutes. The color fades ultimately to colorless. The simple explanation of the process is that there is excess oxygen which in the presence of water oxidizes the NO_2 further to HNO_3 which dissolves in the water, thus auto exhaust to brown cloud to acid rain in a few minutes. Actually depending on the O_2 and NO relative amounts the reaction mixture often contains a significant amount of nitrous acid, HNO_2 which is however also colorless. If this is the case there is always some residual NO and opening the stopcock again later causes a return brown flush. One needs to be careful in the explanation since the reaction in the flask is much too slow to convert NO to NO_2 in the atmosphere, as discussed earlier O_3 and HO_2 are actually the culprits. Also there are very few instances when the NO_2 content of the air causes a significant brownness compared to the particulate scattering of light and its wavelength dependence.

Photochemical aerosol formation A very clean 2–5 L beaker is capped with a suitable inverted watchglass. It is filled with a low concentration of ozone either by effluent from an ozonizer, or better by allowing a mercury lamp to sit inside for ten to twenty minutes while illuminated. The second method has the advantage that the ozone is photochemical, albeit produced by 185-nm photolysis of oxygen, as in the stratosphere. The lamp or ozone source is removed and the room placed in total darkness except for a light beam going through the beaker perpendicular to the audience. I have used a slide projector facing sideways, or an overhead projector with a black cloth covering the lens. A lemon is introduced at this point to symbolize an automobile. A tiny flake (2mm×5mm×0.2mm thick is plenty) of peel of the lemon is removed and dropped into the beaker and the cover replaced. In a few moments whispy smoke should be observed arising from the lemon peel and ultimately filling the beaker.

Lemon peel produces limonene and other large olefinic terpenes in the vapor phase. The olefins react with the ozone to make among other things organic acids which have negligible vapor pressure and are hygroscopic. The trail of "smoke" is the tiny particles so formed. This is a great illustration of how few particles scatter how much light. One can estimate the mass of terpene from the fragment of lemon, and in fact one finds that the mass of ozone (typically 100 ppmv for the photochemical source) is often yet less. The beaker can not be reused without careful cleaning. The "used" surface catalyses the removal of ozone if you try to reuse it. It smells of "old lemon" at the end of the experiment. It is neat to "pour out" the fog into the light beam. The method of detecting small particles is called the Tyndall Effect after its discoverer.

The room as an exponential dilution flask If you ever end up with one of the air monitors described above and it appears to be working it is an interesting exercise to liberate a small quantity of the gas which the monitor detects in the room with the detector. Several electric fans or a number of students beating the air with large pieces of cardboard serve to mix the flask in such a way that steady readings are observed. The readings decay exponentially with a time constant equal to the time constant for the ventilation of the room. Typically this is between ten and twenty minutes. If the data are extrapolated back to zero time, the volume of the room has been measured and the amount of pollutant added is known then the procedure constitutes a calibration of the detector, as well as an interesting illustration of the fact that exponential decay arises naturally. (Many students do not know why natural logarithms are called natural).

Safety

The experiments described are demonstrations. The instructor needs to be knowledgeable about the use of the equipment prior to use. Particular care must be taken with the ultraviolet lamp (exposure to radiation) and the toxic nature of the gases involved. The auto exhaust demonstration uses NO and NO_2, which will be maintained within the flask; and which must be properly disposed of. In preparation of the room as an exponential dilution flask, appropriate calculations should be made so that the exposure to noxious gases is below the appropriate 8 hour OSHA standards, but yet well above the detection limit for the instrument used in the concentration determination.

Acknowledgements

I would like to thank the National Science Foundation for funding under grant NSF-ATM 5-33911.

References

1. Wayne, R. P. *Chemistry of Atmospheres, An Introduction to the Chemistry of the Atmospheres of Earth, the Planets, and their Satellites;* Clarendon Press: Oxford, 1991.

2. Finlayson-Pitts, B. J.; Pitts, J. N. Jr. *Atmospheric Chemistry;* John Wiley: New York, NY, 1986.
3. Seinfeld, J. L. *Atmospheric chemistry and physics of air pollution;* John Wiley: New York, NY, 1986.
4. G. Brasseur and S. Solomon, *Aeronomy of the Middle Atmosphere*, D Reidel, Hingham, MA., 1984.
5. Chapman, S. *Mem. Roy. Meteorol. Soc*, **1930**, *3*, 103.
6. DeMore, W. B., Sander, S. P., Golden, D. M., Molina, M. J., Hampson, R. F., Kurylo, M. J., Howard, C. J., Ravishankara, A. R. *Chemical Kinetics and Photochemical Data for Use in Stratospheric Modeling, Evaluation Number 9*, NASA-JPL Publication 90-1, Pasadena, CA 1990.
7. U.S. National Acad. Sci., 1975, Environmental Impact of Stratospheric Flight.
8. Rowland, F. S. *Ann. Rev. Phys. Chem.*, **1991**, *42* 731.
9. Forman, J. G., Gardiner, B. G., Shanklin, J. D. *Nature*, **1985**, *315*, 207.
10. Solomon, S. *Nature*, **1990**, *347*, 347.
11. Butcher, S. S.; Charlson, R. J. *Introduction to Air Chemistry*; Academic: New York, NY, 1972.
12. Haagen-Smit, A. J., Bradley, C. E., and Fox, M. M., *Industrial and Engineering Chem.*, **1953**, *45*, 2086.

RECEIVED October 1, 1992

CHAPTER 4

Aspects of Surface Science for Emphasis in the Physical Chemistry Curriculum

A. W. Czanderna

The purposes of this chapter are to provide an overview of modern surface science; to provide a brief overview of the roles of compositional analysis, structural determination, and chemical bonding at surfaces in surface science; to provide a brief overview of five widely used methods of surface analysis; and to make suggestions about how the fundamentals of surface science may be taught as a single course or included in undergraduate courses in physical chemistry. Further details about surface science and surface analysis methods can be obtained from literature citations. Both the surface (i.e., the outer monolayer of atoms on a solid) and the interface (i.e., the boundary between two compositionally different solids) are included in the term "surface" when it is used in this chapter.

The rationale for presenting this overview of the most important phenomena in modern surface science related to studying materials and processes is the following: The three phases of interest are the solid (S), liquid (L), and gas (G) phases, none of which is infinite. The boundary region between these phases, i.e., the surface phase, has fundamentally different properties from those of the bulk. The possible boundaries are the S/S, S/L, S/G, L/L, and L/G surfaces. These boundaries are studied in surface science to develop an understanding of phenomena and to develop theories that will permit predicting future events. The emphasis of modern surface science excludes the L/L and L/G interfaces, both of which have been extensively treated as classical physical chemistry of surfaces *(1)*. Some of the broad *topical areas* of study at the S/G, S/L, and S/S surfaces are listed in Table 1. An understanding of these topics is enhanced by applying the modern methods of surface characterization. Since these primarily employ experimental techniques that are used in the vacuum, the topics discussed will be primarily related to the S/G interface. Adequate material is available about the S/L interface *(1)*, although it is currently the least understood interface. Comments about S/S interfaces will be appropriately incorporated into the discussion of S/G interfaces.

For the physical chemistry curriculum, there are two levels to address. The first is a one-semester course in surface science, and the second is how to incorporate aspects of surface science into a standard two-semester course in physical chemistry. For an introductory course, the major challenge for a discipline as broad as surface science is to select the material that can be taught to seniors or to first-year graduate students. The material can be understood by students with three years towards a B.S. in physics, chemistry, materials science, one of the major engineering fields, and from hybrid departments such as engineering physics and applied science. A sequential listing of topical headings for an introductory course in surface science and 50-minute lectures devoted in each area are presented in Table 2. Further details about the material taught under each of the topical headings are presented in the subsequent sections. As will be seen, the main theme of the course is that surface science is primarily an experimental science.

2503–6/93/0043$08.50/0

Table 1. Topical Study Areas at Different Interfaces Between the Solid (S), Liquid (L), and Gas (G) Phases

Interface		Topical Area of Study
Solid A / Solid B	S/S	Corrosion, grain boundary passivation, adhesion, delamination, epitaxial growth, nucleation and growth, abrasion, wear, friction, diffusion, boundary structure, thin films, solid state devices, mechanical stability, creep.
Liquid / Solid	S/L	Wetting, spreading, lubrication, friction, surface tension, capillarity, electrochemistry, galvanic effects, corrosion, adsorption, nucleation and growth, ion electromigration, optical properties, cleaning techniques.
Gas / Solid	S/G	Adsorption, catalysis, corrosion, oxidation, diffusion, surface states, thin films, condensation and nucleation, permeation, energy transfer.

Not all surface scientists will agree with the theme of the course, the topics covered, or the time devoted to each topical area. An entire three-hour course could be offered in many of the *subtopic* areas. While some excellent treatises exist for some of the subtopics, a text completely suitable for the course described does not exist yet. Eight books are particularly helpful *(1-7)* and two books on chemisorption should be considered for supplementary reading *(8,9)*.

A brief overview of aspects of surface science is presented in the next section, and the physical principles of the most important experimental methods are discussed. The subtopical areas in surface science listed in Table 2 are emphasized. Incorporating aspects of these subtopical areas into the standard two-semester course in physical chemistry will be discussed in the final section.

Principal Topics in Surface Science

The principal topics in surface science are summarized in this section and highlighted in Table 2. They are derived from the *fundamental aspects* of surfaces, i.e., surface area and topography; thermodynamics and equilibrium shape; elemental composition; structure and defects, adsorbate interactions, bonding, and location; dynamic surface processes including adsorption, diffusion, and vibration; depth of the surface phase; and electronic distribution at the boundary. Of these aspects, composition, bonding, and structure are the most fundamental parameters that govern the properties of surfaces. For the applied topical areas of surface science (Table 1), surface composition is the most important, which is why most emphasis must be given to it.

Table 2. Principal Topics and Subtopics Covered in 42 Lectures on Aspects of the Fundamentals of Surface Science

No. of 50 min. Lectures	Principal Topic	Main Subtopics
2	Scope, Overview	Historical Overview, Fundamental Parameters, Scope, Physical Adsorption versus Chemisorption
2	Surface Area	BET Area, Porosity
6	Surface Structure, Topography, and Diffusion	LEED, FIM, SEM, surface diffusion, STM, AFM
24	Surface Composition and Composition in Depth	Input-Output Particles, Ions, Neutrals, Electrons and Photons; ISS, SIMS, RBS, AES, XPS
2	Surface Thermodynamics	Thermodynamic Relations, LaPlace and Kelvin Equations, Surface Tension and Free Energy, Depth of Surface Phase, Equilibrium Shape of Surfaces
6	Chemisorption and Nature of Gas/Solid Interactions	Kinetics and Equilibrium of Adsorption, Amount Adsorbed, Nature of Interaction, Ordered Structures, Work Function, Desorption, Vibrational Spectroscopies, Scanning Tunneling Microscopy (STM)

Scope and Overview The breadth of surface science requires restricting the scope of a one-semester course. Of the five types of interfaces, the restriction of interest is first drawn primarily to the S/G interface and second to the S/S interface. The principal applications or topical areas of interest for various interfacial regions (e.g., Table 1) should be *clearly differentiated* from the *fundamental aspects* of surfaces. The latter must be addressed at the introductory level in the physical chemistry curriculum because using one or more of the topical areas (Table 1) as a vehicle for discussing some of the aspects of surfaces may lead to confusion between fundamental understanding and applications of it.

Experimental work conducted to study surfaces is now very extensive; the theoretical treatments are difficult. In science, it is customary to adopt a model based on an ideal situation and to compare the behavior of real systems with the ideal model. What is a realistic view of the boundary at a solid surface? It is not the ideal plane of infinite dimensions, but on an atomic scale it consists of different crystal planes with composition, structure, orientation, and extent that are fixed by the pretreatment of the solid. Imperfections such as an isolated atom, a hole, an edge, a step, a crevice, a corner, and a screw dislocation may also coexist on the surface (see Figure VII-6 in reference 1.). Wide variations in the microscopic topography may also adversely influence the stability of the surface.

Solid surfaces are frequently treated in a variety of ways, e.g., outgassing, chemical reduction, flashing a filament, ion bombardment, cleavage, field desorption, or depositing a thin film to minimize uncertainty about the initial composition of the surface *(10)*. However, impurities may accumulate during some of these treatments at or in the boundary, in trace or larger quantities, and drastically alter the behavior of the boundary. Following a controlled use of the solid, a reexamination of the surface permits evaluation of the influence of that use on the measured properties of the surface. Characterization of the surface before, after, and if possible, during the use of the material is clearly required.

For an overview of characterizing a solid surface, consider these questions: How much surface is there and where is it located? Is the surface real or clean? What solid form does the surface have? What is its topography and structure? What thermodynamic processes occur? How do surface species migrate? What is the equilibrium shape of surfaces? What is the depth of the surface phase? How much gas or liquid is adsorbed and where? What is the nature of the adsorbate-solid interaction? How should these phenomena be studied? The history of studying these effects in gas adsorption on solid surfaces alone leads us to recognize that careful experimentation is the primary necessity in surface science. As each of the characterization questions are briefly addressed in the following paragraphs, the experimental methods deemed most appropriate for studying materials will be indicated.

The concept of a surface is not well defined, especially for most samples of industrial interest. The structure and composition of a material often deviate from their bulk values at depths of only nanometers and in a spatially inhomogeneous manner. As is well known, each method of surface spectroscopy has its own depth and areal resolution and, therefore, presents its unique view of the boundary region of the solid *(11)*. The information available from surface analysis includes identifying the elements present, their lateral distribution, and their depth distribution; structural and topographical information can also be obtained.

Compositional analysis, the primary subject of interest for a course, involves determining three quantities. The elemental identity, i.e., the atomic number, is of primary interest. However, it is also desirable to know the chemical state of the species, e.g., whether it is elemental, oxidized, or reduced. Finally, it is necessary to determine the spatial distribution of the chemical species. An important trend in compositional surface analysis is the demand for greater and greater lateral resolution, e.g., a resolution of 20 nm can now be routinely reached using Auger electron spectroscopy. The study of compositional differences between grains and the grain boundary region is extremely important for deducing how these differences change the properties of the materials.

Structural surface analysis also involves three levels of desired information. For an ideal, atomically flat, single crystal, the structure is specified by the geometry in each cell of the surface unit mesh. Secondly, real surfaces contain defects, such as dislocations, steps, kinks, ledges, and grain boundaries. Finally, a new trend is emerging for determining local atomic order for a particular chemical species. Determining surface structure is essentially limited to studies of solid-vacuum (S/V) or S/G interfaces.

Most studies of chemical bonding at surfaces are related to the S/G and S/L interfaces *(5-9)*. The latter can be probed with ion and electron beams but only with difficulty. The adsorption of gases on solids is a significant branch of surface science. Adsorption is involved in many processes (e.g., corrosion, contamination, and catalysis) of interest to the physical chemist and may be used to measure properties such as the area, acidity, and basicity of surfaces. Prior to discussing surface area, potential energy diagrams for S/G interactions should be introduced and the differences between physical adsorption and chemisorption discussed *(12)*.

To complete the introduction, a generalized set of questions are posed that should be used to describe a method of study or a measurement technique. These are: What is the physical mechanism or process? What is being measured and what can be done with (or learned from) the measured quantity? What is the experimental apparatus and how is the measurement made? What are the advantages or opportunities? What are the problems or limitations? What kind of typical results have been obtained? What is the future potential? These categories are particularly helpful for students who will be confronted with approximately 35 different methods or techniques in the succeeding lectures.

Real and Clean Surfaces, Surface Atom Density, and Solid Forms Real and clean surfaces are introduced by indicating that surfaces do not consist of infinitely smooth planes, but may have steps, edges, vacancies, corners, crevices, dislocations, grain boundaries, etc. The necessity of studying real surfaces in applied science is followed by a more detailed amplification of why clean surfaces must be studied in order to secure an understanding of surfaces and the S/G interaction. The concepts of studying ideal or model systems with restrictions and controls on the many variables are compared with models from various scientific disciplines such as the kinetic theory, a dilute solution, the point particle, harmonic oscillator, or an Einstein crystal.

Broadly speaking, real surfaces are those obtained by ordinary laboratory procedures, e.g., mechanical polishing, chemical etching, industrial processes, etc. Such a surface may react with its environment and be covered with an oxide (generally), possibly with chemisorbed species, and by physically adsorbed molecules from the surroundings. Real surfaces have been studied extensively because they are easily prepared, readily handled, and amenable to many types of measurements. The real surface is the one encountered in most practical applications.

For obtaining clean surfaces, various solid forms are used, such as powders, foils, vacuum-prepared thin films, coatings, filaments, cleaved solids, field emitter tips, single crystals, and polycrystalline solids. These may be formed from metals, semiconductors, and compounds. Clean surfaces, which may be obtained by outgassing, chemical reduction, cleaving, field desorption, ion bombardment and annealing, preparing a thin film, or flashing a filament in ultra high vacuum (UHV) *(10)*, are more difficult to prepare and keep clean. Clean single crystal surfaces are especially useful for comparing theories with experimental results. They can be prepared from ordered single crystals and then perturbed in carefully controlled experiments. Fundamental studies on clean, well-ordered surfaces *(3)* have increased the basic understanding of S/G interactions in the modeling tradition of the kinetic theory, the dilute solution, the point particle, and the harmonic oscillator. Studies with clean, single-crystal surfaces are directly applicable to semiconductor devices *(6)*.

The surface atom density of a typical solid is of the order of 10 atoms/nm^2. The time it takes to form a monolayer of adsorbed atoms on a clean solid surface is given by $t_m = (N/A)(1/\nu\beta)$, where N/A is the surface atom density (atoms/cm^2), ν is the arrival rate of the impinging gas, and β is the sticking coefficient. From the kinetic theory, $\nu = P/(2\pi mkT)^{1/2}$, where P is the pressure, m is the mass of the gas molecule, k is Boltzman's constant, and T is the temperature in K. At a pressure of 10^{-6} Torr (0.133 mPa) and 300 K, ν ranges from 3 to 5 collisions/nm^2-s for the molecular masses of gases in unbaked vacuum systems. When $\beta = 1$, t_m will be 2 to 3.3 s for a typical solid. However, $t_m = 2.5 \times 10^5$ s for $\beta = 10^{-5}$ at the same pressure (0.13 mPa). This shows students that under some conditions three days *may* be required to contaminate a clean surface in a high vacuum. The instructor will also want to discuss the cases if the sticking coefficient is small or if β decreases sharply from the initial sticking coefficient. While β is much less than unity for many gas/solid encounters, especially when a monolayer of adsorbate coverage is approached, this calculation provides the sobering realization that pressures of 10^{-9} Torr or less are necessary in a surface analysis system for general applicability, especially when reactive surfaces are studied. Otherwise, the composition measured may be that formed by reaction with the residual gases in the vacuum (e.g., H_2O and CO in an unbaked system or H_2 and CO in a baked system). A nomograph *(15, Fig. 1.1)* shows the interrelationships of pressure, degrees of vacuum, molecular incidence rate, mean free path, and monolayer formation time when β is unity.

Surface Area Following Langmuir's pioneering work in the 1910s, in which the importance of structure, composition, and bonding to chemisorption was demonstrated, Brunauer, Emmett, and Teller (BET) provided a means for deducing the surface area from multilayer physical adsorption isotherms *(13)*. The BET method for analyzing adsorption isotherms is used routinely by hundreds of laboratories with commercially available equipment. When a significant part of the surface is located internally, hysteresis is observed between the adsorption and desorption branches of the isotherm. The hysteresis results from capillary filling of internal pores. An excellent treatise on the surface area and porosity of solids is available *(14)*, extensive examples from recent studies of this subject have been cataloged *(12)*, and the parameters used to differentiate between physical and chemical adsorption have been tabulated *(12)*. Visual observation of the topographical features of solids is possible using electron microscopic (EM) examination of replicas of surfaces, or a scanning electron microscope (SEM); light scattering, the stylus method, and scanning tunneling microscopy (STM) also provide information about topography. An assessment of the external surface area can be made from SEM or EM photographs, but obviously the internal area is not directly observable.

Thus, qualitative and quantitative answers can be provided to questions about how much surface there is and where it is located, except for S/S surfaces.

Structure and Topography A clear distinction must be made between structure and topography, which are frequently used interchangeably (and incorrectly) in the literature. Topography can be illustrated with SEM photographs showing contouring, hills and valleys, and superatomistic surface features *(16)*. Structure refers to the repetitive spacing of atoms in a surface grating. The results of surface structural determinations show that the lattice spacing in the bulk and at surfaces is the same for most metals, but that halide and many semiconductor surfaces may be relaxed or reconstructed *(3)*. The most common methods of deducing surface structure are by the diffraction (D) of low energy electrons (LEE), either elastically or inelastically (ELEED and ILEED), by reflected high energy electrons (RHEED), and by using the imaging techniques of the field ionization microscope (FIM) *(1-8)*. Visualization of surface atom order can be obtained from scanning tunneling microscopy (STM) and atomic force microscopy (AFM) *(7)*. The value of the LEED, RHEED, and FIM techniques to fundamental surface science is evident in the extensive results that are secured by using them; many additional examples can be found in the journal *Surface Science*.

When considering the determination of structure of single crystal surfaces by LEED, several descriptions are especially good for the coherence phenomena of elastically scattered electrons *(2-8)*. It is important to summarize the principal findings from LEED, but also to note its limitations. A brief description of RHEED and ILEED is also appropriate at this stage *(2)*. For field emission microscopy (FEM) and FIM, the descriptions *(2-8)* can be amplified considerably by using the concise treatment in Gomer's book *(17)*. This book has excellent potential energy diagrams of electrons and ions near an emitting tip. Some outstanding examples of field ion photographs of dislocations, grain boundaries, and vacancies are collected in a chapter by Müller *(18)*. Examples of the utility of both methods for securing qualitative information about the structural specificity of adsorption are also given. The aspects of using FIM for studying surface diffusion processes can be presented qualitatively from published photographs and/or quantitatively by drawing either from review chapters or one of several books.

The incorporation of FIM into a time-of-flight mass spectrometer for carrying out surface analysis by field desorption processes, described in considerable detail by Müller *(19)*, is a natural topic at this point. Furthermore, Block's outstanding method *(20)* of field-ion-mass-spectrometry for studying surface processes and catalytic reactions is worth at least half a lecture period. The importance of new techniques for studying the structural specificity of surface reactions and the short times required for a catalytic reaction adds obvious breadth to the complexity of understanding reactions at the S/G interface.

Surface Composition or Purity and Compositional Depth Profiling The properties of both real and clean surfaces often depend on the presence of chemical groups, e.g., impurities, that are extraneous to the bulk composition. These tend to prevent the self-minimization of the surface energy of the solid. They also influence the growth kinetics, topography, surface diffusion coefficients, and residence time of adsorbed species. Depending on the surface energy, impurities may concentrate at the surface or be incorporated into the bulk. These processes, which involve mass transport to, from, and along the surface, may produce significant time-dependent changes in the properties of the material. Therefore, being able to identify the elemental composition of solid surfaces is *crucial* for both scientific and technological studies.

Until about 24 years ago, experimental techniques for identifying the elements on a surface did not exist. Since then, commercial instruments have become readily available and have been developed for the single purpose of measuring the elemental composition of surfaces. In most surface analysis systems, the sample is introduced from air into an analysis chamber held at 10^{-9} Torr or better via a fast entry air lock. A third intermediary chamber is frequently used because of vacuum design considerations as well as to provide a convenient location for outgassing or processing the sample, or both. All as-received samples introduced from ordinary laboratory air have surfaces contaminated with adsorbates from the atmosphere. Carbon from various adventitious carbon containing gases is the principal contaminant, but O, Cl, S, Ca, and N are also commonly detected. The contamination layer may be removed by ion bombardment, but wary analysts are careful to watch for surface diffusion of contaminants from the unbombarded sample surface into the cleaned area. The basic concepts of surface analysis will be discussed in this section.

The initial presentation of this part of the course focuses on the possible input probes and output particles, and identifies those output particles that do not have widespread application as well as those that will be treated in some detail, i.e., Auger electron spectroscopy (AES), x-ray photoelectron spectroscopy (XPS), secondary ion and neutral mass spectrometry (SINMS), ion scattering spectroscopy (ISS), and Rutherford backscattering spectrometry (RBS). The chapter by Lichtman *(21)*, which is still current enough with the sequence of graphical presentations of ions, electrons, photons, and neutrals as input probes, and what "particles" come out, helps to organize the various methods for identifying elements on the surface. Other experimental probes, either already discussed or to be discussed in subsequent parts of the course, are also indicated from the diagrams. These diagrams have been updated in a recent chapter dealing with ion spectroscopies for surface analysis *(22)*. A simplified version is shown in Figure 1.

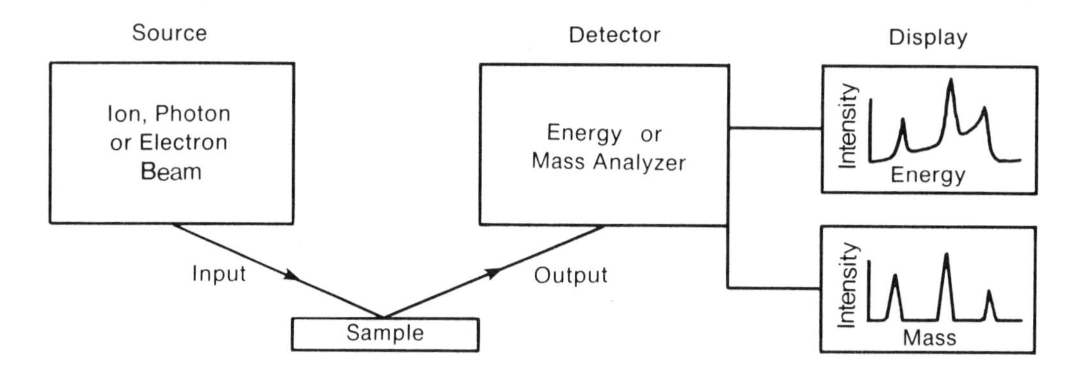

Figure 1. Generalized input and output probes for surface analysis. The electron energy (AES, XPS), ion energy (ISS, RBS), and ion mass (SIMS) are the measured output parameters.

The order of presenting the material in sections E1 through E6 is a personal preference because the physical processes in compositional depth profiling (CDP), secondary ion mass spectroscopy (SIMS), and secondary neutral mass spectroscopy (SNMS) are the same and simplify the transitions in subject material. Some scientists prefer sequences such as XPS, AES, ISS, RBS, SIMS, SNMS, and then CDP. In fact, each of the seven methods can be treated separately.

Compositional Depth Profiling Many studies in science and technology require determining the depth distribution of elements. Surface sensitive probe beams can be used for this purpose in combination with the most commonly used method of ion etching, which is also known as ion erosion, ion milling, sputter etching, compositional depth profiling (CDP), depth profiling, and (atomic) layer-by-layer microsectioning. Ideal and experimental depth profiles are shown in Figure 2 for a multilayer stack of partially oxidized aluminum on partially oxidized silicon. Sputtering, the process of removing material by bombarding the surface with an energetic ion beam (typically 0.5 to 10 keV argon ions), in principle provides a means for atomically microsectioning the solid, and the sputtering time can be

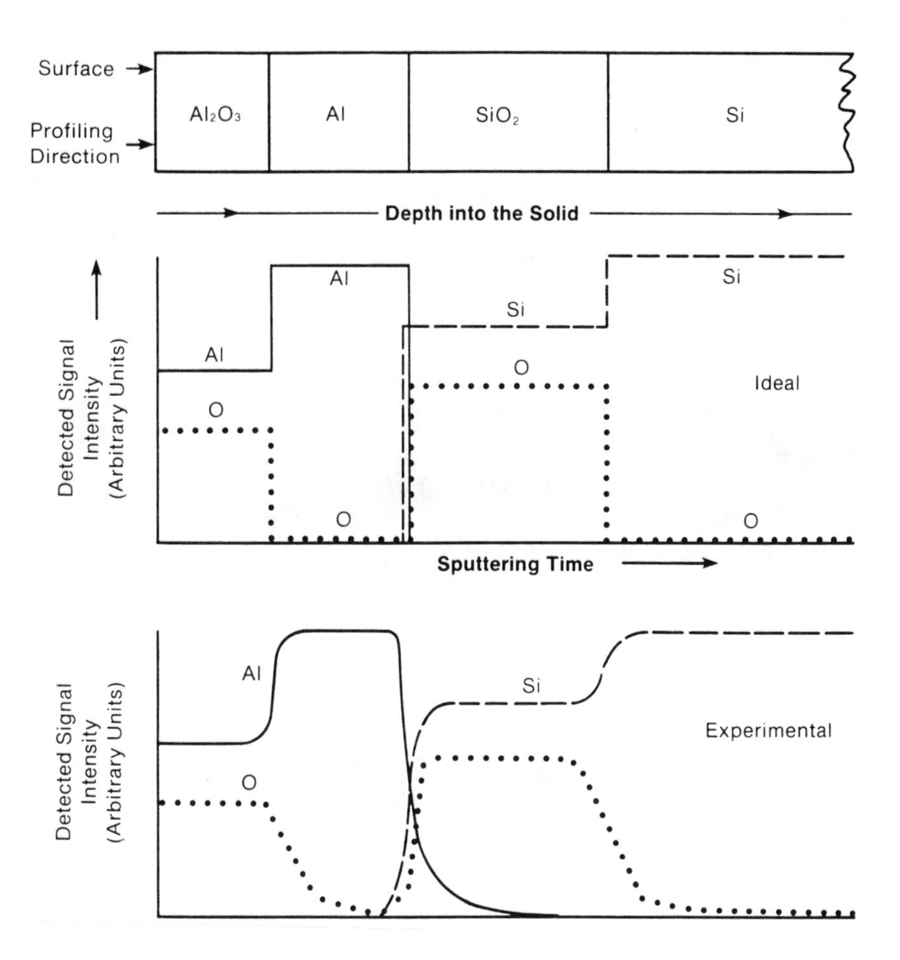

Figure 2. Ideal and experimental depth profiles for thin film multilayers.

related to depth. The principle of microsectioning *(22)* is shown in Figure 3 to indicate the goal of obtaining an analysis at each atomic layer into the solid. An ion beam produces a Gaussian-shaped crater, so the beam is rastered to remove an area of material that may range from 1 to 100 mm^2. The composition is obtained by analyzing the surface in the central portion of the crater.

When a solid surface is bombarded by ions in the keV range, complex processes of energy transfer and interactions occur on the surface and in the near surface region (Figure 4). As a result of the impact, atomic and molecular particles, electrons, and photons are emitted (ejected) from the surface. The ejected particles

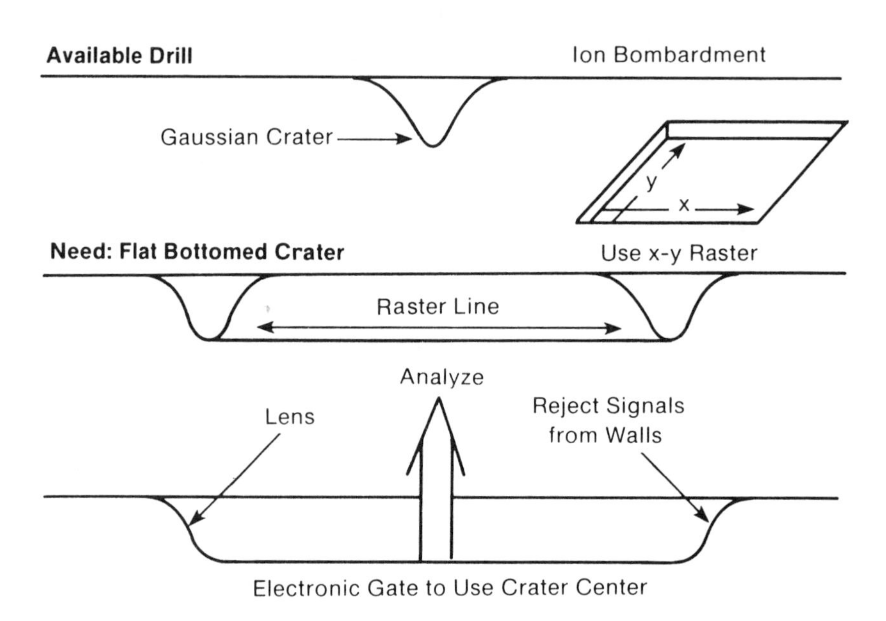

Figure 3. Objectives and practice of depth profiling. The objective of atomic layer microsectioning cannot be achieved because of the Gaussian shape of the ion intensity.

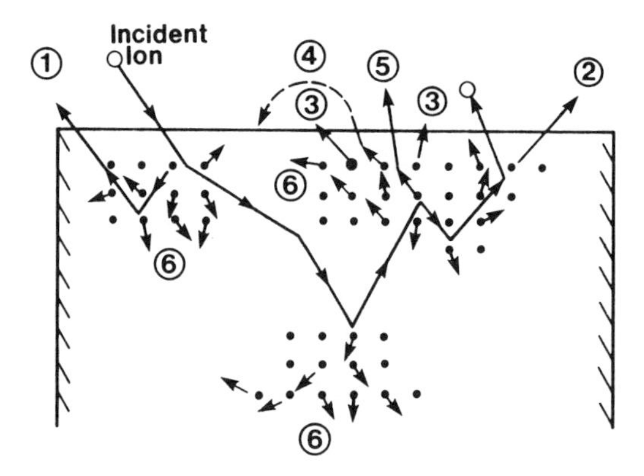

Figure 4. Several of the possible collision processes that occur during ion bombardment. Surface atoms are ejected from being hit from above (1) or below (2); some do not receive enough energy to leave the surface (3) or leave and fall back (4). Occasionally, bulk atoms receive enough energy to escape (5); others, simply receive energy (6).

may be in neutral, excited, or charged (+ or - secondary ions) states and originate from the surface zone. The ejected secondary electrons may result from various surface processes, such as Auger de-excitation and Auger neutralization of the impinging primary ions or emitted secondary particles. They may also result from bulk processes, such as ionization. The emitted photons are generated from de-excitation and neutralization processes above the surface, on the surface, and in the bulk. Energy transfer by ion bombardment also has an effect in the region of primary ion impact, in addition to the obvious sputtering processes. The primary ions will be implanted, whereas surface and near surface atoms will be knocked into the bulk. As the primary ion and recoiling lattice atoms undergo a collision cascade, atoms in the lattice will be displaced both inward and outward in a zone of mixing. As a result, changes in the surface structure, such as defect formation, amorphization of structure, bond breaking, and forming, may occur in the surface damage zone. Energetic neutral particle beams have the same effects on emission processes and changes in the surface zone.

The advantages of sputtering for in-depth analysis are considerable. Physical sputtering is an atomic process, so, in principle, atomic depth resolution is possible. Depth profiling can be performed *in situ* in vacuum, which means surface compositional analysis and sputtering can generally be performed simultaneously. When pure inert gases are used, clean surfaces are produced, and reoxidation and recontamination can be avoided. Reasonable erosion rates are achieved, e.g., 0.1 to 100 nm/min, so microsectioning to depths of 1000 to 2000 nm is accomplished within reasonable periods of time.

The sputtering process itself limits these considerable advantages, however, as we can see by comparing ideal and experimental profiles (Figure 2). A zone of mixing is created from knock-in and knock-out effects, which broadens the depth resolution. Preferential sputtering may occur, and surface roughness may develop. Structure may be destroyed and new chemical states may be formed. Bulk and surface diffusion of the target atoms may be enhanced during sputtering. Matrix effects may change the rate of erosion (e.g., resulting from compositional and structural differences) as the process proceeds. The bombarding gas ions are implanted into the solid; this is particularly serious when reactive gases such as oxygen or nitrogen are used. Redeposition of sputtered material occurs and complicates the analysis when it occurs on the surface being analyzed. Enhanced adsorption of residual gases from the vacuum may also occur, placing more stringent requirements on the vacuum level maintained during in-depth analysis. Further details can be developed from references provided *(22)*.

Ion Spectroscopies Using Ion Stimulation In this section, brief descriptions of ion spectroscopies that depend on ion detection after ion stimulation (Figure 1) are given for SIMS, ISS, and RBS. These are principal topics of all ion spectroscopies for surface analysis *(22)* because they are widely used with commercially available instruments.

In SIMS, a beam of primary ions is directed toward the sample surface, where most of the energy of the ions is dissipated into the near surface region of the solid by a series of binary collisions (Figure 1). As a result of both the momentum (and energy) transfer to the solid and the multiple collisions, surface and near surface atoms and molecular clusters are ejected (sputtered) from the solid. Some of the particles are sputtered as secondary ions, and these may be directed into a mass spectrometer for analysis. A large body of information about secondary ion emission and the applications of SIMS to surface and thin film analysis have been documented *(22,23)* in several conferences on SIMS.

The detected signal of the *i*th element in SIMS requires controlled destruction of the surface and subsurface of the sample to produce a secondary ion current $I_i^{\pm}$. The current for the *i*th element depends on many factors that may be related by the expression

$$I_i^{\pm} = I_p f_i^{\pm} C_i S_i \eta_i \quad (1)$$

where I_p is the incident ion current (ion/s), S_i is the sputtering yield of both ions and neutrals (particles/incident ion), $f_i^{\pm}$ is the fraction of the particles sputtered as ions, C_i is the concentration of the *i*th element (corrected for isotopic abundance) in the sputtered volume, η_i is the collection efficiency of the SIMS instrument, and ± refers to a positive or a negative particle. Most SIMS instruments can be operated in a mode to detect either positive or negative ions, so I_i^+ or I_i^- can be considered separately in Equation (1). C_i depends on the sample selected; the factors affecting the other parameters in Equation (1), i.e., S_i, I_p, and f_i, have been discussed in great detail *(22,23)*.

The most obvious advantage of SIMS compared with other surface analysis techniques is that it identifies elements at a low detection limit on the surface of conductors, semiconductors, and insulators. Other advantages include the ability to detect H and He on the surface, which is physically impossible with AES and impractical with XPS. The outstanding feature of SIMS is a detection limit to as low as 10^{-6} of a monolayer, depending on the sample (that affects S_i and f_i), beam size, and how fast the surface is sputtered away. As little as 10^{-18} g of a sample species may be sufficient to provide a detectable signal. Thus, using care in the instrument and bombardment parameters, signals can be restricted to one to two monolayers. *Static and dynamic SIMS* are terms used to divide the erosion rate of the solid into regimes so that only surface species (static) or species from the surface and bulk (dynamic) are detected. The sensitivity of SIMS to different isotopic masses provides new possibilities for studying corrosion mechanisms, self-diffusion phenomena, and surface reactions involving any exchange of atomic species. Isotopic labeling of both cations and anions could be used for studying reaction mechanisms. Despite the apparent potential, the literature contains relatively little mention of isotopic labeling with SIMS analysis of the surface. The ability to depth-profile and to maintain a constant monitor of the composition is one of the outstanding features of a SIMS apparatus. The obvious application here is to detect, down to about a 0.1 ppm atomic fraction, impurity accumulations at S/S interfaces lying hundreds of atomic layers below the outer surface. Ion microprobes provide capability to image the surface under investigation. This, combined with a lateral resolution as low as 100 nm, provides a capability for analyzing the composition of individual grains in polycrystalline materials and for determining the distribution of elements (bulk or trace) across the surface.

The principal limitation of SIMS is that it destroys part of the sample, preventing it from being analyzed further; there is no chance for a second look at the same spot on the sample. The factors causing large variations in the production of secondary ions make routine quantifications a remote hope unless standards are

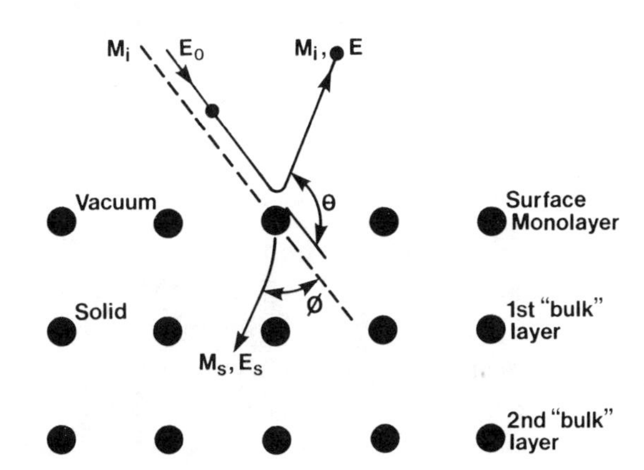

Figure 5. Schematic representation of the scattering of a projectile ion from a surface atom and the recoil of the surface atom M_S with the energy E_S.

used in well-studied systems. Matrix effects, e.g., the variation in the signal of the same element in different chemical environments, can alter the detectability of elements by factors up to 10^5. Finally, good lateral resolution requires small beam sizes, making static SIMS of small spots impossible because the current density is too high.

For SNMS (ions in, neutrals out), it is necessary to post-ionize neutral particles to perform mass spectrometry on them. During the last several decades, sputtering has been accomplished by using ion beams, and post-ionization has been done by the interaction of neutrals with excited atoms, electrons, or photons. In essence, SNMS is very similar to SIMS. A low energy ion beam is used to sputter the surface, and a mass spectrometer is used to analyze secondary particles. However, in SNMS, the true secondary ions are deliberately discarded and the secondary neutrals are post-ionized for subsequent detection. Post-ionization can be achieved by using a low-pressure plasma *(23)* an electron beam *(23)*, or a laser *(24)*. In each case, the ionization cross section is predictable and is largely independent of any matrix effects (although there are variations in the angular and energy distributions of sputtered neutrals). As a result, SNMS is a quantitative technique with a useful narrow range of sensitivities.

For ISS, a collimated monoenergetic beam of ions of known mass (M_1) is directed toward a solid surface, and the energy of the ions scattered from the surface is measured at a particular angle. The energy E_0 of the projectile ion is reduced to E during a collision with a surface atom (M_2), and the intensity of scattered ions is measured and normally presented versus an E/E_0 ratio rather than as an energy loss spectrum. From the number of scattered ions of mass M_1 appearing at particular E/E_0 ratios, information may be deduced about the mass and number density of the various surface species in the target, i.e., the elemental composition of the surface. While a number of review articles have been written, one of the most useful as a teaching aid was published as part of a series of reviews to aid students in the materials sciences *(25)*.

The physical principle for an ion scattered from a surface atom is shown schematically in Figure 5. Using the single binary elastic collision model, i.e., ignoring all interactions with other atoms and applying conservation of energy and momentum, the energy E of the scattered ions of mass M_1 with an incident energy E_0 scattered at a laboratory angle Θ is

$$E = E_0 (1+\alpha)^{-2}[\cos \Theta + (\alpha^2 - \sin^2 \Theta)^{1/2}]^2 , \tag{2}$$

provided that $\alpha \geq 1$, where α is M_2/M_1 and M_2 is the mass of the surface atom. The energy after the collision depends on E_0, M_1, M_2, and Θ. It is customary to represent the experimental data as scattered ion current vs. E/E_0 and to fix Θ and M_1 during the experiment so peaks in the energy loss spectrum E/E_0 correspond to M_2 values deduced from Equation (2), where it was assumed the binary collision model is valid. It was also assumed that the target atoms are stationary and are not coupled to their neighbors, e.g., typical lattice vibration times of about 10^{-13} s are large compared with a collision time of 10^{-15} to 10^{-16} s and the thermal energy/atom at 25°C is only about 0.04 eV compared with typical incident ion energies of 0.5 to 3 keV. For the energies used, the de Broglie wavelengths are small compared with atomic dimensions, so classical mechanics provide an adequate description of the collisions. The signal intensity I_i^+ resulting from the scattering of an ion beam by surface species M_i, can be expressed by

$$I_i^+ = I_0 (N_i - \beta_j N_j) P_i (d\sigma_i(\Theta)/d\Omega)_{\Theta(SC)} \Delta\Omega \tag{3}$$

where I_i^+ is the signal intensity originating from a surface species M_i, I_0 is the primary beam intensity, N_i is the number of scattering centers (of mass M_i) per unit area, β_j is the shadowing factor for the species M_i due to the coverage of another surface species of density N_j, P_i is the probability that an incident ion will not be neutralized, $(d\sigma_i(\Theta)/d\Omega)$ is the differential scattering cross section evaluated at the scattering angle Θ(SC), and $\Delta\Omega$ is the solid angle seen by the detecting system. In Equation (3), I_0 and $\Delta\Omega$ are determined by the instrumental parameters chosen. The solid itself affects N_i, $d\sigma_i/d\Omega$, β_j, and P_i. It can be presumed that signal intensity is directly related to N_i.

For a pure substance with no adsorbed species ($N_j = 0$), the exposed surface atom density depends on the crystal plane exposed. In general, β can be related to all geometric considerations that affect the visibility of the target atom to the probe ion. The Z (atomic number) dependence of $d\sigma_i/d\Omega$ typically increases by a factor of 10 from low to high Z. If a reasonable interaction potential is known or can be assumed, the differential scattering cross section can be derived using the methods of classical mechanics. The screened Coulomb potential, $V(r) = Z_i Z_2 e^2 r^{-1} \exp(-r/a)$ has been used most frequently by ISS users. Tables have been published from which differential scattering cross sections can be derived. The essential point is that the signal intensity is not strongly dependent on different forms of interaction potentials. Cross sections for multiple and subsurface collisions can be ignored because of neutralization processes.

Perhaps the most important, and least understood, parameter in Equation (3) is the probability that an incident ion will not be neutralized. Depending on the energy, only about 0.01% to 10% of the projectile ions are scattered as ions; the remaining ions experiencing binary collisions are neutralized. The projectile ion approaches the surface with a deep potential well available for an electron (–24.4 eV for He^+, –21.6 eV for Ne^+, and –15.5 eV for Ar^+) compared with a work function of about 5 eV for a typical solid. Since the closest approach during a large angle collision encounter is of the order of 0.01 nm, the quantum mechanical probability that electrons will tunnel from the solid to the ion is high. The details of the neutralization process during the approach, collision encounter, and flight from the surface have not yet been established.

Although the experimental facilities one needs vary with the goals for a particular problem, the essential needs for ion scattering experiments are a vacuum chamber, an ion gun, a target, and ion energy analyzer. Ions are produced in a source by electron bombardment of a gas, as the source is held at the desired accelerating potential (~0.3 – 3 keV). Ions are drawn from it by a negatively biased electrode, and the ion beam is formed by an ion focusing lens system. For most applications, only one type of singly charged ion species is desired in the beam, e.g., the ions of ^{3}He, ^{4}He, ^{20}Ne, or ^{40}Ar. Scattered ions are detected by using an energy analyzer either of the sector cylindrical mirror or hemispherical type and an electron multiplier detector. For most instruments, the scattering angle Θ is fixed, but in custom made spectrometers it may be varied.

The main feature of ISS spectra is that only one peak appears for each element, or isotope of each element, as predicted from theory. Thus, six elements will yield only six peaks, a situation that is considerably simpler than for other surface spectroscopies. An ISS spectrum is simple, but the peaks are broad and resolution between neighboring elements in the periodic table is poor, especially for high Z elements.

There are several major advantages to using ISS as a surface sensitive technique in addition to the obvious ability to identify elemental masses on solid surfaces of samples that are conductors, semiconductors, and insulators. Qualitatively, the current detection limits can be as low as 0.01 to 0.001 monolayers for the light elements and 0.001 to 0.0001 monolayers for heavier elements. Depth profiling is routinely accomplished by using the same ion gun for etching and supplying the projectile ions for scattering. Sensitivity to the outer surface layer is the strongest advantage of ISS because the detected signal results primarily, and possibly only, from atoms in the outer monolayer. Structural information can be deduced about the arrangement of two or more elements on single crystalline substrates.

Structural information can also be deduced by using multiple scattering, particularly double scattering. Isotopes of the same element can be detected when all other factors are constant. Chemical information can be deduced about the influence of different chemical environments on the detected signal for elements engaging in quasi-resonant transfer processes.

The strongest limitation of ISS is that the mass resolution is poor, especially for high Z masses. This results from broad peaks (0.02 – 0.03 in E/E_0) and poor energy resolution when M_1 is much less than M_2. Thus, a small concentration of an element with a mass near M_2 cannot be easily resolved. Moreover, the surface under investigation will be damaged by the bombardment from the projectile ions. Even though reduced bombardment energies (~300 to 700 eV), low current densities (0.4 to 40 $\mu A/cm^2$), and pulsed time of flight techniques may be employed, sputtering of the outer monolayer is unavoidable, but may be of no consequence with "static" ISS techniques.

In RBS, a collimated monoenergetic beam of ions of known mass M_1 is directed toward a solid material, and the energy of the ions scattered is measured at a particular angle. The initial energy E_0 of the projectile ion is reduced to E

during the passage through the solid and by the collision with a surface or bulk atom. The intensity of scattered ions is measured and presented in an energy loss spectrum. From the intensity of scattered ions of mass M_1 appearing at the reduced energy E, information may be deduced about the number density, mass, and depth distribution of atoms at the surface or in the bulk. Particle accelerators in the 1 – 3 MeV range became readily available in the late 1940s, and in the 1950s RBS instruments were applied to analyze bulk composition. Thin film and surface analysis with RBS, which were demonstrated in the late 1950s and early 1960s, respectively, have grown dramatically in the last two decades, as evident in a book devoted to the subject that was published in 1978 *(26)*. That book, by Chu, Mayer, and Nicolet, contains a comprehensive bibliography of review articles, books, and categorized applications. RBS is also covered concisely in a recent text *(2)*.

RBS for detecting ions scattered from the surface and bulk atoms of a solid is typically performed with incident energies of 1 to 3 MeV, rather than 0.5 to 3 keV, as in ISS. The energy of the elastically scattered ions is reduced during a binary elastic collision, provided that there are no nuclear reactions. The scattering cross section is greatly reduced at MeV energies, so backscattered particles reach the detector typically after traversing several hundred nanometers into the solid (Figure 6). The ion loses an average energy, which is termed the *stopping cross*

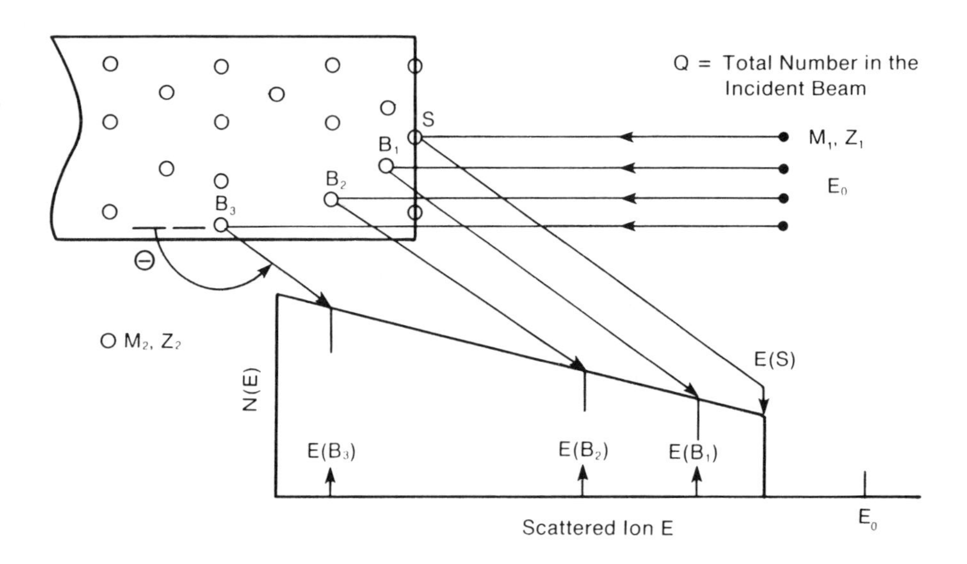

Figure 6. Schematic representation of the backscattering of projectile ions from surface and bulk atoms of M_2 and Z_2 through a scattering angle Θ to form an RBS spectrum.

section, primarily because of collisions with electrons in the solid. The collision encounters are inelastic (quantum mechanical), so a monoenergetic incident ion energy will emerge from the solid with an energy distribution about its initial energy as a result of statistical fluctuations, which is termed *energy straggling*. The energy of a backscattered ion is then reduced by the stopping cross section before and after the binary collision, so the mean energy of the ion provides a measure of the depth at which the collision occurred. Most significantly for RBS, the central force field is well known, which means an expression for the cross section can be derived. About 1 in 10,000 ions will be backscattered, so the analysis is essentially nondestructive. Because very few backscattered ions will have a scattering angle near 170°, beam widths of 1 to 2 mm are required, and the energy resolution is compromised to improve the signal-to-noise ratio.

The probability that an ion is scattered into a particular solid angle Ω is given by the differential scattering cross section $d\sigma/d\Omega$, and is usually called the scattering cross section in the literature. The total number of detected particles Q_D resulting from MeV ion scattering into a solid angle Ω can be expressed by

$$Q_D = Q_0(Nt)\sigma\Omega \tag{4}$$

where Q_0 is the total number of incident particles, N is the number density of particles per cm^3 in the target, t is the thickness, thus Nt becomes the number of particles per unit area in the target. The principal difference between Equations (4) and (3) is the absence of a neutralization probability. Both alpha particles and protons do not undergo charge exchange with a solid until their energy is somewhat less than 1 MeV. Notice also that Q_D is for all particles detected irrespective of their energy, which suggests multichannel techniques will be employed experimentally.

The essential experimental needs for RBS are a particle accelerator (usually 1 to 3 MeV helium or proton ions), a solid state detector, and a multichannel analyzer, aside from the obvious needs for a vacuum and a target. The important properties of particle accelerators (e.g., Van de Graaff generators) are well documented *(2, 26)*. Extensive data are available for assisting in RBS data analysis as a result of nearly 40 years of using accelerators in nuclear physics. These include relative abundances of isotopes, kinetic factors (E/E_0, where E is the energy after a binary collision from an incident energy E_0) for all atomic masses at various scattering angles for both $^4He^{2+}$ and $^1H^+$, stopping cross sections for the elements for various incident ion energies, scattering cross sections between helium and all elements for various incident ion energies, scattering cross sections between helium and all elements for various scattering angles, and the yield from the surface at various scattering angles of ^4He *(2, 26)*. The main features of RBS spectra are that only one peak appears for each mass of an element (or isotope) present, and the peak width is directly related to the thickness of the element. The peak height can be calculated from first principles, so RBS is quantitative. The data can be gathered in a matter of minutes, so the analysis is effectively nondestructive. The dynamics of changes such as interdiffusion can be studied *in situ*.

The outstanding advantage of RBS is that it provides quantitative and nondestructive in-depth compositional analysis. The time needed to acquire data is only a few minutes. The essentially nondestructive nature of RBS allows studying a single sample with materials processing variables such as temperature and time. The technique is more sensitive to high Z elements for which atomic fractions down to 0.001 can be determined. RBS as it is normally practiced is not surface sensitive, and the high vacuum chambers generally used preclude a serious study of surfaces unless a system is designed with ultra high vacuum (UHV) capabilities. When single crystals or epitaxial layers are used, structural information, both at the surface and from the bulk, can be obtained by using channeling and blocking techniques.

As for disadvantages of RBS, different isotopes of the same material are not generally resolvable, and the sensitivity for low Z elements is limited because the scattering cross sections are up to 1000 times smaller than those for high Z elements. Thus, detectabilities of an atomic fraction of 0.1 to 0.01 are typically available for Be through F in the periodic table. Where isotopes are resolvable, there are similar advantages for RBS as those stated for SIMS and ISS but at lower detection sensitivities for the low Z elements and with a lack of resolution between neighboring elements with large atomic numbers. The incident ions are typically in a 1-mm full-width-half-maximum (FWHM) beam, so the lateral resolution is poor compared with that of the other surface techniques. Therefore, lateral sample uniformity is important; SEM sample analysis is also recommended to identify scratches, dust, and other defects. No chemical information can be extracted from the data, however.

Electron Spectroscopies Using X-ray or Electron Stimulation The important and well-developed topic of electron spectroscopy for elemental identification requires effort to keep within six lecture periods. A useful approach is to present plots of i(E), N(E), and dN(E)/dE versus ejected electron energy, as done in a tutorial paper *(27)*. Then, the energy diagrams *(28)* can be used interactively with the students to identify the transitions that produce electrons with energies used for AES, XPS, and other electron spectroscopies. The importance of the elastic, inelastic, and plasmon electrons also should be mentioned and the host of other variants of the principal electron spectroscopies.

The important topics of cross sections, Auger and X-ray yields, Auger transitions for the periodic table, escape depths, effects of angles of incidence and exit on yields, charging, incident beam damage, chemical shifts, etc., can then be covered sequentially. With the tutorial chapter *(27)*, introductory chapters *(28)*, the overview chapter *(29)*, and recent book *(30)*, the instructor has all that is needed for a basic presentation of surface analysis by electron spectroscopy, including typical results, applications, advantages, and limitations.

In XPS, an X-ray source is directed towards the solid surface and the energy of electrons ejected from the solid are measured with an energy analyzer. The ejected electrons have a kinetic energy equal to that of the incident photon energy minus the binding energy of the electron in the solid. Ejected electrons from Auger processes are also detected in XPS and are treated in this section when discussing AES.

The basic XPS principle can be understood by referring to the electronic energy level diagram of Figure 7. An X-ray of energy $h\nu$ has the possibility of ejecting any electron that has a binding energy (BE) of less than $h\nu$ with an intensity that is related to the photoelectric cross section. The zeroth approximation is that every electron orbital level of energy ϵ will give rise to a peak in the XPS spectrum characterized by BE equals ϵ. This is not strictly true, because on removing one electron from an n-electron system, the remaining $n-1$ electrons relax towards the hole created by the loss of the electron. Thus, BE is less than ϵ. In addition, it is this final state relaxation that is directly responsible for the probability of a two-electron process in which an electron is promoted from one of the valence level molecular orbitals to an unoccupied higher valence level simultaneously with the removal of the core electron, becoming greater than zero. This results in two or more peaks, i.e., a main peak and shake-up peak, appearing in the photoelectron spectrum for one ϵ value.

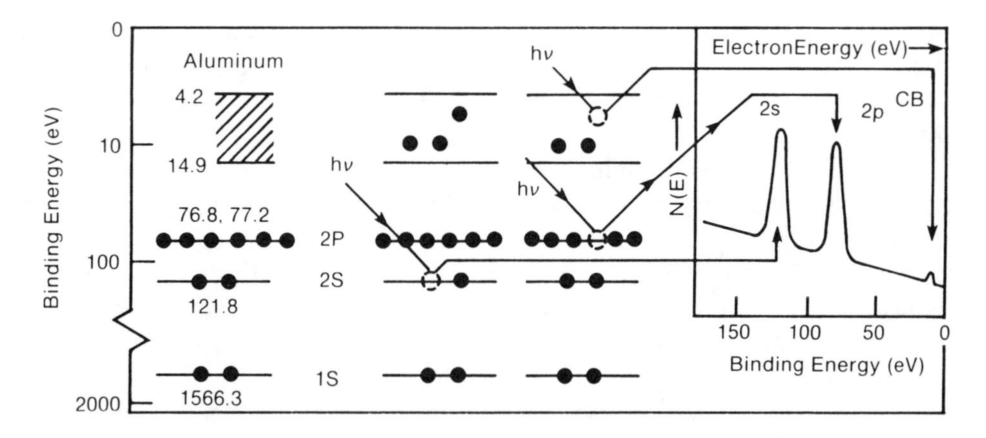

Figure 7. Electron energy levels for Aluminum showing ejection of 2s, 2p, or valence band electrons by photons to produce the intensity shown in an XPS spectrum of $N(E)$ vs. Binding Energy, BE.

Another final state effect that gives rise to an increase in the number of photoemission peaks is multiplet splitting. If the valence levels contain unpaired electrons, as is often the case for transition metal materials, removal of a core electron by XPS results in two possible final states, with the spin of the remaining core electron either up or down. The coupling with the unpaired spins of the valence levels will be different for up or down, and so the final state energies and hence, the experimentally determined BE values, will be different, resulting in multiple peaks in the photoemission process for a given initial state ϵ.

Every element in the periodic table has a unique set of electron binding energies, so qualitative elemental identification is assured in XPS from the measured kinetic energy, $h\nu - \epsilon$. The elemental peak intensities depend on the photoionization cross section that is reasonably well established. The sensitivity of the lowest Z elements increases to $Z = 12$, where it drops dramatically. The repeated drops at higher Z values correspond to successive observations of 1s, 2p, 3d, and 4f electrons as those producing the strongest lines in XPS spectra. All elements that have core levels can be detected, though the magnitude of the cross sections and hence, the relative sensitivities to the different elements, vary by $\sim 10^2$. Quantitative analysis can be done reasonably well *(30)*.

The signal intensity I_i of the ejected photoelectrons for the ith element is given by

$$I_i = F N_i \sigma_i \Theta_i \lambda_e T_A D G \tag{5}$$

where F is the incident photon flux, N_i is the atomic density, σ_i is the ionization cross section, Θ_i is an asymmetry factor, λ_e is the electron mean free path through the material, T_A is the analyzer transmission factor, D is the detector efficiency, and G is a geometric factor.

Electrons ejected from an energy level must escape from the solid into the vacuum for detection without experiencing inelastic collisions. Since X-rays penetrate several hundred nanometers into the solid, it is essential to know the mean escape depth of the ejected electrons. A minimum mean escape depth of about 0.3 to 0.5 nm occurs at a KE of 60 to 100 eV; mean escape depths may range from 0.3 to about 3 nm and depend on the energy of the ejected electron. While this variation must be treated carefully for interpreting data, the escape depth information clearly establishes the sampling depth of identified elements. The escape depth is often used advantageously for nondestructive CDP by varying the collection angle relative to the surface normal for the ejected electrons. At grazing exit, nearly all the electrons are from the outer atomic layer; without ion bombardment, this type of CDP is limited to the escape depth of the electrons.

In addition to atomic identification, information on the chemical environment of the atom is available primarily from the chemical shift phenomenon. Small differences in BE (1 to 10 eV) occur with differences in the chemical environment of the atom. For metallic species, different oxidation states are usually distinguishable in this manner, but for nonmetallic species, such as C, significant variations are observed, depending on the electronegativities of the ligands. Chemical shifts can be handled in some cases by theoretical means, but for most practical situations, a large empirical data base is the preferred means of assigning shifts to particular chemical environments. The other two most common means of providing chemical information are from the shake-up and multiplet splitting phenomena. Since these effects are caused by electron transitions between occupied and unoccupied valence levels and since valence level energies are characteristic of the molecular or chemical state rather than the atomic state, the presence of structure due to shake-up and multiplet splitting provides additional fingerprinting of the chemical environment of an atom.

XPS is the most quantitative (except for the specialized use of RBS) of the surface sensitive techniques, although in practical situations it is often limited by a combination of an unknown depth distribution over the probing depth. It quite obviously provides detection of all elements except hydrogen and helium, all with reasonably well-known cross sections and without strong matrix effects. It is especially sensitive for detecting low-*Z* elements.

Binding energy shifts measured for the same atom in different oxidation states and chemical environments can often provide chemical information. Satellite structures with a well-documented data base on standard compounds also are valuable for extracting chemical information. XPS is the least destructive of the major techniques described in this section. This is because an X-ray source is used, and for related applications with synchrotron or ultraviolet (UV) sources, even less damage is introduced into the sample. Easily degradable materials, especially organic polymers, can be studied routinely with XPS but only selectively using ion or electron excitation sources. As a surface technique, XPS is sensitive to 2 to 30 monolayers, depending on the material. However, by using special procedures, e.g., grazing exit XPS, a sensitivity to the top one or two monolayers is easily realizable.

For depth profiling, XPS has moderate usefulness as a practical matter. This is because the data accumulation times make it difficult to analyze and rapidly ion etch. However, XPS has revealed that changes in the chemical state of some elements occur during ion bombardment. As the other principle limitation, XPS has a poor lateral resolution, but rapid advances are being made towards beam sizes of less than 0.01 mm. Photon sources are not easily confined to exciting small sample areas; when analyzing electrons ejected from small areas becomes possible, the signal-to-noise problem may still require excessive data accumulation times.

In AES, an electron beam is directed towards the solid surface to form vacancies in one of the core electron energy levels and the energy of electrons ejected from the solid are measured with an energy analyzer. The ejected Auger electrons have a kinetic energy approximately equal to that of the energy released from filling a core level minus the binding energy of the Auger electron prior to ejection.

The Auger process can be understood by considering the ionization of an isolated atom under electron bombardment. The potential energy necessary to initiate an Auger electron transition is derived from the energy of a vacancy in an underlying core electron energy level. It is immaterial to the transition process how the vacancy came to exist. When a core level is ionized, the vacancy is immediately filled by another electron, as depicted by the $L_1 \rightarrow K$ transition in Figure 8. The energy $E(K) - E(L_1)$ from this transition can be released as characteristic X-rays (X-ray fluorescence spectroscopy) or be transferred to another electron, possibly in the L_2 level, which is ejected from the atom as an Auger electron with a kinetic energy KE_A. The measured energy for an Auger de-excitation is given approximately by

$$KE_A = E(K) - E(L_1) - E(L_2) - \Phi \tag{6}$$

where Φ is the work function of the analyzer material. This process is termed the KL_1L_2 Auger transition and clearly involves forming a K core vacancy by ionizing a 1s electron, filling the vacancy with a 2s electron in the L_1 level, and ejecting a 2p electron from the L_2 level. Auger transitions involving the KLL, LMM, and MNN levels are the most common, although MMM, NNN, NNO, NOO, and OOO have also been identified and used. The kinetic energies of Auger transitions for all electron energy levels have been tabulated *(2)* or plotted *(32)*.

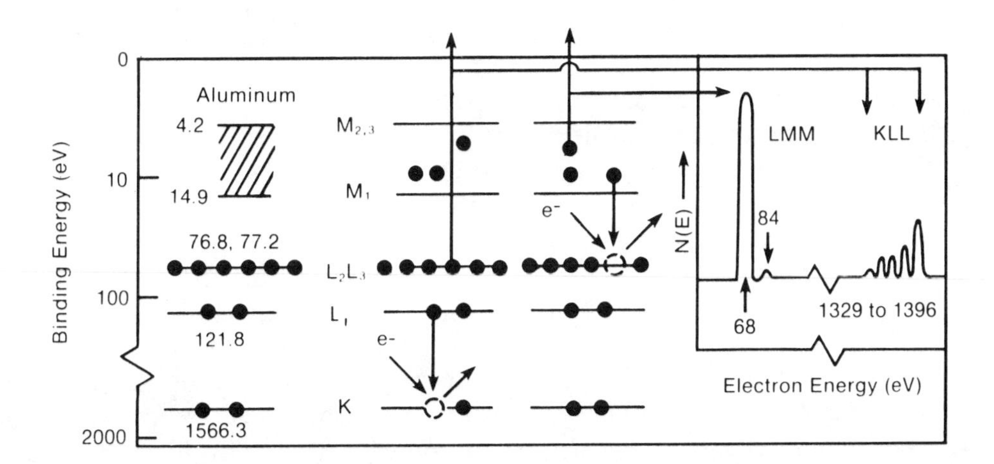

Figure 8. Electron energy levels for Aluminum showing ejection of a K or an L electron by incident electrons to produce the KLL or LMM Auger electron intensity in a spectrum of $N(E)$ vs. Electron Energy.

The Auger electron process clearly involves three electrons: one for forming the core vacancy, one that fills it, and the ejected Auger electron. The value of KE_A is determined by the differences in the energy levels for an atom of a particular Z. The core level electrons can be ejected from the solid by the primary electron, and these also will have characteristic kinetic energies KE_c given by E_p - E_K, where E_p is the energy of the incident electron beam and E_K is the core electron ejected by the primary electron. By varying E_p, the values of KE_c will change, whereas those of KE_A do not. These "characteristic loss electrons" are especially useful for assessing if sample charging is shifting the position of Auger electrons.

The intensity of a particular Auger electron emission, e.g., KL_1L_2, is proportional to the current of ionizing particles passing through the escape volume, the density of atoms in the volume, and the probability an ionizing event will occur in the K level; the latter is the cross section. The Auger current is reduced further by the acceptance angle and transmission of the analyzer as well as the Auger transition probability. The result of rigorous considerations yields a result similar to Equation (5) *(2,31)*.

Quantitative analysis with AES is not as simple as in XPS. First, the process is more complex, involving three electron levels instead of one. The core level photoionization or electron impact cross section has to be folded with the Auger transition probability. The latter is not as well known as photoionization cross sections, which makes quantitative analysis from first principles more difficult. Experimental standards are also less well established. Finally, the usual manner of recording Auger spectra in the first derivative mode ($dN(E)/dE$) and measuring peak-to-peak heights can introduce large errors because of line shape changes; however, areas under AES peaks can now be integrated from output of $N(E)$ versus E.

All the chemical shift, shake-up, and multiplet splitting information of XPS is, in principle, available in AES. The interpretation is more complex, however, because of the three levels involved in the process. Auger chemical shifts have not been nearly so widely exploited as they have in XPS. The empirical data base is still limited, and again the habit of recording in the first derivative mode has obscured the usefulness of the chemical information. The exceptions are usually XPS researchers, who take their X-ray-induced Auger data in the same manner as the XPS data and treat the analysis in a similar manner.

It should be pointed out that the chemical shifts observed in AES are not usually the same as in XPS for the same atoms in the same chemical state. Often, they are larger because of the two-hole nature of the final state in the Auger process. The difference between the XPS and Auger chemical shift has been termed the Auger parameter and is an additional useful guide to the chemical state of the atom *(30)*.

The basic instrumentation for AES measurements includes an electron beam source, a sample, an electron energy analyzer, a detector, a recording system, and a vacuum chamber. The electron beam source will usually have a series of selectable, fixed primary energies in the range up to 10 keV; energies of 2, 3, 5, and 10 keV are widely used. The primary source of electrons may include deflection plates to raster a finely focused beam over a sample area. In recent years, minimum beam sizes have been decreased from nominal 5000 to 200 to 20 nm FWHMs, providing exceptional lateral resolution and increased current densities that require special experimental caution.

Energy analyzers have been (in order of progressively greater use) of the retarding grid, cylindrical mirror, and hemispherical types. Most commercially available instruments now use hemispherical analyzers, so the entire analyzer detection system for XPS and AES is identical. The smaller beam sizes place less stringent requirements on the sample size and topographic conditions for the analysis. However, the widespread coupling of AES and Scanning Auger Microscopy (SAM) with ion etching places initial demands on the initial flatness for securing better profiles.

Auger spectroscopy, which is one of the three most widely used methods of surface compositional analysis, has a number of outstanding advantages. It is the fastest of the common methods for identifying elemental composition, and that feature combined with superb lateral resolution has even made it possible to obtain laterally resolved maps of several elements during depth profiling. AES is especially good for depth profiling at high ion etching rates, again because of its data acquisition speed. AES is sensitive to 2 to 10 monolayers, but unlike XPS, there is no way of changing the escape depth of the detected electrons, except for the characteristic loss electrons and using angle resolved techniques. It also has good sensitivity to the light elements.

The majority of data have been collected in the $dN(E)/dE$ versus E mode because of a high background of inelastically-scattered electrons. However, with an increasing number of studies now securing $N(E)$ versus E data, an increasing amount of chemical information should be extracted from the experimental work. The chemical information appears as different shapes of particular peaks, and this has been largely obscured in the past.

Principal limitations in AES include damage to the sample by the electron beam, including introducing artifacts known as electron stimulated desorption. Severe charging problems may also develop, especially with poorly conducting materials. Negative charge must be supplied to all samples during analysis because of the net ejection of electrons from the sample by Auger processes. There is no sensitivity to atomic masses, so many of the mechanistic processes involving atom transport or exchange cannot be elucidated by isotopic labeling. Finally, there are many possible Auger processes that may occur, especially for the high Z elements. Even for binary alloys, peak interference problems may be encountered that complicate the interpretation of the data.

General Comparisons of SIMS, SNMS, ISS, RBS, AES, and XPS Comparisons of aspects of SIMS, SNMS, ISS, RBS, AES, and XPS have been listed in several tables *(11)*. Summaries of the principal advantages and limitations of each of these techniques are also given in six additional tables *(11)*.

Surface Thermodynamics, Equilibrium Shape, Depth of Surface Phase, and Diffusion

These areas are well organized in at least five texts *(1-4, 33)*. The thermodynamic relations for the surface energy of crystals can be set down in a straightforward manner, and the equivalence between surface tension and surface free energy can be established for isotropic crystals. A rigorous development of the LaPlace and Kelvin equations is followed with a discussion of the implications of the equations on pure solid surfaces.

Surface atoms are in a markedly different environment from that of bulk atoms. Surface atoms are surrounded by fewer neighboring atoms, which are in a surface-unique anisotropic distribution, compared with those in the bulk. The surface phase has a higher entropy, internal energy, work content (Helmholtz energy), and free energy (Gibbs energy) per atom than the bulk phase. For isotropic solids, surface free energy and surface tension are equivalent. At equilibrium, the solid surface will develop the shape that corresponds to the minimum value of total surface free energy. The surface free energy of solids can be calculated for surfaces of different structures. The relative magnitude of surface free energy can be deduced from experiments involving thermal faceting, grain boundary grooving, etc. The low index planes are (usually) the most stable because they have the lowest free energy. All small particles and all large, flat, polycrystalline surfaces, which are characteristic of the type encountered in most applications, possess a relatively large surface excess energy and are thermodynamically unstable.

Based on calculations made for inert gases, ionic halides, and semiconductors, the depth of the surface phase, i.e., that part of the interface with unique properties, may extend to 5 nm. In the latter two cases, substantial deviations in the actual surface structures from those characteristic of a truncated but otherwise bulk solid have been reported from theoretical analyses of LEED intensities from single crystals. The reconstruction or rearrangement of the surface involves movements of over 0.05 nm in the surface atomic plane and possibly in the first bulk layer as well.

Surface atoms are restrained from interatomic motion by nearest neighbor bonding, but these bonds are weaker than those at a corresponding position in the bulk. The potential barriers for surface diffusion are lower than those for bulk diffusion, so less activation energy is required to produce surface diffusion processes. The mechanism of surface diffusion may change with temperature, because the surface population of adatoms, adions, vacancies, ledges, and the like, or other conditions such as structure and ambient atmosphere, may change.

An adequate overview of surface diffusion can be done in one lecture, but three are preferred. The concepts of the diffusion coefficient, mass transport due to capillarity, surface self-diffusion, and thermal grooving appear to be well developed and documented. There are a number of fascinating studies in the literature on using the FEM and FIM to study surface diffusion. A large number of these have been presented in *Surface Science*. Some very interesting results have been published that can serve for a one-lecture presentation *(34)*.

Chemisorption and the Nature of Gas/Solid Interactions

The course is concluded by providing an overview of chemisorption using references 1, 8, and 9 as primary sources and references 6, 33, and 35 as secondary sources. The students are required to review the coverage in section B, and brief mention also can be made of the experimental methods already covered that can be used to study chemisorption. This part of the course applies primarily to the G/S surface, although the general concepts are valid for the L/S surface. Adsorption is the accumulation of a surface excess of two immiscible phases. Except for multilayer physical adsorption, the forces of interaction limit the amount adsorbed to one atomic layer, i.e., up to a monolayer for chemisorption. Mass gain, volumetric, and radiotracer techniques are used to measure the amount of gas adsorbed directly. Indirect techniques require relating the amount adsorbed to the increase or decrease in intensity of some "output particle" such as desorbed gas, scattered ions, ejected Auger electrons, or quanta of radiation. The direct techniques are limited in their applicability but as to the indirect techniques, there is some danger in assuming that the change in signal intensity varies linearly with adsorbate coverage.

Adsorbed gases may bind to the surface nondissociatively or dissociatively, remain localized or have mobility, form two-dimensional surface compounds, or incorporate themselves into the solid, either as an absorbed entity or as part of a new compound. The bonding interaction between a chemisorbed species and the solid depends on the particular geometric configuration, fractional coverage, and electronic interaction. To understand the bonding interaction, a wide diversity of measurement techniques are used, such as infrared (IR) spectroscopy, magnetic susceptibility, electron spin and nuclear magnetic resonance, work function, conductance, LEED, impact desorption, FEM, FIM, electron energy loss (EELS), ultraviolet photoelectron spectroscopy (UPS), and so on. The most significant recent methods using modern vibrational spectroscopy to study molecules on surfaces have been treated *(9)*. Of these, IR-reflection absorption (RA) and EELS are obvious choices because they may be used at various pressures (IR-RA) or in vacuum (EELS). Both provide the same essential information, i.e., the absorption of energy determined by the separation of the quantized vibrational energy levels. For discussing bonding onto single crystals and the arrangement of absorbates, the traditional use of LEED is essential, and impact collision ISS has become popular in recent years. For local bonding order on polycrystalline solids, electron stimulated desorption ion angular distribution (ESDIAD) and surface extended x-ray analysis of fine structure (SEXAFS) should be discussed.

For presenting adsorption isotherms, isobars, and rates of adsorption, the silver/oxygen system is an excellent example for illustrating physical adsorption, chemisorption, adsorption, and (the possibility of) compound formation for a particular isobar as a generalized concept for gas/metal systems *(36)*. Oxygen has two chemisorbed states on silver and has been studied by techniques ranging from LEED, AES, ISS, residual gas analysis (RGA), and work function using single crystals, to FEM on tips cleaned by field evaporation, to measuring the equilibrium adsorption up to 40 kPa microgravimetrically. The results obtained on the silver/oxygen system provide ample opportunity to discuss nondissociative and dissociative adsorption, localized and nonlocalized adsorption, the formation of two-dimensional compounds, absorption processes, homogeneous and heterogeneous surfaces, condensation and sticking coefficients, the basic adsorption equation, and variations of isosteric heats and activation energies with coverage. The Langmuir, Tempkin, and Freundlich isotherms and tests for them make good homework assignments. Then, the basic desorption equations and the processes of isothermal and thermal desorption are treated; again, for convenience the work done on the silver/oxygen and carbon dioxide on oxygen-covered silver systems are used as examples *(36)*. These provide ample opportunity to treat first and second order processes, the interpretation of the pre-exponential, coverage-dependent activation energies of desorption, and thermal desorption from heterogeneous surfaces.

The interaction between a chemisorbed species and the solid is highlighted by a discussion of the geometric configuration, the fractional surface coverage, and the electronic interaction. The understanding that can be gained by using techniques such as IR-RA, magnetic susceptibility, EPR and NMR, contact potential, work function, conductance, impact desorption, FEM, FIM, UPS, EELS can be delivered by the instructor or offered to the students for a series of student seminars. The general concept of surface states, band bending, and the electronic properties of surfaces can be mentioned in passing at this point.

Classes with Different Disciplinary Backgrounds

When teaching students with heterogeneous backgrounds, the instructor has to anticipate difficulties because of the nature of their prior studies (e.g., chemists will not necessarily see the identity between binding energies and potential wells, and engineers who have not been required to take a course in modern physics will have problems with such concepts as work function, the density of states, and tunneling through barriers). The lack of a uniform symbolism for the Gibbs, Helmholtz, internal energy, and enthalpy functions will confuse part of the class, and physicists may become impatient with the seemingly unending breadth of the coverage of the course and apparent lack of rigor about anything. The instructor has abundant opportunities for explaining the same concept in the language of the physicist, the chemist, and the engineer, which also teaches the students to probe the *concepts* behind the words rather than to remain narrowly attached to the language learned in a particular discipline. Furthermore, the content of the course permits teaching an appreciation of basic concepts to engineers and chemists and the importance of applications of science to physicists. As part of my final exams, students have written fascinating answers when asked how they would propose to study a given gas/solid system to reach a particular goal. How would they control the variables? What solid form and what cleaning techniques would be chosen? Why? What measurements should be made to help them obtain the information needed? And how much money do they need?

Including Aspects of Surface Science in Lectures or in the Laboratory

The summary presented in section II produces a satisfactory overview of the fundamentals of surface science. As shown in Table 2, 80% of the course is focused on composition, structure, and bonding. At many institutions, a course of three semester hours at the senior/first-year graduate level cannot be taught, either because of limited staff, qualified staff to develop the course, or enough interested students. However, opportunities abound in which these aspects of surface science can be included in the normal physical chemistry curriculum. Many physical chemistry professors indicate three to six 50-minute periods or the equivalent is about all that can be realistically considered. One approach would be to include the aspects at the appropriate time rather than to set aside a block of time for a "dose of surface science." Another limitation is the $500,000 to $1 million cost for SIMS, XPS, AES, RBS, and ISS instruments. Professors at institutions with these capabilities are encouraged to work with colleagues to design hands-on experiments. The remainder of this section is directed towards professors at institutions with limited resources. The problems suggested, however, are obvious exceptions, and are indicated in Table 3 by reference 1 or 2, chapter (Ch.), and problem (Pr.) numbers.

When treating the kinetic theory, the time it takes to contaminate a surface and the implications of a decreasing mean free path as pressure increases for using ion and electron beams in vacuum (13 mPa and lower) can be discussed and problems assigned. Comments can be made about real and clean surfaces and that several monolayers of absorbed gas exist on all surfaces at room temperature.

During the treatment of thermodynamics and equilibrium, physical adsorption isotherms and surface area problems can be used. Equations for the adsorption isostere can be derived in about five lines to show $\Delta G = \Delta H - T\Delta S$. The concurrence by surface scientists about a surface phase can be included with work on changes of state.

When treating electron energy levels in atoms, the important transitions for XPS and AES (Figures 7 and 8) can also be discussed. This is especially important because the nonradiative Auger yield predominates (at low Z) versus the radiative fluorescence yield when core electrons are ejected to leave an atom in an excited state *(2, Chap. 11)*. The crucial importance of both XPS and AES as methods for measuring the composition of the outer 2 to 10 monolayers should be emphasized, as well as the influence of detecting chemical shifts with XPS for different valence states of atoms and/or different ligands bonded to atoms or ions.

When discussing structure in the solid state, the important results of LEED can be included. If material on DeBroglie wavelengths and the wave-particle nature of electrons have not been taught, the basis for LEED might be more usefully covered with particles and waves and combined with the XPS and AES material. The influence of ion bombardment for depth profiling and SIMS can also be included with structure as a way of analyzing structural composition in depth and also for amorphisizing it.

The similarity of the chemical bonding of molecules/atoms to surfaces and the structure of molecules can be easily discussed together. The use of vibrational spectroscopies *(9)* and the importance of rotational and vibrational energy levels for using infrared absorption for studying absorbed species will clearly enlarge the scope of the solid state to the surface phase for students. An attenuated total reflectance (ATR) attachment for a Fourier transform infrared spectroscopy (FTIR) would permit assigning a laboratory for directly measuring the influence of the adsorption of gases on surfaces. Alternatively, the adsorption of CO on any of a number of adsorbents using transmission IR or FTIR will introduce the basic concepts of chemisorption.

The double layer should be discussed when treating the S/L interface or electrochemical phenomena. Some texts include double layer phenomena in a chapter on surface chemistry, which is acceptable. These chapters frequently discuss the LaPlace and Kelvin equations, which are extremely important for surface tension. These equations serve as a good lead to discussing surface energy and its consequences on the stability of interfaces.

Safety Considerations

For laboratory experiments, most equipment is designed to provide the ability to study surfaces in vacuum or surface reactions in controlled atmospheres. Vacuum

Table 3. Suggested Problem Assignments Related to Aspects of Surface Science

Topic	Reference 1	Reference 2
Clean Surfaces (Kinetic Theory)	Ch. 17; Pr. 1*	Ch. 1; Prs. 1, 2*
Surface Area	Ch. 16, Prs. 11,12 Ch. 8, Pr. 8	--
Composition - XPS	--	Ch. 9; Prs. 2, 3
Composition - AES	--	Ch. 6; Prs. 5, 6 Ch. 11, Prs. 1, 3
Composition - SIMS	--	Ch. 4; Pr. 8
Composition - RBS	--	Ch. 2; Prs. 5, 6; Ch. 3; Pr. 4
Composition - ISS	--	Derive Equation (2)**
Structure - LEED		Ch. 7; Prs. 1, 2, 3
Bonding - Heat of Adsorption	Ch. 16, 4, 36	(See Ref. 8)
Bonding - Desorption	Ch. 17; Prs. 4-7	--
Surface Energy	Ch. 7; Prs. 8, 9	--
Structure - Surface Sites	--	Ch. 1; Prs. 1, 2
Surface Diffusion	Ch. 17; Pr. 10	--

*Also See Real and Clean Surfaces, this chapter.
**See Ion Spectroscopies, this chapter.

systems and gas handling stations provide a range of conditions from ultra high vacuum to partial pressures of a wide range of gases. Reaction rates may be controlled and studied by varying temperature, pressure, and reactant concentrations. Equipment will normally be operated under high vacuum conditions or reduced pressures and will therefore, present all the associated hazards. In case of implosion, there are missile hazards. *Safety glasses must be worn at all times*. During normal preparation for operation, parts of the system may be baked out at temperatures up to 250°C, so precautions against burns must be taken. High pressure gas bottles are connected to and are stored near the apparatus; therefore, care must be taken not to damage the valve or otherwise mishandle gas bottles.

Toxic gases may be used in some studies, and they must be handled and vented carefully. Handling of toxic gases should be done only when another person can be reached by voice command. Ventilation must be on at all times when toxic gases are being used. Specific standards for safe handling and use of each toxic gas must be developed.

Low voltage (120, 208 V) systems are used, and all ordinary precautions against shock must be taken. Liquid nitrogen or dry ice may be used as coolants at times, and vacuum distillation processes may be used. Insulated clothing and eye protection must be worn during coolant transfers. Dewars must be taped to prevent flying glass from any possible implosion. Furnaces may be used to heat samples inside the vacuum. *Thermally protective gloves* are to be worn when handling a hot furnace. Forepump motors operate at temperatures that can cause burns. *Thermally protective gloves* are to be worn when handling a hot pump.

Since different types of samples may be studied, each material may have an associated handling safety hazard. Materials handling data sheets should be obtained from suppliers or other appropriate sources to avoid any unsafe practices.

Acknowledgments

I am pleased to express my appreciation for the helpful comments and constructive criticisms by the students who took my surface science courses at Clarkson University between 1966 and 1977 and at the University of Denver from 1988 to 1991. I also express similar appreciation for those who enrolled in an American Chemical Society short course (Surface Science) between 1983 and 1987, and American Vacuum Society short courses (Fundamentals of Surface Science, Methods of Surface Analysis, and Depth Profiling) from 1975 to 1991.

Literature Cited

1. Adamson, A. W. *Physical Chemistry of Surfaces, 5th edition*, Wiley: New York, NY, 1990.
2. Feldman, L. C.; Mayer, J. W. *Fundamentals of Surface and Thin Film Analysis*, Elsevier: New York, NY, 1986.
3. Somorjai, G. A. *Principles of Surface Chemistry*, Prentice-Hall: Englewood Cliffs, NJ, 1972; *Chemistry in Two Dimensions*, Cornell Press: Ithaca, NY, 1981.
4. Blakely, J. M. *Introduction to the Properties of Crystal Surfaces*, Pergamon Press: Oxford, UK, 1973.
5. Roberts, M. W.; McKee, C. S. *Chemistry of the Metal-Gas Interface*, Oxford U. Press: Oxford, OX2.6DP, UK, 1978.
6. Morrison, S.R., *The Chemical Physics of Surfaces*, second edition, Plenum: New York, NY, 1990.
7. Ertl, G.; Kuppers, J., *Low Energy Electrons and Surface Chemistry*, VCH Verlagsgesellschaft mbH: Weinheim, FRG, 1985.
8. Tompkins, F. C. *Chemisorption of Gases on Metals*, Academic: New York, N.Y., 1978.
9. *Vibrational Spectroscopy of Molecules on Surfaces*, Yates, Jr., J. T.; Madey, T. E., Eds.; Methods of Surface Characterization, Volume 1, Plenum: New York, NY, 1987.

10. Roberts, R. W.; Vanderslice, T. A. *Ultrahigh Vacuum and Its Applications*, Prentice-Hall: Englewood Cliffs, NJ, 1963.
11. Powell, C. J.; Hercules, D. M.; Czanderna, A. W.; in *Ion Spectroscopies for Surface Analysis*, Czanderna, A. W.; Hercules, D. M., Eds.; *Methods of Surface Characterization*; Plenum: New York, NY, 1991, Vol. 2, pp. 417-437.
12. Czanderna, A. W.; Vasofsky, R. *Progress in Surface Science*, **1979**, *9*, 45.
13. Brunauer, S.; Emmett, P.; Teller, E. *J. Am. Chem. Soc.* **1938**, *60*, 309.
14. Gregg, S. J.; Sing, K. S. W. *Adsorption, Surface Area, and Porosity*, 2nd edition, Academic: New York, NY, 1982.
15. Roth, A. *Vacuum Technology*, North Holland: New York, NY, 1976.
16. Wells, O. C., *Scanning Electron Microscopy*, McGraw-Hill: New York, NY, 1974.
17. Gomer, R. *Field Emission and Field Ionization*, Harvard U. Press: Cambridge, MA, 1961.
18. Müller, E. in *Conference on Clean Surfaces*, Johnstone, M.C., Ed.; Ann. N.Y. Acad. Sci.: New York, NY, 1963, Vol. 101, p. 590.
19. Müller, E. W. in *Methods of Surface Analysis*, Czanderna, A. W., Ed.; Elsevier: New York, NY, 1975, p. 329.
20. Block, J. H. Ibid, p. 379.
21. Lichtman, D., Ibid, p. 39.
22. A. W. Czanderna, op. cit. ref. 11, pp. 1-47.
23. Benninghoven, A.; Rudenauer, F. G.; Werner, H. *Secondary Ion Mass Spectrometry*, Wiley: New York, NY, 1987.
24. Becker, C. A., op. cit. ref. 11, pp. 296-346.
25. Helbig, H. F.; Czanderna, A. W. J. Educational Modules in Materials Science and Engineering *(JEMMSE)*, **1979**, *1*, 379-402.
26. Chu, W. K.; Mayer, J. W.; Nicolet, M-A. *Backscattering Spectrometry*, Academic: New York, NY, 1978.
27. Sickafus, E. N. *JEMMSE*, **1979**, *1*, 95-131.
28. Carlson, T. A. *Photoelectron and Auger Spectroscopy*, Plenum: New York, NY, 1975.
29. Riggs, W. M.; Parker, M. J. op. cit. ref. 19, pp. 103-158.
30. *Practical Surface Analysis by Auger and X-ray Photoelectron Spectroscopies*, Briggs, D.; Seah, M., Eds.; Wiley: New York, NY, 1990, 2nd edition.
31. Powell, C. J. in *Quantitative Surface Analysis of Materials*, McIntyre, N. S., Ed.; ASTM: Philadelphia, PA, 1978, STP643, p. 5.
32. Siegbahn, K.; et. al., *ESCA, Atomic, Molecular and Solid State Structure Studied by Means of Electron Spectroscopy*, Almqvist and Wiksells: Uppsala, Sweden, 1967.
33. Zangwill, A. *Physics at Surfaces*, Cambridge U. Press: Cambridge, UK, 1988.
34. Ehrlich, G. in *Chemistry and Physics of Solid Surfaces*, Vanselow, R.; England, W., Eds.; CRC Press: Boca Raton, FL, 1982, Vol. III, pp. 61-77.
35. Hayward, D.O.; Trapnell, B.M.W. *Chemisorption*, Butterworths: London, UK, 1964.
36. Czanderna, A. W. in *Microweighing in Vacuum and Controlled Environments*, Czanderna, A. W.; Wolsky, S. P., Eds.; Elsevier: New York, NY, 1980, pp. 175-232.

RECEIVED October 1, 1992

Why Modernize the Laboratory?

CHAPTER 5

A Consortium-Based Approach to Laboratory Modernization: The Pew Physical Chemistry Project

Colin F. MacKay

The Physical Chemistry Project of the Mid-Atlantic Consortium of the Pew Science Program in Undergraduate Education began with a simple question. "Of the specialized laboratories in the undergraduate core curriculum which is the least developed, which least reflects current practice?" Why did the question come to be asked? It was a response to an initiative of the Pew Foundation. Beginning in 1987 that Foundation defined 8 consortia across the nation, each consisting of a combination of liberal arts colleges and universities. These are the Western Cluster, centered on the Pacific Coast, the Mid-States Science Cluster, the Great Lakes Cluster, the Carolinas/Ohio Science Education Network, the New York State Cluster, the New England Consortium for Undergraduate Science Education, and the Mid-Atlantic Regional Cluster. Each was challenged to prepare proposals describing projects particularly suited to a consortial approach. The consortium which submitted the Physical Chemistry Project, the Mid-Atlantic Cluster, consists of 8 colleges: Bucknell, Franklin & Marshall, Bryn Mawr, Haverford, Swarthmore, Lafayette, Muhlenberg and Vassar, and two universities, the University of Pennsylvania and Princeton University.

The Mid-Atlantic Cluster undertook a range of non-curricular projects. Two of these were broad based grants programs. The first provided summer grants to undergraduates to do research in institutions other than their own. The second offered small grants to encourage cooperative research among faculty members of two or more cluster institutions, with the aim of developing long-term relations between faculty members of different institutions. Undergraduate participation in the proposed work was a strong consideration in grant decisions. There were two programs in important emerging areas in the cluster, neuroscience and computer science. These shared the important underlying aim of creating a mutual support network for the relatively small numbers of faculty in these areas at the colleges. In neuroscience the cluster sponsored postdoctoral grants for graduates of the cluster universities to work at the colleges. In computer science it sponsored summer programs.

However, in many ways the cluster's most ambitious efforts concentrated on three laboratory curriculum development projects, one each in Biology, Chemistry, and Physics. The program in Biology aimed at a unification of molecular and cell biology at the introductory level. Faculty participants produced a new laboratory manual emphasizing modern approaches, which was presented at a specially organized symposium at Princeton in April of 1991. The Physics project concentrated on the development of modern experiments at the intermediate level with special emphasis on optical physics. Participants will present their work at a symposium at the April 1992 meeting of the American Physical Society. Finally there is the project to develop modern materials for the physical chemistry laboratory.

Why Laboratory Development: Why Physical Chemistry?

There are at least two obvious reasons for concentrating on laboratory development. The first is that chemistry is a laboratory centered science. It is in the laboratory that the distinctive activity of that science takes place. The second is that good laboratory programs draw students and poor ones repel them.[1] Why physical chemistry laboratory? It is certainly true that modern techniques used in research in physical chemistry have broad applications and some acquaintance with them is essential in undergraduate chemistry. However, there is a more important reason. Physical chemistry is a crucial course in chemistry programs. Its position in the curriculum and the fact that it is the most mathematical and most abstract of the core courses make it a significant gateway course for majors. Combine these attributes with a laboratory program that has changed little in 20 years[2], that reflects little if any of the progress made in the discipline over that time, and you describe a gateway that many students don't see worth the effort of attempting to pass.

Objectives of a Laboratory Program

In our view laboratory work in the sciences should have several objectives.

- The teaching of basic techniques;
- Reduction to practice of some of the ideas and concepts discussed in the lectures;
- Demonstration of contemporary activity in the field.

The relative proportions of these vary depending on the level of the course, but current research methods should be increasingly represented at the junior and senior levels.

It is in the last area mentioned above that laboratory instruction in physical chemistry most often fails. Spectroscopic techniques in combination with computer acquisition and processing of data are at the heart of modern physical chemistry. As

---

[1] Some of the support for this statement is based on student testimony. However, there is one study that shows that at the high school level a good laboratory program is second only to aptitude in motivating students to choose a chemistry major. See Babu, G, Wystrach, V.P., and Perkins, R., *J. Chem. Ed.,* 1985, *62*, 501.

[2] 20 years may be a bit of an understatement. In preparation for a talk at the 1991 Middle Atlantic Regional Meeting I looked at a laboratory manual written by John Jaspers and published in 1938. Much of the equipment described there still provides the backbone for many physical chemistry laboratory programs.

one example lasers have become ubiquitous[3] because of their special capabilities, particularly the purity of light and the high photon concentration that they provide. Computer-linked spectroscopic systems are now employed to give direct information about molecular properties, to study rapidly reacting systems, and to conduct analyses requiring high sensitivity. It is only in the last few years that we have begun to see a few presentations of laser-based experiments In the *Journal of Chemical Education*. Textbooks are still barren.

This is not just a problem for physical chemistry. The proposed techniques can be applied fruitfully in all areas of chemistry. Indeed the history of the discipline shows that the movement of technical advances from physics through physical chemistry to chemistry at large has been an important element in the development of the whole field. Thus the project planners saw great advantages to introducing undergraduates (who, after all, are in the general education stage of their development as scientists) to modern physical chemistry techniques whatever their ultimate field of specialization.

Aims of the Pew Program

The Pew Physical Chemistry project had several aims:

- demonstrating the value of the consortial approach to the development of ambitious educational projects;
- encouraging the physical chemistry educational community to think boldly about what is possible in the laboratory by creating a benchmark demonstration project;
- creating a nucleus around which those across the nation who share an interest in improving the physical chemistry laboratory program can coalesce;
- improving physical chemistry laboratory instruction at the ten member institutions;
- creating a group of modern experiments built around six or seven instrument packages of varying cost, so that at least some can be afforded by institutions like those in the cluster;
- making the material developed by the group widely accessible.

The project sponsored experiments at both the introductory and advanced level and gave great weight to projects which involved computer control of apparatus and computer processing of data. All experiments were required to be developed to the point at which their use required only the general background shared by all

---

[3] Robert Moore, the Associate Director of the Physical Chemistry Project has recently done a quick sampling of the experimental articles in the Journals of Physical Chemistry and Chemical Physics. 40% used lasers in some way.

physical chemistry faculty, that is that they meet a criterion of transferability. The principle technique for assuring this is discussed below. The joint demands for technical expertise and for transferability made a broad based consortium a particularly advantageous choice for carrying out the program and insured the utility of the experiments devised for institutions ranging from small colleges to major research universities.

The Plan of Execution

No matter how appealing an ambitious curricular development program is, it will fail without a careful plan for its execution. The plan for carrying out the Physical Chemistry project contained several elements.

- *Recruiting Someone With a Full-time Commitment to the Program.* Given the demands made on the time of all the faculty participants, the planners saw it as essential to have someone with a full-time commitment to the program to provide a core around which to build the project's activities. While the original plan envisioned the main role of the holder of the full-time position as providing overall technical support for the program, in actuality the recruitment of Robert Moore from the faculty of Virginia Commonwealth University as Associate Director of the project converted it into a significant leadership position.

- *Incentives for Participation.* Given the research pressures on university faculty members and the increasing emphasis on significant research activity at the liberal arts colleges as a criterion in evaluation of faculty, along with the college's traditionally heavier teaching demands, we felt that we needed significant incentives to encourage participation and so two were built into the program. The first was the opportunity to acquire state of the art equipment, useful both in the teaching laboratory and in research, a powerful incentive for an active scientist. For this the project had a budget of $135,000 with each allocation from those funds to be matched by recipient institutions on a dollar for dollar basis. With the institutional commitment the total equipment budget then was $270,000 for the ten institutions. The second incentive was provision of funds to support released time, to pay summer stipends, and to support student assistants. $169,000 was budgeted in this category. $15,000 was allocated for supporting meetings, communication, and other administrative costs. The original grant period was three years, but it was extended for an additional year.

- *Assurance of Transferability.* Familiarity with a given technique all too often guarantees that the written material provided for those less familiar with that technique will inadvertently omit information essential to its successful performance elsewhere because the designer of the experiment is well past the stage of the beginner, and so assumes a level of familiarity in the area above that which actually exists in the non-expert. To guard against this we used the strategy of designating a faculty member at one institution as the lead developer and asking for a second faculty member at a different institution who lacked expertise in the area to serve as a tester. *This aspect of the program had an important faculty development aim as well: to allow faculty members who wished to develop new expertises the opportunity and time to do so in a strongly supporting environment.* Several of the participants in the program did in fact use the opportunity to explore new areas. Regrettably we had minimal participation from senior faculty members.

- *Orchestrating Progress.* It is essential in a program of this type to provide a series of occasions at which participants will report on their work. These serve as significant incentives to progress. The Pew project had four levels of these with each reaching out to a larger audience. The first level consisted of thrice yearly meetings of the participants. The site of these meetings was rotated among the consortium's colleges so that participants had the opportunity to become familiar with each of them. The second level was a meeting in October 1991 at Haverford College to which physical chemists from both the consortium and from other Delaware Valley institutions were invited. For the third and fourth levels we hoped not only to present our work, but provide a forum for others working on physical chemistry laboratory development first in the Mid-Atlantic area and then nationally. With this goal in mind symposia were organized at the May 1991 Middle Atlantic Regional Meeting, and at August 1991 meeting of the American Chemical Society.

Survey of Accomplishments

With respect to apparatus the work proposed fell into two broad categories: The first consisted of experiments built around instruments purchased under the program; the second of experiments built around apparatus commonly found at many teaching institutions. As might be expected most of the development was under the first category. Perhaps the most ambitious of these projects involved the use of lasers in the undergraduate curriculum. This seemed a natural area in which to work for several reasons. The University of Pennsylvania is a leader in the application of lasers to the solution of chemical problems, and indeed faculty members there had already written a successful grant proposal to the Keck foundation which included a laser for their undergraduate laboratory. The expertise was not limited to Penn. Tom Stephenson of Swarthmore used lasers in his research as did Marianne Begemann at Vassar. Julio DePaula was to join the Haverford faculty and Valerie Walters was to join the Lafayette faculty. Both had considerable laser expertise. An important consideration in recruiting Robert Moore to the project was his broad experience with lasers. Finally, we felt that lasers had developed to the point at which they might be used with relative ease by non-experts in the technology.

Moore, Stephenson and DePaula defined a standard nitrogen dye laser system which met our criteria for use by non-experts and for reasonable cost. Two of the group's lasers were supplied by PTI, one by Laser Photonics. Experiments have been developed which can be used at both the introductory and advanced levels. Lester (Penn), Moore (Associate Director of the Project), Begemann (Vassar), DePaula (Haverford), Walters (Lafayette) and Moog and Hess (Franklin & Marshall) all participated. Contributions to this point are listed in Table 1.

TABLE I. CONTRIBUTIONS FROM THE PEW PHYSICAL CHEMISTRY PROJECT

Marianne Begemann, Vassar
Dye Laser Studies of Plasma and Flame Generated Species.

Julio DePaula, Haverford
Time-Resolved Laser Spectroscopy in the Undergraduate Laboratory.

Michelle Francl, Bryn Mawr
Integration of Modern Computational Techniques and Graphics into the Undergraduate Physical Chemistry Curriculum.

Margaret Kastner, Bucknell
X-Ray Crystallography: Instructional Materials.

Marsha Lester and Robert Moore, U. of Pennsylvania
Experiments in Raman Spectroscopy for the Physical Chemistry Laboratory.

Colin MacKay, Haverford
The Use of an MS or MSD to Study Gas Phase Reactions.

Richard Moog, Franklin & Marshall
New Approaches in Spectroscopy Experiments.

Robert Moore, Haverford
How Expensive are Lasers, Really?

Miles Pickering, Princeton University
Structure-Reactivity Relationships Explored by Cyclic Voltammetry.

Thomas Stephenson, Swarthmore
Fluorescence Probes of Solution Phase Dynamics.

Valerie Walters, Lafayette
Two Photon Absorption Spectroscopy of Phenanthrene.

Brian Williams, Bucknell
The Characterization of Detergents Using Fluorescent Probe

David Van Dyke, U. of Pennsylvania
Two Cyclic Voltammetry Experiments for the Introductory Physical Chemistry Laboratory.

David Van Dyke, U. of Pennsylvania
A Computer-Interfaced Titration Experiment for the Physical Chemistry or Instrumental Analysis Laboratory.

While the Group made strong efforts in developing the use of lasers there were other areas of significant activity. In terms of effort expended, that in the area of uv-visible and fluorescence spectroscopies was second only to that in the laser project. Moog (Franklin & Marshall), Stephenson (Swarthmore), and Williams (Bucknell) all contributed. Some of the experiments required new equipment, some used available apparatus. Also in the area of use of available equipment, MacKay (Haverford) worked on the use of a mass spectrometer or mass selective detector to study gas phase kinetics, and Moore has been developing an experiment using gas chromatography to study thermodynamic properties. Electrochemistry is poorly represented in current physical chemistry laboratory textbooks. VanDyke (Penn), Pickering (Princeton), and Stephenson (Swarthmore) chose to work in this area.

The original focus of the project was on experimental physical chemistry with an emphasis on computer acquisition and processing of data, but two strong proposals opened other possiblities. Michelle Francl (Bryn Mawr) has used her skill in computational chemistry to produce material which requires student interaction, and which results in some striking visualizations of electronic structures. Marj Kastner (Bucknell) will provide videotapes and computer materials giving instruction in X-ray crystallography.

Assessment

A thorough assessment of this program should be done in several years when one can judge how lasting its effect is at the institutions making up the consortium and how successful is the wider movement for improvement in the physical chemistry laboratory program of which it is a part. However, some preliminary judgments in terms of the goals outlined earlier are possible.

- There have been clear advantages to carrying out such a project in a consortial framework. There is the obvious one of numbers. The ten institutions of the consortium muster thirty-six physical chemists, a number even a large state university cannot match. Furthermore, in each at least one physical chemist is responsible for the laboratory program, creating a pool of more than ten faculty members with a direct interest in the physical chemistry instructional laboratory. Thus it is not surprising that 14 faculty members participated in the project. The interactions among them have been synergistic.

- The project has demonstrated some of what can be accomplished with the expenditure of some resources, some effort, and some time. Whether it will encourage those not yet involved in upgrading their physical chemistry laboratory programs to think boldly about what is possible only time will tell, but the early response to our effort is encouraging.

- The goal of providing a nucleus around which those across the nation who share an interest in improving the physical chemistry laboratory could coalesce seems in reach. The organization of and widespread participation in the symposium on physical chemistry laboratory development at the August 1991 4th Chemical Congress of North America held jointly with the national meeting of the American Chemical Society is a promising beginning.

As part of our outreach effort Moore organized a well attended symposium at the November 1991 Southeastern Regional Meeting of the A.C.S. That there will be symposia organized by others at the Rocky Mountain Regional Meeting and at the National Meeting in Chicago indicates that the momentum achieved so far will not be lost. Further, the publication by the American Chemical Society of this symposium volume containing material presented at the New York meeting means that a diverse group of new, modern, physical chemistry experiments will be easily available to the educational community.

- All participants agree that through the provision both of modern apparatus and of experiments which use it an important contribution has been made to the Pew project's goal of improving physical chemistry laboratory instruction at the consortium schools.

- The package of experiments developed speaks for itself. Instruments used range from those commonly available in the instrument pools of many institutions to a carefully defined laser system. We are disappointed that notably absent from the group's output are experiments exemplifying the physical chemist's use of nuclear magnetic resonance and of Fourier transform infra-red spectroscopy. However, there is every reason to believe that suitable experiments will be developed elsewhere.[4]

- The publication of this volume and the various meetings inititated by the group testify that the group's goal of making its materials widely accessible will be met.

Some Comments

As important to success as a willingness of individual faculty members to participate is institutional support. A key here was the organizational structure of the consortium. Two groups oversaw its activities, an Administrative Oversight Group consisting of one administrator from each member institution and a Scientific Advisory Group consisting of one senior scientist from each institution. Both groups set policy in their respective domains and all members of each were kept fully informed of activities. Their familiarity with and interest in the project was a significant source of institutional support.

Such success as the project has enjoyed stems from the enthusiastic participation of many young faculty members and their students. Of the two major incentives for their participation, the chance to acquire new equipment and the opportunity to have released time, the one exhausted first was the equipment fund. This is due at least in part to the difficulties of arranging one semester leaves in small liberal arts colleges with only one or two physical chemists.

---

[4] In the area of nuclear magnetic resonance C. Brad Antanaitis of the Physics Department at Lafayette College has developed an experiment on the "*Biophysics of Electron Transfer in Cytochrome C.*" as part of Mid-Atlantic Cluster's Physics Project. This should be of interest to chemists.

The progression of meetings worked very well, achieving its goals of developing a group spirit and providing both incentives for, and check-points on progress. The demographics of the liberal arts colleges in the Mid-Atlantic Consortium, which are heavily weighted to younger faculty members in physical chemistry, proved to be an important strength, but also caused a problem. The project got off to a slow start because several of the institutions were appointing new staff members. This slow start was certainly of concern to those administering the program for the Pew Foundation. As member institutions filled out their staffs the project took off with the added advantage of the fresh skills and points of view that these new appointees brought.

There are many advantages to carrying out curriculum development projects in the framework of a multi-institutional consortium. For the project itself there is an increased pool from which to draw participants and a greater range of expertises available. Faculty from different institutions with similar interests come to know each other and to know the programs at peer institutions. When a consortium sponsors more than one curriculum development project there are additional benefits. A good idea developed in one curriculum project can be transferred to others. Also there is a natural and healthy competition between disciplinary teams.

One final piece of advice to those who would sponsor curriculum development projects. *On the basis of our experience I would argue that if you want materials usable in a range of types of institutions, consortia such as those which the Pew Foundation has sponsored in which strong liberal arts colleges play a significant role is an excellent vehicle.* This is not surprising since these institutions combine a research active faculty aware of trends in their fields with a strong interest in undergraduate education.

Acknowledgements

The Pew Foundation through its Pew Program in Undergraduate Education has provided the basic support for our work. Joan Girgus, the Director of that program and her assistant, Nancy Porter, have been most helpful. Jerry Gollub, Professor of Physics at Haverford College, was the first director of the Mid-Atlantic Cluster. We have had strong support from the administrations of the participating colleges. Our Administrative Oversight Group has varied in composition. Currently it consists of Bruce Partridge (Haverford), Judith Shapiro (Bryn Mawr), James England (Swarthmore), Jeffrey Bader (Lafayette), Nelvin Vos (Muhlenberg), Eugenia Gerdes (Bucknell), Nancy Dye (Vassar), Susanne Woods (Franklin & Marshall), Walter Wales (U. of Pennsylvania), and Nancy Weiss Malkiel (Princeton). All are senior administrators. The program has benefitted greatly from the contributions of our Scientific Coordinating Commitee, consisting of scientists from the various member institutions. It too has varied in composition. Its current members include Ralph Amado (Physics, Penn), Jan Andrews (Psychology, Vassar), Neal Abraham (Physics, Bryn Mawr), Maitland Jones (Chemistry, Princeton), Frank Moscatelli (Physics, Swarthmore), Margaret Kastner (Chemistry, Bucknell), Robert Chase (Biology, Lafayette), Donald Shive (Chemistry, Muhlenberg), James Spencer (Chemistry, Franklin & Marshall), and Colin MacKay (Chemistry, Haverford).

RECEIVED October 1, 1992

Laser Experiments

CHAPTER 6

A Hands-on Helium—Neon Laser for Teaching the Principles of Laser Operation

William F. Polik

POTENTIAL HAZARDS: Lasers

The emergence of the laser as a scientific tool has profoundly changed the practice of science. Lasers are used in astronomy (lunar ranging), biology (optical trapping of bacteria, fluorescent tagging), electrical engineering (optoelectronics, optical computing), physics (laser-induced fusion, atomic spectroscopy), meteorology (air velocity measurements, wind shear detection), and more. Laser applications beyond scientific research include bar code scanning, laser surgery, surveying, cutting and welding, optical gyroscopes, information retrieval from CD-ROM disks, laser printing, and fiber optic communications. The laser has revolutionized the modern practice of physical chemistry, especially in the areas of molecular spectroscopy (*1-3*) and chemical kinetics (*4*). It has been central to the development of a new branch of physical chemistry called chemical dynamics, in which chemical reactivity is studied with full quantum state resolution (*5*).

The importance of lasers in chemistry has prompted the development of numerous laser-based experiments for the undergraduate physical chemistry laboratory, some of which are detailed in this book. However, the introduction of modern instruments such as lasers into the undergraduate curriculum must be done in a thoughtful manner. Special care must be taken so that students do not just turn knobs or push buttons in pre-arranged sequences. If students are to benefit from exposure to new instrumentation, they must learn the underlying principles of its operation. Moreover, most students learn better and retain new concepts longer if they interact directly with the subject matter. Therefore, the purpose of this laboratory experiment is to teach students the principles of laser operation through a "hands-on" approach, in which students are encouraged to observe and experiment with the internal workings of a simple helium-neon laser. Specifically, students completing the experiments described herein will:

- better understand the physical basis of the lasing process
- be able to identify the components of a laser
- understand how to align a simple laser
- gain practical experience with a relatively safe laser
- gain familiarity with laser diagnostic equipment.

The experimental apparatus consists of a commercial helium-neon laser discharge tube terminated at one end with a mirror and at the other end with a Brewster window. A second laser mirror is held in an adjustable mount external to the laser tube. The laser power supply is an industry standard model, with all high voltage connections insulated. Thus, all components of the laser (laser medium, energy source, and optical cavity) are easily identifiable. After identifying each

2503–6/93/0084$07.25/0

component of the laser and recalling its purpose, the student aligns the cavity by adjusting the external mirror and thereby gains experience making delicate adjustments to the optics of a laser system. The student then makes measurements of the optical cavity stability condition, laser beam polarization, and the cavity transverse and longitudinal modes. These measurements characterize the laser light and reinforce the theory of laser operation. The apparatus has been designed to be sufficiently versatile so as to support these various types of measurements on the same laser system. Through these measurements, the student is introduced to the basic types of laser diagnostic equipment: a power meter, a polarization analyzer, and an optical spectrum analyzer.

As any person who works with lasers will testify, hands-on experience with a laser is absolutely essential to comprehend its operation. The open optical cavity, exposed components, and adjustability of laser optics in this teaching laser afford the student a valuable opportunity to investigate the workings of a simple laser. The experiment is very effective when preceded by a lecture outlining the theory of laser operation and a demonstration of the use of the apparatus. Students may then experiment with the apparatus and make the requested measurements at their own pace. By completing this experiment, students develop a basic understanding of the theory behind laser operation and are well-prepared to understand the operation of more complex laser systems they will encounter in the laboratory, e.g., a pulsed dye laser. Thus, this laboratory experiment is recommended as the initial experiment in a series of experiments which use progressively more complex laser equipment to illustrate modern applications of lasers in chemistry.

Since this teaching laser and the diagnostic equipment consists mostly of commercially available equipment, one does not need to be a laser expert to implement this experiment. The experiment serves as a good starting point for an instructor wishing to introduce laser experiments into the physical chemistry laboratory curriculum. In addition to serving as a teaching tool for physical chemistry, the laser can also be used as a general purpose helium-neon laser in the laboratory or in lecture demonstrations.

Theory

Matter and Radiation Since laser operation has been described thoroughly in many textbooks (*6-17*), only a qualitative overview of the lasing process as it relates to the laser and measurements described here is provided. The mathematical description of the lasing process is omitted in order to make this discussion as accessible as possible.

Although laser light consists of the same electromagnetic radiation as "ordinary" light, it possesses several distinguishing properties: high directionality, extreme brightness, high monochromaticity, and substantial phase coherence. The origin of these properties can be appreciated from a rudimentary understanding of laser action and the physical construction of a laser.

Matter is composed of atoms and molecules, which according to quantum theory exist only in discrete, quantized energy levels. Atoms can change energy levels by absorbing or emitting radiation, such that

$$E_{\text{upper}} - E_{\text{lower}} = h\nu \qquad (1)$$

where h is Planck's constant and v is the frequency of the radiation. Einstein demonstrated in 1917 that there are three interrelated ways in which light and matter interact: absorption, spontaneous emission, and stimulated emission (*18*). These three processes are illustrated schematically in Figure 1. In each process, the

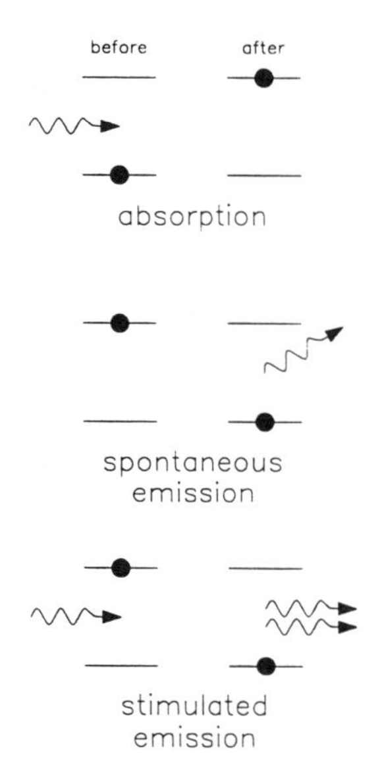

Figure 1. Interaction between radiation and matter: absorption, spontaneous emission, and stimulated emission.

frequency of the emitted light wave must correspond to the energy difference between the initial and final energy levels according to Equation (1). In absorption, an atom is raised to a higher energy level by the presence of a quantum of light, or *photon*, and the photon is absorbed in the process. In spontaneous emission, an atom in an excited state spontaneously falls to a lower energy level, releasing a photon in a random direction. Stimulated emission is analogous to the absorption process, except that the initial state of the atom is higher in energy than the final state and a photon is therefore emitted in order to conserve energy. The stimulated photon is emitted in the same direction and with the same phase as the stimulating photon. A key point in understanding laser operation is to observe that whereas absorption diminishes the intensity of a light beam, stimulated emission amplifies the intensity of a light beam. A laser (*L*ight *A*mplification by *S*timulated *E*mission of *R*adiation) requires that stimulated emission be the dominant interaction between the matter and radiation inside the laser.

The first requirement of a laser is the presence of matter with quantized energy levels which are spaced such that the energy difference corresponds to a frequency in or near the visible part of the electromagnetic spectrum. This matter, which can absorb and emit visible radiation, constitutes the *laser medium*.

Comparison of absorption and stimulated emission in Figure 1 reveals that both processes require the presence of light, but that stimulated emission occurs when the upper quantum state is populated and absorption occurs when the lower quantum state is populated. Thus, in order to insure that the rate of stimulated emission exceeds the rate of absorption, the second requirement for a laser is that a *population inversion* exist, i.e., the number of atoms in the upper state must be greater than the number of atoms in the lower state. Population inversions do not occur in systems at equilibrium, for which the population distribution is described by the Boltzmann relation,

$$\frac{n_{\text{upper}}}{n_{\text{lower}}} = e^{-(E_{\text{upper}}-E_{\text{lower}})/kT} \tag{2}$$

which predicts that lower energy states are always more populated than higher energy states.

Comparison of spontaneous emission and stimulated emission in Figure 1 reveals that both require the upper quantum state of the laser medium be populated, but that stimulated emission occurs when light of the correct frequency is present and spontaneous emission does not require the presence of light. Thus, in order to insure that the rate of stimulated emission radiation exceeds the rate of absorption, the third requirement for a laser is that the excited atoms be in the presence of a high intensity of light, the frequency of which corresponds to the energy difference between the two levels.

The Lasing Process The three requirements for laser operation, a laser medium, a population inversion, and a high intensity of stimulating light, are realized in the physical construction of a laser. The laser medium is a collection of atoms or molecules with quantized energy levels which are contained in the laser. Many different materials have been observed to lase, including solids, liquids, and gases. Since population inversions do not exist at equilibrium, an external energy source is required to disturb the equilibrium situation and populate the upper energy levels. This energy source is often referred to as a *pump*, because it pumps atoms or molecules from lower energy states into excited states. All conventional light sources, in which excited atoms or molecules spontaneously emit radiation, satisfy these requirements of a laser. The technological breakthrough responsible for the invention of the laser was the suggestion by Townes and Schalow in 1958 of a method for surrounding the laser medium with high intensity light (*19*). They suggested the laser medium be placed between two highly reflective mirrors, so that light emitted from the medium would reflect back and forth between the mirrors and bathe the medium in light. This optical configuration for producing high intensities of light is commonly called an *optical cavity*. If one of the highly reflective mirrors slightly transmits light of the laser frequency, light is able to escape the optical cavity in the form of an emitted laser beam. This slightly transmitting mirror is called an *output coupler*. Figure 2 schematically indicates the three components of a laser: laser medium, pump, and optical cavity. Laser action was first demonstrated by Maiman in 1960 with a pulsed ruby laser (*20*). The first example of a continuous-wave laser was the helium-neon laser developed by Javan *et al.* shortly thereafter (*21*).

Figure 2 illustrates the process of laser operation. As a prerequisite for laser activity, the pump creates a population inversion in the laser medium. A few excited atoms or molecules then emit light as they decay to the lower state. At first, this emission is spontaneous because there is insufficient light present to induce stimulated emission. Thus, light is initially emitted in all directions, and most of it is not useful as laser light. Some of the light, however, is emitted down the axis of

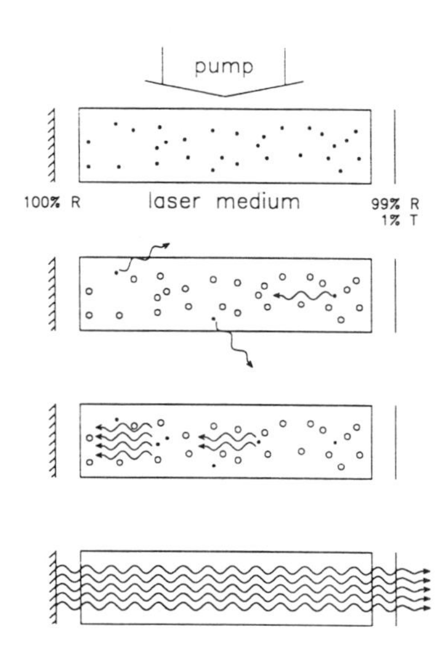

Figure 2. Laser components and operation.

the optical cavity. This light passes by other excited atoms or molecules and induces stimulated emission, thereby producing more light which in turn induces even more stimulated emission. Thus, the intensity of light is amplified as the light passes though the laser medium. The process of stimulated emission builds up as the light oscillates between the two mirrors, until it completely dominates the emission process. A small fraction of the laser light is permitted to escape the laser cavity through the output coupler as the laser beam. If the population of the upper energy level becomes depleted so that a population inversion no longer exists, laser activity stops. Thus, the pump must continuously maintain the population inversion for lasing to be continuous.

The physical construction of a laser is responsible for the distinguishing properties of laser light recognized above. The axial geometry of the optical cavity causes the high directionality of the laser beam along the cavity axis. The extreme brightness of a laser beam results from this high directionality. The monochromaticity results from the existence of sharp transitions between quantized energy levels. If several transitions in a given medium can support laser action, a single one of them can be selected by using mirrors which are optimized to reflect at only the desired wavelength. Finally, the phase coherence of laser light is a direct result of the stimulated emission process.

The Helium-Neon Laser The helium-neon laser used in this experiment is generically depicted in Figure 3, and the atomic energy levels involved in this laser

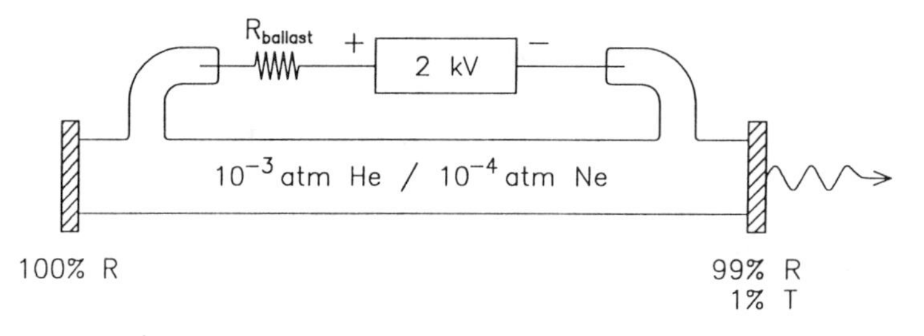

Figure 3. Construction of a helium-neon laser.

are illustrated in Figure 4. In this laser, the laser medium is a mixture of about 1 torr of helium and 0.1 torr of neon contained in a narrow discharge tube. The pump consists of a high voltage DC power supply which creates an electrical discharge through the medium that excites helium atoms from the ground $1s^2$ electron configuration to the 1s2s configuration. This excited state of helium lies very near in energy to the $1s^22s^22p^55s$ electron configuration of neon. Thus, the electronic

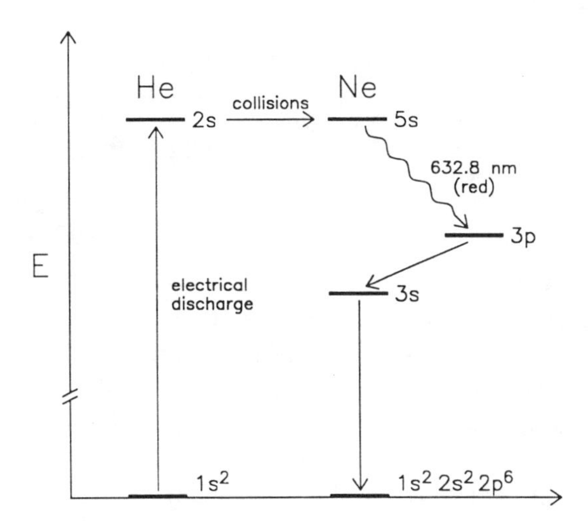

Figure 4. Energy level diagram for helium-neon laser.

energy of excited helium atoms can be transferred very efficiently to neon atoms through collisions. Since only the ground electronic state of neon is populated at room temperature, a population inversion is created between the $1s^22s^22p^55s$ configuration and the lower-lying $1s^22s^22p^53p$ configuration of neon. Laser activity at 632.8 nm can then occur if the gas mixture is contained within a suitable optical cavity. The $1s^22s^22p^53p$ configuration of neon rapidly decays to $1s^22s^22p^53s$, thereby preventing undesirable population build-up in the lower lasing level. Finally, neon atoms return to the ground configuration through deactivating wall collisions.

Methodology

General Description The "hands-on" helium-neon teaching laser consists of three components: 1) a discharge tube containing the helium-neon laser medium; 2) a high voltage power supply that excites the gases in the discharge tube; and 3) an optical cavity consisting of two highly reflective mirrors surrounding the laser medium. These components are mounted on an optical breadboard and are visible to the user. A schematic of the basic apparatus is shown in Figure 5.

One mirror of the optical cavity is sealed directly onto the discharge tube. This concave mirror has a 60 cm radius of curvature and is called a *high reflector* because its reflectivity at the lasing wavelength of 632.8 nm is essentially 100%. The other end of the discharge tube is sealed by a transparent flat window oriented at 56°. At this special angle, known as the *Brewster angle*, this window transmits virtually all light polarized parallel to the plane of incidence. Light polarized perpendicular to the plane of incidence is partially reflected by the Brewster window

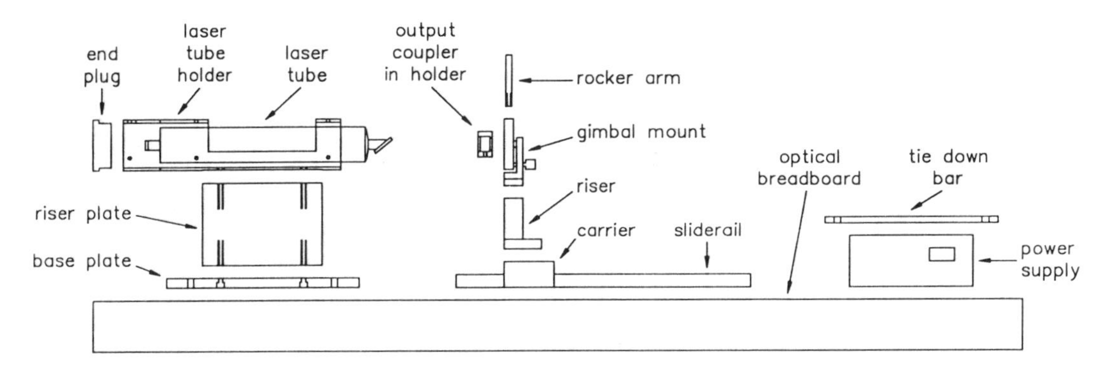

Figure 5. Schematic of "hands-on" helium-neon teaching laser.

and hence escapes from the cavity. If this window were not oriented at the Brewster angle, its reflectivity would introduce a loss in the cavity which would prevent laser operation. The other laser mirror, called the *output coupler*, is a flat mirror held in a gimbal mount on a sliding track so that its orientation and the optical cavity length can be adjusted. The output coupler reflects roughly 99% of the stimulated emission from the laser medium back into the optical cavity and transmits 1% of the stimulated emission as the laser's output beam.

The student should sketch the laser apparatus in a lab notebook, and identify the name and purpose of each component.

Laser Alignment To support laser action, the optical cavity must be aligned so that the light beam inside the optical cavity does not "walk off" the end mirrors after repeated reflections. A systematic search procedure is used to align the output coupler parallel to the high reflector. During this and subsequent procedures, be careful not to touch the Brewster window or output coupler. CAUTION: The emitted laser beam is very intense. Never look directly into a laser beam or into any reflections of a laser beam. Remove watch and jewelry from wrist and fingers to prevent stray reflections.

Turn on the power supply and wait 5 s for the discharge. If the laser has been left in an unaligned configuration, it should not be lasing. Slide the output coupler mount to about 2" from the Brewster window and lock down the carrier onto the sliderail. Adjust the vertical knob so that the output coupler points slightly downward. This setting is easily verified by observing that the vertical post extending upward from the mirror mount slants a few degrees toward the discharge tube. With a delicate touch, grasp the vertical post and rock it back and forth a few degrees. Rocking the output coupler about a horizontal axis sweeps the mirror through a range of angles that includes the correct vertical orientation required for lasing. The horizontal adjustment knob rotates the output coupler about a vertical axis and allows for the correct horizontal mirror orientation required for lasing. Position this knob so that the output coupler points slightly to the left or right of the discharge tube. This position may be verified by viewing the output coupler holder from above the apparatus. It is also possible to use a white business card near the Brewster window to look for light from the discharge tube reflected by the output coupler back toward the Brewster window. A 1–2 cm diameter diffuse circle of light should appear slightly off the optical axis; its center should be located about 1 mm below and to the left or right of the Brewster window.

The search procedure calls for a systematic two dimensional search over all vertical and horizontal mirror positions to locate the lasing position of the output coupler. Make a very small adjustment (1/32 of a full turn) of the horizontal knob to move the output coupler toward the correct horizontal position. Then rock the vertical post so as to pass through the full range of vertical orientations. These two different motions must be performed independently, i.e., in succession. As the mirror is rocked between small horizontal adjustments of the output coupler, watch for a flash of red laser light on the output coupler. If none appears, advance the horizontal knob in the same direction and then rock the post again. Continue this procedure. A smooth systematic search eventually produces a flash of red laser light. Once the flash has occurred, stop adjusting the horizontal knob and allow the vertical post to return to its resting position. Then, advance the vertical knob until lasing is continuous. At this point, "tweak" both knobs for maximum laser brightness. CAUTION: The laser beam cannot hurt one's fingers, but be careful not to look directly into the beam. Verify that the beam and any stray reflections are blocked.

Intentionally misalign the laser by randomly twisting both knobs and repeat the alignment procedure until it is confidently and easily performed. Each student should align the laser at least three times.

Note that the laser beam is much brighter inside the optical cavity than outside the cavity. If 99% of the light is reflected from the output coupler back into the laser cavity, the ratio of light intensity inside the cavity to outside the cavity should be 200:1 (the extra factor of 2 arises from the fact that light is propagating in both directions inside the cavity). If a white business card is placed in the path of the beam outside the cavity, the laser beam is merely blocked at that point. However, if the card is placed in the laser beam path inside the cavity, all lasing activity ceases. Verify and explain this observation.

Laser Cavity Stability Laser activity is sustained only when amplification exceeds the loss of laser light from the optical cavity. If an optical cavity is constructed with small or flat mirrors, diffraction causes some of the laser light to spill out around the edges of the mirrors. Such radiation losses must be minimized for efficient laser activity. The problem of diffraction loss can be minimized by using at least one slightly concave mirror so that light is redirected back toward the optical axis upon each reflection. Mathematically, one must trace the path of a light ray which is initially directed slightly off the optical axis of the cavity. If the ray escapes from the cavity, the cavity is said to be unstable. If the ray is found to remain near the optical axis after multiple reflections, the cavity is said to be stable. One can show that a cavity is stable, i.e., can support lasing, when

$$0 < g_1 g_2 < 1 \tag{3}$$

where $g_i = 1 - L/r_i$, L being the separation between the two mirrors and r_i being a spherical mirror's radius of curvature.

Evaluate the laser stability condition, Equation (3), for the present laser. The high reflector has a radius of curvature $r_1 = 60$ cm and the output coupler is flat, i.e., $r_2 = \infty$. Evaluate g_2. Substitute this value of g_2 into Equation (3) so that the only remaining variable is g_1. Replace g_1 with its definition containing the current value of r_1, and rearrange the inequality so as to find the maximum and minimum lengths L for this laser. Can both limits be tested with the present apparatus?

Unlock the sliding carrier for the output coupler and increase the cavity length. Relock the carrier and reestablish lasing if necessary. Note that lasing is more difficult with increased cavity length. Increase L until the cavity no longer sustains laser activity. At this point the laser cavity has become unstable. How does the final, limiting value of L agree with the theoretical prediction?

Another method of introducing a variable loss into the laser cavity is to insert a thin glass or quartz flat into the laser cavity between the Brewster window and the output coupler. This flat window is placed in a mount which permits rotation about a horizontal axis, thus varying its angle with respect to the laser beam. Adjust the angle of the flat to the Brewster angle, 56°, i.e., parallel to the Brewster window on the discharge tube. CAUTION: This procedure causes a stray reflection to be emitted from the flat toward the ceiling in the vertical plane containing the laser beam; DO NOT LEAN OVER THE LASER OR THE EMITTED LASER BEAM during this procedure. Insert the flat into the optical cavity of the laser, and lock down its magnetic base. Lasing may cease due to slight irregularities or imperfections in the flat. Reestablish lasing with slight adjustments to the output coupler. Rotation of the flat away from the Brewster angle introduces a reflection loss into the cavity. The output laser beam intensity decreases as the angle of the flat increases away from the Brewster angle, and finally lasing stops. Rotating the flat back toward the Brewster angle reestablishes lasing. Remove the flat from the optical cavity of the laser.

The angle of the flat may be used to introduce a calibrated loss into the optical cavity. At the Brewster angle, this loss is theoretically zero. The reflection loss at other angles may be calculated from standard formulas found in optics textbooks. A rough measurement of the gain of the laser medium may be made by rotating the flat until lasing just ceases. At this point the cavity losses (due to the output coupler, diffraction, and reflection from the flat) exactly balance the cavity gain. The advanced student can calculate the cavity gain coefficient in this manner and compare the result to typical accepted values (*16*).

Laser Light Polarization While some helium-neon lasers emit randomly polarized light, the discharge tube in the present apparatus (sealed at one end by a flat window oriented at the Brewster angle) produces a polarized light beam. The reason, as mentioned above, is that at the Brewster angle, the window transmits virtually all light polarized parallel to the plane of incidence. Light polarized perpendicular to the plane of incidence is partially reflected out of the cavity. The plane of incidence is the plane containing both the incident and reflected beams. Discrimination against perpendicularly polarized light during the 200 or so passes that a lightwave makes through the Brewster window results in a vertically polarized output beam.

The theoretical expression for transmission of polarized light through a polarization analyzer is

$$P(\theta) = P_0 \cos^2(\theta) \tag{4}$$

which is known as Malus's Law. In this expression, $P(\theta)$ is the transmitted power when the analyzer is oriented at angle θ and P_0 is the maximum power transmitted by the analyzer when oriented at $\theta = 0°$. Thus, this form of Malus's Law assumes that the plane of polarization of the light beam emerging from the laser defines the angle $\theta = 0°$.

With a sheet of polarizing film, investigate the polarization of the laser beam emitted by the present laser. CAUTION: A stray reflection is generated when polarizing film, or any other optic, is inserted into a laser beam; always insert the optic into the beam at an angle such that the stray beam is reflected in a safe direction, e.g., downward. Insert the polarizing film into the laser beam. Rotate the film in a plane normal to the laser beam and observe the variation in intensity of the light transmitted through the film. In which plane is the laser light polarized? Vertical or horizontal? Does this polarization correspond to light polarized parallel or perpendicular to the plane of incidence of the laser's Brewster angle? Explain by sketching the Brewster window and indicating the polarizations of the transmitted and reflected beams.

Quantitatively measure the degree of polarization of the laser light as follows. Set the laser power meter to 10 mW full scale, and place the detector head in the laser beam. CAUTION: Insure that the reflection from the detector head is safely blocked. Deliberately misalign the output coupler in one dimension to cause lasing to cease, and zero the power meter. Reestablish lasing. Place the polarization analyzer in the laser beam between the output coupler and the detector head of the power meter. Check that the polarization analyzer is mounted such that the maximum transmitted power occurs at angle $\theta = 0°$ and the minimum occurs at $\theta = 90°$. If the maximum transmitted power is less than 3 mW, reset the power meter to 3 mW full scale and rezero the meter. Acquire experimental data for the transmitted power $P(\theta)$ as a function of analyzer angle θ. Take a reading every 10° from –30° to 120°. Plot this data as the ratio $P(\theta)/P_0$ on the y axis versus θ on the x axis. On the same figure, plot the theoretical expression for polarized light according to Equation (4). Is there sufficient agreement between the data and Malus's Law to suggest that the laser beam is polarized? Highly polarized?

Laser Transverse Modes Most helium-neon lasers are designed to emit a laser beam that produces a single spot on a screen. Detailed examination of the cross-sectional intensity profile of such a beam reveals it to have a Gaussian profile, with maximum intensity in the center which smoothly drops off with increasing radius. It is possible for a laser to operate in more complicated *transverse modes*, in which the radiation field within the optical cavity contains both maxima (antinodes) and minima (nodes) in the cross-sectional plane perpendicular to the optical axis. Transverse modes are characterized by two integers, m and n, representing the number of vertical and horizontal nodes. The TEM_{00} mode experiences the minimum diffraction loss, has the minimum divergence, and can be focussed to the smallest possible spot size. For these reasons, it is often the most desirable mode for commercial applications. Higher order modes are larger in diameter and therefore suffer higher diffraction losses.

The present laser discharge tube has a large diameter bore, which makes it capable of operating in several TEM modes. In fact, this laser often operates in a superposition of several TEM_{mn} modes simultaneously in what is called a "multi-transverse mode" condition, thereby obscuring the well-defined individual modes. Since the TEM_{00} mode is the narrowest transverse mode, it is possible to suppress all higher order modes by introducing a small aperture on the optical axis inside the laser cavity. Higher order modes are discriminated against by greater absorption losses on the aperture because they have larger diameters than the TEM_{00} mode.

Slide the output coupler close to the discharge tube and establish laser action. Insert a -1" focal length diverging lens in the output beam of the laser, and display the resulting divergent beam on a white sheet of paper several feet after the lens. The laser may be operating on multiple transverse modes, thereby obscuring distinctive nodal patterns. Slightly adjust the output coupler alignment and/or cavity length to select a single TEM_{mn} mode while suppressing the others. Sketch this mode in your lab notebook and label it with the appropriate integers, e.g., TEM_{12} for 1 vertical node and 2 horizontal nodes. Locate and sketch at least three different transverse modes.

The laser can be forced to operate in TEM_{00} by introducing an aperture of the proper size to discriminate against higher order modes. The easiest way of doing this is to increase the cavity length to $L > 50$ cm. At this length, the bore of the discharge tube acts as the limiting aperture. At shorter cavity lengths, a pinhole of roughly 1 mm diameter inserted on the optical axis near the Brewster window suppresses higher order transverse modes. This pinhole is held in an *xy* translation mount so that it can be precisely positioned on the optical axis of the laser. Obtain TEM_{00} operation by one or both of these methods.

Laser Longitudinal Modes The high reflectivity of the end mirrors in a laser optical cavity causes the light to make many round trips before it escapes though the output coupler. Constructive interference of the light inside the cavity results when the maxima of the waves and their subsequent reflections overlap in space. The electric field oscillations of these lightwaves then add in phase, and the resulting light intensity increases and induces stimulated emission at that wavelength more efficiently. If the laser mirrors are nearly flat, such constructive interference occurs when an integral number of wavelengths is contained in the roundtrip distance of the cavity, i.e., when

$$q\lambda_q = 2L \tag{5}$$

where q is an integer, λ_q is the wavelength, and L is the cavity length. If the wavelength does not satisfy Equation (5), then destructive interference occurs among successive reflections of the light in the cavity, thereby suppressing stimulated emission at these wavelengths. Thus, a laser cavity only supports laser activity at distinct wavelengths, λ_q. Each constructively interfering, oscillating electromagnetic field configuration is referred to as a *longitudinal mode* and is characterized by a unique integer index q. It is often more convenient to express the laser cavity boundary condition of Equation (5) in terms of frequency, ν_q,

$$\nu_q = \frac{c}{\lambda_q} = \frac{qc}{2L} \tag{6}$$

where c is the speed of light. The frequency separation between adjacent modes is therefore

$$\Delta\nu = \nu_{q+1} - \nu_q = \frac{(q+1)c}{2L} - \frac{qc}{2L} = \frac{c}{2L}$$

A laser can operate only at the longitudinal mode frequencies. In a single longitudinal mode laser, only one longitudinal mode frequency occurs within the linewidth of a lasing transition. However, if the linewidth is wider than the spacing between adjacent longitudinal modes, then the laser operates at several closely spaced frequencies, each corresponding to a different longitudinal mode.

The frequency composition of a laser beam can be investigated with an optical spectrum analyzer. A spectrum analyzer is a tunable interference filter that transmits light of only certain frequencies. The instrument consists of two highly reflective, spherical mirrors which form an optical cavity much like a laser cavity, but without the laser medium. Frequencies are transmitted when the cavity contains an integral number of wavelengths during a round trip, while other frequencies are reflected. By electronically varying the separation between the mirrors, the transmission frequency of the spectrum analyzer can be scanned. The spectrum analyzer therefore serves as a high-resolution, scanning, interference filter which is used to analyze the spectral composition of a beam of light.

Align the helium-neon laser and insert a diverging lens into the output beam to observe the transverse mode structure of the laser beam. Insert a pinhole aperture into the cavity or increase the cavity length to $L > 50$ cm so that the laser is operating in the TEM_{00} mode. Remove the diverging lens. Measure the cavity length L with a meter stick. Knowing that λ_q is nominally equal to 632.8 nm for a helium-neon laser, use Equation (5) and the measured cavity length to calculate the a value for q, the number of wavelengths of light contained in the round trip distance of the optical cavity. Be careful to keep the proper number of significant figures in the calculation.

Prepare the Burleigh SA-800-C spectrum analyzer by setting its controls as follows: Photoamp Gain = 12:00 o'clock, Photoamp Bias = 12:00, Dispersion Multiplier = ×2, Dispersion Variable = 9:00, Centering = 12:00, and Sweep = .01 s. (Prepare the Coherent 240-1-B spectrum analyzer controller as follows: Sweep Expansion = ×1, Amplitude = 3:00 o'clock, Offset = 12:00, Risetime = 10 ms.)

Prepare the oscilloscope as follows: Vertical Input = 1 V/div DC, Horizontal Sweep Time = 1 ms/div, and Trigger = positive slope and external source.

Verify that two cables lead from the spectrum analyzer (SA) to the controller, one for adjusting the SA mirror spacing and the other from the SA detector. Verify that two cables lead from the controller to the oscilloscope, one to the vertical input and the other to the trigger input. Turn on the spectrum analyzer and oscilloscope.

If the set-up has not been disturbed since it was last used, a repeating pattern of longitudinal modes should appear as vertical peaks on the oscilloscope screen. If longitudinal modes do not appear, the SA must be aligned. Position the SA so that the laser strikes the SA in the center of the input lens. Orient the SA so that its optical axis is approximately collinear with the laser beam. Translate/orient the SA so that the reflection from the front surface of the SA lens reflects back along the laser beam. Use the *xy* translation and gimbal mount angle adjustments on the SA mount to search for a 1–2 cm diameter red spot being reflected from within the SA back toward the laser. Position this reflection near, but not directly on, the laser output coupler. Several laser modes should now appear as peaks on the oscilloscope screen.

Use the translation and angle adjustments on the SA mount to sharpen up the peaks on the oscilloscope screen. Adjust the oscilloscope vertical amplification as necessary. Observe that if the spectrum analyzer is adjusted to reflect the laser beam directly back into the laser, the spectrum becomes unstable. Excess optical feedback into the laser causes laser instability. Back off slightly from this condition so that the spectrum is stable, the peaks are of maximum height and minimum width, and the baseline is flat.

The Burleigh SA has a free spectral range (repeat distance) of 8.00 GHz. Use the SA Dispersion Variable and Centering controls to adjust the appearance of spectrum so that two identical sets of longitudinal modes appear on the oscilloscope screen and are separated by exactly 8 divisions. (The Coherent SA has a 7.50 GHz free spectral range. Use the Amplitude and Offset controls to separate two sets of modes by exactly 7.5 divisions.) The SA is now calibrated to 1 GHz = 1 ms (one major horizontal division on the oscilloscope screen). If the SA mount controls are adjusted for better sharpness of peaks, the calibration must be verified. Once the SA is calibrated, do not adjust the Dispersion Variable (Coherent SA: Amplitude) control, as this invalidates the calibration.

Use the Centering (Coherent SA: Offset) control to center one set of longitudinal modes on the oscilloscope screen. Adjust the oscilloscope timebase to expand this set of longitudinal modes. If more than one peak is present, the SA is revealing that the laser is not perfectly monochromatic but instead is lasing at as many frequencies as there are peaks. If the laser is operating in a single TEM mode, each peak arises from a different longitudinal mode.

Apply some pressure with your fingers to the optical breadboard near the laser cavity and watch the oscilloscope screen as longitudinal modes sweep through the gain region of the spectrum. Describe and explain this observation in your notebook. What is changing? The laser frequency? The longitudinal mode index *q*? The laser cavity length?

Determine the frequency separation $\Delta\nu$ between adjacent longitudinal modes by measuring the number of oscilloscope divisions between adjacent peaks. Note the current horizontal timebase setting of the oscilloscope, and use the SA calibration to convert this distance to frequency (GHz). Use Equation (7) and the experimentally measured value of $\Delta\nu$ to infer the length of the laser cavity. Is the inferred value consistent with the measured cavity length?

Additional exercises that can be performed with a spectrum analyzer include measuring $\Delta\nu$ for different cavity lengths, observing the frequency spectrum when the laser is not operating in TEM_{00}, and analyzing the polarization of adjacent modes in a randomly polarized laser.

When Finished Turn off the discharge tube, power meter, spectrum analyzer, and oscilloscope. Slide the output coupler to about 2" from the Brewster window and lock down the carrier. *Leave the laser in an unaligned configuration* by twisting both the vertical and horizontal knobs one or two full turns. This allows the next students who work with this equipment to rightfully claim to have aligned a laser from scratch when they have completed the experiment!

Safety

The issue of laser safety has been thoroughly addressed by the American National Standards Institute (ANSI), and several excellent overviews of their findings exist (*22-24*). Continuous visible lasers are classified by their maximum power as being Class 1 ($\leq 10^{-6}$ W), Class 2 ($\leq$ 1 mW), Class 3a ($\leq$ 5 mW), Class 3b ($\leq$ 0.5 W), or

Class 4 (> 0.5 W). Class 1 lasers are considered to be completely safe, even if viewed directly for extended periods. Class 2 lasers are considered to be safe if not stared at directly. Specifically, damage does not occur within the natural aversion time of the eye, 0.25 seconds. Class 3a lasers also pose little risk for momentary direct viewing if the beam is not collected and focused by an optical instrument, e.g., by a surveyor's transit. Class 3b lasers can pose a risk if viewed directly, even without collection optics. Class 4 lasers are the highest power and can cause skin as well as eye damage. A similar classification scheme exists for pulsed lasers, but pulse length and energy become the classifying factors.

The laser described in this article emits between 1 and 6 mW radiation, depending on cavity length and alignment. Thus, according to the ANSI laser classification scheme, it is technically a Class 3b laser; however, its maximum power only slightly exceeds the Class 3a power limit and its operating power often falls within the 3a power level during normal use. Although the laser is relatively safe, the possibility of eye injury does exist from carelessness or misuse. Thus, it is important always to observe proper safety procedures when operating the laser.

Users of the laser must never look into the laser beam or into any direct reflections of the beam. Because the path of the direct beam is very evident, the primary safety concern is to eliminate or predict and block all stray reflections. This is of most importance when inserting or removing optical components from the emitted laser beam. The presence of the laser beam should always be detected indirectly with a white business card, and never directly with the eye. The laser should be placed in a room to which access is limited to the users of the laser. Unfortunately, laser safety goggles which block the 632.8 nm radiation of the helium-neon laser are not practical for this experiment because all aspects of the laser which are to be observed can not be seen through such goggles. The maturity of the user is also an important consideration with this laser. The laser has been used without direct supervision by mature high school and college students after thorough instruction; it is recommended that younger students be directly supervised if they use the laser. Users must maintain a safety conscious attitude at all times.

Data Analysis

The basic calculations requested in the Methodology Section are performed and typical results of the requested measurements are given in this section.

The optical cavity stability condition is given by Equation (3). For the current high reflection, $r_1 = 60$ cm, and flat output coupler, $r_2 = \infty$, Equation (3) reduces to

$$0 \text{ cm} < L < 60 \text{ cm}$$

The lower limit, $L > 0$ cm, cannot be tested with the current apparatus because of the finite length of the discharge tube. However, students can test the maximum length L of the laser by sliding the output coupler along the rail. The maximum cavity length obtained is typically near 57 cm, in close agreement with the theoretical value. The laser power dependence is plotted against cavity length in Figure 6, revealing a sharp power dropoff at cavity lengths over 50 cm.

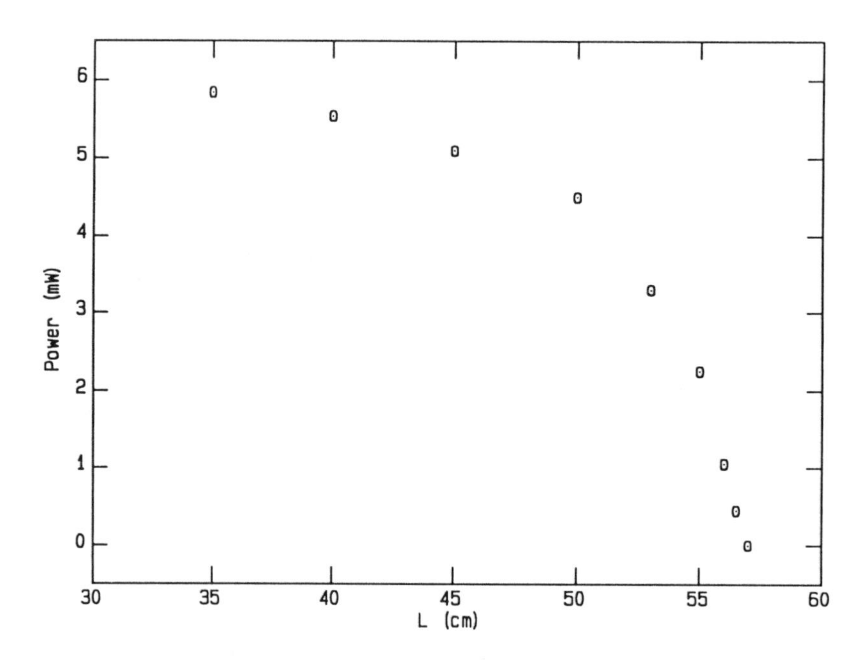

Figure 6. Dependence of laser power on optical cavity length for the "hands-on" helium-neon laser. The theoretical maximum cavity length is 60 cm.

The orientation of the Brewster window results in full transmission of vertically polarized light and partial reflection of horizontally polarized light. Discrimination against horizontally polarized light results in a vertically polarized laser beam. Qualitative analysis of the polarization of the light reflected from the Brewster window onto the breadboard with polarization film reveals that it has a much larger component of horizontally polarized light than present in the emitted laser beam. Student data of the transmission of the highly polarized laser light through a polarization analyzer is plotted with the theoretical expression of Malus's Law in Figure 7.

Several well-defined low order TEM modes can be obtained with the present laser at cavity lengths of 30 to 35 cm. When producing these modes, it is sometimes useful to slightly move the discharge tube (using the nylon screws or the play in the clearance holes mounting the holder base to the breadboard) so that the optical axis strikes the output coupler in a different position. Photographs of several TEM modes obtained with this laser are presented in Figure 8.

The longitudinal mode index q depends on the cavity length. For a cavity length of L = 30–50 cm, q is on the order of 10^6 for a helium-neon laser. Note that the precision with which q can be determined is limited by the precision of the cavity length determination and the laser wavelength determination (632.817 nm is the accepted wavelength for this helium-neon laser). A spectrum analyzer reveals that two to three longitudinal modes are supported when the laser is operated in TEM_{00} with a cavity length of 50 cm. Finger pressure on the breadboard near the laser cavity slightly changes the cavity length, causing longitudinal modes to sweep across the finite linewidth of the spectral transition giving rise to the laser light. These modes have different values of q, but the same wavelength when at the same position in the spectral profile. Figure 9a presents an oscilloscope trace of the

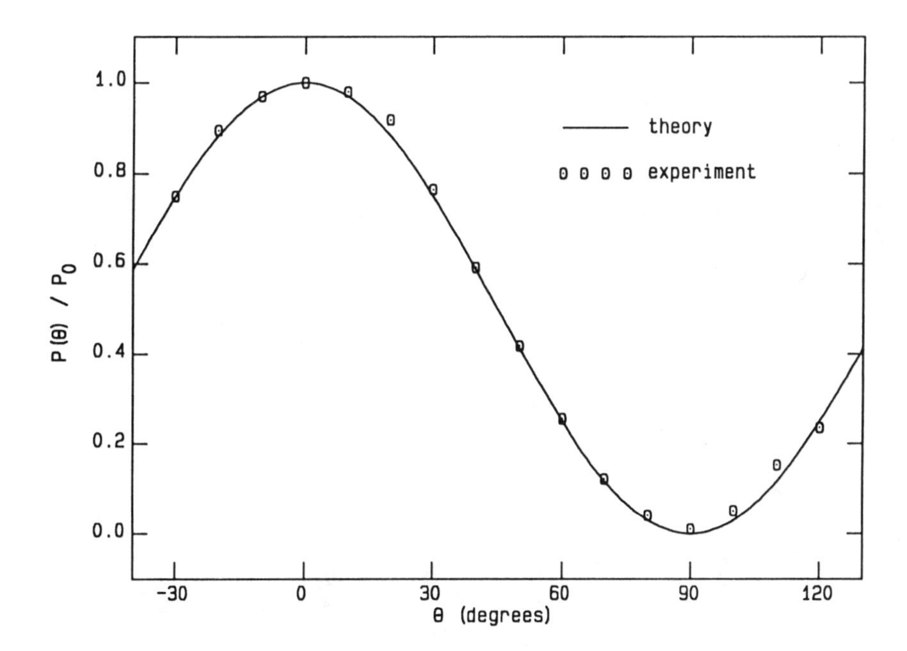

Figure 7. Dependence of laser light transmission on polarizer angle for the "hands-on" helium-neon teaching laser. Data are plotted as points. The solid line represents the Malus's Law theoretical expression for fully polarized light.

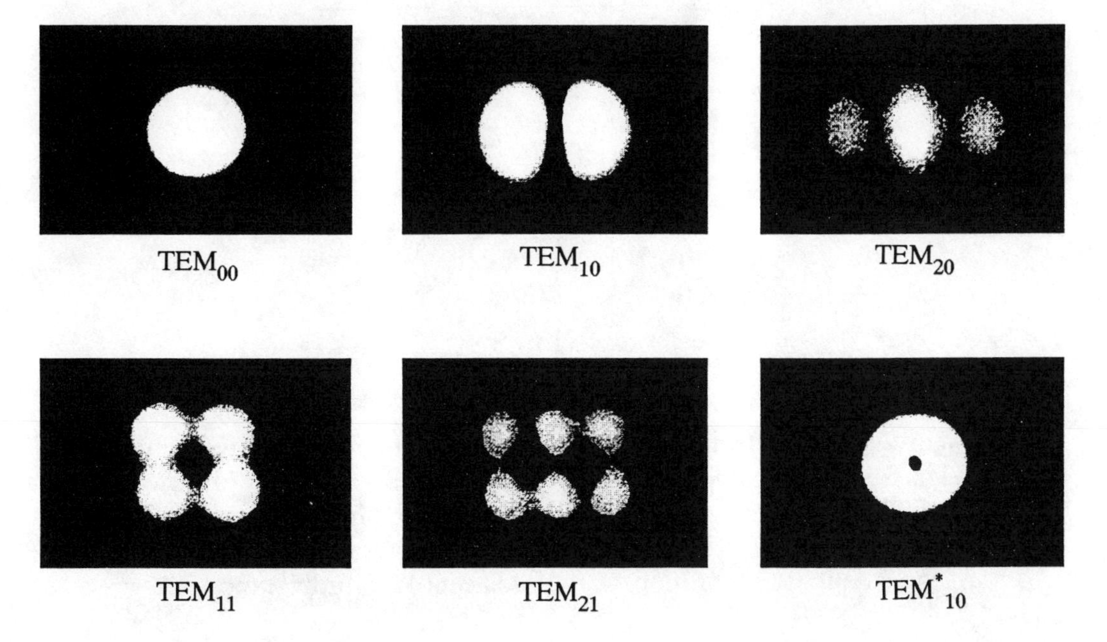

Figure 8. Transverse electromagnetic modes observed with the "hands-on" helium-neon teaching laser. Each mode is labeled TEM_{mn}, where m and n refer to the number of vertical and horizontal nodes, respectively. The TEM^*_{01} mode, or "doughnut mode," is a superposition of the TEM_{01} and TEM_{10} modes. (photography by Richard G. Blair)

output of a spectrum analyzer with an 8 GHz free spectral range which is analyzing the frequency characteristics of the current laser with a cavity length of L = 50 ± 1 cm. Figure 9b expands a set of longitudinal peaks. The cavity length calculated from this measurement is

$$L = \frac{c}{2\,\Delta\nu} = \frac{3.00\times10^{10}\ \text{cm s}^{-1}}{2\ (0.295\times10^{9}\ \text{s}^{-1})} = 50.8\ \text{cm}$$

which is consistent with the stated cavity length.

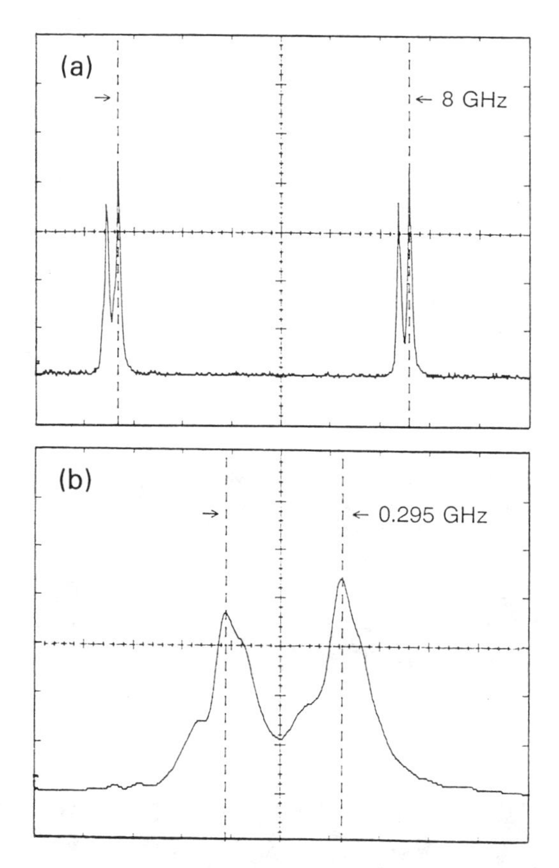

Figure 9. Characterization of the "hands-on" helium-neon teaching laser frequency composition with a spectrum analyzer. (a) Two longitudinal modes are observed with an 8 GHz free spectral range spectrum analyzer. (b) Horizontal expansion permits measurement of the frequency separation, $\Delta\nu$, of adjacent longitudinal modes.

Acknowledgments

It is with great pleasure that I acknowledge the work of Professor John R. Brandenberger of Lawrence University, who is active in introducing lasers into the undergraduate physics curriculum. The apparatus and experiments described in this article are based, in part, on experiences from a NSF Faculty Enhancement Program organized by Professor Brandenberger. These experiences, along with additional experiments suitable for undergraduate physics and physical chemistry laboratories, are detailed in two reports (*25,26*). Financial and equipment support from the Pittsburgh Conference on Analytical Chemistry and Applied Spectroscopy (1989 National College Grants Program), the National Science Foundation (USE-9050487), the E.I. Du Pont and de Nemours Company, and Coherent, Inc. is gratefully acknowledged.

Literature Cited

1. Demtröder, W. *Laser Spectroscopy*; Chemical Physics 5; Springer-Verlag: New York, NY, 1982.
2. Andrews, D.L. *Lasers in Chemistry*; Springer-Verlag: New York, NY, 1986.
3. *Laser Applications in Physical Chemistry*; Evans, D.K., Ed.;Marcel Dekker: New York, NY, 1989.
4. Kovalenko, L.J.; Leone, S.R. *J. Chem. Educ.* **1988**, *65*, 681.
5. Levine, R.D.; Bernstein, R.B. *Molecular Reaction Dynamics and Chemical Reactivity*; Oxford University Press: New York, NY, 1987.
6. Lengyel, B.A. *Introduction to Laser Physics*; Wiley: New York, NY 1966.
7. Beesley, M.J. *Lasers and Their Applications*; Taylor and Francis: London, 1972.
8. Lengyel, B.A. *Lasers*; 2nd ed.; Wiley-Interscience: New York, NY, 1972.
9. O'Shea, D.C.; Callen, W.R.; Rhodes, W.T. *Introduction to Lasers and Their Applications*; Addison-Wesley: Reading, MA, 1977.
10. Verdeyen, J.T. *Laser Electronics*; Prentice-Hall: Englewood Cliffs, NY, 1981.
11. Meyer-Arendt, J.R. *Introduction to Classical and Modern Optics*; 2nd ed.; Prentice-Hall, Englewood Cliffs, NJ, 1984; pp 492-519.
12. Yariv, A. *Optical Electronics*; Holt, Rinehart and Winston: New York, NY, 1985.
13. Eastham, D. *Atomic Physics of Lasers*; Taylor and Francis: London, 1986.
14. Pedrotti, F.L.; Pedrotti, L.S. *Introduction to Optics*; Prentice-Hall: Englewood Cliffs, NJ, 1987; pp 148-175.
15. Wilson, J. *Lasers: Principles and Applications*; Prentice Hall: New York, NY, 1987.
16. Milonni, P.W.; Eberly, J.H. *Lasers*; Wiley-Interscience: New York, NY, 1988.
17. Svelto, O. *Principles of Lasers*; 3rd ed.; Plenum Press: New York, NY, 1989.
18. Einstein, A. *Phys. Z.* **1917**, *18*, 121. Translation can be found in *The Old Quantum Theory*; Permagon: Elmsford, NY, 1967; pp 167-183.
19. Schawlow, A.L.; Townes, C.H. *Phys. Rev.* **1958**, *112*, 1940.
20. Maiman, T.H. *Nature*, **1960**, *187*, 493. Maiman's original paper describing the first laser was turned down for publication by *Physical Review Letters* because it was deemed to be of insufficient interest.
21. Javan, A.; Bennett, W.R.; Herriott, D.R. *Phys. Rev. Let.*, **1961**, *6*, 106 (1961).
22. Mallow A.; Chabot L. *Laser Safety Handbook*; Van Nostrand-Reinhold: New York, NY, 1978.
23. Sliney, D.; Wolbarsht, M. *Safety with Lasers and Other Optical Sources*; Plenum: New York, NY, 1980.
24. Winburn, D.C. *Practical Laser Safety*; 2nd ed.; Marcel Dekker: New York, NY, 1990.

25. *Laser Physics and Modern Optics in Liberal Arts Colleges*; Brandenberger, J.R., Ed.; Lawrence University: Appleton, WI, 1987.
26. Brandenberger, J.R. *Lasers and Modern Optics in Undergraduate Physics*; Lawrence University: Appleton, WI, 1989.

Hardware List

Laser Apparatus Construction

The "hands-on" helium-neon teaching laser described herein is constructed as much as possible from commercially available components. However, several components are not commercially available and must be custom fabricated in a machine shop.

A schematic of the basic apparatus is given in Figure 5. The commercial components with vendors and catalog numbers are listed in Table 1. All components are mounted to an optical breadboard predrilled with 1/4-20 tapped holes. It is important that the breadboard be as flat as possible; hence, a commercial product

Table 1. List of components for "hands-on" helium-neon teaching laser.

Item	Vendor and Catalog Number
1'×4' Optical Breadboard	TMC 74-109-02 or Newport XSN-14
Brewster Window Laser Tube	Melles Griot 05-LHB-570
Power Supply	Melles Griot 05-LPL-379-065
Lead Assembly	Melles Griot LO-7086-1
Output Coupler	Spectra Physics G3802-002
Gimbal Mount (for Output Coupler)	Newport GM-1 (see text for vertical travel stop modification)
Riser (for Gimbal Mount)	Oriel 11950; Newport BP-1
Table Rail and Carrier	Oriel 11422; Oriel 11641
Laser Tube Holder	custom (Figure 10)
End Plug	custom (Figure 11)
Riser Plate	custom (Figure 12)
Base Plate	custom (Figure 13)
Output Coupler Holder	custom (Figure 14)
Rocker Arm (for Gimbal Mount)	custom (Figure 15)
Tie Down Bar (for Power Supply)	custom (Figure 16)

is recommended. Although 1" hole spacing is recommended for maximum flexibility when positioning components, 2" spacing is more economical and suffices. The minimum required breadboard size is 1' × 3'; however, 1' × 4' is recommended to allow room for mounting of laser diagnostic equipment.

The laser tube is a widebore (0.078" inner diameter) helium-neon discharge tube with a 200 mm long gain region. The tube, terminated at the anode (high voltage) end with a 60 cm radius concave mirror and at the cathode (ground) end with a Brewster window, is specified as being capable of producing over 4 mW of multimode 632.8 nm radiation. The laser tube is mounted with its optical axis 4.5" above the breadboard in a custom holder with nylon screws. CAUTION: The laser tube should be oriented in the holder so that the Brewster window faces down. This insures that the stray reflection from the Brewster window is directed down toward the breadboard and not up toward the ceiling. This orientation has the added benefit of keeping the window dustfree. The tube holder is partially cut away so that the discharge tube is open to view. The anode end of the discharge tube is enclosed with a nylon plug so that fingers cannot find their way to the high voltage terminal. Two 0.25" wide slits are provided for viewing the high reflector from the top and for passage of the high voltage power lead from the bottom. The discharge tube is powered with an industry standard power supply, which connects to the tube with a lead assembly with spring clips at the laser end and an Alden connector at the power supply end. The red lead with the insulated ballast resistor is the high voltage lead and is clipped to the discharge tube anode; the black lead is the ground lead and clips to the discharge tube cathode. It is recommended that the lead assembly be firmly clamped to the breadboard near the discharge tube to prevent any cable strain from being transmitted to the tube. For convenience, the power supply is mounted to the optical breadboard with a custom tiebar.

The 15 mm diameter flat output coupler (radius = ∞; transmission = 0.9 ± 0.2%) is set into a 1" diameter custom holder which is in turn mounted in a gimbal mount. The gimbal mount is raised with a spacer so that the center of the output coupler is also 4.5" above the optical breadboard and is mounted onto a carrier which slides along a precision 12" table rail. The table rail is mounted to the breadboard so that the output coupler can be brought to within 1–2" of the Brewster window. A custom rocker arm is added to the top of the gimbal mount so that it can be swept though all vertical angles with slight finger pressure. It is convenient to label the adjustment knobs of the gimbal mount with H and V for horizontal and vertical, respectively. In order to prevent overextension and damage to the retaining spring during this sweeping process, it is highly recommended that the gimbal mount be slightly customized to add a stop to the vertical travel. This is simply done by drilling a 3/16" clearance hole through the lower left corner of the moving side of the mount (widening the existing tapped hole of the GM-1 mount), and drilling and tapping a 6-32 hole opposite this clearance hole on the fixed side of the mount. A 6-32 socket head screw can then be inserted though the clearance hole and screwed in to the point at which vertical travel is to be limited. Several wraps of teflon tape about the screw threads prevents the screw from working loose during use.

Schematic plans for the custom laser tube holder, end plug, riser plate, base plate, output coupler holder, rocker arm, and tie down bar are given in Figures 10 through 16, respectively. In order to assist with the widespread implementation of this experiment, the author is coordinating the manufacture and distribution of complete sets of the custom fabricated parts; please contact the author directly for pricing and availability.

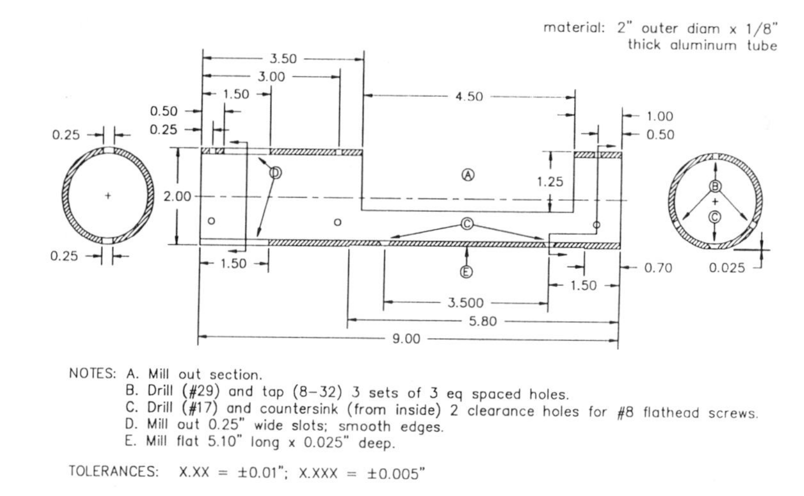

Figure 10. Schematic of custom laser tube holder.

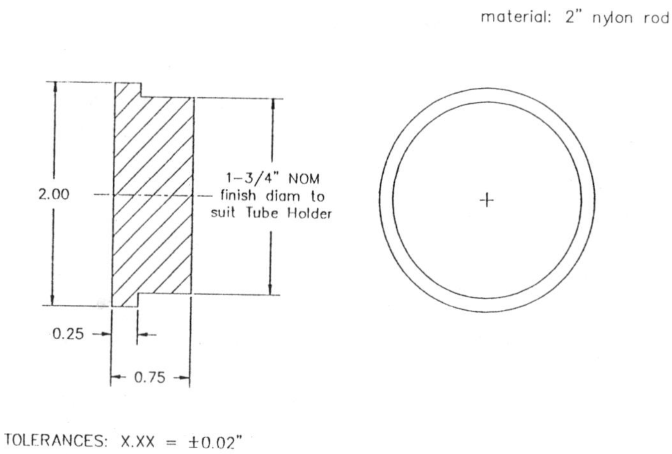

Figure 11. Schematic of custom end plug for laser tube holder.

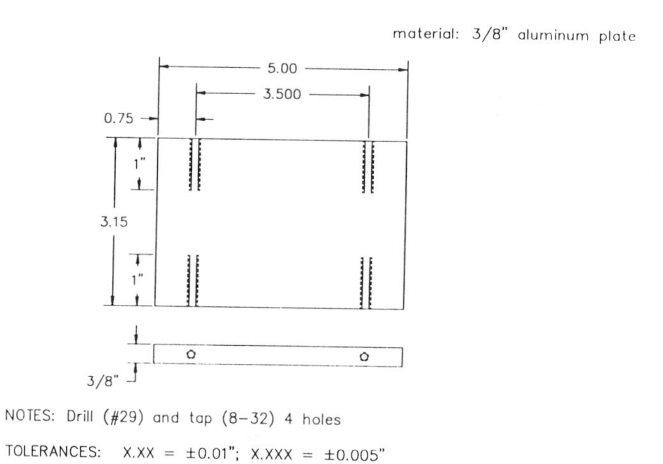

Figure 12. Schematic of custom riser plate for laser tube holder.

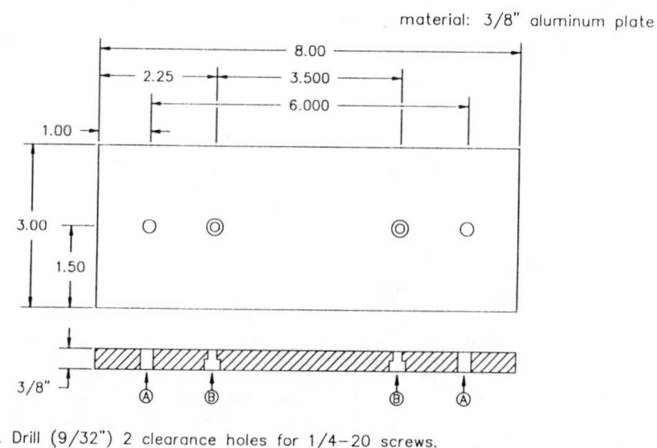

NOTES: A. Drill (9/32") 2 clearance holes for 1/4–20 screws.
B. Drill (#17) and counterbore 2 clearance holes for #8 socket head screws.

TOLERANCES: X.XX = ±0.01"; X.XXX = ±0.005"

Figure 13. Schematic of custom base plate for laser tube holder.

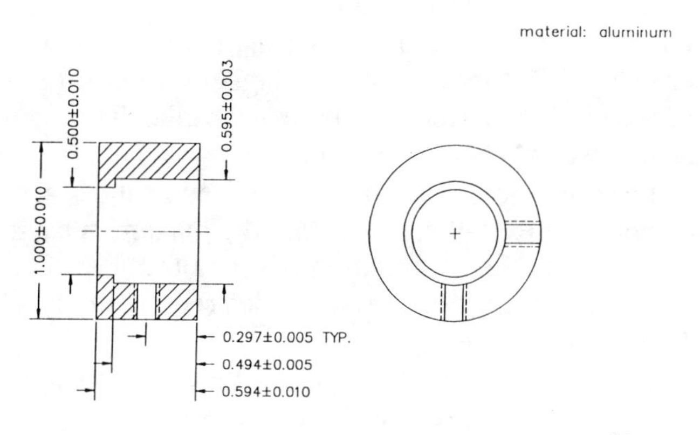

NOTES: Drill (#36) and tap (6–32) 2 holes for setscrews

Figure 14. Schematic of custom output coupler holder.

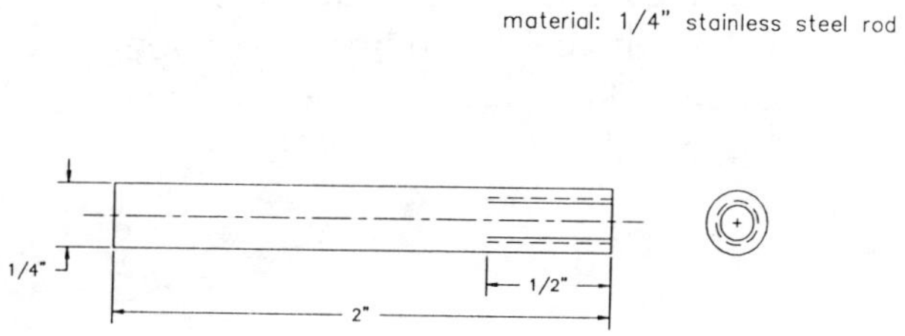

material: 1/4" stainless steel rod

1/4"

1/2"

2"

NOTES: Drill (#29) and tap (8–32) hole.
Insert threaded rod into hole to expose 0.20"–0.25" of thread.

Figure 15. Schematic of custom rocker arm for gimbal mount.

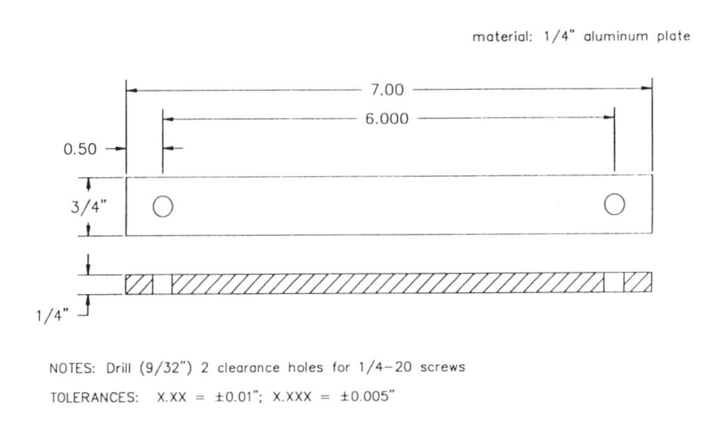

Figure 16. Schematic of custom tie down bar for power supply.

Laser Apparatus Accessories The laser apparatus illustrates the operating principles of laser operation and is a worthwhile component of an instructional laboratory in itself. Added value is gained by acquiring laser diagnostic equipment to characterize the emitted laser beam. The measurements suggested here do not all need to be performed in order for the experience to be worthwhile. They can be phased in as the required diagnostic equipment is obtained.

Some of the laser characterization exercises require a bare minimum of additional equipment. For example, measuring the maximum cavity length requires only a meterstick; observing the polarization of the output beam requires only some polarization film; and observing transverse modes requires only a diverging lens, a lens holder, and a magnetic base.

It is recommended that a power meter be the first piece of laser diagnostic equipment which is obtained. A rotatable Brewster angle window, a polarizer mounted in a rotation stage, and a pinhole mounted in a *x–y* translation mount are the next recommended acquisitions. Investing in good optical mounts is recommended for convenience, protection, and safety reasons. The most costly piece of diagnostic equipment is an optical spectrum analyzer with an associated oscilloscope.

Additional laser diagnostic equipment used in the suggested exercises is listed in Table 2. For maximum flexibility of equipment use, a standardized mounting post diameter of 0.5" is chosen. Vendors which supply the equipment in Tables 1 and 2 are listed in Table 3.

Table 2. List of diagnostic equipment for "hands-on" helium-neon teaching laser exercises.

Item	Vendor and Catalog Number
Meterstick	
Polarization Film	Edmund 38,493
Magnetic Base, Post, and Crosspost Holder	Newport MMB, SP-6, CA-1
-1" Focal Length Diverging Lens and Holder	Newport KBC046, SP-4; Oriel 71594
Power Meter	Coherent 212 or Newport 815-SL
Rotatable Brewster Angle Window	Newport RSX-1, FC-1, SP-4; ESCO R320110
Polarization Analyzer	Newport RSA-1, SP-4; Optics for Research PL-8 or PE-8-VIS
1000 micron Pinhole and XY-Translator	Melles Griot 04-PPM-025, 07-HPH-001, 07-RMN-004
8 GHz FSR Spectrum Analyzer	Burleigh SA-800-C or Coherent 240-1-B
Oscilloscope (for Spectrum Analyzer)	

Table 3. List of vendors.

Burleigh Instruments P.O. Box E Burleigh Park Fishers, NY 14453 (716) 924-9355	Newport Corporation P.O. Box 8020 18235 Mt. Baldy Circle Fountain Valley, CA 92728 (714) 963-9811
Coherent Components Group 2301 Lindbergh Street Auburn, CA 95603 (916) 823-9550	Optics for Research P.O. Box 82 Caldwell, NJ 07006 (201) 228-4480
Edmund Scientific Co. 101 E. Gloucester Pike Barrington, NJ 08007 (609) 573-6250	Oriel Corporation 250 Long Beach Boulevard P.O. Box 872 Stratford, CT 06497 (203) 377-8282
ESCO Products 171 Oak Ridge Road Oak Ridge, NJ 07438 (201) 697-3700	Spectra-Physics Optics Division 1250 W. Middlefield Road Mountain View, CA 94042 (415) 961-2550
Melles Griot 1770 Kettering Street Irvine, CA 92714 (714) 261-5600 (catalog div.) (619) 438-2131 (laser div.) (800) 835-2626	Technical Manufacturing Corporation 15 Centennial Drive Peabody, MA 01960 (508) 532-6330

RECEIVED October 1, 1992

CHAPTER 7

Basic Laser Spectroscopy for the Physical Chemistry Laboratory

Jack K. Steehler

POTENTIAL HAZARDS: High-Pressure Systems, Lasers, High Voltage

Optical spectroscopy is a key experimental tool of modern physical chemistry. With a broad definition of the term "optical", interactions of radiation with matter ranging from low energy nuclear spin transitions (NMR spectroscopy) to high energy transitions of inner shell electrons (X-ray spectroscopy) are included. More traditionally, optical spectroscopy refers primarily to the infrared, visible, and ultraviolet regions. For these spectral regions, the recent widespread availability of lasers for routine use has revolutionized experimental studies *(1)*. At least one third of experimental physical chemistry research utilizes laser light sources. Among the distinctive laser features which have driven this explosion is the high temporal resolution available, with fast pulsed lasers allowing study of molecular structure and reactions on the nanosecond, picosecond, and femtosecond time scales. Additionally, high peak powers of laser sources have opened the door for the application of unusual nonlinear optical effects, such as second harmonic generation (SHG) and coherent anti-Stokes Raman spectroscopy (CARS). The combination of high time resolution and unique spectroscopic tools has provided a wealth of new understanding of fundamental chemical processes.

However exciting the newest, most powerful laser techniques may be, a firm understanding of fundamental photophysical processes is still required. A full understanding of important processes involved in absorption of light and subsequent relaxation is crucial to proper use of advanced techniques. The required level of understanding is definitely nontrivial, and appropriate lecture and laboratory introductions to these topics are essential. While the ability to utilize the most advanced laser techniques is indeed an appropriate goal for graduate training, the basic topics must remain the central focus of the undergraduate experience.

It is also important that the exposure to laser spectroscopy and optical absorption and relaxation processes be as widespread as possible across a variety of educational settings, from research universities to small liberal arts colleges. To that end, a set of basic experiments using affordable instrumentation is needed. The experiments described in this chapter are examples of such experiments, illustrating essential optical processes with affordable instrumentation. The suggested light source is a commercial low power pulsed nitrogen laser.

Theory

The goal of the experiments described here is the development of a comprehensive understanding of the various processes involved in absorption and subsequent relaxation, including various interrelationships *(2)*. Those processes are identified in Figure 1. The energy levels shown include electronic levels (S_0, S_1, S_2, T_1) and vibrational levels within each electronic level. The electronic levels S_n are

2503–6/93/0109$06.00/0

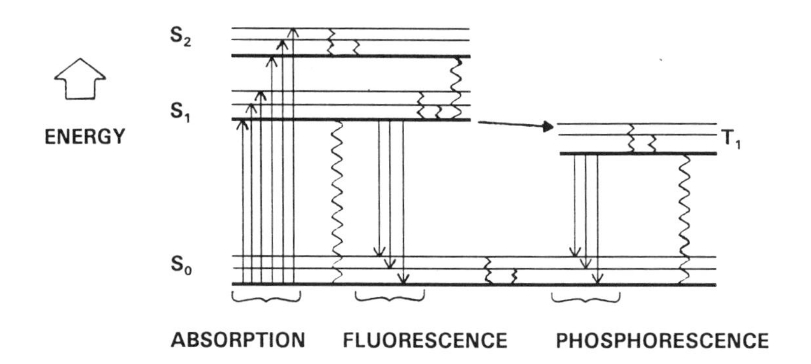

Figure 1. A molecular energy level diagram showing various excitation and relaxation processes. See text for description. The squiggly lines represent nonradiative relaxation.

singlet levels, which have all electron spins paired, while the T_1 level is a triplet level, where unpaired electron spins exist. The key photophysical processes include absorption, vibrational relaxation, intersystem crossing, fluorescence, phosphorescence, energy transfer, and nonradiative relaxation. These terms are defined in the following paragraphs. Each of these many processes involves a characteristic rate, and competition among parallel pathways is an essential consideration.

At room temperature, the majority of molecules will exist in the lowest vibrational level of the S_0 electronic state. Absorption of energy (upward arrow on the diagram) from the incident beam of light causes an excitation into a higher lying singlet electronic state, such as S_1 or S_2. This transition occurs instantaneously. UV-visible absorption spectra represent the wavelength dependence of this excitation process, with only resonant transitions resulting in significant excited state populations. Or to say it another way, only if the length of the arrow matches the gap between energy levels do you get absorption of light. Symmetry rules from quantum mechanics restrict such one photon transitions to those for which a parity change exists (gerade → ungerade). Two photon absorption spectra (where the gap is bridged by the energy of two photons being simultaneously absorbed), may also be utilized, with a requirement for identical symmetries in excited and ground states, a gerade → gerade selection rule *(3)*.

Relaxation, or release of the energy gained by absorption (shown on the diagram as squiggly lines or downward pointing arrows), includes many different possible pathways. The energy associated with the excited molecule may be lost as heat, as light, or by transfer to another molecule. A primary relaxation process is nonradiative relaxation (squiggly lines on diagram), with heat being the final form of the energy. Essentially, the excess energy of the excited molecule becomes thermal motion energy of the surrounding solvent or matrix. Such nonradiative relaxation processes include vibrational relaxation (loss of the energy of molecular vibrations) and heat producing transitions between electronic states (e.g. $S_2 \rightarrow S_1$). The term nonradiative means these processes do not give off light. Energy levels with higher energy than the lowest level of the S_1 manifold typically relax with picosecond lifetimes. Due to the larger energy gap between the lowest level of S_1

and the ground state S_0, the lowest level of S_1 has a longer lifetime, on the order of a few nanoseconds for most organic molecules. In addition to nonradiative relaxation to S_0 (squiggly line on diagram), the S_1 state may emit light (fluorescence, a downward arrow on the diagram), transfer excitation in a resonant fashion to a nearby unexcited molecule (energy transfer, or fluorescence quenching, not shown on the diagram), or may make a transition to a triplet state T_1 of similar overall energy. Energy transfer involves relaxation of an excited electronic state of one molecule, with concomitant excitation of another nearby molecule (energy is transferred from one molecule to another). The conversion to a triplet state, called intersystem crossing, is shown on the diagram as a horizontal arrow pointed to the right and is a quantum mechanically forbidden process of relatively low probability. Intersystem crossing is enhanced by the presence of spin orbit coupling, which mixes the wavefunctions of singlet and triplet states *(4-6)*. Spin orbit interactions (interactions between the spin of the electron and its orbital motion) are enhanced by the presence or proximity of heavy atoms, such as iodine or thallium. Once populated, the T_1 level can relax by nonradiative relaxation (squiggly line on the diagram), energy transfer (not shown on the diagram) or emission of light (phosphorescence, shown as downward arrows on the diagram).

Excited state lifetime The ability to monitor rapidly changing events is a characteristic of laser spectroscopy. Laser pulsewidths of nanoseconds, picoseconds, and femtoseconds are readily available. That pulsewidth, and the response times of detection electronics represent experimental boundaries for the processes which can be studied kinetically. Molecular timescales must also be considered. For events to be monitored by emission spectroscopy, the lowest vibrational state of S_1 is the key energy level. This level may be populated by direct absorption, or by absorption to any higher level followed by picosecond vibrational and electronic relaxation ending at the lowest level of S_1. If pulsed excitation is used, the fluorescence observed from S_1 following the excitation pulse is determined by the finite lifetime of that state. That lifetime is set by the combination of all relaxation processes, since each of these processes depopulates the excited state. Defining a total relaxation rate constant k_{TOT}, we find several contributions to this rate, including k_{FL} (fluorescence), k_{NR} (nonradiative), k_{ISC} (intersystem crossing), and k_{ET} (energy transfer).

$$k_{TOT} = k_{FL} + k_{NR} + k_{ISC} + k_{ET}$$

The observed lifetime, τ_{FL} is $1/k_{TOT}$ and defines the observed emission following the excitation pulse as the first order kinetic decay $I(t) = I_0 \exp(-t/\tau_{FL})$ where I_0 is the initial emission intensity, and $I(t)$ is the intensity a period of time after the excitation pulse. This equation is the integrated form of the first order rate law

$$-(\mathrm{d}N/\mathrm{d}t) = kN$$

where N is the number population of the excited state level.

Similarly, for phosphorescence, a total relaxation rate constant k'_{TOT} is defined, with several contributions, including k_{PH} (phosphorescence), k'_{NR}

(nonradiative), and k'_{ET} (energy transfer). The prime symbols are used to distinguish T_1 processes from similar S_1 processes.

$$k'_{TOT} = k_{PH} + k'_{NR} + k'_{ET}$$

The observed lifetime, τ_{PH} is $1/k'_{TOT}$ and, as above, defines the observed emission following the excitation pulse as $I(t) = I_0 \exp(-t/\tau_{PH})$.

If multiple sample species are excited simultaneously, or if the emitting level is populated by processes involving longlived intermediates, more complicated temporal profiles can be encountered. The simplest example would be an observed temporal profile described as double exponential, involving a sum of two exponential terms *(7)*.

Energy transfer A focus of the main experiment in this chapter is energy transfer, as a means of gaining experience with the kinetic competition occurring among different relaxation pathways. In energy transfer, the excitation energy of one molecule is transferred to a second molecule nearby, making the first molecule return to its unexcited state, with the second molecule becoming excited. Typically, the first molecule is luminescent, while the second molecule is a nonluminescent molecule which relaxes nonradiatively, resulting in quenching of the luminescence. A variety of mathematical models for energy transfer have been used for analyzing experimental data. We will consider three of those models.

The first model, the Stern-Volmer model *(8)*, originally assumed energy transfer occurs at a constant rate, independent of radial separation between donor and acceptor. However, it is also used to describe dynamic collisional energy transfer, with $L_0/L = 1 + k[Q]$. L_0/L is the ratio of unquenched to quenched luminescence intensities, and $[Q]$ is the acceptor (quencher) concentration. Static quenching can also be described by the same formula, with k representing an equilibrium constant for the formation of a nonfluorescent complex between the quencher and the luminescent species.

The second model, the Perrin model *(9)*, assumes a critical distance R_0 exists for energy transfer, with rapid energy transfer (complete quenching) for donor acceptor separations less than the critical distance, and no energy transfer for greater separations. Here $L = L_0 \exp(-v[Q])$, with v being an "effective molar quenching volume" defined as $(4/3)\pi R_0^3$.

The third model, the modified Perrin model *(9)*, assumes a more gradual radial distance dependence (R_0 is reinterpreted as a distance yielding 50% quenching), and the possibility of incomplete quenching for even rather small distances. The resulting equation is $L = L_0 [(1 - \beta) \exp(-v[Q]) + \beta]$, where β is the residual luminescence present even at high $[Q]$.

Room Temperature Phosphorescence (RTP) Phosphorescence is utilized less frequently than fluorescence, in both practical analytical spectroscopy and for physical chemistry research *(10)*. Due to the forbidden nature of the triplet to singlet transition, the lifetime of the T_1 state is relatively long, typically milliseconds to seconds. During this time, nonradiative relaxation competes effectively with emission processes. Unless major efforts are made to minimize nonradiative relaxation, no phosphorescence is seen. Two approaches are used to enhance phosphorescence. The first is the use of low temperatures, with samples typically frozen in liquid nitrogen. The resulting rigid matrix restricts the molecular motions which transfer excitation energy as heat to the surrounding solvent. Thus, more phosphorescence is seen. However, the complexity and inconvenience of cryogenic spectroscopy have prevented the widespread use of this method.

The second approach to minimizing nonradiative relaxation is called room temperature phosphorescence *(11-14)*. The phosphorescent molecules are adsorbed onto a filter paper surface. The molecules become trapped within the solid matrix, and their motions are restricted. In turn, the nonradiative relaxation is significantly reduced, and phosphorescence is observed, even at room temperature. Additives which further restrict motion by packing into the filter paper voids further enhance the light emission.

Phosphorescence is also enhanced by agents which enhance the rate of intersystem crossing (ISC) *(4,15)*. As ISC is enhanced, the relative numbers of excited molecules in S_1 which cross over to T_1 is increased. In RTP, the usual procedure includes the addition of a heavy atom species (e.g. Pb, Tl, or I) along with the phosphorescent organic molecule. It must also be understood that the enhancement of the rate of ISC, which increases the population of T_1, also affects the lifetime of the T_1 state. Intersystem crossing is also part of the nonradiative relaxation of the T_1 state (the T_1 to S_0 transition), and heavy atom additives will also enhance this rate, decreasing the T_1 lifetime. This effect on the T_1 lifetime is the primary focus of the main experiment in this chapter. However, an understanding of all relaxation pathways and rates is required.

Methodology

Experiment 1 - The Speed of Light The time scale of laser experimentation is typically nanoseconds or shorter. Yet undergraduate students do not automatically understand this time scale. In order to provide a frame of reference for the timescale of molecular photochemistry, an introductory experiment which measures the speed of light can be performed.

In this experiment, a pulsed nanosecond laser is sent the length of a laboratory room, bounced off a mirror, and reflected back to a detector near the laser. The time delay is measured, as is the physical distance travelled (typically 50 feet), providing the information needed to calculate the speed of light.

A low power pulsed nitrogen laser (λ = 337 nm, pulsewidth = 3 ns) is used, as shown in Figure 2. The layout is set up by the instructor, in advance, along a back wall of the laboratory, or totally contained on top of a long laboratory bench. This arrangement, along with appropriate laser beam blocking shields, minimizes the student exposure to the laser beam. The detector used is a PIN photodiode, connected to an oscilloscope with an appropriate terminating resistor. Measurements of fast electrical signals can require careful selection of a terminating resistor to prevent "ringing" of the signals, and to ensure a sufficiently short electronic RC time constant *(16)*. A bandwidth of at least 100 MHz is preferred for the oscilloscope, which may be analog or digital. Figure 3 shows typical results, with a calculated speed of light within 2% of the expected value, 2.998×10^8 m/s in vacuum, 2.987×10^8 m/s in air.

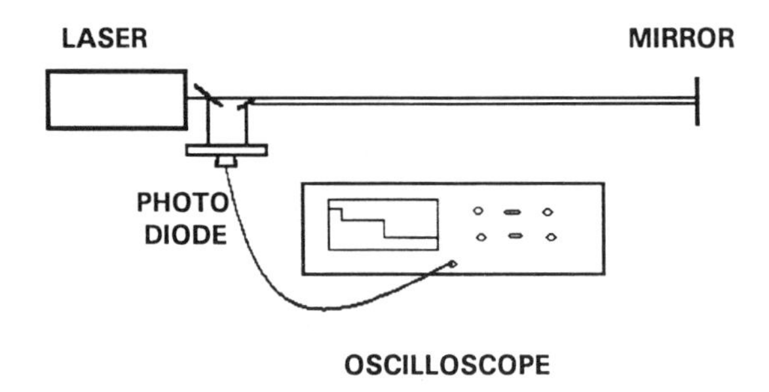

Figure 2. The experimental setup for measurement of the speed of light.

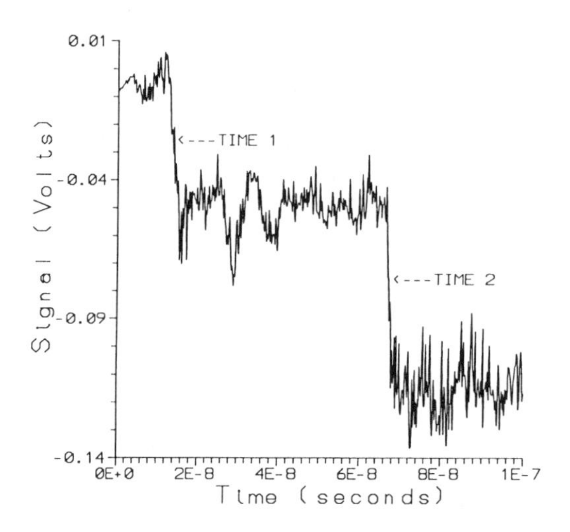

Figure 3. Experimental results for the determination of the speed of light. At time 1, the initial laser pulse hits the photodiode. At time 2, the laser pulse which has travelled to the end of the room and back hits the photodiode.

From this experiment, an appreciation of the time scale of molecular events is gained. A useful rule of thumb is that one foot is the distance light travels in one nanosecond, and 0.3 millimeter is the distance light travels in one picosecond.

Experiment 2 - Room Temperature Phosphorescence Lifetimes The goal of this experiment is the determination of the most appropriate models to describe the effect of a particular heavy atom (I^-) on the phosphorescence lifetime of a particular compound (2-naphthoic acid). As noted in the literature *(17)*, different heavy atoms and different luminescent molecules may yield different results, so the student should not overgeneralize from the results of this experiment. Direct measurements of phosphorescence lifetimes are required, rather than the relative intensity measurements described in the theory section. Since the heavy atom affects both the step which populates T_1 and the step which depopulates T_1, an intensity measurement represents a combination of both processes. The population step increases intensity, and the depopulation step decreases intensity. Thus a measurement of the depopulation step alone, through the lifetime measurement, is required.

Sample preparation involves dissolving appropriate chemicals, deposition of appropriate volumes onto the filter paper surface (needed for RTP), and several drying steps. The sample chosen is 2-naphthoic acid. A solution of 1.0×10^{-1} M is prepared in 1.0 M NaOH. The NaOH enhances solubility, and also serves as a "packing agent" on the filter paper surface, enhancing phosphorescence significantly *(17-19)*. Solutions of KI are also prepared, as a source of the heavy atom, iodine, in concentrations of 0.2, 0.4, 0.6, 0.8, and 1.0 M. Filter paper (Whatman #1 CHR) squares, 2 cm × 2 cm, are precut, and washed in methanol to remove possible luminescent contaminants. A 5 μL portion of the 2-naphthoic acid solution is spotted onto the center of each of six filter paper squares. A 5 μL portion of each of the five KI solutions is then spotted onto the same position on one of the paper squares, on top of the 2-naphthoic acid. The remaining paper square is spotted with a 5 μL water blank. The six samples thus have the same amounts of 2-naphthoic acid and varied amounts of KI codeposited on the filter paper surface. The samples are air dried for 20 minutes, and then dried for 30 seconds in an air blower set on ~200 °C. Samples are kept in a dessicator until use. If time and interests allow, the experiment may be repeated with a lighter halogen, such as KBr or KCl. Alternatively, replicate samples of the I^- experiment can be used to provide additional data points for the data analysis.

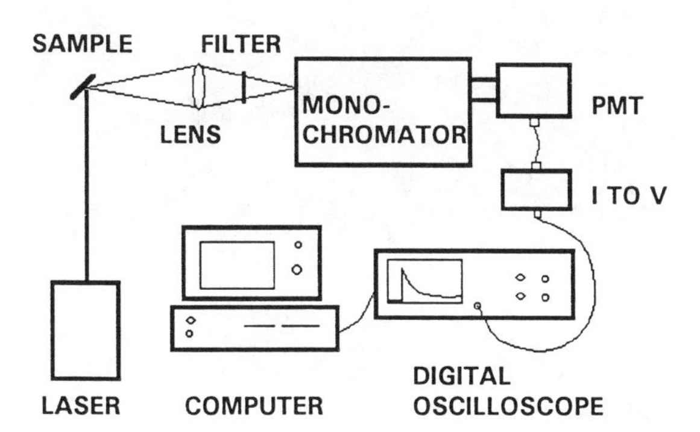

Figure 4. The experimental setup for the room temperature phosphorescence lifetime measurements.

The experimental setup is shown in Figure 4. The nitrogen laser delivers 3 ns pulses with 40 kW peak powers, at 337 nm. Note: the peak absorption of 2-naphthoic acid is at 290 nm *(20-22)*, but sufficient absorption occurs at 337 nm. The samples are held in a flowing stream of dry nitrogen during the experiment, to prevent quenching by water or atmospheric O_2 *(23-24)*. A lens collects the phosphorescence and directs it towards a monochromator, preceded by a UV blocking glass filter. The monochromator is set at 518 nm. Photomultiplier detection, current to voltage conversion, and data collection and averaging complete the setup. As mentioned in experiment 1, for fast time scale experiments some care is needed to ensure rapid photomultiplier and electronic response times *(16,25)*. However, this experiment occurs on a long enough time scale that such effects are less important. The data collection and averaging may be done on a digital storage oscilloscope, or with a boxcar averager. The digital oscilloscope will yield faster data acquisition, since all time windows are observed simultaneously. Computer interfaced instruments are recommended, to allow quantitative fitting of phosphorescence lifetime data, rather than visual approximations from analog data presentations. Tracings from displays of analog oscilloscopes might also be used, though the small signal levels expected here do not make this the preferred choice. A typical data collection averaging 1000 laser pulses is shown in Figure 5. Similar decay curves are collected for each of the six samples, beginning with the sample treated with the highest amount of I. Note: The phosphorescence signal for 2-naphthoic acid in the absence of any I^- is very weak, and may not be measurable. Literature values of phosphorescence lifetimes for 2-naphthoic acid range from 730-909 ms *(21, 26-27)*, but significantly shorter lifetimes are expected here, in the presence of the NaOH and KI.

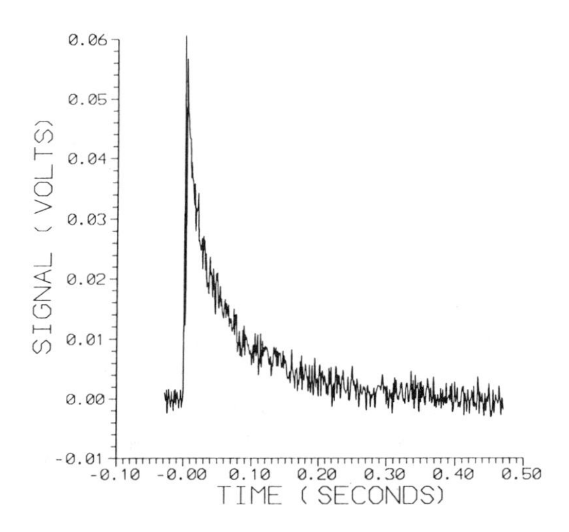

Figure 5. Experimental data for the determination of a phosphorescence lifetime. The sample is 2-naphthoic acid with 0.6 M KI codeposited.

Safety

All laser experiments require appropriate eye protection. For the nitrogen laser suggested here, the laser wavelength is 337 nm. Laser safety goggles which block this wavelength are required. Note that ultraviolet blocking eyewear not designed for work with high peak power lasers is not sufficient and is not recommended. The optical setup recommended in this chapter is an open configuration, with components mounted on an open breadboard layout. Users should ensure that stray reflections and the main excitation beam itself are appropriately blocked by flat black blocking devices which prevent such beams from leaving the experimental table. Users should also ensure that no optics (including the sample itself) are moved while the laser is on. Such movement can generate unexpected and uncontrolled reflections and must be avoided. Additional laser safety precautions can be found in the literature *(28)*.

Chemically, normal precautions to avoid skin exposure to solutions and solvents should be used. Of the chemicals used in these experiments, methanol is toxic, and sodium hydroxide is caustic and can cause chemical burns. Safety goggles are required at all times during these experiments, including all sample preparation and cleanup steps.

Data Analysis

Experiment 1 - The Speed of Light The speed of light is $\Delta x/\Delta t$, where Δx is the pathlength difference between the two pulses of light hitting the detector, and Δt is the measured difference in time of arrival for the two pulses. An error analysis to establish the measurement uncertainty is also appropriate.

Experiment 2 - Room Temperature Phosphorescence Lifetimes The first step in data analysis is the determination of the phosphorescence lifetime. The digitized data may be fit to exponential decay equations in a variety of ways *(7, 27)*. The simplest is to plot ln (I) vs. t, and perform a linear fit to the straight line portion of the data. The slope is $-1/\tau$. If the initial portion of the data is nonlinear, indicating multiple exponential behavior, only the shortest lifetime should be determined and used. A variety of additional methods for fitting the decay data may also be used, but the simple method described here will be adequate for this experiment. A set of typical results would include lifetimes ranging from 86 ms for $[I^-] = 0.2$ M to 32 ms for $[I^-] = 1.0$ M.

The second step in data analysis is consideration of the three models for energy transfer. We are seeking the best model, and the constants derived from it. In all cases, the concentration of I^- in the deposition solution (0, 0.2, 0.4, 0.6, 0.8, 1.0 M) will be used for the data analysis, while recognizing that a more complex system exists on the filter paper upon drying.

For the Stern-Volmer model, a plot of $1/\tau$ vs. $[I^-]$ should be linear, following the equation

$$1/\tau = 1/\tau_0 + k_2[I^-],$$

where k_2 is the bimolecular rate constant for the interaction of triplet state 2-naphthoic acid and I^-. The slope is used to determine k_2.

For the Perrin model, a plot of ln (τ_0/τ) versus $[I^-]$ should be linear, with τ_0 being the lifetime in the absence of I^-. The equation is $\ln(\tau_0/\tau) = v[I^-]$. The slope yields a value for v. Note: If τ_0 cannot be measured experimentally due to low signal intensities, a value extrapolated to $[I^-] = 0$ on the Stern Volmer plot may be used.

Finally, for the modified Perrin model, a plot of $[\ln(\tau/\tau_0)-\beta]$ versus $[I^-]$ will be linear, following the equation $[\ln(\tau/\tau_0)-\beta] = -v[I^-] + \ln(1-\beta)$. Here the slope is $-v$. Note that this plot requires knowledge of β, in addition to experimentally determined quantities. For this two parameter model (v and β), several approaches can be taken. First, reasonable estimates of β can be used, and varied until the most linear plot results. An appropriate initial estimate of β would be $\tau([I^-]=1$

$M)/\tau_0$. Secondly, simultaneous optimization of the two variables can be used. Methodologies such as a grid search optimization or simplex optimization may be used *(29)*. The most readily accessible method should be used.

Depending on the chemical system used, any or all of the models may adequately describe the energy transfer process. For example, a literature study of fluorescence lifetimes of 2-naphthalenesulfonate showed all three models were adequate when light halogens were used, but only the modified Perrin model fit the experimental data for I^- *(9)*.

In addition to making the necessary plots, students should discuss the agreement of their data with the proposed models, and should infer numerical values for the constants inherent in each model. Discussion of the conceptual meaning of those numerical results should also be included in the lab report. Finally, since the observed phosphorescence intensities run counter to intuition (more I^- yields brighter phosphorescence despite the enhanced relaxation rate), a summary of the multiple roles played by I^- in the experiment should be included.

Acknowledgments

The author gratefully acknowledges the invitation to participate in the production of this book, and the contributions of the many undergraduates who have explored time resolved laser spectroscopy with the author over the last seven years. The research projects of Sharon Lau and David Michie were particularly important in developing the experiments included here.

Literature Cited

1. Andrews, D. L. *Lasers in Chemistry, 2nd Ed.*; Springer-Verlag: Berlin, 1990.
2. Steinfeld, J.I. *Molecules and Radiation, 2nd Ed.*; MIT Press: Cambridge, MA, 1985; p. 287.
3. Reference 2, pp. 415-426.
4. Vo-Dinh, T. *Room Temperature Phosphorimetry for Chemical Analysis*; John Wiley: New York, 1984; p. 50.
5. Weissbluth, M. *Atoms and Molecules*; Academic Press: New York, NY, 1978; p. 648.
6. Reference 2, p. 50.
7. Demas, J. N. *Excited State Lifetime Measurements*; Academic Press: New York, NY, 1983.
8. Demas, J. N.; Demas, S. N. *Interfacing and Scientific Computing on Personal Computers*; Allyn and Bacon: Boston, MA, 1990; p. 399.
9. White, W.; Seybold, P. G. *J. Phys. Chem.* **1977**, *81*, pp. 2035-2040.
10. Skoog, D. A. *Principles of Instrumental Analysis, 3rd Ed.*; Saunders College Publishing: Philadelphia, PA, 1985; pp. 243-244.
11. Reference 4.
12. Hurtubise, R. J. *Anal. Chem.* **1989**, *61*, pp. 889A-895A.
13. Hurtubise, R. J. *Solid Surface Luminescence Analysis*; Marcel Dekker: New York, NY, 1981.
14. Schulman, E. M.; Walling, C. *J. Phys. Chem.* **1973**, *77*, pp. 902-905.

15. Suter, G. W.; Kallir, A. J.; Wild, U. P.; Vo-Dinh, T. *Anal. Chem.* **1987**, *59*, pp. 1644-1646.
16. Reference 7, pp. 112-126; 188-191.
17. Reference 4, pp. 50-64.
18. Niday, G. J.; Seybold, P. G. *Anal. Chem.* **1978**, *50*, pp. 1577-1578.
19. McAleese, D. L.; Dunlap, R. B. *Anal. Chem.* **1984**, *56*, pp. 2246-2249.
20. Asafu-Adjaye, E. B.; Su, S. Y. *Anal. Chem.* **1986**, *58*, pp. 539-543.
21. Schulman, E. M. *J. Chem. Educ.* **1976**, *53*, pp. 522-524.
22. Marzzacco, C. J.; Deckey, G.; Halpern, A. M. *J. Phys. Chem.* **1982**, *86*, pp. 4937-4941.
23. Reference 4, pp. 58-61.
24. McAleese, D. L.; Freedlander, R. S.; Dunlap, R. B. *Anal. Chem.* **1980**, *52*, pp. 2443-2444.
25. Lytle, F. E. *Anal. Chem.* **1974**, *46*, pp. 545A-557A.
26. Goeringer, D. E.; Pardue, H. L. *Anal. Chem.* **1979**, *51*, pp. 1054-1060.
27. Nithipatikom, K.; Pollard, B. D. *Appl. Spectrosc.* **1985**, *39*, pp. 109-115.
28. Sliney, D.; Wolbarsht, M. *Safety with Lasers and Other Optical Sources*; Plenum: New York, NY, 1980.
29. Walters, F. H.; Parker, L. R. (Jr.); Morgan, S. L.; Deming, S. N. *Sequential Simplex Optimization;* CRC Press: Boca Raton, FL, 1991.

Hardware List

Laser: Laser Science model VSL-337 nitrogen laser.

Digital oscilloscope: Hewlett Packard model 54503A.

PIN photodiode: United Detector Technology type 10DP/SB.

Monochromator: Any 1/4 meter focal length monochromator. Smaller monochromators or even colored glass filters monitoring all phosphorescence may also be suitable. Note: other phosphorescence spectrometers with lifetime measurement capabilities are also suitable for this experiment.

Photomultiplier tube and power supply: Almost any photomultiplier tube and power supply would be suitable. Pacific Instruments is a good source of PMT housings and power supplies.

Mounts and optics: Suitable sources of low cost optics and mounts are Melles Griot and Newport Corporation. Homemade mounts and optical tables are quite appropriate.

RECEIVED October 1, 1992

CHAPTER 8

Three Applications of a Nitrogen-Laser-Pumped Dye Laser in the Undergraduate Laboratory: From Spectroscopy to Photochemistry

Julio C. de Paula, Jeffrey Lind, Matthew Gardner, Valerie A. Walters, Kristen Brubaker, Mark Ledeboer, and Marianne H. Begemann

POTENTIAL HAZARDS: High-Pressure Systems, Lasers, High Voltage

Many current problems in physical chemistry are being solved with the help of the laser. The monochromaticity, coherence, and power of this light source have revolutionized high-resolution spectroscopy. Also, the advent of pulse shortening techniques has made lasers the tools of choice for the study of short-lived reaction intermediates; nowadays, picosecond transients may be probed with laser systems that are available commercially.

The majority of undergraduate laboratory courses across this country do not introduce students to the uses of lasers in physical chemistry. Two reasons are typically given for this curricular deficiency. First, there is some apprehension among faculty about the cost of laser and detection equipment. Second, many believe that the successful implementation of laser-based experiments requires special expertise among the faculty.

The arguments above are compelling indeed, especially when we consider small departments with limited budgets. Fortunately, these concerns are becoming less and less important, as recent advances in laser and computer technology have resulted in the commercial availability of equipment that is both affordable and easy to use.

It is our purpose to present a series of experiments built around a nitrogen-laser-pumped dye laser, a short focal length monochromator, a photomultiplier tube, a boxcar integrator, an oscilloscope, and a personal computer. We describe three configurations of this basic equipment, which was used by our students in elementary and advanced studies of kinetics and spectroscopy.

Two of our experiments illustrate principles of electronic spectroscopy, but in ways that exploit the unique characteristics of the laser. For example, we describe a spectroscopic study of atoms in the gas phase that revolves around the optogalvanic effect. In this experiment, developed at Vassar College, principles of quantum mechanics are reinforced by studying a system, in this case the plasma, that has industrial applications. Plasmas are used routinely in the electronics industry, for example.

Another advantage of laser optogalvanic spectroscopy (LOGS) is its simplicity. Because the absorption of a photon is measured indirectly, sophisticated and expensive photon detection techniques or equipment, such as photomultiplier tubes and monochromators, are not needed. Nonetheless, this project does provide an opportunity for the student to obtain experience with atomic spectral analysis and familiarity with basic concepts of electronics.

Our molecular spectroscopy experiment, developed at Lafayette College, also introduces a technique not encountered in traditional undergraduate laboratory

2503–6/93/0120$08.75/0

courses: two-photon absorption spectroscopy. The study of two-photon spectra is important because different selection rules govern the probability of absorption of one or two photons by a molecule. Consequently, some electronic states can be observed more clearly in two-photon spectroscopy than in one-photon spectroscopy.

In our experiment, we use a focussed beam from the nitrogen-laser-pumped dye laser to deliver the high photon densities necessary for the observation of strong two-photon absorption signals from the aromatic hydrocarbon phenanthrene. Data from one and two photon spectroscopic measurements are combined to arrive at a thorough description of the electronic structure of phenanthrene. The experimental results are then checked against the theoretical predictions from a simple Hückel molecular orbital calculation.

The pulsed nature of the output from the nitrogen-laser-pumped dye laser suggests that the laser spectrometer can be operated in the time domain. Indeed, with a slight modification in the gated integrator settings, lifetimes of excited states or reaction intermediates may be determined in the nanosecond to millisecond time scale. With this feature in mind, the group at Haverford College designed a kinetics/photochemistry experiment that exploits the nanosecond time resolution of the instrument.

We chose to study Ru(II) *tris*(α-diimine) complexes, because they have been touted as useful photosensitizers for the catalytic splitting of water, the main goal of artificial photosynthesis. In order to gain insight into the structural factors that improve the efficiency of a photosensitizer, electron transfer reactions between the excited states of several Ru(II) *tris*(α-diimine) complexes and $Fe(H_2O)_6^{3+}$ are studied by quantitating the decrease in the excited state lifetime with increasing quencher concentration (the Stern–Volmer relation). The kinetic data are analyzed by least squares methods (linear and nonlinear) and the reaction mechanism is interpreted in light of current theories of electron transfer reactions in the condensed phase.

We are devoted not only to the implementation of interesting laser-based experiments, but also to facilitating the transfer of this technology to other programs. To this end, we have used versatile instrumentation that is not only affordable, but also easy to install, use and maintain. Transferability is being tested further by the groups at Haverford and Lafayette, who have exchanged experiments. We also hope to show that, in addition to providing a cost-effective and user-friendly means of revitalizing physical chemistry laboratory courses, the laser and detection system used in these experiments is sophisticated enough to represent an important addition to a small department's research capabilities.

The following discussion begins with general comments about instrumentation and laser safety. Subsequently, the three experiments mentioned above will be described in detail, with separate sections on theory, methodology, and data analysis. Lastly, a detailed hardware list will be provided.

General Methodology

The Lasers Novices in laser spectroscopy are referred to the excellent monograph by Andrews *(1)* for a general introduction to lasers and a more detailed explanation of the detection equipment. We do point out, however, that one need not be an expert in lasers and electronics to implement the experiments described in the subsequent sections.

We recommend a nitrogen-laser-pumped dye laser (hereafter referred to as *nitrogen/dye laser* for the sake of convenience) system because it provides short, high-energy pulses of high-resolution, tunable radiation for under $25,000. Alternative systems include dye lasers pumped by Nd:YAG or excimer lasers. These are considerably more expensive, however, and produce very energetic and, consequently, very hazardous radiation. Therefore, nitrogen laser based systems provide a good balance between safety, cost-effectiveness, and versatility.

The group at Vassar used a EG&G Model 2100 nitrogen/dye laser system. As this product is no longer available, the groups at Haverford and Lafayette have tested products from Photon Technologies International (PTI, South Brunswick, NJ). We describe the PTI system in detail, with the purpose of identifying performance features that are compatible with the projects outlined in subsequent sections. Naturally, any system with similar specifications may be used successfully (see Hardware List).

The nitrogen laser used in our experiments is a PTI PL2300 laser, capable of delivering 1.4 mJ pulses of 600 ps duration at repetition rates that vary from single shot to 20 Hz. This laser is very easy to install and use. Installation consisted of uncrating the laser, connecting it to the 120 V line, and connecting a nitrogen cylinder to the appropriate inlet. After minor pressure adjustments, the specified ultraviolet output was obtained. Routine maintenance of this instrument is also easy; it consists of regular changes of gas supply, and periodic cleaning of the spark gap and the laser cavity.

The nitrogen laser pumps a PTI PL202 laser, which consists of oscillator and amplifier stages and delivers pulses of about 500 ps duration. Depending on the choice of dye, wavelengths ranging from 360 to 900 nm may be obtained from this laser. For blue and green dyes, we observe about 10% conversion of the nitrogen output into dye output. High spectral resolution (0.04 nm) is specified by the manufacturer for the grazing incidence cavity.

All of the optical components of this dye laser are aligned at the factory and do not require further adjustments. The dye cuvettes sit on pre-aligned holders. Dye changing consists of simply emptying and cleaning the supplied 1-cm quartz cuvettes and filling them with the dye solution of choice.

Detection for Emission Spectroscopy The basic instrumentation required for the emission experiments described below is schematized in Figure 1. In our design, a very small portion of the dye laser energy is diverted by a microscope slide to a Hammamatsu S1721 photodiode, which was wired in house according to manufacturer's instructions. Most of the laser energy enters a PTI MP1 sample compartment. The compartment consists of a cuvette holder for 1-cm cells, two premounted supracil quartz lenses, one iris diaphragm, and two holders for two-inch square optical filters. The first lens (25 mm diameter, 70 mm focal length) focusses the dye laser output into the sample, and the other (38 mm diameter, 60 mm focal length), at 90° to the first, collects scattering or emission from the sample. One of the filter holders and the iris are located near the entrance port of the sample compartment, before the focussing lens. The second holder is located between the collecting lens and the entrance slit of the monochromator. The sample compartment was mounted directly to the monochromator by PTI.

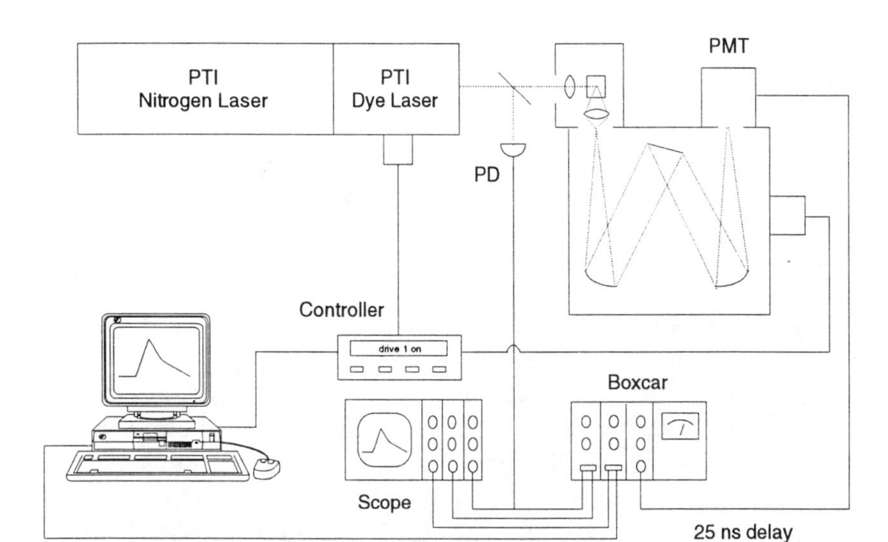

Figure 1. Schematic diagram of the nitrogen-laser-pumped dye laser and emission detection equipment used in our experiments. Abbreviations: PD, photodiode; PMT, photomultiplier tube.

The PTI 001 single monochromator has a focal length of 0.25 m and is equipped with a 1200 grooves/mm grating. Scanning of the monochromator and dye laser gratings can be done manually or electronically, via the PTI 01-5010 controller. The photomultiplier tube (PMT) is a side-window Hammamatsu R928, operated at room temperature. The tube has a nominal rise time of 4 ns. PTI markets this PMT in a compact housing that contains the power supply and an amplifier. In response to our request, PTI included a switch in this housing that enables the user to choose between amplified or "direct" (minimal time constant) operation. The former mode of operation is useful when acquiring spectral data, where amplification of weak signals may be desired, while the latter mode is necessary for nanosecond time domain operation. PTI shipped the entire sample compartment/monochromator/PMT unit preassembled on a breadboard. The unit was ready for use upon receipt.

There are a variety of ways to process the PMT signal, depending on the nature of the experiment. The equipment recommended here allows for the acquisition and averaging of both spectral and time-domain data. Our design uses a Tektronix 2225 50 MHz, or similar, oscilloscope for real-time display of the PMT signal; this feature helps in the optimization of the signal-to-noise ratio. The main signal processing unit is a Stanford Research Systems (SRS, Sunnyvale, CA) Model 250 boxcar integrator. This unit is triggered by the signal from the Hammamatsu photodiode. The time-dependent PMT signal is integrated over time slices (or gates) that can be as short as 2 ns or as large as 15 μs; moreover, the gate may be either stationary or scanned.

For the acquisition of spectra, where either the monochromator or the dye laser grating is scanned, the gate may be wide and is kept stationary. For a given monochromator focal length, the spectral resolution of the system increases with decreasing slit widths, and with increasing groove density of the grating. The resolution found in our two-photon spectra is representative of the performance of our system.

For the processing of time-domain data, the gratings are fixed and the gate is scanned over the desired time range. The PMT signal is delayed by about 25 ns with respect to the photodiode signal because the electronics of the boxcar integrator impose a 25 ns delay between the arrival of a trigger pulse and the activation of the gate. The time resolution of our system was estimated by measuring the apparent width of a dye laser pulse with the monochromator/PMT/integrator setup. The 500 ps pulse is seen as a 40 ns pulse by our detection equipment. This reflects broadening of the laser pulse by the electronics in the PMT power supply and in the boxcar integrator.

A SRS Model 245 Computer Interface Module controls various integrator functions, including gate scanning, and sends digital data to the microcomputer, either via a RS232C or a GPIB interface. SRS provides a software package that controls data acquisition and contains some data analysis capability, including linear and nonlinear regression. Documentation of SRS hardware and software is very good. The gated integrator manual describes in detail the integrator settings required for a variety of experimental configurations. The software runs under a user-friendly graphical interface on PC clones. No math coprocessor is required, although the data analysis routines will run faster with one. Dot-matrix printers, laser printers and HPGL plotters are supported, but the output is not of high quality. However, data can be exported in ASCII format and fed into data analysis or graphics programs for further processing. We experimented with XT and 386SX machines, and found that data acquisition with the SRS software via an RS232C line ran more reliably on the latter system.

A SRS Model 240 DC-300 MHz current amplifier is optional. This unit amplifies the direct output of the PMT by as much as a factor of 125, without broadening the transient significantly. We recommend this current amplifier if nanosecond time-domain data of weak signals are to be analyzed.

The groups at Lafayette and Haverford Colleges purchased the components for the afore-mentioned system in the Spring of 1990. The total hardware cost, which excludes optional items discussed above, was approximately $40,000.

Detection for Optogalvanic Spectroscopy A N_2/dye laser was used in these experiments because it allows for the acquisition of spectroscopic and time-dependent data. The laser used in this study, an EG&G 2100, provided a maximum of only 30 μJ/pulse. If more powerful pulses are used, such as those from the PTI system described above, then care should be exercised to prevent saturation of strong optogalvanic transitions. The spectroscopy part of this experiment may be done with a chopped continuous-wave (cw) dye laser and lock-in amplifier, instead of the pulsed laser and boxcar averager. However, a cw laser may not be used in the two-photon and time-resolved luminescence experiments described in later sections.

LOGS is a very appealing technique for the undergraduate laboratory because it does not require extensive detection equipment. As shown in Figure 2, changes in the discharge impedance are measured by monitoring the voltage across a resistor in series with it. With the commercially available hollow cathode lamps (Buck Scientific, Inc., East Norwalk, CT) used in this work, the resistor, and coupling capacitor were 2.5 kΩ and 10 nF, respectively. The lamps require about 300 V to ignite (Bertan high voltage power supply 210-03R), but can be operated at slightly

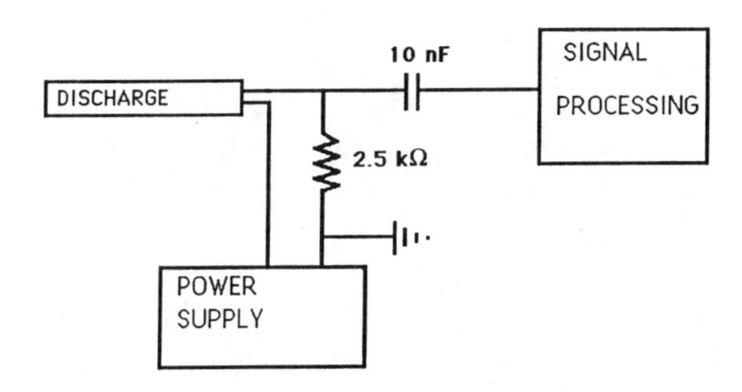

Figure 2. Schematic diagram of the optogalvanic detection circuit.

lower voltages during continuous operation. Typically, spectra are obtained with a discharge current of about 25 mA. The laser output is directed into the center of the discharge and the resulting optogalvanic signal is displayed on a 50 MHz oscilloscope, such as the Tektronix 2225, and directed to a boxcar averager/gated integrator and a chart recorder, such as the Philips PM 8272, for hardcopy. More elaborate designs would use a computer to control data acquisition and display, as described above.

GENERAL SAFETY CONSIDERATIONS

Before attempting any of the experiments described in this chapter, we strongly recommend that instructor and students become thoroughly familiar with laser safety practices. The discussion below is not comprehensive and serves merely as an introduction to those aspects of laser operation that represent hazards. Further study of laser safety is imperative.

Standards for laser safety are delineated clearly in the American National Standards Institute's Standard Number ANSI Z136.1 "Safe Use of Lasers", whose latest version was published in 1986. Copies of this document may be purchased from the Laser Institute of America, 12424 Research Parkway, Suite 130, Orlando, FL 32826. ANSI Z136.1 is a highly technical document that laser novices may find difficult to understand. The excellent monograph by Winburn *(2)* explains the standards in more accessible language and is an appropriate companion to ANSI Z136.1-1986.

In general terms, laser hazards are associated with exposure to high voltages and with biological effects of laser light and laser dyes on tissue. Throughout the discussion that follows, it is important to remember that even the most thoughtful safety measures can only *minimize* the risk of exposure, as it is impossible to completely eliminate such risks in a laboratory setting.

High Voltages The PTI PL2300 N_2 laser was designed in such a manner as to minimize exposure to high voltage areas under normal operating conditions. Care must be exercised, however, when opening the console for maintenance. High voltage areas are marked clearly by the manufacturer. Of course, maintenance is to be carried out by the instructor, who must follow the manufacturer's instructions very carefully.

Electrical hazards may also be found in the detection equipment. In the LOGS experiment, the detection circuit should be enclosed in an electrical component box. When turning off the power supply, sufficient time should be allowed before handling the hollow cathode lamp or detection circuit, so that any residual voltage can discharge.

The PMT assembly manufactured by PTI and used in the emission experiments does not pose a high risk of exposure to high voltages, when operated under normal conditions. If purchasing components from other vendors, however, please follow the recommended safety guidelines and place high-voltage power supplies as far away from high traffic areas as experimentally feasible.

Tissue Damage by Laser Radiation Laser light can inflict severe damage to skin and to the eye. Damage as severe as blindness may result from even the briefest exposure of unprotected eyes to direct or reflected laser beams. Consequently, all personnel occupying the same room where the laser is operating must wear specially designed laser goggles that attenuate radiation generated by the laser. To prevent burns, skin exposure to laser radiation should also be avoided.

Detailed studies have determined damage thresholds for continuous-wave (cw) and pulsed lasers and the results are summarized in ANSI Z136.1-1986. Exposure limit data are very useful in calculating the maximum safe operating power of a laser and the optical density (OD) of protective eyeware *(2)*.

A number of companies, such as Spectra Physics (Mountain View, CA), Uvex Winter Optical (Smithfield, RI), and Glendale Protective Technologies (Woodbury, NY), market laser goggles that comply with ANSI Z-136.1-1986. When ordering, it is important to specify the mode of operation of the laser (cw or pulsed) and the wavelength range for which protection is desired. For a pulsed laser system, such as the one described in this chapter, the energy per pulse, the pulse width, and the pulse repetition rate should also be specified.

In order to minimize further the risk of eye and skin injury, we recommend that the laser and detection systems be placed in a room that is isolated from other student or faculty activity. If this is not possible, then the laser system and its immediate area should be enclosed by curtains or movable partitions. We also recommend that the laser and detection system be kept just slightly above waist level, and that all optics external to the laser cavity and the sample compartment be enclosed within cardboard, wood, or metal boxes whose interiors are painted black. This applies especially to the microscope slide and photodiode combination.

Chemical Hazards All of the experiments described below use a dye laser. If the students are to mix their own dye solutions, then it is incumbent upon the instructor to warn the class of the potential carcinogenic activity of laser dyes. It is imperative that the mixing of dye solutions be done in a hood, and that rubber gloves be worn during the operation.

LASER OPTOGALVANIC STUDIES OF PLASMA-GENERATED SPECIES

THEORY *General* Optogalvanic spectroscopy is based on the optogalvanic effect, which occurs when the absorption of laser light by a plasma generated species induces an impedance change in the plasma *(3,4)*. By monitoring the discharge impedance (or voltage) as a function of laser wavelength, one measures indirectly the absorption of photons and obtains an optogalvanic spectrum. The spectrum produced is the same as the laser absorption spectrum, but with spectral features whose relative intensities may be quite different. There are numerous mechanisms that may be responsible for the optogalvanic signal, depending on the species and transition involved. In most cases, it is assumed that the cross section for collisional ionization is different for the upper and lower level of the species, resulting in a change in discharge impedance upon excitation *(5)*. Normally, when this mechanism is in effect, the plasma impedance decreases upon absorption of the photon. However, other possible mechanisms by which the photon energy can be coupled to that of the charged species in the plasma, and thereby produce an optogalvanic signal, exist *(6)*. For example, processes such as non-resonant laser photoionization increase the probability of autoionization, and changes in the cross section for associative ionization or Penning ionization may be important.

The experiments described here are among the simplest in optogalvanic spectroscopy. Typically, they are completed in one, or possibly two, four hour laboratory periods. The experiments involve the acquisition of the excited state electronic spectrum of neon or argon generated in a commercially available hollow cathode lamp, and measuring the time dependence of the optogalvanic signal as a function of transition. An added bonus that comes with developing these experiments is the ability to use the optogalvanic spectrum of neon or argon as a means of calibrating the wavelength of dye lasers.

Several other LOGS experiments that use a more complex discharge cell are possible, although they are perhaps more suitable for advanced independent projects. If a hollow cathode discharge is designed that allows for the input of several gases, then the spectra of open shell free radicals can be studied. For example, NH_2 can be observed in a N_2/H_2 plasma, and CN can be observed in a C_2N_2 plasma *(6)*. The He_2 molecule, which has a repulsive ground state, can be detected easily in a He discharge of sufficiently high pressure *(6)*. Optogalvanic spectroscopy has also been used to determine the photodetachment threshold for I^- generated in a discharge of I_2 vapors *(7)*. This could be reproduced as an advanced physical chemistry laboratory experiment. Experiments that investigate optogalvanic spectroscopy as a plasma diagnostic are also feasible. For example, it has been shown that the Stark splitting and broadening of Rydberg transitions in helium, which are observed in optogalvanic spectroscopy, can be used to determine the electric field strength *(8)* and electron temperature *(6)*, respectively, in certain plasmas.

Atomic Spectroscopy The excited states of neon and argon are described by using j–l coupling, a coupling case intermediate between Russell–Saunders (L–S) and j–j coupling *(9,10)*. Most students become fairly adept with the coupling of angular momenta and the determination of term symbols in atoms that follow the Russell–Saunders coupling rules. However, whether they truly understand the physical basis for what they are doing mathematically is somewhat questionable. By studying the neon or argon spectra, students obtain good understanding of coupling of angular momenta and of selection rules for atomic transitions.

For a one-electron atom, one only needs to consider the spin angular momentum, s, and orbital angular momentum, l, of a single electron in order to determine the total angular momentum, j, of the atom. In a many electron atom, one must consider the spin and orbital angular momenta of all the open shell electrons. The total angular momentum of the atom will depend on how the individual spin and angular momenta are coupled together. It is the relative orientation of the individual electronic angular momenta with respect to one another that determines the term values (energies) corresponding to a particular electronic configuration. There are basically two schemes that are used to couple individual spin and angular momenta together, Russell–Saunders (or *L–S*) coupling and *j–j* coupling. In the first scheme, the orbital angular momenta of all the electrons l_1, l_2, ... , l_k are coupled together to form a total orbital angular momentum, *L*. Likewise, the individual total spin angular momenta s_1, s_2, ... , s_k are coupled together to form a total spin angular momentum, *S*. *L* and *S* are then coupled together to form the total angular momentum, *J*, for the atom. In the *j–j* coupling scheme, the orbital and spin angular momenta of each electron are coupled together to form a total angular momentum, j_i, for each electron: $l_1, s_1 \rightarrow j_1$; $l_2, s_2 \rightarrow j_2$; .... ; $l_k, s_k \rightarrow j_k$. The j_1, j_2, ... , j_k are then coupled together to give the total *J* for the atom. There is no definite *L* and *S* for this type of coupling and therefore, *L* and *S* are not good quantum numbers and the selection rules $\Delta L = 0, \pm 1$ ($\Delta l = \pm 1$ for the electron undergoing a transition) and $\Delta S = 0$ do not hold. The selection rule $\Delta J = 0, \pm 1$ and that $J = 0$ cannot connect with $J = 0$ still hold. Also, even terms must combine with odd terms (the Laporte rule).

Russell–Saunders coupling is used for atoms in which the interaction energy of the l_i with one another and of the s_i with one another is stronger than the interaction between the individual l_i and s_i for each electron. If the interaction energy between the individual l_i, s_i pairs is larger than that among the different l_i and the different s_i, then *j–j* coupling is used. The interaction of the l_i with one another and the s_i with one another is electrostatic in nature, whereas the l_i, s_i pairs couple together because of the magnetic moments associated with the angular momenta. The latter phenomenon is called spin–orbit coupling. Thus, Russell–Saunders coupling is used when the electrostatic energy is greater than the magnetic spin–orbit energy, and *j–j* coupling is used when the magnetic spin–orbit energy is greater. For most of the lighter elements ($Z \leq 40$), Russell–Saunders coupling is valid. However, for heavier atoms and particularly for the excited states of heavier atoms, *j–j* coupling becomes more appropriate. Even for light atoms in excited states, the interaction of the spin and orbital angular momenta of the excited electron with the rest of the electrons may be weaker than their interaction with each other. In these cases, a coupling scheme intermediate between Russell–Saunders and *j–j* coupling may be appropriate.

The lowest excited states of neon correspond to the electron configuration $1s^2 2s^2 2p^5 3s^1$. This configuration gives rise to four levels which, in Russell–Saunders coupling terms, would be designated as $^3P_2$, $^3P_1$, $^3P_0$, and $^1P_1$ levels. In this notation, the superscript designates the spin multiplicity (2S + 1) of the state, the subscript designates the state's total angular momentum quantum number, *J*, and the letter designates the quantum number *L* (a P state corresponds to $L = 1$). However, as mentioned above, neon is described more accurately by *j–l* notation. In this coupling scheme, the $1s^2 2s^2 2p^5$ core is described by Russell–Saunders coupling and results in two core states labeled as $2p^5(^2P^o_{3/2})$ and $2p^5(^2P^o_{1/2})$. Odd states are labeled with the superscript "o". The excited electron's orbital angular momentum, *l*, is then coupled to the total angular momentum, *j*, of the core electrons, resulting in a *j–l* coupling value. The total *J* value for the state is obtained by coupling the excited electron's spin angular momentum to the *j–l* coupling value. Thus, the first four excited states arising from the $2p^5 3s^1$ configuration are designated as the $3s[3/2]^o(2,1)$ and the $3's'[1/2]^o(0,1)$ levels. The unprimed and primed values of *n* and *l* for the excited electron indicate whether the core state is $2p^5(^2P^o_{3/2})$ or $2p^5(^2P^o_{1/2})$, respectively. The value of the *j–l* coupling is given in square brackets and are listed from lowest to highest energy. Note that the same number of states and the same values for the total angular momentum are obtained as when using Russell–Saunders coupling. Most of the strongest transitions that are observed in the neon optogalvanic spectrum between 495 and 680 nm originate in one of the 3s or 3′s′ levels and terminate in one of the 3p or 3′p′ levels. Lines that originate in the 3p and 3′p′ levels and terminate in the 5s, 5′s′, 6s, 6′s′, 4d, and 4′d′ levels are also observed. By taking the appropriate differences between transition frequencies, the separation between the four 3s and 3′s′ levels, the four 5s and 5′s′ levels, and the four 6s and 6′s′ levels can be determined. A comparison between the separation of these energy levels in neon (or argon) with that of an atom that follows pure Russell–Saunders or *j–j* coupling *(9,10)* is made. From this comparison, the changes in multiplet splitting when progressing from pure Russell–Saunders to *j–j* coupling is observed for a *ps* configuration. A *ps* configuration is an electron configuration in which there is one unpaired electron in a p orbital and one in a s orbital. When Russell–Saunders coupling holds, this splitting is small compared to the energy difference of states differing in *L* or *S* but having the same electron configuration. For example, the splitting between the $^3P_2$, $^3P_1$ and $^3P_0$ mentioned above would be small relative to the energy separation between them and the $^1P_1$ level. As *j–j* coupling is approached, this splitting becomes comparable to the separation between levels with different *L* or *S*.

Time Dependence of the Optogalvanic Signal. The time dependence of the optogalvanic signal provides a snapshot of how the plasma responds to the perturbation in excited state density caused by the absorption of a photon. The system will return to equilibrium on a time scale that depends on the states involved *(6)*. In this part of the experiment, students will investigate qualitatively the time dependence of the optogalvanic signal. It is observed that the time dependence of a rare gas transition originating in a radiatively metastable level is quite different from that originating in a nonmetastable level. In neon and argon, two of the four lowest excited states (the 3s and 3′s′ levels) are radiatively metastable due to ΔJ selection rules and have collision-free lifetimes on the order of tens to hundreds of seconds. The nonmetastable levels have lifetimes on the order of nanoseconds *(3)*. Transitions from the metastable levels are particularly strong, but transitions originating in the nonmetastable levels can be observed easily also. All signals have an initial fast rise, corresponding to a decrease in discharge impedance, that peaks at 4–5 μs after the trigger pulse. The decrease in plasma impedance results from additional electrons produced by ionization of the excited states populated by the laser and from super-elastic collision of thermal (slow) electrons with the excess excited states. The decrease in impedance is followed by an increase in discharge impedance that is a maximum at 10–15 μs after the trigger signal. The plasma impedance increases because the laser has depopulated plasma-excited states, which play an important role in maintaining the excited state density in the plasma. In essence, the laser causes a momentary rapid decrease in plasma impedance due to ionization and super-elastic collisions, but these processes convert the laser-excited states to ground state species, resulting in a net decrease in the plasma excited state density. The portion of the signal corresponding to an increase in plasma impedance is much stronger for transitions from nonmetastable levels, indicating the importance of metastable states in maintaining the equilibrium plasma excited state density. A weak second decrease in plasma impedance is sometimes observed at long times (about 30 μs) after the trigger signal. The probable cause for this portion of the signal is the excess ions themselves, which are produced by the laser but which take a longer time to drift to the cathode and, thereby, cause a decrease in the plasma impedance.

PROCEDURES *Atomic Spectroscopy* In the first part of the experiment, students obtain the excited state electronic spectrum of the hollow cathode fill gas, neon or argon. Depending on the amount of time available, all or part of the region from 670 to 495 nm is studied. On the order of eighty lines are observed in this spectral region. The dyes used to cover the complete wavelength range are absolute ethanol solutions of Rhodamine R640 and R6G, Coumarin C540 and C500. To obtain the spectrum, the laser grating is scanned, with the boxcar gate set at a fixed delay relative to the laser trigger signal. A portion of the neon spectrum is shown in Figure 3. This spectrum was obtained with a gate duration of 5 μs, a delay from

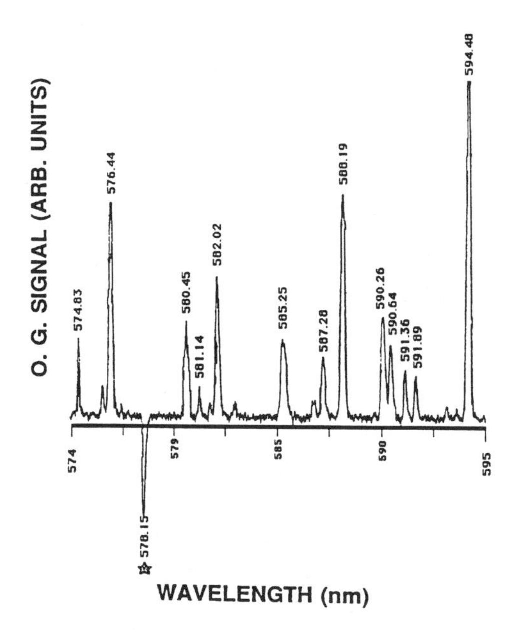

Figure 3. Neon optogalvanic spectrum. The transition marked with an asterisk is assigned to copper.

trigger of about 200 ns, and an input time constant of 10 μs. The laser was scanned at 0.02 nm/s with a pulse repetition rate of 10 Hz. By simultaneously observing the optogalvanic signal and the gate output from the boxcar on the oscilloscope, students establish operating parameters for the boxcar averager. The students record the wavelength of all lines observed and, with the use of various reference books *(11,12)*, assign as many as possible to known transitions of the rare gas. Lines that are not assigned to the rare gas may be assigned to the cathode material itself. A copper hollow cathode lamp was used to obtain the spectrum shown in Figure 3 and a transition assigned to copper is indicated with an asterisk.

Time Dependence of the Optogalvanic Signal. In the second part of the experiment, students obtain the time dependence of the optogalvanic signal. To do this, the laser frequency is set at the center of a transition and the boxcar gate is scanned at a fixed rate for a fixed period of time after the laser trigger pulse. As shown in Figure 4, the shape of the optogalvanic signal as a function of time is significantly different for transitions originating in radiatively metastable levels (Figure 4a) compared with those originating in nonmetastable levels (Figure 4b). By observing the signal on an oscilloscope, students set the boxcar scan range, and, by measuring the signal shape for various time constants, gate durations, and scan times, they can determine the

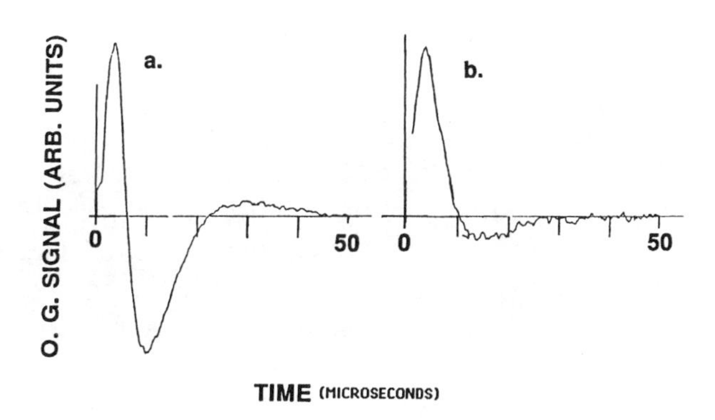

Figure 4. Time dependence of optogalvanic signals in neon. (a) lower level is metastable; (b) lower level is nonmetastable.

parameters for best signal recovery. The signals shown in Figure 4 were obtained with a gate duration of 100 ns, a scan range of 50 μs, a scan time of 200 s, and a laser pulse repetition rate of 10 Hz. The input time constant was 100 μs. If possible, it is useful to delay the optogalvanic signal relative to the trigger signal in order to completely capture its initial fast rise. In this case, approximately 30 feet of coaxial cable were inserted in the optogalvanic signal channel in order to delay the signal as much as possible. Students are expected to determine what transitions to study in the above manner and to see if the results shown in Figure 4 are general for the gas studied.

If transitions are observed in atoms sputtered from the cathode, their time dependence can also be studied. The transition observed in copper (Figure 3) is identified *(13)* as the $^2D_{3/2}$ - 4p $^2P^o_{1/2}$ transition ($3d^9 4s^2$ - $3d^{10}4p$). The lower level of the transition is radiatively metastable. The time dependence of the copper transition shows no fast portion corresponding to a decrease in plasma impedance and the portion of the signal corresponding to an increase in plasma impedance occurs at much shorter times, 4–5 μs, after the laser trigger pulse. The time dependence of the copper optogalvanic signal explains why it shows an opposite sign from the neon transitions in Figure 3.

DATA ANALYSIS As mentioned previously, students assign as many lines as possible to transitions in the rare gas studied. The most helpful sources of help in this task are American Institute of Physics Handbook *(11)* and Bashkin and Stoner *(12)*. The latter reference presents excellent energy level diagrams.

The students calculate the splitting between the four 3s and 3′s′ levels by taking the appropriate differences between transition frequencies. If enough transitions are observed, then they can calculate the same splitting in the 5s, 5′s′, and 6s, 6′s′ levels. They are asked to compare the observed splitting with that for ps configurations in other atoms and to discuss the physical basis for the various coupling cases. In discussing the time dependence of the optogalvanic signals, students are asked to explain qualitatively the basis for the observed differences.

TWO-PHOTON ABSORPTION SPECTROSCOPY OF PHENANTHRENE

THEORY *Two-Photon Spectroscopy.* Electronic absorption spectroscopy is useful in structural determinations of molecules in excited electronic states. Transitions from the ground electronic state to higher energy electronic states can result from the absorption of one or more photons of electromagnetic radiation by a single molecule. Several important features of two photon absorption spectroscopy that are pertinent to the observation and identification of excited electronic states of molecules are described below. Excellent, yet relatively uncomplicated, treatments of two photon molecular spectroscopy, providing much more detail, can be found in references *14-17*.

1. Simultaneous absorption of two visible photons by a single molecule results in excitation to electronic states far into the uv region. UV-vis spectrophotometers typically do not permit observation of states much higher in energy than 50,000 cm^{-1}. Although states at even higher energies are observable with two photon absorption, this experiment focusses on an energy region that is readily observed with a UV-vis spectrometer, in order to demonstrate the complementary relationship between the one and two photon techniques.

2. Different selection rules apply in one and two photon spectroscopy. These selection rules vary according to the symmetry properties of the molecule. The probability of a one photon transition from the original state (o), with wavefunction ψ_o, to the final state (f), with wavefunction ψ_f, is proportional to the square of the electric dipole transition moment;

$$\mathbf{M}^{i}_{of} = \int \psi_f\, \mathbf{r}_i\, \psi_o\, d\tau \quad (i = x, y, z) \tag{1}$$

In this equation, $\mathbf{r}_i$ is the electric dipole moment operator with components in the *x*, *y*, and *z* directions. For a transition to be allowed, the integral must be nonzero. Simple inspection of a character table (*vide infra*) indicates whether or not the integral is zero by symmetry. The allowed excited state symmetries in a transition from a totally symmetric ground state are those that correspond to the x, y, and z vectors in the character table. The transition tensor, $\mathbf{S}^{ij}_{of}$, determines whether a two photon transition is allowed, and is proportional to the product of two electric dipole transition moments,

$$\mathbf{S}^{ij}_{\mathrm{of}} \propto \mathbf{M}^{i}_{\mathrm{ok}}\mathbf{M}^{j}_{\mathrm{kf}} \qquad (i, j = x, y, z) \tag{2}$$

where k refers to an intermediate (virtual) state. Again, inspection of the character table reveals which transitions are allowed in two photon absorption. In this case, the allowed excited state symmetries correspond to the quadratic functions of the x, y, and z vectors (i.e., x^2, xy, $x^2 - y^2$, etc.)

For molecules with a center of symmetry, an electronic state is gerade (g) if its wavefunction is symmetric with respect to inversion, or ungerade (u) if it is not. In one photon absorption, only g → u and u → g transitions are allowed. In two photon absorption, only g → g and u → u transitions are allowed. Hence, in centrosymmetric molecules, one would expect to see entirely different electronic states with the two techniques. These so-called parity selection rules do not apply to phenanthrene which has no center of symmetry, but may be useful in extensions of this experiment to other aromatic hydrocarbons, as described below.

3. Another advantage of employing two photon spectroscopy is that measurements of band intensities using circularly and linearly polarized radiation can be used to identify those states that are classified as totally symmetric. Such polarization measurements in one photon spectroscopy do not yield symmetry information.

The excited electronic states of the aromatic hydrocarbon, phenanthrene, are investigated in this experiment. Three experimental techniques are used; the most novel of which is two photon absorption spectroscopy in the form of two photon laser-induced excitation fluorescence. The other techniques are one photon absorption and dispersed fluorescence. These spectra permit determination of the energies of excited electronic states and, also, vibrational frequencies within these states.

Symmetry of Phenanthrene. As inferred above, this experiment makes use of the symmetry properties of phenanthrene. The three symmetry elements present in phenanthrene can be seen in Figure 5. There is one C_2 proper axis of rotation and two planes of symmetry, σ_{zy} and σ_{zx}. Hence, phenanthrene has relatively low symmetry, belonging to the C_{2v} symmetry point group.

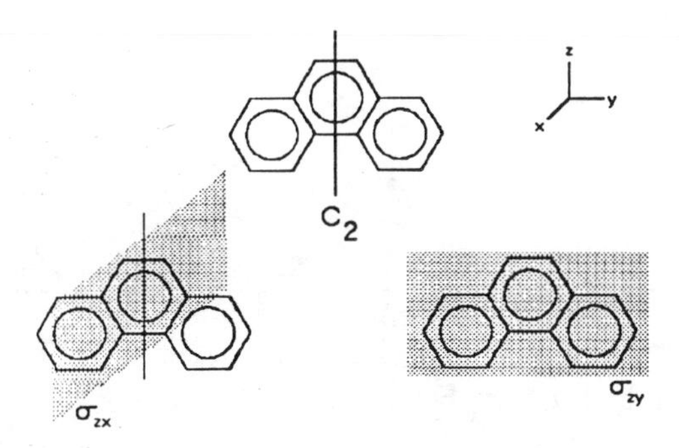

Figure 5. The symmetry elements of phenanthrene.

The character table for the C_{2v} group is shown in Table I. The electronic states of phenanthrene can be classified as belonging to the A_1, A_2, B_1, or B_2 irreducible representations. The totally symmetric representation in this point group is A_1. The allowed transitions for phenanthrene can be determined from this table. Consideration of the above selection rules reveals that one photon transitions to states of A_1, B_1, and B_2 symmetry are allowed, and two photon transitions to states of A_1, A_2, B_1, and B_2 symmetry are allowed.

TABLE I. Character Table for the C_{2v} Point Group

C_{2v}	E	C_2	σ_{zy}	σ_{zx}		
A_1	1	1	1	1	z	x^2, y^2, z^2
A_2	1	1	-1	-1		xy
B_1	1	-1	-1	1	x	xz
B_2	1	-1	1	-1	y	yz

Theoretical Predictions. The symmetry properties of molecules are also useful in simplifying theoretical calculations of these states. Hückel molecular orbital calculations (*18, 19*), employing the symmetry properties of phenanthrene, can be performed as part of this experiment or the results of these calculations can be used as additional background information in a purely spectroscopic investigation. The latter approach is taken here. These calculations provide the symmetries and approximate energies of the molecular orbitals of phenanthrene from which the electronic states are formed. For simplicity, we focus on the four lowest energy excited electronic states of phenanthrene and the calculated energies and symmetries of these states are shown in Table II. The Hückel molecular orbital

TABLE II. Energies and Symmetries of the Four Lowest Energy Electronic States of Phenanthrene Calculated from Hückel Molecular Orbital Theory

Excited Electronic State Energy (in cm^{-1})	Symmetry
31,100	B_2
35,200	A_1
35,200	A_1
39,300	B_2

calculations were performed with Hückel parameter β = 73 kcal/mol, the value used for the related molecule, naphthalene *(19)*. As seen in Table II, two of the four states have A_1 symmetry, and the other two states have B_2 symmetry. Transitions to states of A_1 and B_2 symmetry are allowed in both one and two photon absorption.

Spectroscopic Investigation. The theoretical results shown in Table II guide the experimental investigation of phenanthrene by suggesting the wavelength range within which to look for the states. All four excited states can be observed by using a *combination* of one photon (UV-vis) absorption spectroscopy, and two photon laser-induced excitation fluorescence. Although transitions to all four states are allowed in both one and two photon spectroscopy, the relative intensities are sufficiently different that different sets of electronic states are observed with the two techniques. Polarization measurements are carried out to determine if the states observed with two-photon absorption are totally symmetric. Several of the excited electronic states exhibit vibrational structure and the frequencies (usually expressed as wavenumbers in units, cm^{-1}) of the excited state vibrations can be measured. For a comparison of ground and excited state vibrational frequencies, a dispersed fluorescence spectrum yielding the ground state frequencies is also obtained. Frequency shifts between ground and excited states are known to give specific information on the geometry differences in the two states.

Related Aromatic Hydrocarbons Phenanthrene was chosen for this investigation of aromatic hydrocarbons for several reasons. First, only moderate introduction to group theory is required to gain an appreciation of the symmetry properties of phenanthrene. Second, transitions to some of the same electronic states are allowed and actually observed in one and two photon spectroscopy, emphasizing that they are just different forms of an already familiar technique; electronic spectroscopy. Finally, the selection rules can be determined from the electronic state symmetries alone, i.e., there is no need to invoke vibronic coupling to explain the existence of certain bands in the two photon spectrum. More complicated concepts can be illustrated in additional investigations of other aromatic hydrocarbons, for example, anthracene and naphthalene. For naphthalene, Hückel calculations employing the symmetry properties of the molecule, are described in great detail in reference *19*. Both anthracene and naphthalene possess a center of symmetry, hence, the previously mentioned parity selection rules pertain. Both molecules exhibit substantial vibrational structure in their absorption and dispersed fluorescence spectra (*20*). The two photon spectra of anthracene (*21*) and naphthalene (*22*) exhibit bands that are vibronically allowed, allowing introduction of this concept. Polarization measurements can be made on these vibronically allowed states, keeping in mind that the measurements indicate when the *vibronic* symmetry is totally symmetric. (The vibronic symmetry is the same as the electronic symmetry when the vibration is totally symmetric, as has been assumed for phenanthrene.) Additional experiments on the deuterated derivatives of these aromatic hydrocarbons, to investigate the relative effects of deuteration on the vibrational frequencies in the ground and excited states, might also be interesting.

PROCEDURES *Sample Preparation* Spectra are obtained of solutions of phenanthrene in methanol of spectrophotometric quality. Both phenanthrene and methanol can be purchased from Aldrich Chemical Company, Milwaukee, WI. **SAFETY NOTE:** Phenanthrene is listed as combustible, an irritant and possible mutagen,which should be handled with rubber gloves and in a well-ventilated hood. Additional safety information can be found in the Material Safety Data Sheet, supplied with purchase of the phenanthrene. Similar considerations apply to anthracene and napthalene.

For the two photon spectra, 10^{-3} M solutions are satisfactory. For these spectra, it is very important to filter the phenanthrene solution before filling the cuvette. This removes dust and, therefore, greatly reduces the amount of scattered light.

One Photon Spectrum A UV–vis spectrum of phenanthrene in a solution of methanol constitutes the one photon absorption spectrum. Due to significant differences in band absorption coefficients, at least two sample concentrations are required to see all the regions of structure in the wavelength range predicted by theory. Since only wavelength information is required, the concentrations are not crucial. Concentrations that yield absorbances of less than 1 for the bands in a given region with a good signal to noise ratio, are appropriate.

Two Photon Spectrum The experimental apparatus for obtaining the two photon fluorescence spectra is as described previously. The output from the nitrogen pumped dye laser is focussed with a 0.5 inch diameter, 1.0 inch focal length quartz lens into the cuvette containing the phenanthrene solution. A phenanthrene molecule within the region of high photon density can absorb two identical incident photons and emit a single photon of higher energy than the incident photon. The wavelength of dye laser radiation for two-photon absorption must be twice the wavelength that would be required for one photon absorption. To obtain the spectrum, the fluorescence intensity is monitored as the dye laser wavelength is scanned. The monochromator is set to 368 nm where the fluorescence is most intense. The slits and PMT voltage are set such that a good signal-to-noise ratio is observed, without saturating the detector (about 600-700 mV for the setup described previously).

TABLE III. Suggested dyes and their wavelength ranges.

Region	Dyes
480–550 nm	Coumarin 500
610–670 nm	Sulforhodamine 101
650–700 nm	DCM

The suggested dye laser wavelength regions and the dyes covering these regions are shown in Table III. The second and third dye regions overlap and are used to record the multiple vibrational bands of a single electronic state. The dyes can be purchased as ready made solutions from PTI, or in solid form from Eastman Kodak Co. (Rochester, NY).

Polarization Measurements In these measurements, the intensity of a two photon absorption band is measured with circularly polarized light, and also with linearly polarized light of the same intensity. The dye laser is set to the wavelength maximum of a vibronic band. Relatively inexpensive sheet polarizers (Oriel Corporation, Stratford, CT) are sufficient to polarize the radiation. It is recommended that the circularly polarized measurement be obtained first. To "clean up" the linear polarization of the laser radiation, a linear dichroic sheet polarizer is placed after the laser and before the sample. A circular sheet polarizer or 1/4 wave sheet retarder is then placed between the linear polarizer and the sample to obtain circularly polarized radiation. With the sample removed, the intensity of the radiation is measured at a photodiode placed after the cuvette holder. With the sample in place, the intensity of the PMT signal is recorded. This gives I_{circ}.

To obtain the measurement with linearly polarized radiation, the circular sheet polarizer is removed. Now, the intensity of the radiation at the photodiode in the absence of sample, should be greater than before. (Every optic in the beam causes some reduction in power.) The laser power must be adjusted so that the photodiode measurement is exactly the same as when the radiation was circularly polarized. This can be done by turning down the discharge voltage of the laser, or by placing a neutral density filter in the beam path. (A neutral density wheel with variable optical densities would be ideal.) Placing the sample in the beam path, and measuring the new PMT signal gives the I_{lin} value. The ratio of I_{circ}/I_{lin} gives the polarization ratio, Ω.

Dispersed Fluorescence The dispersed fluorescence spectrum of phenanthrene is obtained by pumping the solution with the 337 nm radiation from the N_2 laser directly. For this spectrum, the wavelength of the monochromator is scanned. To physically make the rearrangement, one merely pushes the dye laser off to the side, and switches the grating control to the emission monochromator. (Note that this spectrum will justify the collection of fluorescence at 368 nm in the two-photon excitation spectra, as this is where the emission maximum occurs.)

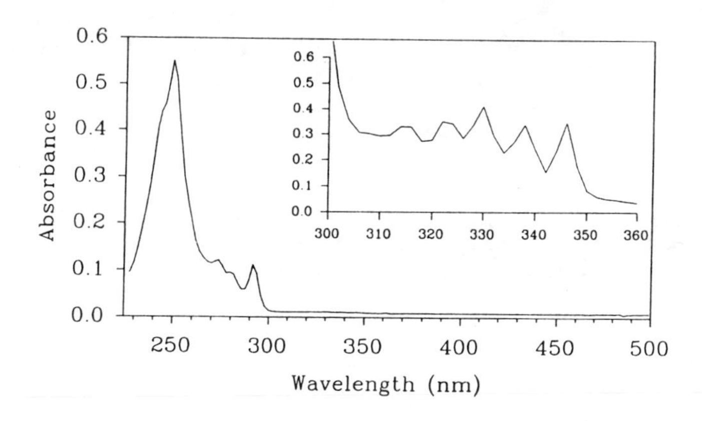

Figure 6. UV–vis spectrum of a solution of phenanthrene in methanol.

DATA ANALYSIS A UV–vis spectrum of phenanthrene, obtained with diode array detection, is seen in Figure 6. (Scanning instruments may provide higher resolution.) This spectrum encompasses the wavelength range within which the four low energy states are expected to lie, but only three states are observed. Two of these states exhibit vibrational structure, as evidenced by sets of bands with similar intensities that are separated by energies on the order of vibrational energies. The lowest energy state, seen in the inset, shows five vibrational bands, and a second state of intermediate energy shows three vibrational bands. The band at 250 nm shows no vibrational structure, but is thought to correspond to a third electronic state.

To determine the vibrational frequencies, the band wavelengths (in nm) are first converted to units of cm^{-1}. The frequency in cm^{-1} of the lowest energy peak in a set of bands is then subtracted from the frequency in cm^{-1} of each of the higher energy peaks. The lowest energy peak is assumed to be the origin band resulting from absorption from the v=0 level in the ground state to v=0 in the excited state. For the lowest energy state seen in the UV–vis spectrum, the vibrational frequencies are found to be 684, 1401, 2154, and 2945 cm^{-1}, as seen in Table IV.

TABLE IV. Vibrational Frequencies (in cm^{-1}) of the Ground and First Excited States of Phenanthrene.

Ground State			First Excited State	
Literature (*23*)	Dispersed fluorescence	Literature (*23-25*)	1-photon spectrum	2-photon spectrum
408	334	395		
714	700	671	684	698
833				
1043				
1355	1321	1376	1401	1370/1424
		1407		
1450		1446		
1580		1519		
1606				
	2054	2115	2154	2014
	2706	2842	2945	2700

The two photon excitation fluorescence spectra acquired with the suggested dyes provide the electronic state energies of two excited states and vibrational frequencies in these states. Figure 7 shows the spectrum of one of these states. Note that this state is allowed in one photon spectroscopy and would be expected between 250 and 265 nm. However, it was not observed by this technique because it was obscured by nearby strong one-photon transitions *(15,23)*. This is the fourth predicted state.

The vibrational frequencies of the two photon states and the ground state, can be determined from the two photon spectra and dispersed fluorescence spectrum, respectively, as illustrated above for the one photon spectrum. The vibrational frequencies for the lowest state seen in two photon spectroscopy (which is the same state as the lowest state seen in one photon spectroscopy), and the vibrational frequencies of the ground state are given in Table IV. Typically, the difference in frequency of a given vibrational mode in the ground state and an excited state is less than about 10%, and the data were correlated on this basis.

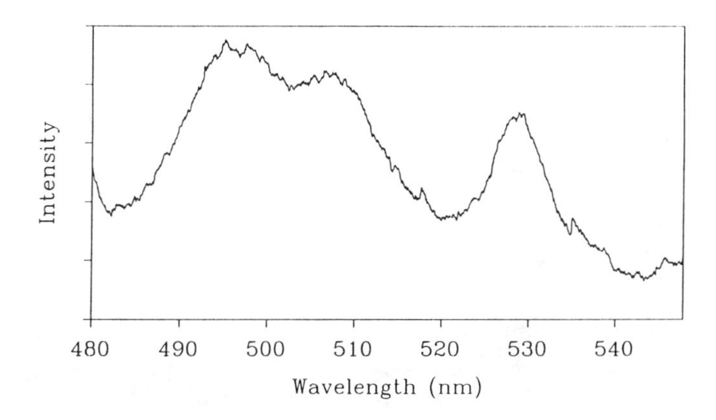

Figure 7. Two photon fluorescence excitation spectrum of a 10^{-3} M solution of phenanthrene in methanol.

Literature values are included for comparison and agree with our data to within about 25–30 cm^{-1}. Importantly, the direction of the frequency shifts is the same as in the literature. The vibration at 1355 cm^{-1} (literature value), apparently due to a ring breathing mode, shifts up on going to the first excited state. This frequency increase is opposite of what one would expect for a π-π^* transition, as the antibonding character generally leads to a weakening of the bonds, resulting in lower frequencies of the vibrational normal modes. (This expected trend can be modified, however, if the degree to which the various stretches and bends comprise the normal mode differs between the ground and excited state.) The vibration at 714 cm^{-1} (literature value), due to a combination C–C–C bend and C–C stretch, decreases on going to the excited state. It should also be emphasized that due to the relatively low spectral resolution, the observed bands in all the spectra may be due to several vibrations of similar energy, and the above determined energies can be thought of as pertaining to the most intense vibration in each group.

The symmetry information from the polarization measurements can be evaluated by determining Ω, the ratio of the band intensities I_{circ} / I_{lin}. In a two photon transition to a totally symmetric state, Ω is less than 3/2. In a transition to a state of other symmetries, Ω is equal to 3/2. Both two photon states are found to be totally symmetric, implying that the other two states seen only in the one photon spectrum, have B_2 symmetry.

PHOTOCHEMISTRY OF RUTHENIUM(II) tris-(α-DIIMINE) COMPLEXES

THEORY Ruthenium(II) *tris*(α-diimine) complexes have been used extensively as photosensitizers in solar energy conversion systems (*26,27*). Upon photoexcitation, these Ru(II) complexes can reduce many substances; in principle (though not in current practice), the power "stored" in such a photo-generated redox pair can be used to split water into H_2 and O_2, two energy rich molecules. These processes are summarized below, where L represents the diimine ligand and Q represents the oxidant:

$$RuL_3^{2+} + h\nu \rightarrow {}^*RuL_3^{2+} \quad (3)$$

$$^*RuL_3^{2+} + Q \xrightarrow{k_q} RuL_3^{3+} + Q^- \quad (4)$$

$$2\ RuL_3^{3+} + H_2O \rightarrow 2\ RuL_3^{2+} + 1/2\ O_2 + 2H^+ \quad (5)$$

$$Q^- + H^+ \rightarrow Q + 1/2\ H_2 \quad (6)$$

In other words, solar energy is used to make fuels from abundant and inexpensive substrates, such as H_2O. Although the chemistry of such a solar cell is straightforward, there are some technological barriers to be overcome before ruthenium(II) complexes can form the basis of a commercial photovoltaic device *(26)*.

The project described in this section is designed to give insight into the chemical and physical factors that improve the photochemical efficiency of Ru(II)-diimine complexes. The basic strategy consists of manipulating the structure and the redox potentials of the complexes by using 2,2′-bipyridine (bpy) or *o*-phenanthroline (phen) ligands that exhibit a variety of substituents (Figure 8). These bidentate ligands form *tris* complexes with Ru(II) that have rough D_3 symmetry.

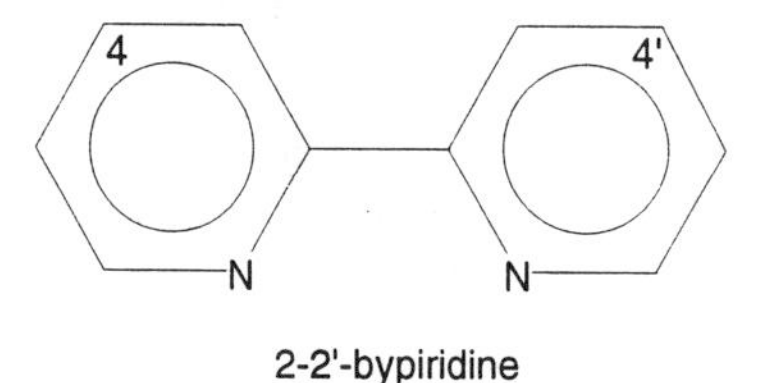

2-2'-bypiridine

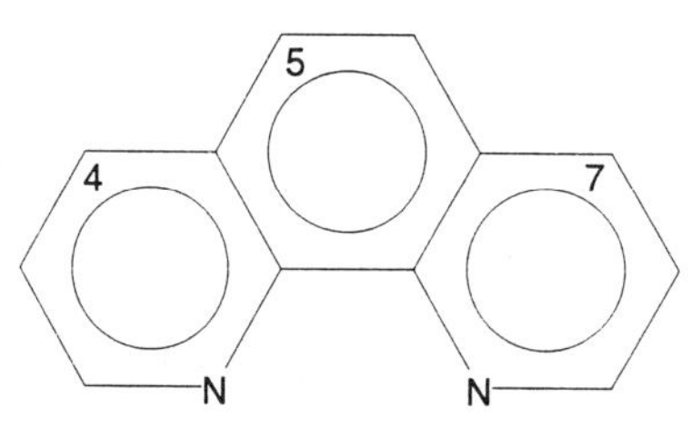

1,10-phenanthroline

Figure 8. Structures of 2,2′-bipyridine and 1,10-phenanthroline, showing the ring position numbers where the ligands used in this study are substituted. The other ligands are : 4,4′-dimethyl-2,2′-bipyridine, 4,7-dimethyl-1,10-phenanthroline, and 5-chloro-phenanthroline.

Before embarking on a discussion of the factors that affect photochemistry, it is useful to consider the electronic structure of Ru(II) *tris*-(α-diimine) complexes.

The following brief discussion will use $Ru(bpy)_3^{2+}$ as example, but the principles are completely transferrable to substituted bipyridyl and phenanthroline ligands. More thorough treatments of the problem are given by Watts *(27)* and by Krausz and Ferguson *(28)*.

The electronic absorption spectrum of $Ru(bpy)_3^{2+}$ at room temperature shows a broad feature at 450–460 nm, with a shoulder to the blue *(27)*. Several lines of evidence indicate that this transition consists of a metal-to-ligand charge transfer (MLCT) from a molecular orbital that has mostly Ru(II) d orbital character to a molecular orbital that has mostly ligand π orbital character. The MLCT excited state can be thought of as having partial Ru^{3+}-bpy^{-} character.

The dipole moments of MLCT states of Ru complexes with bpy and phen ligands have been measured, by quantitating the solvent dependence of the absorption spectra *(29,30)*. These experiments indicate that the MLCT excited states have rather large dipole moments, a result that is not expected if the electronic excitation were distributed more or less evenly among all three ligands. It has been proposed, therefore, that the charge transfer is localized within the molecule, possibly onto only one of the diimine ligands.

Ru(II) diimine complexes are highly luminescent, due to emission from the MLCT excited state. At room temperature and in aqueous solution, $Ru(bpy)_3^{2+}$ shows strong luminescence at about 600 nm *(27)*. The lifetime of this excited state under the same conditions is 600 ns *(31)*. The red luminescence can be quenched by energy transfer or by electron transfer to acceptor molecules:

Energy transfer: $$^*RuL_3^{2+} + Q \rightarrow RuL_3^{2+} + Q^* \quad (7)$$

Electron transfer: $$^*RuL_3^{2+} + Q \rightarrow RuL_3^{3+} + Q^- \quad (8)$$

A variety of inorganic and organic species may act as quenchers. The rate constant for the quenching reaction is obtained by analyzing the luminescence data. Typically, the luminescence quantum yield, ϕ, of the donor, in our case the Ru(II) complex, is monitored as a function of the concentration of the quencher. The quenching rate constant, k_q, is determined from the so-called Stern–Volmer relation:

$$\phi_0 / \phi = 1 + \tau_0 k_q[Q] \quad (9)$$

where ϕ_0 and τ_0 are the luminescence quantum yield and the luminescence lifetime of the donor in the absence of acceptor, respectively. The quantum yield of luminescence is proportional to the luminescence lifetime, so that the Stern–Volmer relation may also be written as:

$$\tau_0 / \tau = 1 + \tau_0 k_q[Q] \quad (10)$$

The equation above suggests that k_q may be obtained from an experiment where the luminescence lifetime of the donor is measured at different quencher concentrations. A plot of τ_0 / τ versus [Q] is expected to be linear, with slope $K_{SV} = \tau_0 k_q$, where K_{SV} is the Stern–Volmer rate constant.

Determining k_q does not give any insight into the mechanism of quenching. For a given system, k_q may reflect a combination of energy and electron transfer processes. There are some criteria, however, that govern the relative efficiencies of energy transfer and electron transfer. When applied to simple systems, these criteria help determine the predominant mechanism of quenching.

According to Förster, efficient energy transfer depends on a number of factors, which are summarized lucidly by Stryer *(32)*. Some of these are geometrical in nature; for example, the efficiency of energy transfer increases with decreasing distance between the donor and the acceptor. For a given distance, energy transfer will depend strongly on the relative energies of the excited states of the donor and acceptor. Efficient energy transfer is expected to occur if the excited state of the donor is higher in energy than the excited state of the acceptor, on which the excitation energy will reside ultimately. In practical terms, this means that the luminescence spectrum of the donor molecule must be blue-shifted relative to the absorption spectrum of the acceptor molecule. Quantum mechanical considerations also stipulate that there must be some measurable overlap between the luminescence spectrum of the donor and the absorption spectrum of the acceptor.

Förster's theory of *energy* transfer has been tested on a number of systems and has survived in more or less its original form. On the other hand, theories describing *electron* transfer reactions in the condensed phase are still being debated and refined. Hopfield *(33)* and Marcus *(34)* have contributed considerably to our understanding of electron transfer processes. In qualitative terms, they postulate that rates of electron transfer depend on: (i) the distance between the donor and acceptor; (ii) the degree of quantum mechanical coupling between the molecular orbitals of donor and acceptor; (iii) the free energy change, ΔG^0, for the reaction; and (iv) the reorganization energy, λ, of the donor, acceptor, and the medium. The latter parameter refers to the energy cost incurred by molecular rearrangements that must result from the transfer of charge along a finite distance. Generally, it is expected that rates of electron transfer will increase with decreasing donor-acceptor distances, and that the maximum rate will be observed when the reaction is activationless, i.e., when ΔG^0 is large and negative, and when $\Delta G^0 = -\lambda$. In other words, the rate is optimized when the standard free energy change for the reaction is matched exactly by the energy required for reorganization of the donor, acceptor, and solvent molecules.

The considerations above provide a basis for the design of an experiment on the photochemistry of Ru(II) diimine complexes. The laser spectrometer described previously is used to obtain lifetime data for the complex as a function of concentration of a quencher, such as $Fe(H_2O)_6^{3+}$. To our knowledge, this donor-quencher system has been studied by steady-state luminescence and flash photolysis *(35)*, but not by time-resolved luminescence. By obtaining the luminescence spectrum of the complex and the absorption spectrum of the quencher, the relevance of energy transfer processes in the quenching mechanism is evaluated. Likewise, the importance of electron transfer may be estimated by considering the postulates of Marcus theory.

The experiment may be run in a variety of ways. At Haverford College, we divide the class into groups of two or three students, with each group being assigned a different ligand. The results from each group are posted and the student is asked to provide explanations for any observed trends in k_q with ligand structure. Alternatively, the entire class may work on one complex only, but each group is to obtain data at different temperatures. In this latter version of the experiment, data interpretation would include an assessment of the mechanism of quenching and a determination of the Arrhenius parameters: the activation energy and the pre-exponential factor.

PROCEDURES *General* At Haverford College, this project is part of our junior-level integrated laboratory course. The students synthesize the complexes, ascertain purity by absorption spectrometry, and carry out the luminescence quenching measurements. While the synthesis is fairly straight-forward, we recognize that it is not an appropriate part of a physical chemistry laboratory course. In order to facilitate the incorporation of this project into a more conventional physical chemistry curriculum, we make two suggestions:

(a) The instructor may synthesize the compounds and make them available to the students in pure form. Consequently, the project may be carried out exactly as described in this section.

(b) The instructor may purchase $Ru(bpy)_3Cl_2 \cdot 6H_2O$ from G.F.S. Chemicals (Powell, Ohio) and use this compound only. The project's goals are then shifted to an evaluation of energy vs. electron transfer as the main mechanism of quenching by Fe(III), and to a determination of the Arrhenius parameters for the reaction. We recommend that the instructor check the purity of the purchased compound, according to the method proposed below. Better than 95% purity can be obtained through recrystallization of the crude complex from hot water *(36)*.

Synthesis and Purity Checks Bipyridyl and phenanthroline complexes of Ru(II) were synthesized according to the procedures of Broomhead and Young *(36)*, without modification. $RuCl_3 \cdot 1\text{–}3H_2O$ was purchased from Alfa Products (Ward Hill, MA) and the ligands (see Figure 9) were purchased from G.F.S. Chemicals. All synthetic procedures should be carried out in a well-ventilated hood.

We have used a spectrophotometric method for determining the purity of the Ru(II) complexes. The method exploits the fact that the MLCT transition is characteristic of the complex only; no other bands from unreacted $RuCl_3$ or free ligand appear in the 400–500 nm region.

Typically, an approximately 50 μM aqueous solution of the complex is made in a volumetric flask by weighing accurately a small amount of oven-dried complex and diluting to the mark. The absorption spectrum of this solution is taken with a 1-cm cuvette and the absorbance at the peak of the MLCT transition is determined. Care must be taken to keep the absorbance of this solution between 0.4 and 1.0, as the experimental uncertainty increases sharply outside this range. By application of the Beer–Lambert Law, an apparent value of the extinction coefficient of the complex is calculated. The extent to which the literature and the observed values agree is a measure of the purity of the compound. The extinction coefficients reported by Lin *et al.* *(35)* were used as the theoretical values in this calculation.

Luminescence Quenching Experiments The students should prepare two stock solutions, one containing complex and another containing $Fe(H_2O)_6^{3+}$ (made from $FeCl_3 \cdot 6H_2O$). Both solutions should be made in 0.1 M HCl, in order to prevent the formation of ferric hydroxide and other dimeric ferric species.

From the stock solutions and 0.1 M HCl, at least six solutions (4–5 mL each) are made, each having the same concentration of complex (about 4–5 μM) and a known concentration of quencher (a range of 0–3 mM is appropriate). Immediately

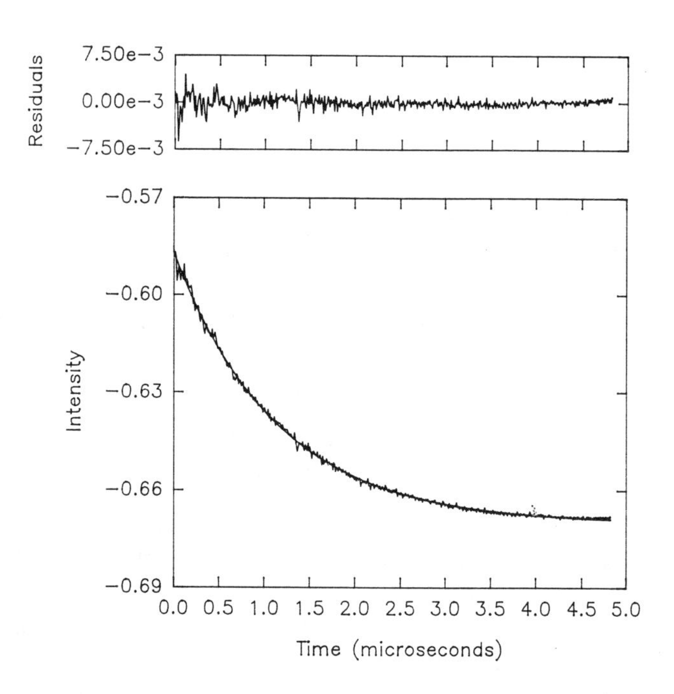

Figure 9. Luminescence decay curve of Ru(II) *tris*(5-chloro-1,10-phenanthroline) in 0.1 M HCl, at 25 °C.

prior to the luminescence measurement, enough solution is transferred to a 1-cm quartz fluorescence cell, the cell is capped with a rubber septum, and the contents are purged with nitrogen or argon for about 10 min. Purging removes O_2, which is a powerful quencher of Ru(II) diimine luminescence *(31)*. Special care must be taken during this purging step, so as to minimize the risk of pressure building up inside the cuvette. Over-pressurization may result in the spraying of solution containing 0.1 M HCl.

Luminescence lifetime measurements are made with the system described previously. The dye of choice for this experiment is Coumarin 440 (Exciton, Dayton, OH), which exhibits high gain in the 420–460 nm region, where the MLCT transitions for the chosen complexes occur. The dye laser grating should be set to the wavelength of the MLCT band for the complex. The emission monochromator should be set to about 600 nm initially.

As the best decay data are observed at the emission maximum, the luminescence spectrum of the complex must be acquired. The task is made simple by the fact that most of the ruthenium complexes used in this study emit at around 600 nm. Therefore, a decay curve should be observed on the oscilloscope at this wavelength, regardless of the identity of the complex. A stationary boxcar gate should be set to a value that ensures integration of about 90% of the decay curve. Once these settings are made, scanning the emission monochromator remotely over a 500–700 nm range will afford the luminescence spectrum. The signal-to-noise ratio and spectral resolution may be optimized by adjusting the entrance and exit slits of the emission monochromator.

With the emission monochromator set to the emission maximum of the complex, lifetime data are obtained for all six purged solutions. Here, the boxcar gate should be scanned in time over a range that encompasses at least two half-lives.

The gate width should be set to a value that ensures at least 400 data points over this time range. We recommend that the sample containing no quencher (τ_0 determination) be run first, as it will exhibit the longest lifetime of the data set. As an example, a portion of the decay curve for Ru(II) *tris*-(5-chloro-1,10-phenanthroline) is shown in Figure 9.

The last piece of experimental data required for this project is an absorption spectrum of the quencher, $Fe(H_2O)_6^{3+}$. The information will be useful in an evaluation of the quenching mechanism.

DATA ANALYSIS *Least-Squares Fitting of the Decay Curves* The luminescence of the Ru(II) complexes used in this study decays with first-order kinetics:

$$d[{}^*RuL_3^{2+}] / dt = -k_{obs} [{}^*RuL_3^{2+}] \qquad (11)$$

The decay curve may be fitted to the following equation:

$$I(t) = B + I(0) \exp(-k_{obs}t) \qquad (12)$$

where $I(t)$ is the emission intensity as a function of time, $I(0)$ is the intensity at t = 0, and B is the background intensity, or the intensity at infinite time. Fitting of the data to Equation 12 should employ a least-squares method. The students at Haverford College use the program KINFIT (On-Line Instrument Systems, Jefferson, GA), which fits data to a number of kinetic schemes. The software runs on PC clones (with or without a math co-processor) and the user can fit with either a successive integration or a Levenberg-Marquadt algorithm. KINFIT provides extensive fitting diagnostics: standard deviation of the fit, a Durbin–Watson factor for the fit, a plot of residuals, and a plot of the autocorrelation function. In lecture, the students are introduced to these concepts and are instructed in the use of the program.

Figure 9 shows the results of a Levenberg–Marquadt fit to Equation 12 of the luminescence decay of Ru(II) *tris*(5-chloro-1,10-phenanthroline). The calculated lifetime is 1.13 ms, the standard deviation of the fit is 8.8×10^{-4}, and the Durbin–Watson factor is 0.95. These data underscore the importance of using more than one criterion to analyze the fit. For the data in Figure 9, the standard deviation is small, which is indicative of a good fit, but the residuals are not perfectly random. This is shown by the small oscillations in the plot of residuals and by the Durbin–Watson factor, which should be greater than one for random residuals.

There are several alternatives to KINFIT. Some plotting and statistical software packages, such as SigmaPlot (Jandel Scientific, Corte Madera, CA), support least squares fitting of user-defined functions. If the course budget does not allow for the purchase of software, the instructor may write a fitting program based on published algorithms, such as those described by Bevington *(37)*. Bevington provides complete FORTRAN codes, which can be used as is or can be translated into any other language, such as BASIC. It is also possible to use Bevington's code as the basis for an electronic spreadsheet macro. Regardless of the calculational method, we recommend strongly that an analysis of residuals be included in the evaluation of the fit results.

Stern–Volmer Analysis After obtaining the luminescence lifetimes, the student should fit the data to the Stern–Volmer relation, eq 10, by standard linear regression methods. This can be done with an electronic calculator or with a number of software packages, such as SigmaPlot or any electronic spreadsheet. Figure 10 shows the results for the Ru(II) *tris*-(5-chloro-1,10-phenanthroline) / Fe(III) system. The dashed lines define the 90% confidence interval for the calculated line.

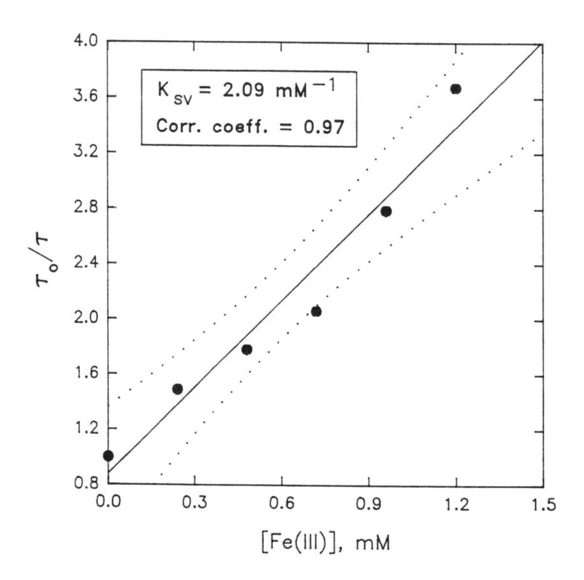

Figure 10. Stern–Volmer plot for the quenching of luminescence of Ru(II) *tris*(5-chloro-1,10-phenanthroline) by $Fe(H_2O)_6^{3+}$ in 0.1 M HCl, at 25 °C.

Data Interpretation As stated earlier, we ask different groups of students to determine k_q values for the RuL_3^{2+}/Fe(III) system, where the ligands are differently substituted forms of bpy or phen. Data from the entire class are pooled and interpreted as a whole by each student.

First, the student is asked to determine which quenching mechanism predominates: electron or energy transfer. This is done as suggested in the *Theory* section. The absorption spectrum of $Fe(H_2O)_6^{3+}$ shows no features in or immediately beyond the 600-650 nm region, where the RuL_3^{2+} complexes luminesce. This observation argues for electron transfer as the main mechanism of quenching.

The second part of the interpretation is concerned with the rationalization of any observed trends in k_q values with respect to ligand structure. Table V summarizes the data obtained by the class who took our integrated laboratory course in the Spring of 1991.

The trends are not strong, but can be understood in terms of the postulates of Marcus theory. The ΔG^0 for these reactions are large and negative *(35)*, so that we can assume that the donor-acceptor distance is one of the most important determining factors of the rate. In other words, the closer the two metal centers get during a collision, the higher the rate of electron transfer. With this in mind, one may expect k_q to be smaller for substituted bpy and phen than for the corresponding unsubstituted ligand. In fact, the opposite is true.

TABLE V. Summary of Luminescence Quenching Data (Spring of 1991)

Ligand	k_q ($M^{-1}s^{-1}$)
2,2′-bipyridine	2.12×10^8
4,4′-dimethyl-2,2′-bipyridine	1.06×10^9
1,10-phenanthroline	1.14×10^9
4,7-dimethyl-1,10-phenanthroline	1.62×10^9
5-chloro-1,10-phenanthroline	1.73×10^9

One must remember, however, that the MLCT state, from which the electron is donated, has considerable L^- character, so that a fair amount of electron density is donated from the metal to the periphery of the molecule. If a peripheral substituent can conjugate with the π system of the ligand, the extent of intramolecular charge separation will increase. The dipole lengths of the MLCT states in this series of complexes support this conclusion *(30)*. Hence, conjugation of the substituent with the aromatic system pulls electron density away from the donor, Ru(II), and places it in the immediate vicinity of the acceptor, Fe(III), very effectively. This may be viewed as shortening the effective donor–acceptor distance, which leads to an increase in the electron transfer rate.

Of the substituents studied in this project, Cl can donate electron density to the π system via resonance, and the methyl group can hyperconjugate with the π system. Both effects will explain qualitatively the observed trends in the rate of electron transfer with ligand structure.

Safety

The laser dyes used in these experiments are generally of unknown toxicity, although some are suspected carcinogens. The methanol used as a solvent is moderately toxic, and inhalation and skin contact should be avoided. Details of the Ruthenium compound's toxicity are unknown.

Concluding Remarks

The versatility of the laser-based spectroscopic system described above is evident from the thematic scope of the proposed experiments: with one system, it is possible to explore introductory and advanced topics in atomic, molecular spectroscopies, and photochemistry. Yet, we outlined only some of the basic uses of the instrumentation. Two other experimental configurations are suggested below.

The laser/monochromator/PMT system can be used as the basis for a flash photolysis instrument. With the laser providing an actinic beam, the only additional piece of instrumentation required for the upgrade would be an intense, continuous source of white or monochromatic light, which would impinge on the sample at 90° relative to the laser pulse and would act as the probe beam. The PTI sample compartment is ideal for this purpose, because the cuvette holder may be illuminated from three directions. Light absorbed by the probe beam may be analyzed by the boxcar/PMT system described above, without modifications.

Low-resolution Raman spectroscopy may also be carried out on the system described in this chapter. The group at Lafayette College used the nitrogen laser as the excitation source and acquired a Raman spectrum of liquid carbon tetrachloride with a 0.25 m single monochromator and a PMT operating at room temperature. Although the spectral resolution was low, the experiment demonstrates the sensitivity of our system. Data of higher resolution may be obtained by substituting a 0.25 m or higher focal length double monochromator for the 0.25 m single monochromator.

Acknowledgements

The authors would like to acknowledge financial support from the Pew Science Mid-Atlantic Consortium, which is funded by The Pew Charitable Trusts. JdP acknowledges additional support from the Howard Hughes Medical Institutes. Special thanks go to the integrated laboratory students of the last two years at Haverford College, particularly: Scott Wasserman, Gustavo Arrizabalaga, Marc Neff, Brian Roe, Auren Weinberg, Jessica Weiss, Michael Massiah, Seamus McElligott, Deborah Gross, Nadine Srouji, and Kaustuv Bannerjee, for the data shown in Table V.

Literature Cited

1. Andrews, D. L. *Lasers in Chemistry*; Springer-Verlag: Berlin, Germany, 1990.
2. Winburn, D.C. *Practical Laser Safety*; Marcel Dekker, Inc.: New York, NY, 1990.
3. Zalewski, E.F.; Keller, R.A.; Engleman, Jr., R. *J. Chem. Phys.* **1979**, *70*, 1015-1026.
4. Lawler, J.E. *Phys. Rev. A* **1980**, *22*, 1025-1033.
5. Demtroder, W. *Laser Spectroscopy*; Springer-Verlag; Berlin, Germany, 1982.
6. Pfaff, J.; Begemann, M.H.; Saykally, R.J. *Mol. Phys.* **1984**, *52*, 541-566.
7. Webster, C.R.; McDermid, I.S.; Rettner, C.T. *J. Chem. Phys.* **1983**, *78*, 646-651.
8. Doughty, D.K.; Lawler, J.E. *Appl. Phys. Lett.* **1984**, *45*, 611-913.
9. Herzberg, G. *Atomic Spectra and Atomic Structure*; Dover Publications, Inc.; New York, NY, 1944.
10. Karplus, M.; Porter, R.N. *Atoms and Molecules*; W.A. Benjamin, Inc.; California, 1970.
11. *American Institute of Physics Handbook* 3rd Ed.; Gray, Dwight E. Ed.; Mcgraw-Hill: New York, NY, 1972.
12. Bashkin, S.; Stoner, J. O. Jr.; *Atomic Energy Levels and Grotrian Diagrams* Vol I; North-Holland, 1979.
13. Shenstone, A.G. *Phil. Trans. R. Soc. A* **1948**, *241*, 297-322.
14. McClain, W.M. *Acc. Chem. Res.* **1974**, *7*, 129-135
15. Goodman, L. and Rava, R.P., *Acc. Chem. Res.* **1984**, *17*, 250-257.
16. Friedrich, D.M. *J. Chem. Educ.* **1982**, *59*, 472-481.

17. Johnson, P.M., *Acc. Chem. Res.* **1980,** *13*, 20-26.
18. Yates, J.; *Hückel Molecular Orbital Theory*; Academic Press: New York, NY 1978.
19. Cotton, F.A; *Chemical Applications of Group Theory*; Wiley-Interscience: New York, NY 1971.
20. Berlman I.B.; *Handbook of Fluorescence Spectra of Aromatic Molecules,* 2nd ed.; Academic Press: New York, NY 1971.
21. Dick, B.; Hohlneicher, G. *Chem. Phys. Lett.* **1981,** *83*, 615-621.
22. Dick, B.; Hohlneicher, G. *Chem. Phys. Lett.* **1981,** *84*, 471-478.
23. Dick, B.; Hohlneicher, G. *Chem. Phys. Lett.* **1983,** *97*, 324-330.
24. Warren, J.A.; Hayes, J.M.; and Small, G.J. *Chem. Phys.* **1986,** *102*, 323-336.
25. Sethuraman, V.; Edelson, M.C.; Johnson, C.K.; Sethuraman, C.; and Small, G.J. *Mol. Cryst. Liquid. Cryst.* **1980,** *57,* 89.
26. Willner, I.; Mandler, D.; Maidan, R. *New J. of Chem.* **1987,** *11*, 109-121.
27. Watts, R. J. *J. Chem. Educ.* **1983,** *60*, 834-843.
28. Krausz, E.; Ferguson, J. in *Progress in Inorganic Chemistry*; Lippard, S. J., Ed.; J. Wiley & Sons: New York, NY, 1989, Vol. 37; pp. 293-390.
29. Kober, E. M.; Sullivan, B. P.; Meyer, T. J. *Inorg. Chem.* **1984,** *23*, 2098-2104.
30. de Paula, J.C., manuscript in preparation for submission to *J. Chem. Educ.*
31. Demas, J. N. *J. Chem. Educ.* **1976,** *53*, 657-663.
32. Stryer, L. *Ann. Rev. Biochem.* **1978,** *47*, 819-846.
33. Hopfield, J. J. *Proc. Natl. Acad. Sci. USA* **1974,** *71*, 3640-3644.
34. Marcus, R. A.; Sutin, N. *Biochim. Biophys. Acta* **1985,** *811*, 265-322.
35. Lin, C.-T.; Bottcher, W.; Chou, M.; Creutz, C.; Sutin, N. *J. Am. Chem. Soc.* **1976,** *98*, 6536-6544.
36. Broomhead, J. A.; Young, C. G. in *Inorganic Syntheses*; Fackler, J. P., Jr., Ed.; J. Wiley & Sons: New York, NY, 1982, Vol. 21, pp. 127-128.
37. Bevington, P. R. *Data Reduction and Error Analysis for the Physical Sciences*; McGraw-Hill: New York, NY, 1969, pp. 204-246.

Hardware List

The following hardware was used in the above experiments:

Lasers: Photon Technologies International PL2300 nitrogen laser; Photon Technologies International PL202 dye laser.

Monochromator: Photon Technologies International Model 001 (0.25 m focal length, f/4, 1200 grooves/mm grating); Photon Technologies International 01-5010 Digital Drive Unit (also controls the grating from PTI PL202 dye laser).

Detectors: Hammamatsu R928 Side-Window Photomultiplier Tube (in PTI 01-512 housing and power supply); Hammamatsu S1721 Photodiode.

Signal Processing: Tektronix 2225 50 MHz Oscilloscope; Stanford Research Systems Model 250 Gated Integrator;

Stanford Research Systems Model 280 Power Supply and Display Module;
Stanford Research Systems Model 245 Computer Interface Module;
Stanford Research Systems Model 265 software for controlling SRS 245 and SRS 250;

Any PC-Compatible microcomputer (286 or above; math co-processor optional), with at least one serial port, or GPIB interface board;

A dot matrix printer or HPGL plotter.

Optional: Stanford Research Systems Model 240 DC-300 MHz Current Amplifier;

Laser Photonics, Inc. (Orlando, FL) markets a nitrogen/dye laser that is similar in design, performance, and price to the PTI system. Less expensive systems may be put together, but with some sacrifice of performance and versatility. Laser Science, Inc. (Cambridge, MA) and Laser Photonics manufacture inexpensive, albeit not very powerful, nitrogen/dye laser systems, which should be appropriate for the lifetime and optogalvanic studies described above.

Other modifications in detection and signal processing are also possible. For example, if no spectral data are desired, then the monochromator may be replaced with an appropriate combination of color and notch filters. Also, a fast photodiode may substitute for the photomultiplier tube, with attendant loss of dynamic range and sensitivity.

The experienced spectroscopist will find it cost-effective to replace the PTI MP1 sample compartment with optics and mounts marketed by companies such as the Newport Corporation (Fountain Valley, CA), the Oriel Corporation (Stratford, CT), and Melles-Griot (Irvine, CA). Volume II of the Oriel catalog has useful suggestions on how to configure luminescence detection systems from their components.

Laser Science, Inc. markets an inexpensive boxcar integrator, but the gates cannot be scanned for lifetime determinations. On the other hand, a fast storage oscilloscope may be used to process transient signals. Within this scheme, a scope camera may be used to afford hard copy of the data.

RECEIVED October 1, 1992

Flash Photolysis of Benzophenone

Patrick L. Holt

POTENTIAL HAZARDS: Lasers, High Voltage

Since its development by Porter and co-workers in 1950 *(1)*, flash photolysis has been widely used to investigate fast chemical reactions on time scales ranging from femtoseconds to seconds. This technique has been used to identify and determine the properties of highly unstable chemical species, species that could not be detected by other means. Results from these experiments have also enhanced our understanding of chemical reactivity and structure. In a typical flash photolysis experiment, a sample is exposed to a brief flash of light. Molecules in the sample absorb photons of light and are excited to higher energy states. A second light source, often called the probe, is then used to monitor what happens to the resulting states. In modern physical chemistry research, lasers are often used for both the excitation source and the probe source. Pulsed laser systems are available that emit flashes of light that last for only a few nanoseconds (10^{-9} s), picoseconds (10^{-12} s), or even femtoseconds (10^{-15} s). The short-duration, high-intensity light pulses emitted from these systems are ideal for exciting large numbers of molecules and for investigating extremely rapid processes.

Obviously, flash photolysis has had an enormous impact on many areas of modern physical chemistry research, making it imperative that students be exposed to this technique in the undergraduate lab. Indeed, many elegant flash photolysis experiments have been described *(2-6)*, but few make use of the modern technology found in a typical research lab. In this chapter, we will describe a flash photolysis experiment that utilizes laser and computer technology to investigate the photochemical and photophysical behavior of benzophenone, a system that exhibits kinetic behavior on time scales ranging from picoseconds to milliseconds.

Theory

Photophysics of Electronically Excited States When a molecule absorbs a photon of light and is promoted to an electronically excited state, it may become unstable and experience a number of changes to lose or redistribute its excess energy. As these changes occur the molecule is said to undergo decay or relaxation. The processes that occur may ultimately lead to a permanent chemical change in the system or they may end with the molecule returning to its original state.

Excited state relaxation pathways may be either radiative, or nonradiative in nature. Radiative decay involves the emission of the photon from the excited state, leaving the molecule in a lower energy electronic state. If the emission occurs without a change in the total electronic spin of the molecule, it is called fluorescence. If a spin change does occur, then the emission process is referred to as phosphorescence. Typical fluorescence lifetimes range from 10^{-12} s to 10^{-6} s whereas phosphorescence lifetimes may be as long as minutes. Phosphorescence occurs over a much longer period of time because it involves a change of spin, a process which is generally rather slow.

In nonradiative transitions, energy transfer or redistribution occurs without the emission of photons. Both internal conversion and intersystem crossing are examples of intramolecular nonradiative relaxation. These relaxation pathways involve a change in the electronic state of the system in which excess electronic energy is redistributed internally as vibrational and rotational energy. No change in electronic spin occurs during internal conversion whereas intersystem crossing is accompanied by a change in spin. Excess vibrational and rotational energy also may be transferred through collisions to surrounding molecules in a process known as thermal degradation. In the solution phase, complete thermal degradation, in which all of the excess vibrational and rotational energy is lost, generally occurs in less than a nanosecond. Intermolecular electronic energy transfer is generally not so efficient; the large energy gaps involved may not be easily accommodated by surrounding molecules.

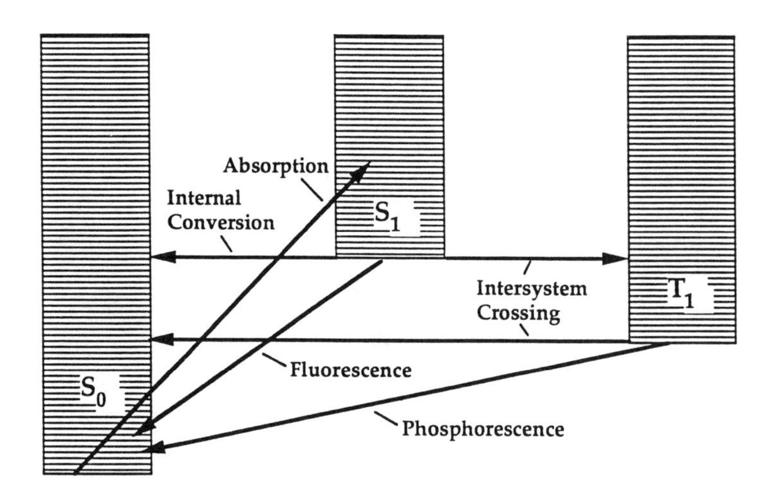

Figure 1. Jablonski diagram depicting photophysical processes.

Various types of radiative and nonradiative decay are illustrated in the Jablonski diagram shown in Figure 1. The lowest energy state or ground state is referred to as S_0. The S is used to indicate that all the electrons in the molecule are paired and that the total electronic spin is zero; this state is said to be a singlet state. The ground state molecule absorbs a photon and is promoted to the S_1 state. This state is also a singlet state because the absorption of a photon does not change electron spin. The subscript indicates that this state is the first or lowest excited singlet state. In the solution phase, excess vibrational energy imparted to S_1 during the absorption process is lost through thermal degradation. The S_1 state, now in its lowest vibrational level, may subsequently fluoresce and return to S_0. It also may undergo internal conversion to form S_0 or undergo intersystem crossing to generate a triplet state, represented here as T_1. (The terms singlet and triplet refer to the multiplicity of the system which is given by the expression $2S+1$, where S is the total spin quantum number. When S = 0, the multiplicity is equal to one, hence the name singlet. When S = 1, the multiplicity is equal to three, corresponding to a triplet.) As before, excess vibrational and rotational energy is lost rapidly through collisions. In the case of the triplet, the molecule will then remain trapped in the T_1 state until it emits a photon as phosphorescence or decays through intersystem crossing. Both transitions involve a change in multiplicity and leave the molecule in the S_0 state.

(A word of caution here: Jablonski diagrams can be misinterpreted with the relative positions of the electronic states mistakenly taken to represent differences in molecular geometry. The horizontal positions of these states are purely a matter of convenience; they do not indicate any particular structural differences. In addition, it is important to note that according to the Franck–Condon principle the molecular geometry will not change during a radiative transition. It is only after the transition that any shifts in geometry may occur.)

Of course, there are many systems that exhibit more complex photophysical behavior than is described here. For example, photochemical change may also occur in which excited molecules isomerize, dissociate, or react with other molecules. More elaborate discussions of electronic relaxation may be found in a number of physical chemistry and photochemistry texts *(7-12)*.

Benzophenone Photochemistry Benzophenone photochemistry has been the subject of intense scientific scrutiny for a number of years *(13-24)*. The lowest excited singlet state (S_1) of benzophenone lies roughly 26,200 cm^{-1} above the ground state (S_0). Following excitation to S_1, benzophenone undergoes intersystem crossing to form the triplet state (T_1) in less than 100 picoseconds *(20,21)*. In many solvents, the triplet relaxes almost completely through intersystem crossing to the ground state with a solvent-dependent lifetime on the order of few microseconds. However, in the presence of alcohols, triplet relaxation is much faster; Beckett and Porter reported a lifetime of 60 ns for benzophenone triplets in isopropanol *(15)*. The accelerated relaxation rate is due to an additional decay path in which the benzophenone abstracts a hydrogen atom from a solvent molecule to form the protonated ketyl radical, $C_6H_5\dot{C}OH$. This process is temperature dependent *(16)* and is facilitated by the weaker C–H bonds alpha to the hydroxyl group *(15)*. Solvent radicals formed in this process may then donate a hydrogen to another benzophenone molecule to form a second ketyl radical. The protonated ketyl radicals formed in these solvents may subsequently dimerize to form benzopinacol. In basic solutions, the deprotonated benzopinacol anion may form through the reaction of a protonated ketyl radical with a deprotonated ketyl radical. In both cases, the dimerization reaction occurs on the millisecond time scale. The complete reaction scheme is

$$\phi_2CO\ (S_0) + h\nu \rightarrow \phi_2CO^*\ (S_1) \qquad (1)$$

$$\phi_2CO^*\ (S_1) \rightarrow \phi_2CO^*\ (T_1) \qquad (2)$$

$$\phi_2CO^*\ (T_1) \rightarrow \phi_2CO\ (S_0) \qquad (3)$$

$$\phi_2CO^*\ (T_1) + ROH \rightarrow \phi_2\dot{C}OH + \dot{R}'OH \qquad (4)$$

$$\phi_2CO\ (S_0) + \dot{R}'OH \rightarrow \phi_2\dot{C}OH + \dot{R}'O \quad (5)$$

$$\phi_2\dot{C}OH \rightleftharpoons \phi_2\dot{C}O^- + H^+ \quad (6)$$

$$2\ \phi_2\dot{C}OH \rightarrow \underset{HO\quad}{\phi_2C} - \underset{\quad OH}{C\phi_2} \quad (7)$$

$$\phi_2\dot{C}OH + \phi_2\dot{C}O^- \rightarrow \underset{HO\quad}{\phi_2C} - \underset{\quad O^-}{C\phi_2} \quad (8)$$

where ϕ represents C_6H_5. Reaction 7 dominates in acidic solutions and reaction 8 dominates in basic solutions.

In our investigations of the reaction kinetics of this system, we will ignore reaction 2 because it occurs so rapidly that the generation of T_1 is essentially instantaneous on the time scales explored in this experiment. In the absence of candidates for hydrogen abstraction (like isopropanol), triplet decay is governed by reaction 3 and is obviously first-order with respect to the T_1 concentration:

$$d[T_1] / dt = -k_3 [T_1] \quad (9)$$

Integration of equation 9 gives the time dependence of the triplet concentration:

$$[T_1]_t = [T_1]_0 \exp(-k_3 t) \quad (10)$$

Here, $[T_1]_t$ is the concentration of triplets at time t, $[T_1]_0$ is the initial triplet concentration, and k_3 is the first-order rate constant. The lifetime, τ, of T_1 is equal to the reciprocal of the rate constant:

$$\tau = 1 / k_3 \quad (11)$$

The lifetime is actually the average time of existence for a T_1 molecule and should not be confused with the half-life.

In the presence of alcohols, triplet decay is produced by both reactions 3 and 4. The corresponding rate equation is

$$d[T_1] / dt = -k_3 [T_1] - k_4 [T_1] [ROH] \quad (12)$$

Since the alcohol is present in excess, the rate equation may be reduced to a pseudo first-order expression:

$$d[T_1] / dt = -k' [T_1] \quad (13)$$

where $k' = k_3 + k_4 [ROH]$. The triplet decay is then given by

$$[T_1]_t = [T_1]_0 \exp(-k' t) \quad (14)$$

The observed rate constant k' is larger than k_3 and, consequently, the presence of alcohols will accelerate triplet decay.

The protonated ketyl radical, formed in reactions 4 and 5, disappears slowly in reaction 7. The disappearance is second order and is given by

$$d[\phi_2\dot{C}OH] / dt = -k_7 [\phi_2\dot{C}OH]^2 \quad (15)$$

The time dependence of the ketyl radical concentration is then given by

$$[\phi_2\dot{C}OH]_t^{-1} = k_7 t + [\phi_2\dot{C}OH]_0^{-1} \quad (16)$$

The rate of decay of the deprotonated ketyl radical via reaction 8 is second-order overall. The corresponding rate expression is

$$d [\phi_2\dot{C}O^-] / dt = -k_8 [\phi_2\dot{C}OH] [\phi_2\dot{C}O^-] \quad (17)$$

The equilibrium expression for reaction 6 may be rearranged to yield

$$[\phi_2\dot{C}OH] = [\phi_2\dot{C}O^-] [H^+] / K_{eq} \quad (18)$$

Substitution of this into equation 17 gives

$$d[\phi_2\dot{C}O^-] / dt = -k_{obsd} [\phi_2\dot{C}O^-]^2 \quad (19)$$

where $k_{obsd} = k_8[H^+]/K_{eq}$. Hence, the reaction kinetics appear to be second order with an observed rate constant k_{obsd} that is directly proportional to $[H^+]$. The integrated expression for this reaction is

$$[\phi_2\dot{C}O^-]_t^{-1} = k_{obsd}\, t + [\phi_2\dot{C}O^-]_0^{-1} \quad (20)$$

A summary of rate expressions and typical lifetimes for various species in the benzophenone system is presented in Table 1.

Table 1: Properties of Various Species in the Benzophenone System.

Species	λ_{max} (nm)	Lifetimes	Decay Rate
S_0	382		
S_1	530	10 – 50 ps	$-k_2[S_1]$
T_1	525	50 ns – 20 μs	$-k_3[T_1]$
$\phi_2\dot{C}OH$	545	μs – ms	$-k_7[\phi_2\dot{C}OH]^2$
$\phi_2\dot{C}O^-$	630	μs – ms	$-k_8[\phi_2\dot{C}O^-][\phi_2\dot{C}OH]$ or $-k'[\phi_2\dot{C}O^-]^2$

Experimental Methods

As mentioned earlier, the technique of flash photolysis will be used to investigate the photodynamics of the benzophenone system. In these experiments, a short pulse of light is used to initiate the series of events discussed above. Light from the probe source is directed through the sample and onto a photodetector (either a photomultiplier tube or a photodiode). Since different species generated during the photo-induced process have different absorption spectra, it is possible to follow reaction kinetics by measuring the intensity of the transmitted probe light as a function of time. Specifically, the appearance and disappearance of individual species can be monitored by adjusting the output of the probe source to a wavelength that is absorbed strongly only by that species. Listed in Table 1 are wavelengths corresponding to maxima in the absorption spectra for important species in the benzophenone system.

The two experiments described below investigate different time domains of the benzophenone system. The first experiment uses a pulsed nitrogen laser and a digital storage oscilloscope to examine the decay of the benzophenone triplet in the presence of different solvents. The second experiment uses an inexpensive setup with a simple photographic flash unit, a helium–neon (HeNe) laser, and a computer equipped with an analog-to-digital conversion board to study the second-order decay of the ketyl radicals.

Benzophenone Triplet Decay ***Apparatus*** For investigations of the nanosecond relaxation of the benzophenone triplet, it is necessary to use an excitation source that generates pulses of light that last for a few nanoseconds or less. In addition, the light source must emit light at a wavelength that is absorbed strongly by ground-state benzophenone molecules. Pulsed nitrogen lasers, which provide a 337 nm output, will work well as the excitation source. An energy of output of 1 mJ per pulse is adequate for this experiment.

A continuous-wave Xe lamp is a reasonable probe source for this experiment. The benzophenone triplet has an absorption maximum at 525 nm so it is desirable to eliminate other wavelengths produced by the lamp. Filtering may be accomplished by using a low-resolution monochromator or specially designed bandpass filters.

The sample solution is placed in a 1-cm path length, ultraviolet-grade cuvette. The nitrogen laser beam and the lamp output are oriented to pass through the sample at right angles to each other. A photomultiplier tube (PMT) is used to detect the intensity of the probe light transmitted by the sample. The PMT output should be monitored with a digital storage oscilloscope that has a response of 100 MHz or better. Depending on the sensitivity of the oscilloscope, it may be necessary to amplify the photomultiplier tube output with a current-to-voltage amplifier. To measure the true time dependence of the PMT output, the input impedance of the oscilloscope should be 50 Ω. For data analysis, the oscilloscope trace can be photographed or ported to a computer. Any of a number of spreadsheet programs may be used to analyze the data. A block diagram of the experimental setup is shown in Figure 2.

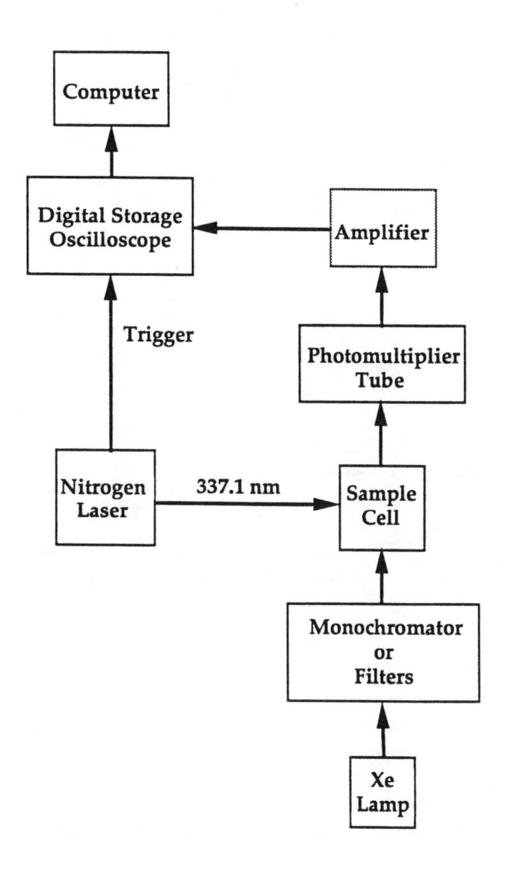

Figure 2. Block diagram of setup to monitor benzophenone triplet decay.

Procedure The procedure for conducting these experiments is relatively simple. Prepare benzophenone solutions with a variety of solvents exhibiting different degrees of susceptibility to hydrogen abstraction. Generally 0.01 M solutions of benzophenone in benzene, methanol, and isopropanol will generate good results. Before each experiment is run, solutions must be purged of oxygen which can react with (quench) T_1 to generate S_0. Remove oxygen by bubbling nitrogen through the sample cell for fifteen to thirty minutes. After this is done seal the cell and position it at the point where the excitation and probe beams intersect. Eliminate any external lighting that may contribute to unwanted background noise. Trigger the oscilloscope without firing the nitrogen laser to record the probe intensity baseline. (With some oscilloscopes, it may be possible to begin recording data before the flash unit fires. If this is the case, baseline data and flash data may be recorded on the same scan, eliminating the need for a separate baseline scan.) Set the oscilloscope to trigger when the nitrogen laser fires. Fire the nitrogen laser and record the oscilloscope trace. Make adjustments to the vertical scale and horizontal time base so that the trace corresponding to triplet decay occupies as much of the screen as possible. Record the baseline and fire the nitrogen laser again. Repeat this procedure until satisfactory data have been obtained.

Ketyl Radical Decay ***Apparatus*** Investigations of the millisecond kinetics of ketyl radical decay may be done with the same setup as discussed above. However, it is possible to do these experiments with a less complex and less expensive flash photolysis system that still utilizes laser and computer technology. The experiment described below concentrates on the decay kinetics of the deprotonated form of the ketyl radical. A diagram of the apparatus is shown in Figure 3.

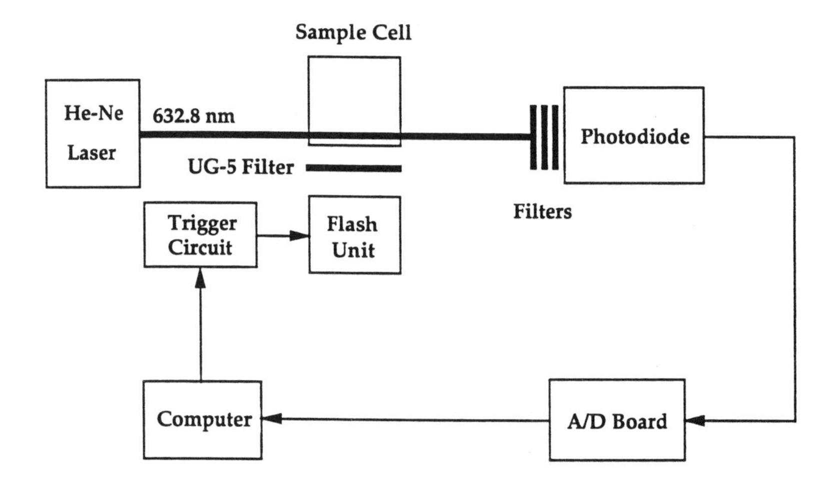

Figure 3. Block diagram of setup to monitor deprotonated ketyl radical decay.

This system uses a standard photographic flash unit as the excitation light source. If the clear ultraviolet-absorbing plastic piece that covers the flashlamp is removed, these flash units can provide excitation wavelengths throughout the visible and into the ultraviolet regions of the spectrum. Typical flash durations range from a few hundred microseconds to a few milliseconds. This is certainly short enough to study the kinetic behavior of ketyl radicals.

A red HeNe laser may be used for investigations of the deprotonated ketyl radical which has an absorption maximum at 630 nm; the laser output is at 632.8 nm. A 0.5 mW HeNe laser provides adequate power for these experiments.

The sample is placed in a 5-cm path length cylindrical sample cell. The flash unit and the HeNe laser are configured so that the excitation light and probe beam pass through the sample at right angles to each other. The flash unit should be positioned as close as possible to the curved surface of a 5-cm-long cylindrical sample cell. The HeNe laser beam should be aligned to travel parallel to the long axis of the cell just inside the curved surface adjacent to the flash unit, where excitation-induced effects are expected to be the largest. A photodiode may be used to monitor the intensity of transmitted probe light; the intensity of the transmitted light is high enough that a photomultiplier tube is not necessary. As the circuit in Figure 4 shows, the photodiode, operating in the photoconductive mode, is reverse biased with a 9 V battery. The rise-time of the detector is less than 20 ns. Ground glass plates (Rolyn Optics #55.3000) may be placed in front of the photodiode to scatter the incoming HeNe laser beam, making the detector less sensitive to small variations in the beam direction and reducing the light level to ensure that the

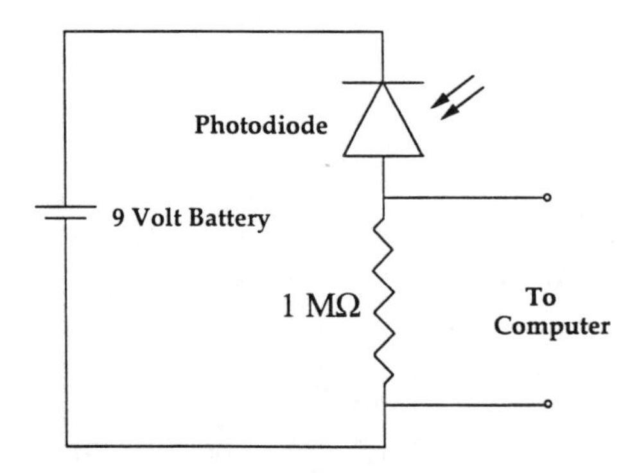

Figure 4. Photodiode circuit diagram.

detector response is linear with respect to beam intensity. In addition, a number of 3-mm thick Schott Glass OG570 filters may be used to block extraneous ultraviolet light produced during the flash pulse. To reduce background interference from flash-generated visible light, the flashlamp should be covered with a Schott UG2 filter that absorbs visible light and transmits ultraviolet light.

The flash experiment may be automated with any computer system equipped with a multifunction interface card that is capable of 12-bit analog-to-digital conversion and digital output. The photodiode output is connected to an analog input channel and the analog-to-digital conversion rate is adjusted to measure the output signal as a function of time. In order to monitor millisecond reaction kinetics, the maximum conversion rate should be on the order of 10 kHz or higher. The flash unit is triggered through a digital output channel using the circuit shown in Figure 5 *(25)*. The digital output is connected to the base of a 2N2222 transistor

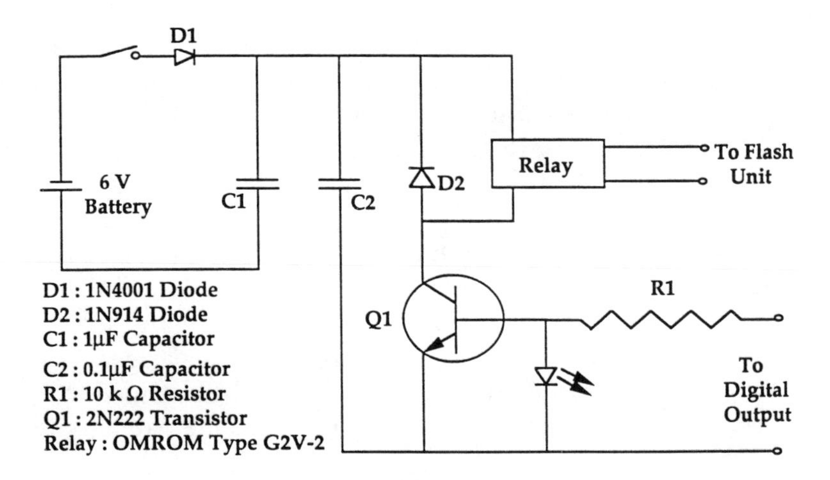

Figure 5. Flash gun trigger circuit.

which is used to activate a one amp relay (OMROM Type G2V-2). The relay is connected to the terminals of the flash gun and when the digital output is set high the relay contacts close causing the flash unit to fire. The rectifier/filter configuration of capacitors C1, C2, and diode D1 filter power from the 6V battery (four 1.5 V type C batteries will do the job). Diode D2 is placed across the relay coil to bypass the reverse current produced when the relay is switched off. A light emitting diode is used to indicate when the circuit was activated.

Software to drive most data acquisition boards may be obtained commercially. Generally board manufacturers offer software packages that contain routines for all of the low-level interface tasks. These routines may be accessed through a number of programming languages including Pascal, FORTRAN, Basic, and C. An excellent discussion of hardware programming may be found in a recent text by Demas and Demas *(26)*. This text has a chapter devoted entirely to programming the Data Translation DT2801 data acquisition board using the Pascal programming language.

Procedure Following the procedure set forth by Goodall et al. *(4)*, prepare a stock solution of 5×10^{-3} M benzophenone in isopropanol. Mix equal volumes of this solution with NaOH solutions to make at least five different solutions ranging in pH from 11 to 13. Since hydrogen abstraction occurs with a quantum yield approaching unity, these solutions should be stored in the dark. Before running a flash photolysis scan, each solution should be purged with nitrogen for fifteen to thirty minutes or subjected to a series of freeze–pump–thaw cycles while attached to a vacuum line to remove dissolved oxygen. The longer time scale of this experiment makes it particularly critical that oxygen be removed to avoid serious quenching effects.

Place the sample cell in the the path of the HeNe laser beam and align the setup as described in the previous section. Eliminate any external lighting that may contribute to unwanted background noise. Connect the photodiode output to an oscilloscope (analog or digital). Adjust the HeNe laser and the photodiode position to maximize the signal on the scope.

If the flash scans are to be recorded with a digital storage oscilloscope, first obtain a HeNe laser intensity baseline. (As discussed earlier, it may be possible to begin recording data before the flash unit fires, eliminating the need for a separate baseline scan.) Set the oscilloscope to trigger when the flash unit fires. Fire the flash unit and record the oscilloscope trace. Make adjustments to the oscilloscope settings to obtain a satisfactory trace. Record the baseline and fire the flash unit again. Repeat this procedure until satisfactory data have been obtained.

If the flash experiment is completely automated as discussed in the previous section, connect the photodiode output to the input of an analog-to-digital conversion board. Run a program that obtains a HeNe laser intensity baseline, triggers the flash unit, and obtains the kinetic data. The baseline data and flash data may be obtained in separate scans or, if the program is designed to begin data acquisition before the unit flash is fired, these data may be obtained in a single scan.

Perform several scans on each solution and store the data on disk for future analysis.

Safety

Considerable caution must be exercised when working with the high intensity light output of lasers. Goggles designed to block ultraviolet light should be worn at all times when the nitrogen laser is used. Never look directly into any laser beam regardless of its power output (even when wearing eye protection). Be aware of reflections from any surface in the beam path; check for reflections and block any that could be hazardous. Serious eye damage can result if these precautions are not taken!

Photomultiplier tubes require high voltage power supplies. These supplies should be switched off when not in use. Never connect or disconnect a cable when the power supply is turned on.

Some care must be taken when making benzophenone solutions, particularly when handling benzene, methanol, and sodium hydroxide solutions. All of these chemicals are moderately toxic. Benzene is a human carcinogen, teratogen, and strong eye irritant. Sodium hydroxide solutions are caustic.

Data Analysis

Benzophenone Triplet Kinetics A typical oscilloscope trace is shown in Figure 6. The trace represents the time dependent output the photomultiplier tube. Note that the PMT output is negative; higher light intensities generate a more negative output. Initially the signal may go to more negative values due to the detection of scattered

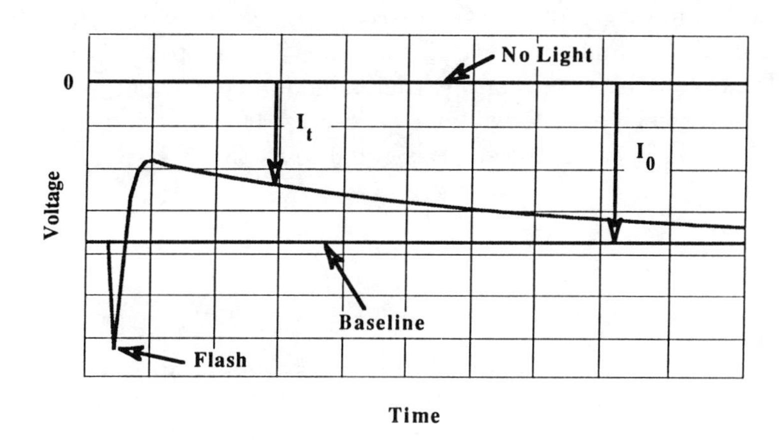

Figure 6. Typical oscilloscope trace.

light from the nitrogen laser excitation pulse. Ignore early data where this scattered excitation light is an obvious perturbation. Measure the transmitted intensity I_t at a number of different times and use the expression

$$A_t = \log (I_t/I_0) \tag{21}$$

to compute the absorbance A_t at each of these times. I_0 is the baseline intensity.

In benzene solutions, the benzophenone triplet undergoes first-order decay. Since it is proportional to concentration, the absorbance decays exponentially with time:

$$A_t = A_0 \exp(-k_3 t) \tag{22}$$

To extract the rate constant k_3 simply plot ln (A_t) versus time (t) and do a least squares fit. The slope is equal to $-k_3$.

With alcohol solvents, the analysis is complicated by the presence of ketyl radicals. These species also absorb light at 525 nm so the sample absorbance depends on their concentration as well as that of the triplet. The oscilloscope trace should appear to have two kinetic components, one fast component corresponding to triplet decay and one slow component corresponding to ketyl radical decay. In fact, the ketyl radical reacts so slowly that its decay can be considered negligible on the time scale of triplet relaxation. Consequently, the second component should appear to be constant with a magnitude represented by A_∞. It can be shown then that the difference between A_t and A_∞ decays exponentially with time,

$$A_t - A_\infty = (A_0 - A_\infty) \exp(-k't) \tag{23}$$

where k' is observed rate constant first introduced in equation 14. To determine k', plot ln ($A_t - A_\infty$) versus time and do a least squares fit. The slope is equal to -k'.

Deprotonated Ketyl Radical Kinetics Unlike the output of the photomultiplier tube, the photodiode output is positive; higher light intensities will produce a more positive signal. Use equation 21 to compute the absorbance at a number of different times. To determine k_{obsd}, plot $1/A_t$ versus t as indicated by equation 20 above and do a least squares fit. Since $A = \epsilon c L$, the slope is equal to $k_{obsd}/\epsilon L$ where L, the sample cell path length, is 5 cm and ϵ, the extinction coefficient, is 5000 M^{-1} cm^{-1}. Compute k_{obsd} for each pH. Now plot k_{obsd} versus $[H^+]$ and do a linear regression. Since $k_{obsd} = k_8 [H^+] / K_{eq}$, the slope is equal to k_8 / K_{eq}. Use $K_{eq} = 6 \times 10^{-10}$ M to compute k_8.

Discussion

Compare your results with literature values and account for any discrepancies. How would your data be affected by the presence of dissolved oxygen in the solutions? How would you expect your results to be affected by the use of different alcohols as solvents?

Design an experiment to investigate dimerization of the protonated ketyl radical. What would you use as your probe source? Would you expect the results to be pH dependent?

Acknowledgments

I would like to thank Victor Mendez for his help in putting together the millisecond flash photolysis experiment. I would also like to the thank the Camille and Henry Dreyfus Foundation for providing the funds to purchase the nitrogen laser.

Literature Cited

1. Porter, G. *Proc. R. Soc. London Ser. A* **1950**, *200*, 284-299.
2. Yamanashi, B. S.; Nowak, A. V. *J. Chem. Educ.* **1968**, *45*, 705-710.
3. Blake, J. A.; Burns, G.; Chang, S. K. *J. Chem. Educ.* **1969**, *46*, 745-746.
4. Goodall, D. M.; Harrison, P. W.; Wedderburn, J. H. M. *J. Chem. Educ.* **1972**, *49*, 669-674.
5. Kozubek, H.; Marciniak, B.; Paszyc, S. *J. Chem. Educ.* **1984**, *61*, 835-836.
6. Hair, S. R.; Taylor, G. A.; Schultz, L. W. *J. Chem. Educ.* **1990**, *67*, 709-712.
7. Atkins, P. W. *Physical Chemistry*; Fourth Edition; W. H. Freeman and Company: New York, NY, 1990; pp. 509-512.
8. Barrow, G. M. *Physical Chemistry*; Fifth Edition; McGraw-Hill Book Company: New York, NY, 1988; pp. 758-765.
9. Levine, I. N. *Physical Chemistry*; Third Edition; McGraw-Hill Book Company: New York, NY, 1988 ; pp. 752-756.
10. Birks, J. B. *Photophysics of Aromatic Molecules*; Wiley-Interscience: New York, NY, 1970.
11. Turro, N. J. *Modern Molecular Photochemistry*; The Benjamin/Cummings Publishing Company: Menlo Park, CA, 1978.
12. Wayne, R. P. *Principles and Applications of Photochemistry*; Oxford University Press: New York, NY, 1988.
13. Porter, G.; Wilkinson, F. *Trans. Faraday. Soc.* **1961**, *57*, 1686-1691.
14. Bell, J. A.; Linschitz, H. *J. Am. Chem. Soc.* **1963**, *85*, 528-532.
15. Beckett, A.; Porter, G. *Trans. Faraday. Soc.* **1963**, *59*, 2038-2050.
16. Topp, M. R. *Chem. Phys. Lett.* **1975**, *32*, 144-149.
17. Tsubomura, H.; Yamamoto, N.; Tanaka, S. *Chem. Phys. Lett.* **1967**, *1*, 309-310.
18. Godfrey, T. S.; Hilpern, J. W.; Porter, G. *Chem. Phys. Lett.* **1967**, *1*, 490-492.
19. Ledger, M. B.; Porter, G. *J. Chem. Soc., Faraday Trans. 1* **1972**, *68* , 539-553.
20. Hochstrasser, R. M.; Lutz, H.; Scott, G. W. *Chem. Phys. Lett.* **1974**, *24*, 162-167.
21. Greene, B. I.; Hochstrasser, R. M.; Weisman, R. B. *J. Chem. Phys.* **1979**, *70*, 1247-1259.
22. Bhattacharyya, K.; Das, P. K. *J. Phys. Chem.* **1986**, *90*, 398 -3993.
23. Hoshi, M.; Shizuka, H. *Bull. Chem. Soc. Jpn.* **1986**, *59*, 2711-2715.
24. Naguib, Y. M. A.; Steel, C.; Cohen, S. G.; Young, M. A. *J. Phys. Chem.* **1987**, *91*, 3033-3036.
25. Mendez, V. M. *M.S. Thesis*; Trinity University: San Antonio, TX, 1990.
26. Demas J. N.; Demas S. E. *Interfacing and Scientific Computing on Personal Computers*; Allyn and Bacon: Boston, MA, 1990.

Hardware List

Listed below are hardware components necessary to execute each experiment. Potential vendors and model numbers are also included where appropriate.

Benzophenone Triplet Decay

1. Nitrogen Laser

A number of companies manufacture nitrogen lasers that are adequate for this experiment. Photon Technology International (PTI) and PRA, a branch of Laser Photonics, both sell nitrogen lasers that produce short pulses with pulse energies of a millijoule or more.

2. Digital Storage Oscilloscope

Both Tektronix and Philips manufacture good digital storage oscilloscopes. The scope should have at least a 100 MHz bandwidth and should be capable of sampling data at 100 megasamples per second or better. These scopes can be provided with a GPIB or RS-232 interface to permit the downloading of data to a computer.

3. Photomultiplier Tube

Hamamatsu and Burle, among others, sell versions of the 1P28 photomultiplier tube. This is an inexpensive PMT with adequate sensitivity and response for this experiment. It is also advantageous to buy a PMT housing with the voltage divider network already installed; Products for Research is a good source for these housings.

A high-voltage power supply is required to operate the PMT. Several companies, including those listed above, sell power supplies. Bertan Associates (Model 602C-15N) sells a cheap, bare-bones power supply that may be used for this experiment.

As mentioned earlier, it may be necessary to amplify the PMT output. Phillips Scientific sells a DC - 300 MHz amplifier that has a gain of ten (Model 6950).

4. Xenon Lamp

PTI sells modules that may be assembled for use as the probe source in these experiments. The L1 illumination system includes as 75 W xenon lamp and a 0.25 m monochromator. This is an extremely versatile system that can be used for a wide variety of experiments.

5. Computer

The apparatus can be interfaced to both DOS and Macintosh computer systems.

Ketyl Radical Decay

1. Photographic Flash Unit

Almost any inexpensive flash gun will do for this experiment. Just remember to remove the plastic piece from in front of the flash to allow UV light to reach the sample.

2. Helium-Neon Laser

Red helium-neon lasers may be obtained from a number of sources. Edmund Scientific is a good supplier of inexpensive helium-neon lasers. Powers on the order of 0.5 mW are adequate for this experiment.

3. Photodiode

Inexpensive photodiodes may be obtained from a number of sources. The PIN-6D-I photodiode form United Detector Technology has adequate sensitivity for this experiment.

4. Data Acquisition Boards

There are a huge number of suppliers of data acquisition boards for both PC and Macintosh systems. The board should be capable of 12-bit analog-to-digital conversion at rates of 10 kHz or better. In addition, it should have a number of digital input/output channels.

5. Computer

The apparatus can be interfaced to both DOS and Macintosh computer systems.

RECEIVED October 1, 1992

CHAPTER 10

Laser Photooxidative Chemistry of Quadricyclane

Joseph J. BelBruno

POTENTIAL HAZARDS: High-Pressure Systems, Lasers

Lasers have opened many new vistas in chemical research. Experiments exploiting the unique properties of lasers include high resolution spectroscopy, the development of multiphoton spectroscopic techniques, the revitalization of nearly abandoned techniques (e.g., Raman spectroscopy) and the first true observation of bond selective photochemistry.*(1)* More conventional laser photochemical studies have routinely employed the unique features of lasers to both modify known photochemical processes and to open new reaction pathways. The properties unique to laser systems include: extraordinary brightness, directionality, spatial and temporal coherence and monochromaticity.*(2-3)* Each of these characteristics is addressed below in comparison to the properties of conventional light sources.

An ideal laser exhibits an infinitely narrow wavelength distribution. In practice, this is not attainable, but nonetheless, the wavelength distribution of real lasers may be quite small, depending upon the complexity of the laser system. Even for lasers with relatively broad wavelength distributions, optical components are available to narrow the range of emitted frequencies. *Monochromaticity* is defined by the spread in wavelength, $\Delta\lambda$, or the corresponding spread in frequency, $\Delta\nu$, of the light source. For a white light source, such as a tungsten bulb, the bandwidth of the visible emission covers the entire visible spectrum, $\Delta\lambda \sim 300$ nm. Of course, the bandwidth of the conventional source is much greater than 300 nm, since the emission of a thermal source extends well into the infrared region of the electromagnetic spectrum. An argon ion laser typically exhibits a bandwidth of approximately 0.01 nm, with minimal bandwidth narrowing optical components added to the system. The bandwidth of even a simple dye laser, the most useful laser to the chemist since its output wavelength is readily varied, will routinely equal or better this value. Dye lasers with bandwidths of the order of 0.0001 cm^{-1} are not uncommon. From a photochemistry viewpoint, the monochromaticity of a laser leads to a concentration of the entire radiative output into the spectral region that is most effective in driving the chemical change.

Directionality is a laser property that is striking in its uniqueness and has been observed by anyone who has attended a laser light show. Observation of the laser output yields a spatially narrow, well-defined beam of light which appears not to expand with increasing distance from the laser exit port. *Directionality* is technically specified in terms of beam divergence, that is, a measure of how rapidly the beam spreads after exiting the laser. A typical red HeNe laser used for alignment or pointing has a divergence of 1 milliradian (recall that 2π radians = 360°). Divergence is defined as twice the angle that the beam edge makes with the

2503–6/93/0166$06.00/0

center of the beam. This may seem confusing, but if one considers the exit port of the laser as a point source, the divergence is the angular spread of the beam as it leaves the exit port. Simple trigonometric relationships will prove that a divergence of 1 milliradian implies an increase of 1 mm in the diameter of the beam for every 1 m traveled from the exit port. This proof is left to the student (see Data Analysis, Question 1). Note that the comparison of directionality properties between a laser and a conventional light source is straightforward. A light bulb, for example, has no directionality at all. One may readily conclude that directionality is unique to lasers.

It is obvious that lasers are "bright" sources; however, the technical definition of this property is more complex. *Brightness* is defined in terms of lumens/m^2-steradian, where the lumen is an intensity unit based upon the response of the human eye to visible light. The lumen is included in a set of units known as photometric units. An alternative unit to the lumen is the watt (a radiometric unit), but the exact choice of unit is not critical, especially since the intent is to compare the laser to a conventional light source. It is important to note that brightness is defined in terms of a value per steradian. A steradian is a measure of the solid angle over which the light is emitted. Solid angle is measured over a sphere (i.e., in three dimensions) and the steradian is a unit analogous to the radian unit for a circle (measured in two dimensional space).*(2)* Perhaps the most interesting comparison, and one that clearly puts the brightness of a laser into perspective, is with the sun. The sun has a brightness of 1.5×10^5 lumens/m^2-steradian. It is crucial to note that the sun emits in all directions or over 4π steradians. A 1-milliwatt HeNe laser with a beam diameter of 1 mm (beam area of 7.9×10^{-3} cm^2) and a solid angle of 10^{-6} steradians (recall that the divergence is very small and approximately 1 milliradian) has a brightness of 2×10^7 lumens/m^2-ster or 100 times brighter than the sun.

The final property among the four originally cited is coherence. Lasers are highly coherent in both space and time. *Coherence* may the most difficult laser property to describe in a simple manner. In the ideal laser, all of the emitted photons appear to have the same phase and to have originated at the same point in space. Therefore, the measurement of the phase of the radiation, for example, at any one point in space and time allows for prediction at any other point. As with monochromaticity, real laser systems do not exhibit the ideal characteristic. Coherence is critical for many laser applications, but since the photochemical experiments described in this laboratory do not depend on the coherence of the dye laser employed, further discussion of this property shall not be pursued. The interested reader is referred to Reference 2 for additional details.

The photochemical experiment described herein utilizes the monochromaticity, directionality and brightness of the laser, *in combination*. One intent of the exercise is to compare the relative efficiencies of conventional and laser photochemistry. A second is to explore the mechanism of a particular photochemical reaction, the methylene blue sensitized photoreaction of quadricyclane with acetic acid, using laser radiation.

Theory

Lasers/Lamps. The normalized intensity distribution obtained from a typical commercial tungsten-halogen lamp is shown in Figure 1 for the visible and near

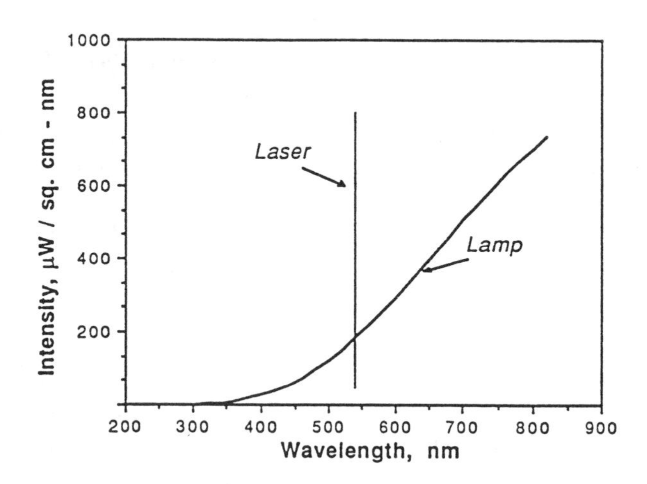

Figure 1. Typical intensity distribution from a tungsten-halogen source.*(4)* For comparison, the output of a dye laser with a bandwidth of 0.01 nm is superimposed on the Figure. The absolute value of intensity scale does not apply to the laser output.

infrared region of the spectrum. The spectral curve continues on to approximately 2.5 µm. As might be expected from the theory of blackbody radiation, the maximum in the distribution function is in the infrared region of the spectrum, with the result that a large fraction of the total output of the source is observed as "heat" and is not useful for the purposes of initiating photochemistry in the visible wavelength region. As a comparison, it is to be noted that the output of a 0.01 nm bandwidth dye laser appears to have no width on the scale used in Figure 1. For a tungsten-halogen lamp with a total output of 500 W (note that we have switched to the more common radiometric units), the output across the visible region of the spectrum would be at most 15% of the total. In this particular case, Figure 1 indicates that the output in the visible region, at a wavelength of 540 nm and for the sample located 6 cm away from the source, is of the order of 0.2×10^{-3} W cm^{-2} nm^{-1}. Note that for thermal sources, the intensity incident on the sample depends critically on the distance between the source and the sample. In fact, the intensity at the sample decreases as the inverse of the square of the separation distance. A 1 mJ pulsed dye laser (with a bandwidth of 0.05 nm and a beam diameter of 1 cm) at the same wavelength operated with a 15 Hz repetition rate and having a 20 ns pulse duration produces an *average* intensity of 0.4 W cm^{-2} nm^{-1}. The laser source is approximately 1000 times brighter than the lamp even though the actual irradiation period is only approximately 0.3 ms (20×10^{-9} s $\times$ 15 s^{-1}) per 1 s of laboratory time, i.e., the actual irradiation period for the laser is only 10^{-5}% of the total elapsed time. We might expect, therefore, that a laser-initiated photochemical experiment would proceed more rapidly than the corresponding lamp driven reaction.

The directionality of the sources is also relevant to the discussion of relative photochemical efficiency. The lamp emits in all directions, i.e., over 4π steradians. Therefore, the intensity calculated above must be divided by 4π to account for the divergence of the source, since only a small fraction of the emission is in the direction of the target solution. The dye laser has a divergence of only a few milliradians. The result is that the emission is typically over a solid angle of 10^{-4} steradians or less. For the purposes of this experiment, all of the laser radiation is directed toward the sample solution.

Finally, the importance of monochromaticity in photochemistry is obvious. This property of light was implicitly included when the brightness of the two sources was discussed above. The laser produces radiation only at the frequency needed to initiate the photochemistry. The lamp emits at all frequencies between the near ultraviolet and the mid infrared. Although in liquid phase photochemical reactions, the absorption bands are much broader than the bandwidth of the laser, the fact that the laser is monochromatic increases the brightness per unit wavelength and, hence, the efficiency of the photopumping, as has already been shown. In gas phase photochemistry the absorption bands, especially in molecular beam experiments, are sufficiently narrow so that the bandwidth of the laser permits selective excitation of a particular quantum state in the photoinitiation process.

Quadricyclane Photochemistry. Quadricyclane has been a target of a number of photochemical studies, beginning with its initial synthesis *(5)* in 1958. In particular, a large body of literature describes the photoisomerization of norbornadiene to quadricyclane, as well as photo- and thermal initiation of the reverse process. *(6-10)* Norbornadiene has been observed to isomerize to quadricyclane by means of direct uv irradiation at 222 nm, the lowest energy electronic absorption band for this molecule.*(6)* In studies reported in the early literature, the reverse isomerization reaction was thermally driven, since the lowest energy electronic absorption of quadricyclane lies in the vacuum uv and only the tail of this absorption band extends as far as 200 nm. Quadricyclane, but not norbornadiene, has been observed to quench the fluorescence of aromatic molecules.*(9)* The quenching is coincident with isomerization to norbornadiene. The photochemical literature describing these two molecules is sufficiently large so as to define the potential surfaces relevant to the isomerization reaction. These surfaces are represented schematically in Figure 2.*(10)* In general, the conclusions were that excitation to the first excited singlet

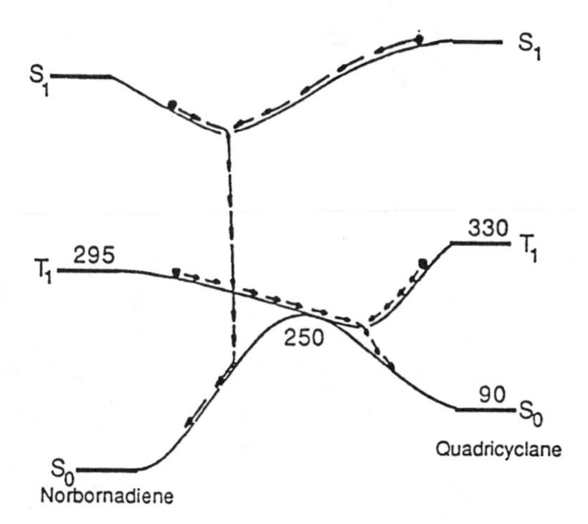

Figure 2. Schematic potential energy surfaces for the interconversion of norbornadiene and quadricyclane. Energies shown are in kJ/mol. Reproduced, with permission, from reference 10.

surface connecting the two isomers favors production of the norbornadiene isomer, while excitation to the lowest triplet surface leads to production of quadricyclane. Norbornadiene is the lower energy isomer, lying ~90 kJ mol^{-1} below quadricyclane. Both isomers have been employed in a wide range of photochemical synthetic studies. The volume of such literature is quite large and, since the concern in this report is with one particular photo-oxidation process, the reader is referred to the literature to satisfy any additional general interests. The particular reaction of interest here is the photo-oxidation of quadricyclane, sensitized by methylene blue in acetic acid. The mechanism of this process was originally reported to consist of solvent attack on an intermediate formed by the interaction of the quadricyclane with singlet oxygen during irradiation with a 500 W halogen lamp.*(11)* However, methylene blue may also lead to photochemical reactivity by an electron transfer mechanism, regardless of the presence of oxygen, and this was the focus of more recent re-examinations of the reaction in question.*(12-13)* The more recent studies have led researchers to postulate a mechanism involving this electron transfer process. In the absence of oxygen, visible light is absorbed by the methylene blue, creating an excited state dye molecule. This excited state is assumed to accept an electron from quadricyclane to produce the quadricyclane radical cation as shown in Scheme 1. Furthermore, the mechanism of the reaction from the quadricyclane

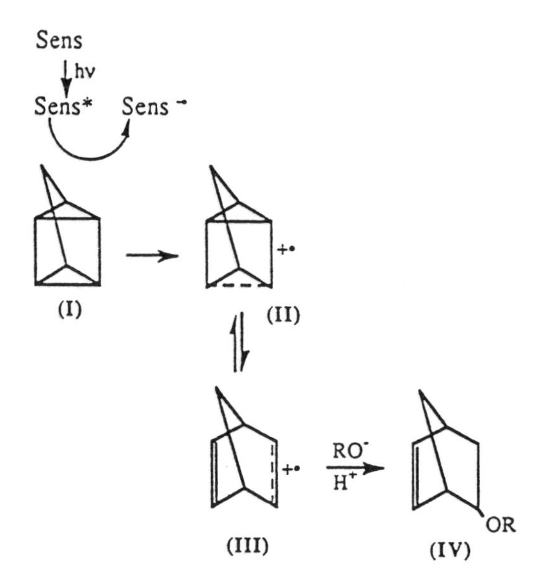

Scheme 1. Photochemical reaction of quadricyclane with acetic acid.

radical cation (**II**) to the final oxidation product (**IV**) is postulated to proceed through the formation of the norbornadienyl radical cation (**III**). The original experiments, in which a 500 W lamp was the photolysis source, resulted in the formation of a substantial quantity of norbornadiene. However, for the laser photochemical studies described in this experiment, the exclusive product was 5-norbornen-*exo*-2-yl acetate (**IV**), as shown in Scheme 1. The mechanistic step from the norbornadienyl cation (**III**) to the final oxidation product would require a nucleophile in the reaction solution. Reactivity has been observed *(11-13)* with both methanol and acetic acid as the nucleophile precursors. There is no valid reason to assume identical mechanisms for the direct uv initiated reaction, which also generates (**IV**) among a large number of products, and the sensitized process (indeed, they must be different!). However, the net effect of the sensitizer is that a reaction which would normally be driven by absorption in the less accessible ultraviolet region of the spectrum is shifted into the visible spectrum.

The intention of this laboratory exercise is two-fold: (1) to explore the properties of lasers and their use in photochemistry and (2) to explore the mechanism of the sensitized photo-oxidation reaction described above. The properties of the laser, relative to a lamp source, are observed by means of a comparison of photochemical efficiency and the calculations in the Data Analysis section of the exercise. In addition, the experiment provides an in-depth introduction to the field of liquid phase photochemistry and mechanistic studies.

In liquid phase studies, the use of added solutes as scavengers or quenchers is common. This laboratory proposes the use of several of these substances in order to choose between the $^1O_2$ sensitized process and the electron transfer mechanism. Complete details regarding the application of these additional solutes are given with the Methodology, but for example, the yield of oxidation product and/or the depletion of quadricyclane as well as the formation of any side product(s), in the presence or absence of oxygen is studied. Singlet excited state methylene blue may undergo intersystem crossing to the triplet state and sensitize the conversion of ground state oxygen into singlet oxygen in a spin-allowed reaction. If the yield of product and the rate of product formation are unchanged in the presence of added oxygen, and the standard conditions are defined as oxygen-free, then the $^1O_2$ sensitized reaction is not significant and additional support is obtained for the electron transfer mechanism. Added oxygen may, alternatively, cause a second mechanism to be competitive with the electron transfer process, but the observation of any change is significant mechanistic data.

Often, the removal of oxygen from a reaction mixture is incomplete rendering the standard conditions described above invalid. An alternative to the added oxygen experiment (and, for a thorough study, an adjunct) is the use of a singlet oxygen quencher and standard conditions in which the reaction solutions are air-saturated. One of the most effective and most commonly employed quenchers *(14)* is 1,4-diazabicyclo[2.2.2]octane known by the common name DABCO. The DABCO chemically removes any singlet oxygen generated by the methylene blue sensitized reaction and, if singlet oxygen is required in the reaction mechanism, should slow down or halt the photochemical reaction. As with all mechanistic studies, alternative interpretations are possible in the DABCO quenching experiments and the student needs to consider all of the collected data and all of the alternative reaction channels.

Finally, the nature of the reaction mechanism may also be directly probed by intercepting the electron transfer process. Quadricyclane has an reduction potential of -0.91 V. This reduction potential implies that quadricyclane is easily oxidized and, as an electron donor, is one of the most efficient saturated hydrocarbons. Since the reduction potential for the excited state of methylene blue is -0.25 V, the electron transfer process is highly favorable and would be expected to go to completion with a ΔG of -65 kJ mol^{-1}. One efficient method to test whether or not the electron transfer mechanism is valid is to introduce a co-solute with a lower reduction potential than quadricyclane. One such molecule, which would not absorb the radiation incident on the sample, is *trans*-stilbene. The reduction potential of the latter is −1.49 V, so that *trans*-stilbene would be the preferred electron donor and one would expect a decrease in the production of the 5-norbornen-*exo*-2-yl acetate (**IV**) product. The exact magnitude of the reduction would be concentration dependent.

Careful examination of the reaction mixture concentrations given in the Methodology section indicates that the acetic acid is present in large excess. Since, under the conditions described below, only a single photo-oxidation product and (no norbornadiene) is observed, the nucleophillic addition to the norbornadienyl cation must be considerably faster than neutralization of the cation and one may assume that it is formation of the quadricyclane radical cation that is the rate determining step. Under this assumption, measurement of the concentration of quadricyclane as a function of laser irradiation time, or the number of pulses, is sufficient to monitor the overall rate of the photo-oxidation reaction since production of the cation and subsequent conversion to the isomeric norbornadienyl cation inevitably lead to product. Conventional sampling/quenching techniques and NMR analysis of the reactant concentration as a function of time would yield the rate constant for this process. All such measurements should be carefully conducted under identical laser energy conditions, since the brightness of the source has already been shown to be a critical parameter in any photochemical study.

Safety

****Laser goggles must be worn at all times during this experiment.**** Always be aware of the laser beam path and the presence of others while adjusting the path. Keep in mind that reflections of the laser beam from polished surfaces are as dangerous as the beam itself.

Quadricyclane is a flammable reagent and glacial acetic acid which is caustic, irratating to the eyes, and inhalation causes irritation of the respiratory tract. Use the precautions appropriate, respectively, to all flammable or caustic reagents and heed general safe laboratory practices when handling these materials. Sample solutions should be mixed and stored in a fume hood. The laser dyes must be handled with care, since some are suspected carcinogens, and, in general, these dyes have not been studied well for adverse effects on humans. Chloroform ($CHCl_3$) is moderately toxic, and a suspected human carcinogen.

Methodology

The experiments contained in this section constitute the "basic" experiment and require two lab periods (approximately 8-10 hours). The time frame may be shortened by requiring only selected experiments from this group. All of the studies that follow are based on an identical mixture of reactants. The standard reaction mixture consists of 10 mL of glacial acetic acid, 0.3 g of quadricyclane and 10 mg of methylene blue. Variations in this solution, such as the addition of scavengers or quenchers, are noted as required below. A typical photochemical experiment involves the use of approximately 5 mL of this mixture; however, the exact volume is dependent upon the choice of sample cell.

Confirmation of Sensitizer Driven Chemistry. The first experiment is intended to demonstrate that the sensitizer is the driving force for the observed photochemistry, regardless of the irradiation source. This is most readily observed by means of the uv-vis spectrum of the standard sample solution in comparison with that of the same solution minus the dye. The presence of the large absorption band with a maximum at ~600 nm for the standard reaction mixture indicates that this solution will be photo-active in the presence of visible light. As has already been noted, quadricyclane has minimal absorption at wavelengths greater than 200nm, so that radiation between 300 and 700 nm will not be sufficient to initiate the oxidation reaction.

Laser vs. Lamp Efficiency. The next experiment is intended to compare the relative efficiencies of laser and incoherent sources in promoting the methylene blue sensitized photo-oxidation. Approximately 5 mL of solution are charged into a 10mL round bottom flask fitted with a rubber septum. The flask is suspended in a flowing water bath and the solution is deoxygenated. This is accomplished by flowing nitrogen or argon through the solution via a syringe needle for 5 minutes (Note that a second needle must be inserted through the septum to provide an exit for the gas). When deoxygenation is complete, the output of a 500 W halogen lamp located approximately 6 cm from the sample is directed at the solution from above. The irradiation is allowed to proceed for 2 h, during which time the initial laser-driven photochemistry experiments may be begun. Although a round bottom flask would be adequate for the laser experiments as well, reflections are minimized by the use of a typical 1 cm pathlength cuvette. The reaction cell is charged with the standard solution and deoxygenated as described above prior to insertion into the beam path of an unfocused dye laser. The exact wavelength of the dye is unimportant, but should lie within the absorption band of methylene blue. Coumarin 540A is an efficient laser dye for both nitrogen and excimer pumped dye lasers. The maximum in the emission curve for this dye lies at approximately 540 nm and the use of that wavelength is recommended. Pulse energies of less than the 1 mJ used in this description of the photochemistry are acceptable, provided the laser irradiation time is extended to compensate for the reduced brightness. The reaction may be allowed to proceed for 1 h at 15 Hz for a total of ~1 msec irradiation time. The two solutions undergo identical workup as outlined in the original literature.*(12-13)* Neutralization by addition of 1 M $NaHCO_3$ is followed by extraction with $CHCl_3$ and the extract is dried over $MgSO_4$. Analysis of the products of the two reactions is performed by 60 MHz ^1H NMR using TMS as a reference. The reaction product will exhibit a complicated spectrum at lowfield ($\delta = 1.2–1.8$), but characteristic resonances are readily observed [$\delta = 6.25$ (m,1H), $\delta = 5.98$ (m,1H), $\delta = 4.65$ (bs,2H) and $\delta = 2.87$ (d,2H)] in addition to the acetyl resonances. (This shorthand notation for NMR results is widely used. Any given band may be described by two symbols, (x, yH). In this notation, yH corresponds to the integrated number of protons and x is a description of the type of band, i.e., singlet (s), doublet (d), triplet (t), multiplet (m) and broad (b).) Optical or infrared spectroscopic techniques may be employed in place of or in addition to NMR; however, the NMR spectrum is less complicated than the optical spectrum and is recommended. If UV-vis spectroscopy is employed, one may utilize the absence of any absorption bands at wavelengths greater than 200 nm for quadricyclane and the presence of such bands in the wavelength region around 220 nm for substituted norbornadienes as the basis for analysis. In the infrared spectrum, quadricyclane exhibits absorption bands at 3050 cm^{-1} and 2990 cm^{-1} with equal intensity. The substituted norbornadiene shows absorption at 3040 cm^{-1} and 2990 cm^{-1}, with the latter considerably weaker than the former. In addition, a characteristic olefinic absorption occurs at approximately 725 cm^{-1}. Questions appearing in the Data Analysis section of this report may be used to explore the efficiency and properties of the two experiments.

Mechanistic and Kinetic Studies. The experiments described above are intended to be illustrative of the properties of laser light. Those that follow, using only the laser source, are intended to be used in an exploration of the reaction mechanism and to estimate the rate of depletion of the starting material. The kinetics of the laser driven reaction, as well as the relevant underlying assumptions, have already been discussed. The evaluation of the phenomenological rate constant for the photo-oxidation requires analysis of the reaction mixtures as a function of time. The absolute depletion rate of quadricyclane may be obtained by comparison of the NMR signal with a standard curve obtained from a set of appropriate concentrations. The concentrations should be developed with the consideration that irradiation with a 1 mJ pulse at 15 Hz for 1 h will produce approximately 50% conversion to the oxidation product. Concentration curve data are also essential to the calculation of quantum yields (see Data Analysis).

Mechanistic details may be extracted by means of added solutes which interact with some step in the photo-oxidation process. The Theory section describes the three possible photochemical experiments: (1) deliberate saturation of the reaction mixture with oxygen; (2) the addition of DABCO to the air-saturated and Ar- or N_2-saturated mixtures and (3) the addition of *trans*-stilbene as an alternative electron donor. For each of these experiments, the basic procedure is that outlined above. In the case of the alternative electron donor, the concentration should be comparable to that of quadricyclane; however, due to solubility considerations the stilbene should initially be dissolved in the quadricyclane with subsequent addition of the acetic acid. The DABCO concentration is less critical, but should be of the order of 10^{-3} M. In neither instance is it necessary to completely quench the particular reaction, but to merely modify the quantum yield sufficiently so as to be certain of the effect of the added solute.

Data Analysis

(1) The "Introduction" section of this experiment described various laser properties in comparison to conventional light sources. It is essential that any potential photochemist to be familiar with these quantities. The student is asked to confirm some of the values used in the text without derivation.

- For a laser beam with an initial diameter of 1 mm, show that a divergence of 1 milliradian produces an increase of 1 mm in diameter for each meter the beam travels.
- Show that a 1 mJ per pulse dye laser operated at 15 Hz yields an average intensity of 0.4 W $cm^{-2}nm^{-1}$ for a pulse width of 20 ns and a bandwidth of 0.05 nm.

- Show that a laser beam with an area of π mm^2 at a point 2 m away from the laser exit port subtends a solid angle of $\sim 10^{-6}$ steradians, if the divergence of the beam is 1 milliradian.

(2) The student should compare the absorption spectra (with and without the sensitizer) and comment on the differences observed, especially in terms of the requirements for photoinitiation of the chemistry. The difference in the efficiency of the two irradiation techniques (laser vs. tungsten-halogen lamp) is striking and the student should discuss the results in terms of the laser properties described in the introduction to the experiment.

(3) The student should discuss the mechanism of the reaction in light of the results of the quenching experiments described in the theory and methodology sections. How do these experiments confirm and/or refute possible mechanisms?

(4) The measurement of the concentration of quadricyclane, either from NMR or optical techniques, may be used to determine the overall rate of quadricyclane reaction. Since the laser chemistry yields a single product, this is the overall rate of the reaction. The usual analysis techniques may be employed to determine the phenomenological rate constant for quadricyclane depletion. The student should first show that the reaction between quadricyclane and acetic acid is expected to be pseudo-first order in quadricyclane, since the acetic acid is present in a fifty fold excess relative to the quadricyclane.

(5) For a 1 cm pathlength sample cell and 1 mJ/pulse incident radiation, all of the laser light incident on the sample is absorbed. This observation should be used to make an estimate of the quantum yield for the laser initiated reaction.

Appendix *(Additional Suggested Experiments)*

The experiment may also be modified to become the focus of an open-ended "project" laboratory. In that case, it may be utilized in one of the comprehensive or "super" laboratories described at the symposium.

Fluorescence Quenching. One possible extension of the photochemical kinetics described here involves the measurement of the fluorescence quenching of methylene blue by quadricyclane. Variation of the quencher concentration allows for the determination of Stern-Volmer kinetics and given the lifetime of the excited singlet state of the dye *(15)*, the quenching rate constant may be determined. This rate constant may be compared to both a theoretical value based on the Gibbs energy, ΔG^0, for the electron transfer step *(16)* and the phenomenological rate constant. These experiments would require one additional laboratory period and the availability of a fluorescence spectrometer. The same laser employed in the photochemical studies may be used, along with a filter to isolate the emission wavelength and a photomultiplier-oscilloscope detection system, to determine fluorescence lifetimes and yields. Alternatively, a conventional apparatus may be employed.

Other Oxidation Reactions. It is possible to vary the nucleophile in the photo-oxidation reaction. The literature *(11-13)* contains reports on the formation of methoxy alcohols if the reaction is studied in methanolic solutions rather than acetic acid. If the analogous reaction in methanol is to be attempted, the sample solution ratios should be maintained, with direct substitution of methanol for acetic acid. Although the experimental procedure is identical to that described above, the literature seems to indicate that a complex mixture of products is possible. The analysis procedure, in this case, would be considerably more complicated. The reader is referred to the original reports for complete details.*(11-13)*

Other Sensitizers. An extension of the electron transfer interception process described above would involve the substitution of a different dye that absorbs in the visible region of the spectrum, but exhibits a less favorable reduction potential. Rose Bengal has been employed in such studies and a significant literature on its application to photochemistry exists.*(17)* The application of Rose Bengal to these studies would require the student to re-examine the UV-vis spectrum of the reaction mixture to ensure that the laser wavelength is appropriate for the new system. Since only a small section of the basic experiment is repeated, this procedure involves approximately 1-2 h of additional laboratory time.

Synthesis. Synthesis of authentic samples of the products for either nucleophile is another potential extension the experiment described herein. If the aceylated product (**IV**) is available through such a synthetic extension of this lab, the rate of formation of product may be directly determined and the nature of the products directly confirmed.

Acknowledgments

The author acknowledges the financial support of the National Science Foundation ILI Program (CHE-8852168), the Pew Charitable Trusts (through the New England Consortium for Undergraduate Science Education), The Hewlett Foundation and Dartmouth College.

Literature Cited

1. Bronikowski M.J.; Simpson W.R.; Girard B.; Zare R.N. *J. Chem. Phys.* **1991**, *95*, 8647.
2. O'Shea D.C.; Callen W.R.; Rhodes W.T. *An Introduction to Lasers and their Applications*, Addison-Wesley Publishing Co., 1978.
3. Andrews D.L. *Lasers in Chemistry*, Springer-Verlag, 1986.
4. BelBruno J.J. Ph.D. Thesis Rutgers University (1980); see also specifications for various intensity tungsten-halogen lamps: Oriel Corporation Vol. II (1989).
5. Cristol S.J.; Snell R.L. *J. Amer. Chem. Soc.* **1958**, *80*, 1950.
6. Dauben W.G.; Cargill R.L. *Tetrahedron* **1961**, *15*, 197.
7. Hammond G.S.; Turro N.J.; Fischer A. *J. Amer. Chem. Soc.* **1961**, *83*, 4674.

8. Hammond G.S.; Wyatt P.; DeBoer C.D.; Turro N.J. *J. Amer. Chem. Soc.* **1964**, *86*, 2532.
9. Murov S.; Hammond G.S. *J. Phys. Chem.* **1968**, *72*, 3797.
10. Turro N.J.; Cherry W.R.; Mirbach M.F.; Mirbach M.J. *J. Amer. Chem. Soc.* **1977**, *99*,7388.
11. Kobayashi T.; Kodama M.; Ito S. *Tetrahedron Lett.* **1975**, *16*, 655.
12. Hatsui T.; Takeshita H. *Chem. Lett.* **1990**, 1253.
13. Hatsui T.; Takeshita H. *J. Photochem.* **1991**,*57*, 257.
14. Turro N.J. *Modern Molecular Photochemistry*, Benjamin-Cummings Publishing Co., 1978.
15. Nicol M.F.; Hara Y.; Wiget J.M.; Anton M. *J. Mol. Struct.*, **1978**, *47*, 371.
16. Rehm D.; Weller A. *Isr. J. Chem.* **1970**, *8*, 259.
17. Akasaka T.; Ando W. *J. Amer. Chem. Soc.* **1987**, *109*, 1260.

Hardware List

The conventional photochemical experiments may be performed using a standard 500 W Hanovia halogen lamp; however, a 300 W flood light has been shown to yield identical results since only the output in the visible region of the spectrum (500-650nm) is useful in the initiation of the chemical reaction. The laser experiments are readily accomplished with a simple nitrogen-pumped dye laser, although a Lumonics 861S/300EPD excimer-dye laser combination was used in development of this exercise. At least two manufacturers are competing to supply reasonably priced nitrogen laser systems for academic use. As noted in the introduction, bandwidth is not a critical parameter in these studies. Also required for analysis of the results is a simple ultraviolet spectrophotometer and a 60 MHz ^{1}H nuclear magnetic resonance spectrometer. Product analysis may also be accomplished via infrared or ultraviolet spectroscopic analysis using standard instrumentation, as noted in earlier sections of the exercise. However, IR spectroscopy records bands from a large number of internal modes of both reactants and products and the uv absorption bands are near the cutoff wavelength of most laboratory spectrometers. The NMR technique is preferred and has high sensitivity coupled with a simpler spectrum.

RECEIVED October 1, 1992

CHAPTER 11

Multiphoton Ionization Spectroscopy of Cesium Atoms

Charles S. Feigerle and Robert N. Compton

POTENTIAL HAZARDS: High-Pressure Systems, Lasers, Vacuums

Prior to the discovery of the laser, optical spectroscopy was viewed nearly exclusively as involving only single photon absorption and emission. The development of the modern pulsed laser, which continues to be pushed to ever higher peak powers and shorter pulse lengths, provided the tool for extending spectroscopy to the multiple photon regime. The growth of one multiple photon technique, multiphoton ionization (MPI), has been explosive. In the 25 years since results of the first multiphoton ionization experiment were reported *(1)*, many of the elements *(2-3)* and a large number of stable and transient molecules *(4)* have now been studied by multiphoton ionization spectroscopy. One reason for the rapid growth in popularity of MPI is that the products are charged particles which can be mass analyzed, energy analyzed, and detected with unit efficiency. Furthermore, one may choose the laser wavelength so that ionization proceeds in two or more steps: first by excitation to a real bound intermediate state, and followed by ionization from this bound excited state. MPI then has the sensitivity afforded by charged particle detection and the structure specificity of traditional bound-bound spectroscopy as well. Such a process is refered to as resonance enhanced multiphoton ionization (REMPI).

It has become customary to assign a particular REMPI scheme by the number of photons required to reach a resonance (m) together with the number of photons required to ionize (n) the resonantly excited state as a (m + n) REMPI. A (2 + 1) REMPI process is contrasted with direct one-photon ionization and non-resonant n-photon MPI in Figure 1. It is sometimes useful to perform one of the steps with a different wavelength and in that case the distinct photon step is represented by a prime. For example, if in Figure 1, one laser were used to excite the atom and a second color used to ionize the excited intermediate, the process would be a (2 + 1′) ionization. With sufficient colors available, it is possible to separate the multiphoton ionization process into a sequence of one-photon bound-bound transitions coupled with a final bound-free ionization step. Since all photon steps in this process are resonant, ionization by this mechanism is refered to as Resonance Ionization Spectroscopy or RIS.

The sensitivity and spectral selectivity of REMPI and RIS have made these techniques suitable for the study of a wide variety of chemically important problems, including: discovery and characterization of excited states of molecules *(5-6)*, photofragmentation and photoionization from excited states *(7)*, measurement of the rotational and vibrational product state distributions from chemical reactions *(8)*, and investigation of the dynamics of gas-surface interactions *(9)*. Since MPI experiments are usually performed using pulsed lasers, mass and energy analysis of

2503–6/93/0178$06.00/0

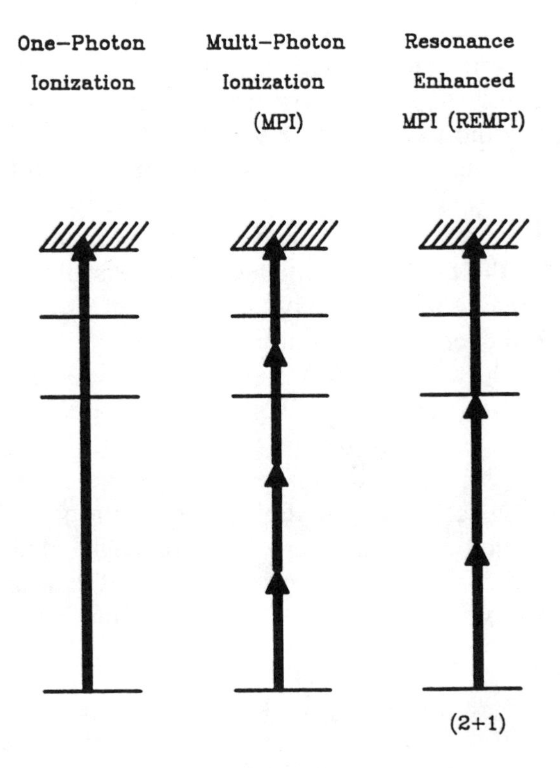

Figure 1. Schematic representation of three different mechanisms for ionizing an atom or molecule are shown: one-photon ionization, 4-photon ionization, and 3-photon resonance enhanced ionization. The hatched lines represent the continuum, the region above the ionization potential for formation of a positive ion plus an electron. The length of the arrows are proportional to the photon energy used.

the products can be readily obtained using time-of-flight techniques. Mass analysis is particularly useful for identifying when photofragmentation is taking place in addition to ionization *(7)*. Electron energy analysis is useful for characterizing the dynamics of the MPI process and to identify the internal energy content of the final ion products *(4)*. More recently, MPI has been performed on weakly bound complexes and clusters formed in free jet expansions, allowing the study of intermolecular interactions and condensation *(4)*.

Maria Göppert-Mayer first introduced *(10)* the concept of a two-photon transition in 1931. Since that time, the theory of MPI has been developed to a high degree of sophistication. It has even been experimentally demonstrated that the sensitivity of REMPI (or RIS, in particular) is so high as to be able to detect a single atom *(2,11)*. Yet, in spite of this impressive list of accomplishments, MPI is seldom taught at an undergraduate level. Rarely are undergraduate students given the opportunity to actually perform a multiphoton ionization experiment. In this article we present an introduction to multiphoton ionization spectroscopy and describe an experiment on resonance enhanced multiphoton ionization of cesium (Cs). This experiment provides a modern alternative to the discharge spectrum of hydrogen, an experiment frequently used in physical chemistry laboratory. The spectrum obtained retains nearly the same simplicity as that for the hydrogen atom, yet has the added advantage of demonstrating spin-orbit coupling and quantum defects in atomic stucture. The MPI experiment is performed using a nitrogen-pumped tunable dye laser, an optical heat pipe, and a gated integrator for signal averaging. With the exception of the heat pipe, which can be constructed from commercially available parts or purchased as a unit, the laser and electronics are available in many undergraduate physical chemistry laboratories.

The spectrum which is recorded is easily assigned as involving intermediate excitation to ns and nd excited states of Cs. For low values of n it is also possible to resolve the fine-structure splittings of the nd levels. Analysis of the data, by fitting the line positions to a Rydberg formula, provides an extremely accurate determination of the ionization potential of the cesium atom. The experiment also provides a probe of multielectron interactions through comparison of quantum defects for the s- and d-electron Rydberg series, quantities which are also obtained from a fit of the line positions measured in the spectrum.

Theory

Multiphoton Excitation The transition moment matrix element

$$\langle f | \mu \cdot \mathbf{E} | i \rangle \tag{1}$$

connecting an initial state, i, with a final state, f, plays a central role in the theory of single-photon dipole-allowed optical transitions. Here, μ represents the dipole moment operator and **E** the electric field vector for linearly polarized light. Analysis of the properties of equation 1 provides the basis for establishing selection rules, for deriving Franck-Condon factors, and in general provides the theoretical basis of traditional optical spectroscopy. A similar entity, the multiphoton transition moment, can be derived for n-photon absorption. In general, all n-photon transition moments (n = 1, 2, ...) can be obtained from a time-dependent perturbation theory analysis of the probability of finding an atom or molecule in an excited state after interaction with an electromagnetic field *(12-13)*. The simplest of the higher order moments is that for a two-photon transition. Since the multiphoton ionization of Cs experiment we describe in this paper involves two-photon excitation to a resonant intermediate state, we will focus our discussion of the theory of multiphoton absorption to the two-photon case.

In the perturbation theory analysis, single-photon absorption is a first order process where the mechanism which populates the excited state is a direct one-photon transition from some initially populated level. However, other pathways to the excited state can be envisioned. For example, a two-photon absorption pathway is contrasted with that for single photon absorption in Figure 2. At first glance, the two-photon pathway appears to be no different than sequential one-photon absorption to the intermediate level, v, followed by a second one-photon transition from the intermediate level to the final excited state. On the contrary, it is well

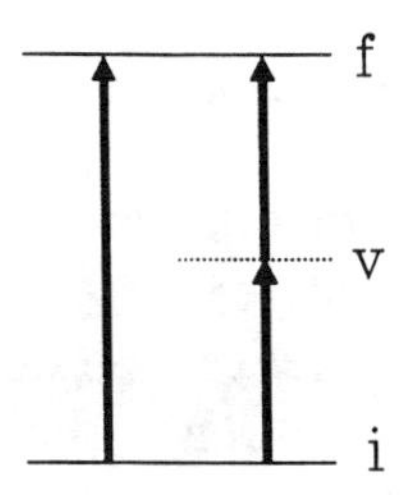

Figure 2. Energy level schematic which contrasts a one-photon and two-photon transition between the states $f \leftarrow i$. The two photon transition proceeds through the non-stationary state virtual level, v.

established that there need not be any bound stationary state at the position of level, v, for two-photon absorption to the excited state to occur! Rather, the "virtual" level, v, is a non-stationary state created by interaction of the light wave with the atom or molecule and consisting of contributions from all possible bound and continuum states according to their transition moment amplitudes and detuning from resonance.

This expansion state description of the level, v, is embodied in the second order time dependent perturbation theory result for the two-photon transition moment,

$$K_{f \leftarrow i} = \sum_{j} \{ \langle f | \mu \cdot \mathbf{E} | j \rangle \langle j | \mu \cdot \mathbf{E} | i \rangle \, / \, (\omega_{ji} - \omega + i\Gamma/2) \} \tag{2}$$

Here, $\hbar\omega_{ji}$ is the energy difference between quantum states i and j, ω is the photon frequency, Γ is the combined width of states i and j, and the summation is over all bound and continuum states, j, of the system. The unique properties of the two-photon transition can be predicted from analysis of equation 2 in which each term in the sum is a product of two single-photon transition moments and the states which these matrix elements couple are restricted by the usual one-photon dipole selection rules. The overall selection rule for two-photon absorption from $f \leftarrow i$ is then the result of one-photon selection rules imposed on the individual transitions to and from the expansion states, j. For two-photon excitation of ground state Cs, with its valence electron in a 6s orbital, the only expansion states allowable by the LaPorte rule, $\Delta \ell = \pm 1$, are those formed from excitation of the electron to an np orbital. A second application of this parity restriction yields ns and nd orbitals as the only possibilities for the final state, f, within the dipole excitation approximation. An equivalent "LaPorte" rule for two-photon transitions is then found to be $\Delta \ell = 0, \pm 2$. As will be shown in a later section, the multiphoton ionization spectrum of Cs is dominated by ns and nd Rydberg series resonances when using 600 - 700 nm excitation.

The above discussion of the selection rules for two-photon excitation of ground state Cs has been based upon treating Cs as a "hydrogenlike" (one-electron) atom. This approximation can be used to understand the major features of the optical spectrum of Cs and the other alkalis because these atoms consist of a single electron outside of a closed shell in their ground state electron configuration. However, the majority of atoms have more than one electron in their valence shell leading to an increased importance of electron-electron correlation and angular momentum coupling in determining their energy levels and optical spectra. The Laporte rule we invoked for analysis of the alkalis is a specific case of the more general requirement that parity change in a dipole allowed transition. Parity specifies whether the wavefunction is even or odd when the coordinates for each particle in the atom are replaced by their negative, i.e., $(x,y,z) \rightarrow (-x,-y,-z)$. For a state derived from a single configuration, its parity can be determined from whether $\sum \ell_i$ is odd or even, where odd parity states are denoted by a naught (∘) in the state symbol. The sum is over the angular momenta of all the electrons in the atom. Even when no distinct configuration exists for a state, it can still be characterized as even or odd parity. Since parity remains a quantized property of multielectron atoms, both the one and two-photon versions of this $\Delta \ell$ restriction will still apply to multielectron atoms as long as the states involved are derived from predominantly one electron configurations. Restrictions on other quantum numbers also exist. What quantum numbers are necessary to further specify the states involved in the transition will depend on the strength of the fine and hyperfine interactions in the atom and the ability of the spectroscopic technique to resolve them. The laser used in this experiment can resolve fine structure in some of the levels of Cs, but is unable to resolve their hyperfine structure. The quantum numbers of interest are then the total electron orbital angular momentum, L, the total electron spin, S, and the total electron angular momentum, J. The selection rules on L, S, and J for one and two-photon dipole excitation *(14)*, in the limit that the states are described by Russell-Saunders (LS) coupling, are contrasted in Table 1. The new feature produced for the two-photon case is the possibility of change of angular momentum by two units. The $nd\ ^2D_{3/2}$ or $nd\ ^2D_{5/2} \leftarrow 6s\ ^2S_{1/2}$ absorptions which are prominent in both the two photon absorption and the (2 + 1) MPI spectrum of Cs are good examples of such transitions.

MPI Signal Estimate While the transition moment analysis is well suited for predicting the selection rules and approximate strengths for two-photon transitions,

Table 1. Selection Rules for Russell-Saunders (LS) Coupled States of an Atom

	ΔL	ΔS	ΔJ
one-photon	0, ±1 (0 ↮ 0)	0	0, ±1 (0 ↮ 0)
two-photon[a, b]	0, ±1, ±2 (0 ↮ 1)	0	0, ±1, ±2 (0 ↮ 1)

[a] Slight differences exist in the selection rules depending on whether the two photons are identical (same polarization and frequency) or not (see Ref. 14). This table refers to identical photons.
[b] That $\Delta L = \pm 1$ and $\Delta J = \pm 1$ are allowed initially appears to violate the two-photon Laporte rule, $\Delta \ell = 0, \pm 2$. However, angular momenta add vectorially and the Laporte rule specifies the change in angular momentum for a single electron which must ultimately be coupled to the angular momenta of the remaining electrons in the atom. Of course, for two photon excitation of ground state Cs, the total L for the remaining electrons is zero and the only possibilities again are $\Delta L = 0, \pm 2$. This is accounted for by the restriction (0 ↮ 1).

a simpler approach can provide insight into the practical requirements for creating a two-photon excitation. One way to address the problem is to ask the question, "what is the probability that two photons will collide with an atom within the lifetime of the virtual state"? We can estimate this probability, P, as the product of the photon flux, F (photons cm^{-2} sec^{-1}), times the virtual state lifetime, τ (sec), times the effective cross sectional area of the atom, σ (cm^2 $photon^{-1}$), to give

$$P = F \tau \sigma \tag{3}$$

This equation differs from the equation for the N-photon transition rate,

$$R = \sigma_N F^N \tag{4}$$

in that the rate of an N-photon transition per atom, R(sec^{-1}) will depend on the laser flux to the Nth power times the generalized N-photon cross section, σ_N (cm^{2N} sec^{N-1}). This difference in laser power dependance in equations 3 and 4 arises because in equation 3 we assume that the virtual state is already prepared at zero time.

To calculate the photon flux requires specification of the power output and focusing conditions of the laser light. The nitrogen-pumped dye laser used in these experiments produces 400 psec pulses with greater than 100 μJ pulse energy. For 650 nm light, this yields around 8×10^{23} photons sec^{-1} during the pulse. To obtain the photon flux we divide the number of photons sec^{-1} in the pulse by the cross sectional area of the laser beam. The laser beam is focused to a smaller cross sectional area than is directly available from the laser in order to increase the flux and subsequently the two-photon absorption rate. For a focused beam, the flux is highest at the beam waist, where the diameter, d, can be approximated as the product of the inherent divergence, θ, of the multimode laser beam times the focal length, f, of the lens

$$d = f \theta \tag{5}$$

With a 0.5 milliradian divergence and f = 75 mm, the cross sectional area of the beam at the waist is about 44×10^{-6} cm^2, yielding a laser flux in the experiment of 1.9×10^{28} photons sec^{-1} cm^{-2}.

We obtain an estimate of the lifetime, τ, of the virtual state from the energy-time formulation of the Heisenberg uncertainty principle,

$$\Delta E \; \Delta t \sim \hbar \tag{6}$$

The energy deficit, ΔE, is the detuning of the virtual state from some one-photon allowed stationary state and we take the uncertainty in time, Δt, to be the "lifetime" of the intermediate (virtual) state. For example, if we take the energy deficit to be about a 0.5 eV or $\Delta E \approx 4000$ cm^{-1}, then the lifetime of the intermediate resonance obtained from equation 6 is $\tau = 1.3 \times 10^{-15}$ sec. Finally, taking 0.1 nm as a typical radius of an atom, the cross section can be estimated as $\sigma = 3 \times 10^{-16}$ cm^2. Combining our estimates for the flux, cross section, and lifetime then gives a probability in equation 3 of 7×10^{-3} that an atom in the laser volume would absorb two photons. (We can extend our analysis to estimate the number of atoms that would be excited within the laser pulse in a typical experiment. If we take a laser volume of ~ 10^{-6} cm^3 and a number density of ~ 3.5×10^{16} atoms/cm^3 (1 Torr), then using the excitation probability of 7×10^{-3} obtained above yields approximately 2×10^8 excited atoms per laser shot.) One concludes that the nitrogen-pumped dye laser has the ability of producing high enough photon flux for frequent "simultaneous" interaction of two photons with an atom or molecule.

Cs Multiphoton Ionization A large number of multiphoton ionization resonances, due to two-photon absorption of ground state Cs to excited *n*s and *n*d Rydberg states followed by ionization from these excited states, are accessible using the laser dye DCM. A representative illustration of the two-photon excitation to the 10s and 9d levels of Cs is shown in the energy level schematic given in Figure 3. Two mechanisms contribute to ionization of these excited states. Ionization can occur through absorption of a third photon giving rise to an overall (2 + 1) ionization process (see Figure 1). As explained earlier, the 2 denotes the number of photons required for the resonant absorption step and the 1 gives the number of photons involved in subsequent ionization. This mechanism is believed to be the dominant source of ionization for the *n*d excited states. An alternative mechanism provides the majority of ion signal from the *n*s states. It is well known that the cross section for photoionization of the excited *n*s states is considerably smaller than that of the *n*d states in this wavelength region. Ions are instead produced from excited *n*s states through ionizing collisions of the type

$$Cs(6s) + Cs(ns) \rightarrow Cs_2^+ + e^- . \quad \text{Associative ionization}$$

$$\rightarrow Cs^+ + Cs + e^- . \quad \text{Collisional ionization}$$

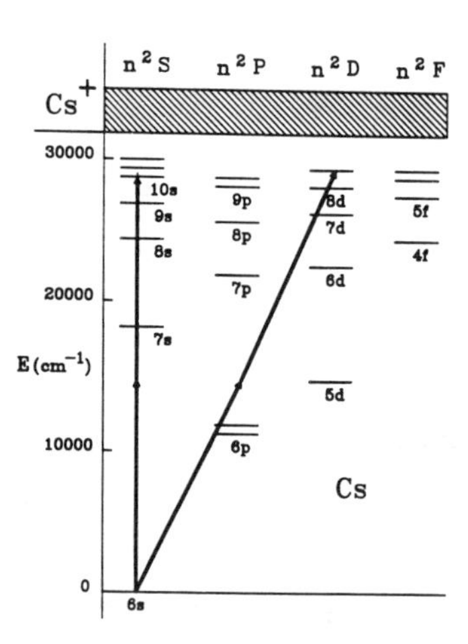

Figure 3. Energy level diagram for Cs showing the s, p, d, and f series. Two photon excitation of the 6s ground state to ns and nd excited states is dipole allowed. These excited states are either collisionally or photoionized to produce a MPI spectrum.

Associative ionization is important at low values of n whereas collisional ionization dominates at high principal quantum numbers. Of course, such collisions can also contribute to the *n*d signal as well. In general, ionization via three-photon non-resonant multiphoton ionization is orders of magnitude lower in intensity under the conditions of the experiment. Therefore, tuning the laser across one of the ns or nd two-photon excitations yields a sharp ionization resonance on top of a weak structureless background.

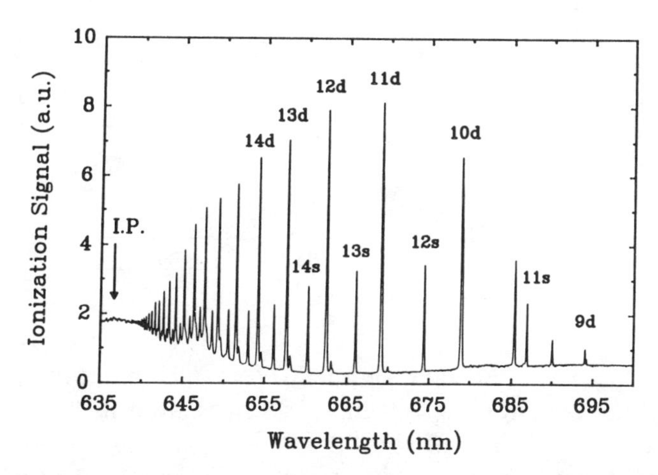

Figure 4. Multiphoton ionization spectrum of Cs recorded using the heat pipe. The spectrum is seen to consist of a d and s Rydberg series which converge at the ionization potential of Cs.

An example of a multiphoton ionization spectrum of Cs, obtained by scanning a dye laser from 700 to 635 nm, is shown in Figure 4. Two Rydberg sequences are observed in the spectrum. The more intense sequence is due to ionizations via two photon excitation of the 6s electron to nd orbitals. In all, twenty-one d-resonance lines are observed spanning $n = 9 - 29$. Each of these lines contain the unresolved $nd\ ^2D_{3/2,5/2}$ fine structure levels which are actually resolved for $n = 9$. Interspaced between these intense lines are a weaker set of resonances belonging to ionization via two photon excitation of the 6s electron to higher ns orbits. Fifteen s-resonances are observed spanning $n = 11 - 25$. Assignment of these lines is very straightforward, simply requiring comparison of twice the photon energy with the known excited state energy levels for Cs *(15)*. Furthermore, these assignments give direct evidence for the resonant step being two-photon absorption, as excitation to the s- and d-levels is parity forbidden in one-photon excitation by the Laporte rule ($\Delta \ell = \pm 1$).

Closer inspection of the 697 to 677 nm region of the spectrum shown in Figure 5 reveals both finer detail in some of the resonances and additional lines due to new ionization pathways. The spin-orbit splitting of the 9d ^2D term is easily observed and measurement of the spin-orbit splitting is even possible. In contrast to the regions between other pairs of d-Rydberg resonances, where only a single s-resonance is observed, the region between the 9d and 10d shows 3 resonance features in addition to "normal" ionization through the 11s. As indicated in Figure 5, two of these correspond to one-photon excitation of the Cs 6s to the 5d $^2D_{3/2}$ and $^2D_{5/2}$ excited states followed by two-photon ionization. While other members of the d-Rydberg series were observed via a (2 + 1) ionization process, these resonances occur by (1 + 2) ionization. Resonance excitation to the 5d $^2D_{3/2,5/2}$ states proceeds in one photon through a quadrupole transition. Quadrupole transitions occur through interaction of the electric field of the light with the quadupole, rather than the dipole, moment of the charge distribution and have selection rules that

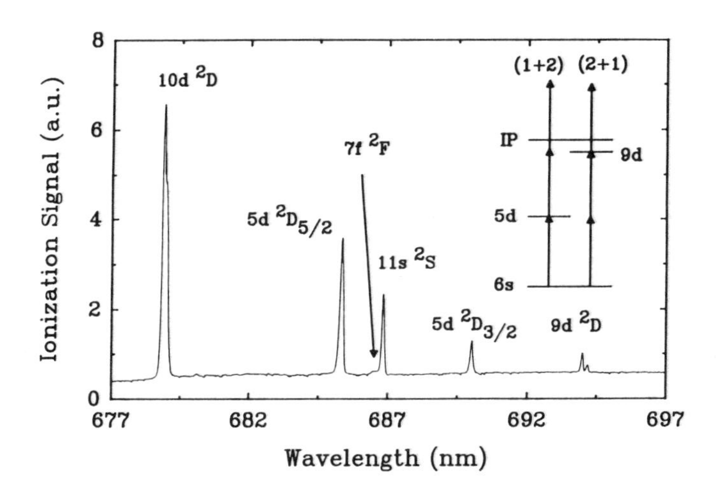

Figure 5. Expanded view of the 677 to 697 nm portion of the multiphoton ionization spectrum of Cs presented in Figure 4. The fine structure splitting of the 9d ^{2}D term is seen to be resolved. (1+2) MPI through the 5d ^2D_J levels are also observed where the excitation step occurs via a quadrupole transition.

allow changes of angular momentum of up to ±2. The transition probability for quadrupolar excitation is typically 10^{-4} to 10^{-5} times smaller than for dipolar excitation and hence quadrupole transitions are usually not prominent in absorption spectra. In the present case, the (2 + 1) and (1 + 2) ionizations through the d-level intermediates are of comparable intensity, demonstrating how modern high powered pulsed lasers provide sufficient sensitivity to observe dipole-forbidden transitions. This 697 to 677 nm region of the spectrum is also a beautiful illustration of the *n* dependence of the spin-orbit splitting. The spin-orbit splitting of the 5d ^{2}D term is here seen to be large compared to the splittings observed for the 9d (keep in mind that the wavelength scale for the 5d excitation represents one rather than two photon excitation of the 9d). Finally, the weakest peak in this region is due to ionization via excitation to the 7f ^{2}F term, made possible by the nearness of the resonant one-photon quadrupole transition to the 5d ^{2}D from which a second photon can produce a dipole transition to bring the Cs atom to the 7f excited state.

The astute observer will also note in Figure 4 the appearance of small resonances at slightly lower energy (longer wavelength) than the 11d through 16d states. These correspond to two photon excitation to *n*p states (n = 12 – 17) which is parity forbidden in a two-photon transition. Such transitions have been reported for sodium *(16)* at high density but are not totally understood. Their occurance may be due to electric field mixing of opposite parity states *(17)*. The electric field can be from the bias electrode or from the presence of electron-ion pairs from MPI *(18)*.

Rydberg series have long been one of the most accurate ways of determining an ionization potential. This is achieved by fitting the transition energies of the Rydberg series to the formula

$$E_{n,\ell} = E_\infty - R_{Cs} / (n - \delta_\ell)^2 \tag{7}$$

using a non-linear least squares routine. Both s- and d-Rydberg series are observed in the spectrum presented in Figure 4 and each are analyzed separately. In this analysis $E_{n,\ell}$ is the energy of the s(ℓ=0)- or d(ℓ=2)-level with principal quantum number n, E_∞ is the ionization potential of Cs, R_{Cs} is the Rydberg constant for Cs (109,736.862 cm^{-1} for ^{133}Cs), and δ_ℓ is the angular momentum dependent quantum defect. Both E_∞ and δ_ℓ are derived from a fit of the observed spectrum to equation 7. The success of this formula in describing the energies of Rydberg levels is a result of Rydberg states of alkalis being approximately one electron systems. Thus the formula for the levels closely resembles the energy equation for the hydrogen atom. Differences in Rydberg states of Cs and states of the hydrogen atom are "lumped" into the angular momentum dependent quantum defect δ_ℓ. The quantum defect δ_ℓ, can be thought of as a dimensionless number that converts the quantum number n into an effective quantum number n^* ($= n - \delta_\ell$) which makes the energy level "hydrogenic". The magnitude of the quantum defect represents the degree of departure from a pure hydrogenic orbital. Those orbitals which penetrate the core to a larger degree will experience a greater interaction with the core electrons resulting in a larger quantum defect. Analysis of the s- and d-Rydberg series thus provides a method for probing how orbitals of differing angular momentum sample the ion core.

Methodology

The apparatus used in the experiment consists of a dye laser (PRA Laser #LN107) pumped by a 2 Megawatt pulsed nitrogen laser (PRA Laser #LN1000), optics to direct and focus the laser beam, a Cs heat pipe, a vacuum (~10^{-3} Torr) and gas handling system for cell preparation, and ionization detection electronics. A block diagram depicting this experimental arrangement is shown in Figure 6. The

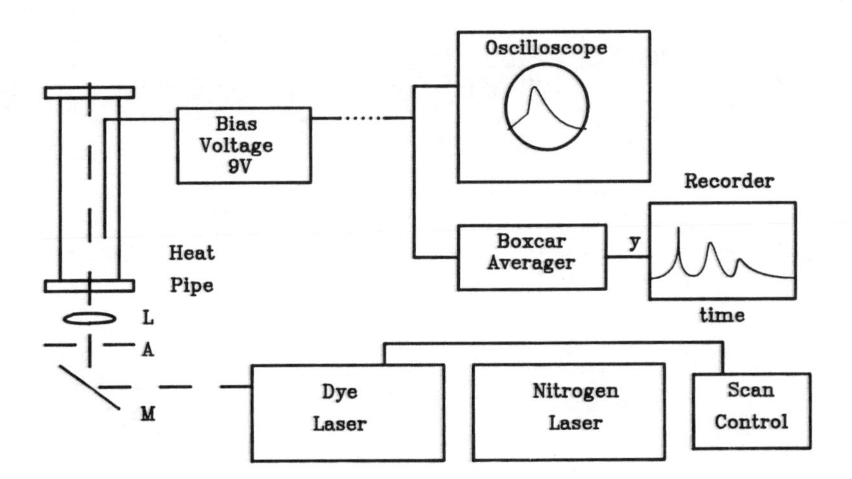

Figure 6. Block diagram of the apparatus used to perform multiphoton ionization experiments. The dashed lines represent the laser beam path and solid lines the electrical connections. L = lens; A = aperture; and M = mirror.

experiment can be conveniently laid out on a standard 4′ × 8′ laser table with room to spare for additional experimental set-ups. The nitrogen laser is simple and relatively safe to operate and students can be quickly trained in its proper use. On the other hand, tuning the optical elements in the dye laser requires more expertise than is generally available from students attempting to use the dye laser for the first time. This does not impose any significant restriction since a single dye, DCM, covers the entire wavelength range of the Cs MPI spectrum they will record and students are therefore not required to change dyes or adjust the optics in the dye laser cavity to perform the experiment. Dye changes and subsequent adjustments to the dye-laser can be done by the instructor or an assistant prior to the laboratory period. The lasing wavelength of the dye-laser is stepper motor controlled through a digital programmer which is easily operated by students for recording MPI spectra.

In our experimental arrangement, the output of the dye-laser is directed into the Cs heat pipe using a flat aluminum mirror. The mirror is aligned so that the laser beam passes cleanly through the input and output windows of the heat pipe. An iris is inserted into the beam path to spatially filter residual amplified spontaneous emission (ASE) from the laser output and the beam is focused into the center of the heat pipe using a 10 cm focal length plano-convex quartz lens. This procedure is usually sufficient to observe an MPI signal on an oscilloscope when the dye laser is tuned to one of the resonances. Further improvements in the strength of the signal can be obtained from adjusting both the dye laser and external optics once the initial MPI signal is observed. When the laser and optics are optimized, peak heights of > 100 mV are easily obtained for ionization through the d-levels. These signals are typically more than 20 times greater than those that are observed when the laser is tuned off resonance.

Since the vapor pressure of Cs is only 1×10^{-6} Torr at room temperature, a heat pipe *(19-20)* is used to heat the sample and increase the vapor pressure to a few Torr. The term "heat pipe" refers to any device which transfers heat by convection of an active or working medium. The working medium in this case is Cs and the latent heat of vaporization is transferred by evaporating liquid Cs in the center of the heat pipe and condensing its vapor in the cooled region of the tube. Figure 7 gives a simplified diagram of a working heat pipe. The pipe is configured with windows so that a laser beam can enter and exit the heat pipe, passing through the dense metal vapor. A rare gas such as Ar is used to confine the metal vapor and prevent fogging of the windows. In a properly working heat pipe, the Cs vapor is confined to the central heated region by collisions with buffer gas at the boundary of the heated zone. Cesium which is condensed at this boundary is recirculated back to the heated region by capillary action using a tube made of stainless steel mesh which acts as a wick. Definition of the heated region is further aided by external water cooling at the ends of the wick. A wire for collecting the charge produced by ionization is situated inside of the tube but placed off axis so as to not block the laser beam from passing through the heat pipe. The wire is connected to a BNC type electrical feedthrough to provide easy monitoring of the ionization current. An external 9V battery provides the biasing to draw either electrons or positive ions formed from ionization to the collection wire.

Assembly and filling of the heat pipe is done by the instructor, making sure that the pipe is first leak free and then taking care when filling to minimize the exposure of the Cs to air. This can be achieved by placing a sealed ampule of Cs (2 g is sufficient for a small heat pipe) into the heat pipe, and breaking the ampule

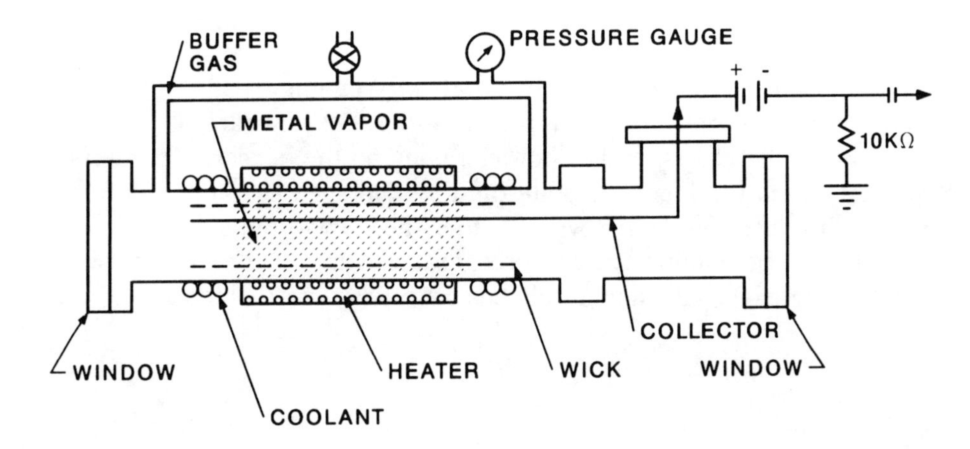

Figure 7. Diagram of the heat pipe used in the Cs multiphoton ionization experiment. The laser beam passes through the length of the pipe and is focused in the metal vapor to produce ionization. The ions or electrons that are produced are drawn to the collector where their current pulse can be measured.

while flowing Ar gas out the open end. The heat pipe is then sealed, evacuated to remove any air contamination, refilled to 3 Torr of Ar, and heated to 200 ^{o}C to distribute the Cs uniformly over the wick. After allowing the heat pipe to cool, the window at one end of the heat pipe can be removed while again flowing Ar to minimize air exposure and the glass pieces from the ampule can be removed. When evacuated and refilled to the desired pressure of Ar, the heat pipe is now ready for use. A leak tight heat pipe can be operated for over a year without cleaning and replacing the Cs.

For carrying out the MPI measurements, the heat pipe is typically maintained at a fixed temperature ($\pm 1^{o}$) between 200 – 300 ^{o} C with a Ar buffer gas pressure of ~3.5 Torr. The MPI signals which are obtained in the Cs experiment are large enough (>100 mV) to observe directly on a 1 MΩ impedence input of a 50 MHz oscilloscope without further amplification. An illustration of such a signal is given in Figure 6. Recording the intensity of this signal as a function of wavelength then yields a multiphoton ionization spectrum. The pulse height of the MPI signal varies from one laser shot to another by typically 25–50% due to smaller variations in laser pulse intensity. The MPI signal is particularly sensitive to these variations since the ionization rate depends on the laser power raised to the number of photons required to ionize the atom or molecule (see equation 4). The effect of these intensity variations is to introduce noise into the spectrum. This noise can be reduced significantly by integrating the signal over a narrow window near the peak and averaging the resultant over a few (3 – 10) shots of the laser using a gated integrator and boxcar averager. Boxcar averagers are commercially available or can be constructed out of easily obtainable electronic parts *(21)*. A Stanford Research Systems Model SR250 boxcar averager was employed in our laboratory. Signal can also be recorded using an electrometer in order to measure the "average" current created by the pulsed laser or using a lock-in amplifier tuned to the laser pulse repetition rate. The output of the boxcar averager or other signal recording device, is then directed to the y-input of a x–y recorder and/or digitized for storage in a computer.

An initial survey scan of the spectrum spanning 635 – 700 nm can be obtained in about 10 min using a nitrogen laser repetition rate of 5–10 Hz, a dye-laser scan speed of 1.0 Å/s, and 3 shot averaging on the boxcar. To obtain a better quality spectrum at the highest resolution, the scan speed should be lowered to 0.2 Å/s and 10 shot averaging employed. Under these conditions, a 200 Å segment of the spectrum can be obtained at 0.4 Å resolution in about 17 min. The starting and ending wavelengths for all scans should be accurately recorded for use in subsequent analysis.

Safety

- Laser goggles must be worn at all times when the nitrogen or dye-laser are in operation. The focussed laser beam has a higher energy density, thus extreme care must be exercised for both the beam and its reflections.
- Laser dyes should be treated as suspected carcinogens. Safety goggles and gloves should be worn during preparation of dye solutions and filling of dye cells.
- Cesium is extremely reactive with water and many other solvents, producing a flammable and explosive reaction. Protect from exposure to water, alcohols, and other oxygen containing solvents.

Data Analysis

1. Prepare a table with the following columns: λ(air), Vacuum Correction, λ(vac), $\Delta\lambda$(calibration), λ(measured), $E_{n,\ell}$(measured), $E_{n,\ell}$(true), ΔE, Assignment. Instructions for filling in the columns are given in the following.
2. Record the peak positions of all peaks observed in the spectrum in the column titled, λ(air).
3. For the most accurate determination of the ionization potential and quantum defects from the data, two corrections should be made to the raw spectral line lists prior to fitting to the Rydberg formula. The wavelengths measured need to be converted from the air wavelength that is read off the dial of the dye laser, to the vacuum wavelengths that are propagated through the heat pipe. The difference in these arises from the difference in the index of refraction of air (1.00029 @ 0° C and 1 atm pressure) and the index of refraction in vacuum (1.00000). In the range of 600 to 700 nm, the (+) correction to the wavelength is between 1.6 and 1.9 Å depending on λ. Tables of this correction can be found in any edition of the CRC Handbook of Chemistry and Physics *(22)* under "Index of Refraction of Air". Write values of the vacuum correction obtained from the CRC Handbook in the indicated column of the table and convert the measured line positions to vacuum wavelengths.

4. The laser should also be calibrated against some known standard in the wavelength range in which it is to be operated. The transition energies of the 5d $^2D_J \leftarrow 6s\ ^2S$ doublet are known *(15)* to be 14,596.84232 ± 0.00020 cm^{-1} and 14,499.25837 ± 0.00020 cm^{-1} for the J = 5/2 and 3/2 components respectively. These lines are observable in the spectra as (1 + 2) MPI signals and thus provide highly accurate and convenient calibrations for the remaining (2 + 1) MPI lines in the spectrum. Identify the 5d $^2D_{5/2,3/2} \leftarrow$ 6s ^2S transitions in the spectrum and compare their measured vacuum wavelengths to those predicted from the above known transition energies. Determine a calibration correction based upon the observed difference, insert it into the table under $\Delta\lambda$(correction) and convert λ(vac) to λ(measured).
5. The transition energies in wavenumbers for the two-photon excitation part of the (2 + 1) MPI lines of the spectrum can now be calculated from $E_{n,\ell}$(measured) = 2 / λ(measured). Record $E_{n,\ell}$(measured) in the table in units of cm^{-1}. Compare $E_{n,\ell}$(measured) with the atomic energy levels *(15)* for Cs. Assign all of the peaks observed in the spectrum and record the assignments in the table.
6. For most students, a Rydberg constant is first encountered in the equation for the energy levels of the hydrogen atom. The Rydberg constant for hydrogen differs from that for Cs due to differences in the reduced masses of H^+ and e^- and Cs^+ and e^-. Calculate the Rydberg constant for Cs, R_{Cs}, and compare with the Rydberg constant for hydrogen, R_H.
7. Fit the measured transition energies for the peaks that make up the s(ℓ=0) Rydberg series to equation 7 to determine the ionization potential of Cs and the s-electron quantum defect. Do the same for the d(ℓ=2) Rydberg series. The fit requires a non-linear least squares routine to simultaneously extract the ionization potential and the quantum defect. Alternatively if a non-linear least squares routine is not conveniently available, linearize the equation by plotting $(E_\infty - E_{n,\ell})^{-1/2}$ vs. n. The plot should have a slope of $1/(R_{Cs})^{1/2}$ and an intercept of $\delta_\ell/(R_{Cs})^{1/2}$. In this case extract both s- and d-electron quantum defects and R_{Cs} from the plot. Compare your results with literature values *(15)*.
8. Compare the quantum defects measured for the s- and d-Rydberg series. Discuss their differences based upon the radial dependence of the radial distribution functions for s- and d-electrons of the same n in hydrogen.

Acknowledgments

Acquisition of the equipment utilized in these experiments was made possible through a grant by the National Science Foundation, #CHE-8852247, and through support by Science Alliance, a state of Tennessee Center of Excellence.

Literature Cited

1. Veronov, G.S.; Delone, N.B. *JETP Lett.* **1965,** *1*, 66.
2. Hurst, G.S.; Payne, M.G.; Kramer, S.D.; Young, J.P. *Rev. Mod. Phys.* **1979,** *1*, 767.

3. Fassett, J.D.; Moore, L.J.; Travis, J.C.; Devoe, J.R. *Science* **1985,** *230*, 262.
4. Compton, R.N.; Miller, J.C. in *Laser Applications in Physical Chemistry;* Evans, D.K., Ed.; Marcel Dekker, Inc.: New York, NY, **1989**, pp. 221-306.
5. Petty, G.; Tai, C.; Dalby, F.W. *Phys. Rev. Lett.* **1975,** *34*, 1207.
6. Chen, P.; Cheyska, W.A.; Colson, S.D. *Chem. Phys. Lett.* **1985,** *121*, 405.
7. Zander, L.; Bernstein, R.B. *J. Chem. Phys.* **1979,** *71*, 1359.
8. Winkler, I.C.; Stachnik, R.A.; Steinfeld, J.I.; Miller, S.M. *J. Chem. Phys.* **1986,** *85*, 890.
9. Hayden, J.S.; Diebold, G.J. *J. Chem. Phys.* **1982,** *77*, 4767.
10. Göppert-Mayer, M. *Ann. Physik* **1931,** *9*, 273.
11. Hurst, G.S. *J. Chem. Ed.* **1982,** *59*, 895.
12. Peticolas, W.L. *Ann. Rev. Phys. Chem.* **1967,** *18*, 233.
13. Steinfeld, J.I.; *Molecules and Radiation*; MIT Press: Cambridge, MA, **1985**.
14. Bonin, K.D.; McIlrath, T.J. *J. Opt. Soc. Am.* **1984,** *B1*, 52.
15. Weber, K.H.; Sansonetti, C.J. *Phys. Rev. A* **1987,** *35*, 4650.
16. Burkhardt, C.E.; Ciocca, M.; Garver, W.P.; Leventhal, J.J.; Kelley, J.D. *Phys. Rev. Lett.* **1986,** *57*, 1562.
17. Klots, C.E.; Compton, R.N. *Phys. Rev. A* **1985,** *31*, 525.
18. Zhang, J.; Lambropoulos, P.; Zei, D.; Compton, R.N.; Stockdale, J.A.D. *submitted to Z. Fur Physik D.*
19. Vidal, C.R.; Cooper, J. *J. Appl. Phys.* **1969,** *40*, 3370.
20. Vidal, C.R.; Haller, F.B. *Rev. Sci. Instrum.* **1971,** *42*, 1779.
21. Malmstadt, H.V.; Enke, C.G.; Crouch, S.R.; *Electronics and Instrumentation for Scientists*; Benjamin Cummings Inc.: **1981**, pp. 422-426.
22. Lide, D.R.; *CRC Handbook of Chemistry and Physics*; CRC Press, Inc.: Boca Raton, FL, 1990.

Hardware List

The major pieces of equipment used in the experiment have already been identified in the methodology section of this paper. The most unique piece of equipment that is used in the experiment is the optical heat pipe. Using Figure 7 as a guide to the overall construction, the heat pipe can be constructed from the following list of parts.

Parts List:

1 ea. - 1.33" Vacuum Nipple; for example MDC Vacuum Products Part #402000. (used for main heater tube.)

2 ea. - 1.33" Vacuum 4-way Cross; for example MDC #405000. (attach one to each end of nipple to provide ports for windows, gas inlets and electrical feedthrough.)

2 ea. - 1.33" Vacuum Flange to 1/4" VCR Fitting Adapter; for example MDC #414006. (attach one to one port of each 4-way cross for matched gas line inlets. See Figure 7.)

2 ea. - 1.33" Vacuum Flange mounted Pyrex viewport; for example MDC #450000. (attach one to each 4-way cross for laser windows.)

1 ea. - 1.33" Vacuum Flange mounted electrical feedthrough with BNC connector; for example MDC #630000. (Attach to open port of one of the 4-way crosses. Used to hold charge collection wire and provide external connection to collector signal lead.)

1 ea. -	1.33" non-rotatable blank vacuum flange; for example MDC #110000. (attach to remaining open end of 4-way cross. Used to blank off and seal.)
2 ea. -	Bolt sets for 1.33" Vacuum Flange; for example MDC #190001.
2 pkgs. -	OFHC copper gaskets for 1.33" Vacuum Flange; for example MDC #191000.
2 ea. -	3/4" I.D. × 2" long, 1/2 section cylindrical RL-series ceramic heater with hole for thermocouple; Thermcraft, Inc., Winston-Salem, NC 27117. (to be placed around 1.33" Vacuum Flange Nipple.)
1 ea. -	Temperature Controller; for example Omega Engineering, Inc. Part #CN5001K2.

Additional Construction Tips:

A wick will need to be constructed for the interior of the pipe. It can be made of 304 stainless steel wire cloth (80 mesh × 0.0055" wire size. Available from McMaster Carr, Inc. or Small Parts, Inc.) which has been rolled and spot welded into a cylinder which will just slip inside the vacuum nipple. The length and placement of the wick should approximately match the heated zone. A collector wire will also need to be made which will extend over the heated zone to the feedthrough. This is easily accomplished by simply bending a piece of 1/16" stainless steel welding rod into an elongated L of the proper size. The collector should lie to the side of center so as to not obstruct the laser from passing through the tube. A cylindrical connector is used to attach the welding rod to the feedthrough. This connector can be simply machined or purchased from MDC Vacuum Products or other vacuum hardware manufacturers.

When wrapping the heater around the tube, care needs to be taken to ensure that the heater does not short to the tube body. The outside of the heater should be wrapped with additional high temperature insulation and covered with aluminum foil to minimize heat loss. Cooling coils will need to be attached to the heat pipe at the inside ends of the two 4-way crosses. Two to three wraps of 1/4" copper tubing, soldered directly to the stainless steel body is adequate. The water inlet to the two coiling coils can be plumbed in series. Very little heating of the water takes place under even slow water flow conditions and tygon tubing is adequate for connection to the copper coils.

The heat pipe will need to be connected to a vacuum and gas handling system through the VCR fittings indicated in the above list. The vacuum and gas handling system is used to both pump out the tube and to fill it with a few Torr of Argon. The vacuum (10 mTorr) provided by a simple two-stage rotary vacuum pump is sufficient. In order to monitor pump out and fills, the system should be equipped with a pressure measurement device such as a piranni gauge capable of measuring from mTorr to atmospheric pressure.

Notice: To prevent possible shocks, the heat pipe should be electrically grounded to earth.

RECEIVED October 1, 1992

CHAPTER 12

Picosecond Laser Spectroscopy

Weining Wang, Andrew R. Cook, Keith A. Nelson, and J. I. Steinfeld

POTENTIAL HAZARDS: Lasers

Introduction to laser spectroscopy

Laser spectroscopy is now widespread throughout all areas of chemistry. Examples of applications include the study of organic, inorganic, and biological photochemistry, polymer and biopolymer dynamics, molecular energy levels and dynamics, liquid crystalline nonlinear optics, environmental pollutants, the human genome, and countless others. For most laboratory applications, there are several features of coherent laser light that may be exploited to achieve goals which would be difficult or impossible to achieve with incoherent light. First, the intensity levels that can be reached conveniently with laser light are far higher than those otherwise available. This facilitates the study of many processes even if the laser does nothing more to the sample than ordinary light, simply because the amount of sample excitation is far greater. For example, the amount of a photochemical reaction intermediate or product present at a given time may become sufficient for spectroscopic characterization. Alternatively, processes may occur under laser irradiation which almost never occur otherwise. For example, simultaneous absorption of more than one visible or infrared photon by a sample is but one example of a *nonlinear optical process* (discussed further below) which may occur under laser irradiation and which may lead to new kinds of excited states or photochemical reaction products.

Another important feature of some lasers that is commonly exploited in the lab is the highly monochromatic character of their output. This is useful for very accurate determination of molecular energy levels through *high-resolution spectroscopy*. To understand this, it is important to realize that no light beam is perfectly monochromatic, i.e., composed of just one wavelength. If a grating is used to separate the different wavelengths of sunlight or lamplight, and a very narrow slit is used to reject most wavelengths, resolution of about 10^{-2} cm^{-1} can be achieved. (Note that the intensity of the incoherent light within this bandwidth is generally very low!) On the other hand some commercial lasers can produce intense output with a spectral width of a few Megahertz (e.g. 3 MHz = 10^{-4} cm^{-1}) and spectral widths on the order of 1 Hz (less than 10^{-10} cm^{-1}) have been achieved with specially designed lasers. Narrow laser spectral widths can be used to resolve closely spaced lines of a molecular absorption spectrum, or to determine the widths of narrow absorption features. This yields improved information about molecular rotational or other energy levels which may be poorly resolved in conventional spectroscopy.

A third important capability of some lasers is production of short, intense light pulses. These are useful for characterization of rapid events such as

2503–6/93/0194$06.75/0

excited-state relaxation or photochemical change through *time-resolved spectroscopy*. Using conventional flashlamps, pulses of duration on the order of 1 microsecond (10^{-6} s) can be generated and so events on this time scale can be observed. Commercial lasers with output pulses in the 1-100 picosecond range (1 ps = 10^{-12} s) have been available for several years, and pulse durations of under 100 femtoseconds (1 fs = 10^{-15} s) are now available commercially. The shortest pulses ever generated are less than 10 fs in duration. Note that a single electromagnetic cycle of visible light takes about 2 fs, so the shortest pulses are only as long as several optical wavelengths. Femtosecond pulses have been used for time-resolved observations of molecular vibrational oscillations and the breakage of chemical bonds.

In this experiment picosecond laser pulses are used to conduct time-resolved spectroscopy of molecular excited states in organic dye solutions. The excited-state transport, trapping, and relaxation processes which you are studying are important in most organic molecular liquids and crystals, and are particular important in the initial steps of photosynthesis. Students gain some experience in the use of pulsed lasers and the associated spectroscopic tools, and in thinking about the kinds of dynamical processes that may follow irradiation of a complex molecular system.

Dynamics of Organic Materials

Electronic Excited State Dynamics Absorption of ultraviolet or visible light by a molecule produces an electronic excited state. Many processes of chemical interest can be initiated in electronic excited states. Photochemical events include photodissociation, photoisomerization, and photoionization. Electronic excited states which do not undergo chemical change may still undergo any of a wide range of dynamical processes including transfer of energy to another molecule, fluorescence, or nonradiative relaxation. Any of these processes returns the initially excited molecule to its ground electronic state. Some of the possible pathways from the excited state are indicated in Figure 1. The ground and optically excited states of most organic molecules are nondegenerate (singlet) states, and so are labeled S_0 and S_1 respectively.

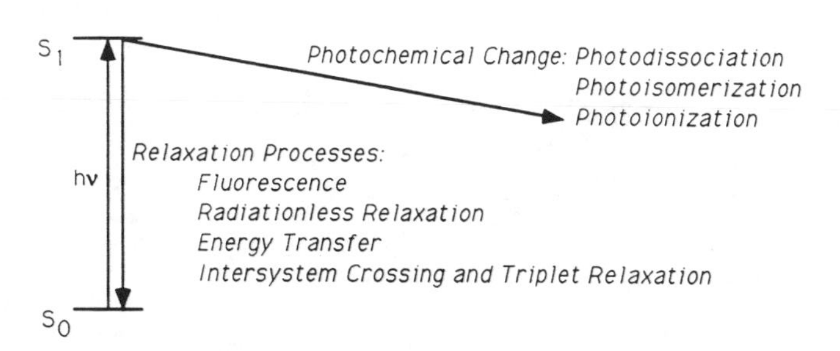

Figure 1. Photochemical processes

Primary Events in Photosynthesis The main subject of this experiment, electronic energy transfer, was first considered in detail in an effort to understand photosynthetic conversion of solar energy. In the first event of photosynthesis, light is absorbed by a chlorophyll molecule to produce an electronic excited state. However, the reactions through which chemical energy is ultimately produced do not take place at the "antenna" chlorophyll molecule that absorbs sunlight. The leaves of a plant are filled with antenna chlorophylls, but have relatively few "reaction centers" at which conversion to chemical energy takes place. How does the electronic excited-state energy get from the initially excited chlorophyll to the reaction center, which is typically on the order of 100 Å away? The energy migrates, through a mechanism first described theoretically by Förster, from chlorophyll to chlorophyll until it "lands" on a reaction center as illustrated schematically in Figure 2.

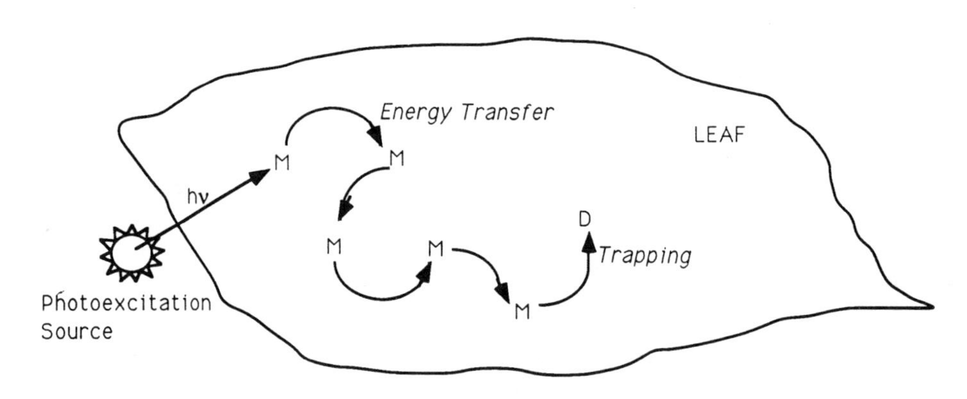

Figure 2. Primary photoexcitation in photosynthesis.

The antenna and reaction center are labeled M and D respectively because the former is a chlorophyll monomer while the latter includes a chlorophyll dimer. (The 1989 Nobel Prize in Biology was awarded for the x-ray determination of the reaction center structure in a photosynthetic bacterium.) The reaction center has a slightly lower electronic energy, so once the energy lands there it becomes trapped and cannot return to the network of antenna chlorophyll molecules. The situation is illustrated schematically in Figure 3.

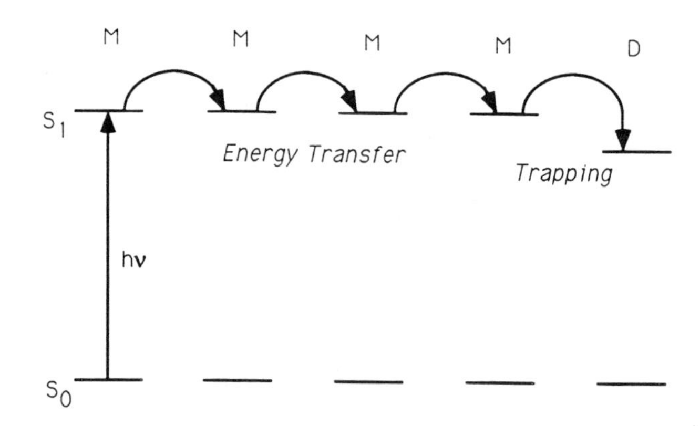

Figure 3. Energy trapping.

Organic Dye Solutions Organic dyes are strongly absorbing and fluorescing molecules. Ordinarily, absorption is followed by fluorescence with a quantum yield of nearly unity and a lifetime of about 5 nanoseconds (5×10^{-9} sec). However, if electronic energy transfer can lead to trapping on a nearby site which does not fluoresce, the lifetime can become extremely short and the fluorescence effectively quenched. Organic dyes can be used as "sensitizers" which absorb light and transfer the energy efficiently to other organic molecules, semiconductors, etc.

Most aromatic molecules have a weak attraction for each other through van der Waals forces, and organic dyes are no exception. In dilute solution, the dye molecules are present as monomers almost exclusively. However, as the dye concentration is increased, weakly bound dimers form. At high concentration, *i.e.*, around 10^{-2} M and above, a substantial fraction of the dye molecules are complexed to others in weakly bound dimers. Just as in the case of chlorophyll, the dimer has a lower excited-state energy than the monomer. (Can you explain this very general trend?) The dimers therefore act as traps for electronic excited-state energy. A concentrated dye solution mimics the leaf in that there are many monomers among which energy is transferred, and a substantial number of dimers into which energy is eventually funneled. In the biological system, the trapping of energy allows it to stay at the reaction center long enough to initiate chemical change. In the dye solution, the dimer is not involved in any chemical reactions but rather undergoes rapid radiationless relaxation back to the ground electronic state. In this manner the electronic excited-state energy is dissipated, and fluorescence is quenched.

In this experiment 100-ps "excitation" laser pulses are used to excite dye molecules in solution. The absorption of the excited states is monitored with variably delayed "probe" laser pulses. In this manner the lifetime of the excited states is determined. As the dye concentration is increased, the rate of trapping and radiationless relaxation increases and the quantum yield decreases. By measuring the excited-state lifetime as a function of dye concentration, the dynamics of energy transfer and trapping in this simple system are determined. The results are of direct applicability in understanding the primary events of photosynthetic systems.

The Mechanism for Energy Transfer Electronic energy transfer is extremely common, but may appear to be mysterious since neither mass nor charge, but only energy, is transferred (unlike electron transfer, by contrast). A classical analogy can be made to a system (like a molecular crystal lattice) of masses and springs. Imagine groups of masses (molecules) bound together with very stiff springs and weakly interacting with each other through very loose springs. If one group starts vibrating, its connections to other groups leads them to vibrate. Eventually the group that was vibrating initially is hardly moving, and other groups vibrate in turn. In the case of electronic energy transfer, molecules are held together through strong electronic forces and interact with each other through weaker (van der Waals) electronic forces. When one molecule is excited, its electron cloud is distorted and other molecules nearby "feel" a change in their interactions with it. Eventually the energy migrates to neighbors, much like vibrational energy. In each case energy, but no mass or charge, is transferred.

Nonlinear Optics In this experiment you will also have the chance to see several effects are observed which illustrate *nonlinear optics*. Nonlinear optics refers to any optical effect whose intensity increases nonlinearly with the intensity of incoming light. For example, fluorescence is generally a linear optical effect: twice as much light in yields twice as much fluorescent light out. By contrast, a laser is a nonlinear optical device: there is no output when the intensity of light pumping the laser is low, and a threshold must be crossed to produce laser output. In *frequency-doubling*, laser light with a wavelength of 1.06 μm (near-infrared) enters a nonlinear optical crystal, and light at 532 nm (green) comes out. The intensity of green light is proportional to the square of the near-IR intensity. The effect is impressive visually since the light which is incident upon the crystal is invisible to the eye, and bright green laser light comes out. This experiment is itself also a nonlinear optical effect. Three laser beams cross inside a sample, and a fourth beam comes out whose intensity increases as the cube of the total incident intensity. Other nonlinear optical effects, including optical generation of ultrasonic waves inside your sample, may be observed.

Picosecond Spectroscopy: Theory and Basic Methods

Students should come to the lab with a good basic understanding of how time-resolved spectroscopy and the "transient grating" experiment work. Here we provide a brief description. A good general reference is the book by Fleming *(1)*. Additional details may be found elsewhere *(2-6)*.

Excitation and Probe Pulses Typically, the experiment involves an "excitation" pulse which passes through the sample at time t = 0 and a "probe" pulse which is delayed by an amount of time which can be varied. In this experiment the excitation pulse is absorbed by the sample to produce electronic excited states, and this initiates the sequence of events which will be monitored. The probe pulse is either absorbed by the excited states, as shown in Figure 4, or by the ground state. As relaxation

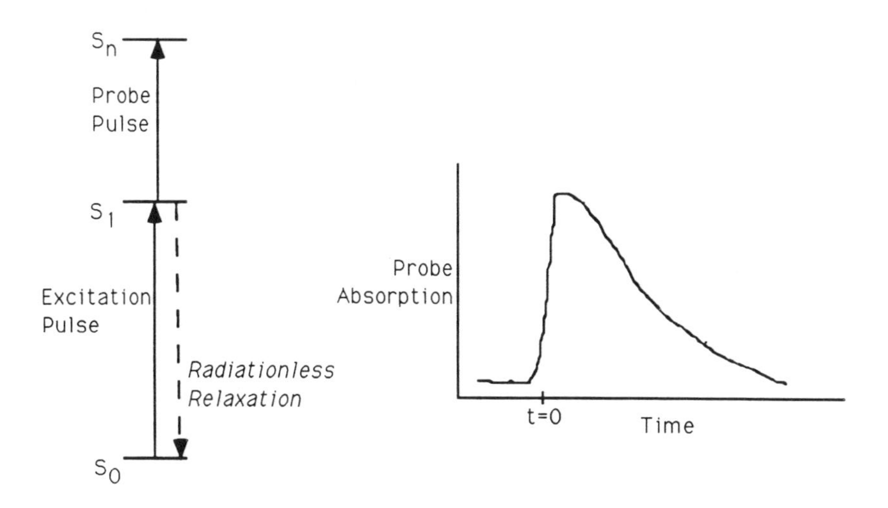

Figure 4. Pump-probe experiment.

occurs, the population of excited states decreases from its initial amount and the population of ground states recovers. If the probe pulse is absorbed by the excited states, then the amount of its absorption decreases. If the probe pulse is absorbed by the ground state, then its absorption increases as the ground state population recovers. The experimental mechanism and typical time-dependent data are sketched in Figure 4.

The sketch of data indicates that probe absorption is measured as a function of delay between excitation and probe pulses. The delay is actually quite simple to control: the probe pulse is directed into a retroreflector whose position can be varied, and then into the sample. (Figure 5) The excitation-pump pulse sequence is repeated many times, with the delay of the probe pulse slightly different each time. The absorption of the probe pulse at each delay is recorded to produce data like that in Figure 4. In this experiment, a total temporal range of about 10 ns is covered. This means that a mechanical delay of the retroreflector is required which can vary over a range of about 1.5 meters (optical delay of 3 m). The pulse duration

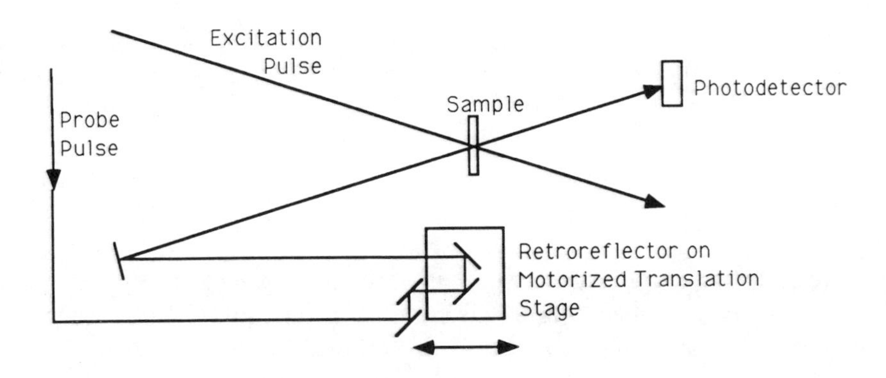

Figure 5. Pulse delay setup.

is about 100 ps, so there should be at least this resolution in the delay. Thus the individual steps by which the retroreflector is translated should be no bigger than 1.5 cm. In practice, it is usually best to increment the delay by much smaller steps (e.g. 1 mm).

Notice that in this experiment, no fast electronic devices are needed to achieve fine time resolution. The time resolution is determined only by the laser pulse duration and by the resolution (i.e., step size) of the probe pulse delay. In fact positions can be controlled to much better than 1μm if necessary, so only the pulse duration limits the experimental time resolution.

The Transient Grating Experiment This experiment provides the same information as the transient absorption measurement illustrated above, but in a manner which can be advantageous in terms of signal/noise and other considerations. Instead of a single excitation pulse, two time-coincident excitation pulses are crossed inside the sample at t = 0 as illustrated in Figure 6.

The interfering excitation pulses create a "grating" pattern of electronic excited states, with a maximum excited state concentration at the interference maxima and no excited states at the interference minima. The pattern persists until the excited states are gone, so by monitoring the grating decay the excited-state lifetime can be determined. The grating is easily monitored by *diffraction* of the variably delayed probe pulse. This works because the excited molecules have

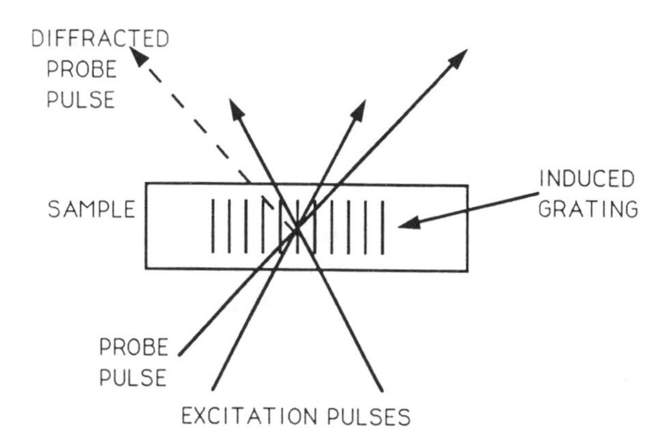

Figure 6. Transient grating.

different optical properties (e.g., different absorption spectra) than those in their ground states. The sinusoidal pattern of excited state concentration therefore acts like a *transient grating* which can diffract a probe pulse incident at the Bragg angle for diffraction. The lifetime of the excited states is determined by the time-dependent decay of diffracted signal intensity. Transient grating data from electronic excited states appear similar to that sketched above for transient absorption.

There are several advantages in the transient grating experiment as compared to transient absorption. The most important is that the signal is a coherent laser spot in a unique spatial direction. In this experiment the signal should be visible to the eye in a lit room! Even rather weak signals can usually be detected and measured accurately because they are measured against a "dark" background. That is, essentially no light other than signal strikes the photodetector. By contrast, in transient absorption measurements the change in intensity of the probe pulse which is transmitted through the sample must be measured. There is always a bright laser spot incident on the photodetector, and (often small) changes in its intensity must be determined.

As an aside, note that the transient grating experiment can be used to measure many other events besides electronic excited-state relaxation. For example, if energy transfer is very rapid, it will wash out the grating pattern since excited states will move out of the grating peaks and into the nulls. The decay of diffracted signal intensity then gives the rate of energy transfer. (Organic dye molecules do transfer energy rapidly, but their lifetimes are too short for energy migration over the grating fringe spacing (usually several μm). It is also possible to generate ultrasonic waves in the sample with the crossed excitation pulses, and to detect the coherent acoustic oscillations by diffraction. This will be discussed further below. Molecular rotational motion, mass diffusion and thermal diffusion, and other responses can also be observed in transient grating experiments.

Experimental Methodology

Solution and Sample Preparation The main part of the experiment will consist of transient grating measurements on solutions of the organic dye Rhodamine 6G

(R6G) in a 1:1 (by volume) ethanol/glycerol solvent mixture. A 5 x 10^{-2} M solution is prepared and then diluted down to 5 x 10^{-5} M. All solutions should be prepared in advance. Small volumes of the solutions will be sufficient, but be sure to keep them sealed overnight to avoid evaporation of the ethanol.

For dilute solutions, experiments can be done in 1-mm or 2-mm spectrophotometer cuvettes. However, with concentrated solutions the green excitation light will all be absorbed within as little as a few µm. In these cases it will be most convenient to "sandwich" a few drops of concentrated solution between two microscope slides which can be held together with clamps. A thin teflon spacer can be inserted at one end of the "sandwich" so that the thickness of the solution trapped between the slides is greatest near the spacer and least at the opposite end. To optimize signal, it is best to find a spot at which most (about 90%) of the green excitation light is absorbed.

Optical Alignment of the Experimental System A detailed description of the laser system is given later; its main elements are summarized here. The heart of the laser system is a Spectron 903 mode-locked and Q-switched Nd:YAG laser. Its output consists of a sequence of 100-picosecond pulses. The wavelength is 1.06 µm, and frequency-doubling yields pulses at 532 nm (green light). One of these pulses is selected and used for excitation in this experiment. The rest of the pulses are used to pump a picosecond dye laser whose output is a single yellow (560-nm) pulse. This is used as the probe pulse in the experiment. These pulses (green excitation, yellow probe) are used to align the experiment. A number of parameters should be optimized before collecting data.

Excitation Pulses The green pulse is split into two identical pulses with a 50% reflector. The two pulses must be crossed in the sample, and must be time-coincident. Schematically, the pulses are directed as indicated in Figure 7.

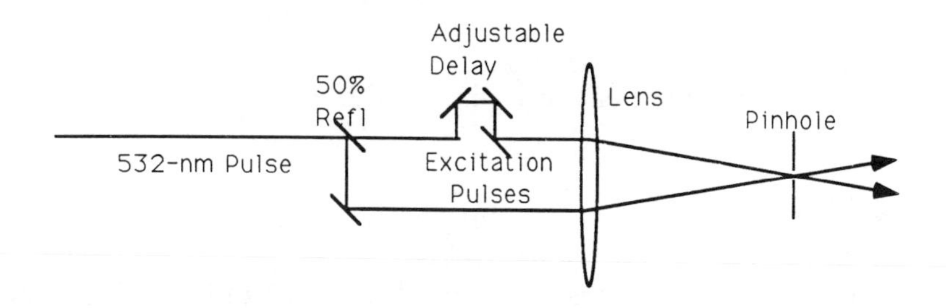

Figure 7. Schematic illustration showing temporal and spatial overlap of the two excitation pulses. The adjustable delay is used to equalize the two path lengths. The two beams are crossed inside the pinhole (along with the probe pulse), and then the sample is inserted in place of the pinhole.

The pinhole is a 100-µm hole drilled through a thin metal foil. **Please note that the pinhole can be damaged by the focussed pulses!** Be sure to **insert several filters** into the green beam before it is split so that its full energy is not incident on the pinhole. The adjustable delay of one of the excitation pulses should be optimized before starting work on the system.

Probe Pulse There are three critical alignments necessary for the probe pulse. First, it must be crossed with the excitation pulses inside the sample. Second, it must be incident on the sample at the Bragg angle for diffraction off the grating formed by the excitation pulses. Third, its alignment along the motorized delay line must be optimized. The experiment including aligned excitation and probe pulses is illustrated schematically in Figure 8.

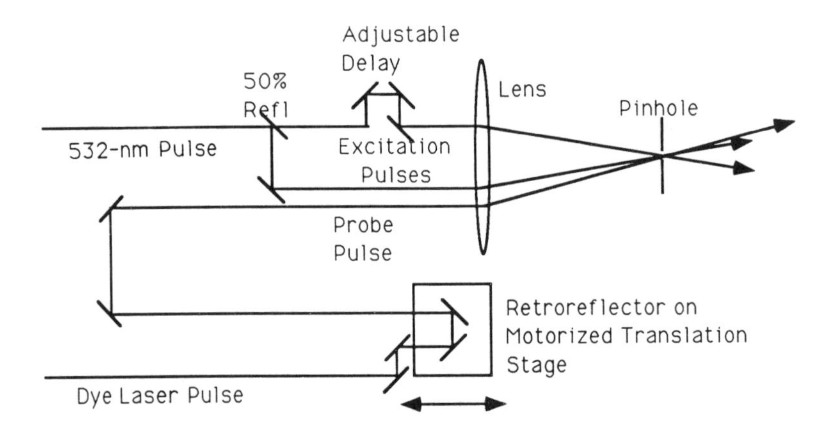

Figure 8. Schematic illustration of the excitation and probe pulses aligned for the transient grating experiment. The excitation pulses are time-coincident, and all three beams are crossed in the pinhole. The probe pulse in aligned along the motorized delay line and is incident on the pinhole at the Bragg angle for diffraction. To do the experiment, the pinhole is replaced by the sample and diffracted light is observed.

The probe beam alignments are discussed below in the order that they may easily be checked.

Aligning the delay line The position of the probe beam as it enters the sample (or pinhole) should not move as the motorized delay line is run. This means that the probe beam must be precisely parallel to the direction along which the translation stage travels. The reflector which steers the beam toward the retroreflector is the one which must be adjusted critically. If the alignment is very poor, then the probe beam will miss the retroreflector entirely as it is moved far back along the delay line. The probe beam should be directed through the pinhole as indicated above, with the retroreflector at the very front of the delay line. The retroreflector should then be moved toward the back of the delay line. If alignment is not perfect, then the probe beam will no longer go precisely through the center of the pinhole. (This is apparent from the deterioration of the interference pattern in the light transmitted through the pinhole.) Aim the probe beam through the pinhole by adjusting the mirror which directs the probe at the retroreflector. Then move the retroreflector all the way to the front of the delay line, and adjust the position of the pinhole so that the probe beam goes through the center. Once again move the retroreflector all the way back, and adjust the mirror which directs the probe at the retroreflector. Then return the retroreflector to the front of the delay line, and adjust the pinhole position. The procedure continues this way (usually only one or two iterations are needed) until the probe beam does not move when the retroreflector is moved. Another way of doing this is to aim the probe beam at a "target" such as a cross on a piece of paper, and observe by eye any movement of the probe beam as the delay line is moved. The procedure is the same: adjust the position of the target when the retroreflector is in front; adjust the alignment of the mirror directing the beam along the delay line when the retroreflector is in back; and repeat until no movement of the beam can be detected. If a target is used, place it somewhere before the lens which focuses the probe beam into the pinhole. This way the probe spot and its movements are big enough to see easily by eye.

Adjusting the Bragg angle The angle of incidence θ_p of the probe beam for Bragg diffraction depends on the angle θ_e between the excitation beams and on the excitation and probe wavelengths λ_e and λ_p respectively, as shown in Figure 9.

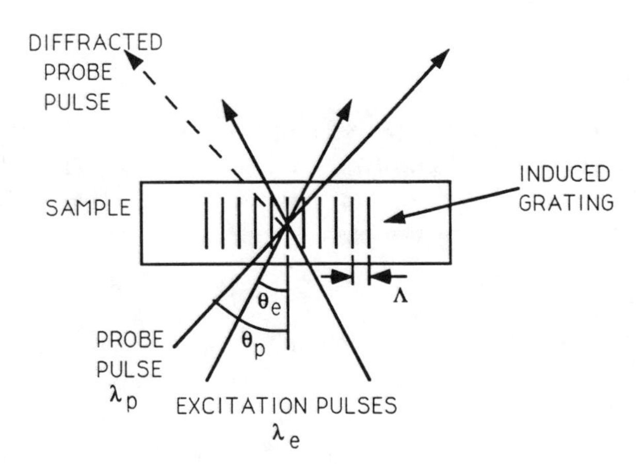

Figure 9. Schematic illustration of the transient grating experiment indicating the excitation and probe wavelengths and angles of incidence, and the grating spacing, Λ.

The grating spacing, Λ, and other parameters are related as follows.

$$\Lambda = \lambda_e/2 \sin \theta_e = \lambda_p/2 \sin \theta_p \qquad (1)$$

Through this relation the grating spacing and the Bragg angle can be calculated once the angle between excitation pulses is set. In practice, the angles between beams are small (a few degrees) and so the separation between beams has the same ratio as the wavelengths, as illustrated in Figure 10.

The Bragg angle is set by translating the probe beam reflector such that the relation $d_e/d_p = \lambda_e/\lambda_p$ is held, where the distances d_e and d_p are defined in Figure 10. Note that it is not necessary that the probe beam be level with the excitation beams; it can be somewhat above or below the plane containing the excitation beams, as shown in the inset.

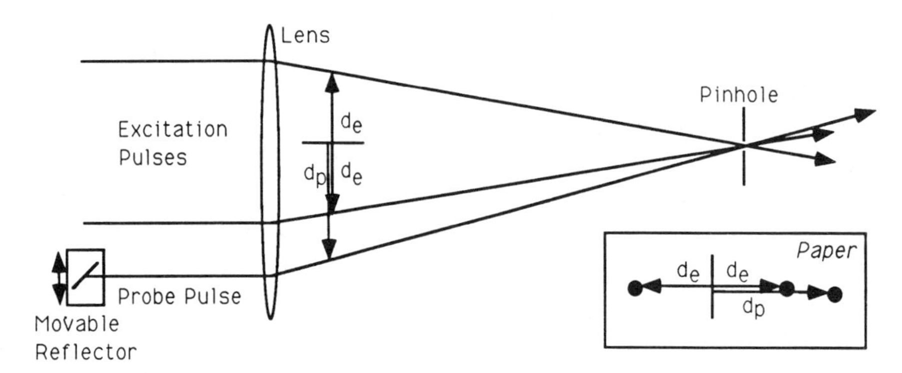

Figure 10. Schematic illustration of the excitation and probe pulses indicating the distances between pulses. The distances are measured from the midpoint between the two excitation pulses. From this point, the distances to the excitation and probe pulses are in the ratio of the excitation and probe wavelengths. The inset shows the laser spots as observed on a card inserted after the lens. The Bragg angle can be set by translating the reflector which directs the probe beam into the lens.

Crossing the probe and excitation beams The two excitation beams and the probe beam should be roughly parallel before the lens so that the lens focuses them all to the same point. The pinhole is placed at that point, and transmission of each beam is optimized by making fine adjustments to the final excitation and probe beam reflectors.

Looking for Transient Grating Signal When the excitation and probe beams are properly aligned, the pinhole should be replaced by a sample. The three incident beams should be absorbed partially and transmitted partially through the sample. The three transmitted beams will appear as in the inset to the Figure 10 on a card inserted after the sample. Diffracted signal should appear as a fourth spot, positioned symmetrically to the transmitted probe beam as shown in Figure 11.

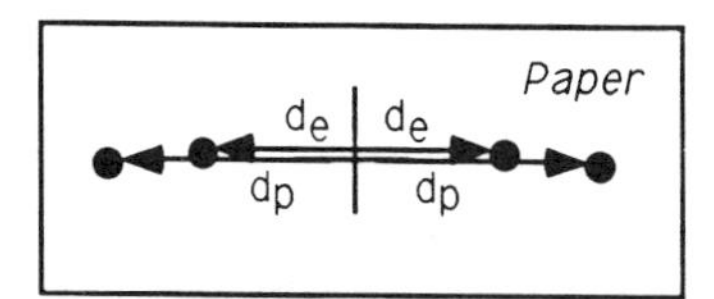

Figure 11. The transmitted excitation and probe beams and diffracted signal as observed on a card inserted after the sample. The signal appears as a fourth spot which is symmetrical with respect to the transmitted probe beam.

If signal does not appear as soon as the sample is inserted, try adjusting the position of the probe retroreflector along the motorized delay line. The signal is strongest when the probe pulse is time-coincident with the excitation pulses, i.e., at $t = 0$. Signal vanishes at earlier times (shorter delays) and decays gradually at longer times. If no signal can be found, check with a pinhole to make certain that the three beams are crossed. Make sure to place the sample precisely at the location of the pinhole. When signal is found, it can be optimized through small adjustments of the beams and the sample position.

Recording Transient Grating Data The diffracted signal should be aimed into a photodetector, and to the extent possible any scattered light other than signal should be excluded. The output of the photodetector is measured by a lock-in amplifier whose output is read by a personal computer. The computer also controls the motion of the delay line. Thus diffracted signal intensity is measured as a function of probe pulse delay.

Experiments to be Conducted The main part of the experiment consists of transient grating experiments on Rhodamine 6G solutions. Experiments on a malachite green solution should also be conducted. If possible, more than one angle between excitation pulses should be used for malachite green. This will demonstrate optical generation of tunable ultrasonic waves. Be sure to measure each angle accurately. These values are needed for subsequent calculations.

Many additional samples could be tried, and at least some should be tried if time permits. Suggestions follow.

(1) Liquid crystals. Nearly any isotropic liquid crystal will produce strong transient grating signal due to molecular orientational alignment. The excitation pulses align the molecules at the grating peaks but not at the nulls. This results in transient grating signal which decays due to orientational relaxation (return to equilibrium). Rotational diffusion times of molecules in liquid crystals and ordinary liquids can be determined this way.

(2) Fluorescent rulers and drawing tools. Many drawing tools including triangles, straight edges, etc. are doped with a highly fluorescent dye (often R6G) to avoid shadows around the edges. Transient grating data from these samples will yield the excited-state lifetimes under conditions where movement is highly restricted.

(3) Jello®! Orange, Lemon, or Strawberry Jello may produce transient grating signals. At one time Rhodamine dyes were used as food colorings in jello, and in fact the jello laser (similar to the dye laser you will use) has been demonstrated *(7)*. Be sure to prepare (cook) this sample in advance!

(4) In general, any material of reasonably good optical quality can be examined. It is helpful (but not essential) for the sample to absorb green light, even weakly. Optical quality can be gauged by trying to read fine print through the sample. If you can do this with the print about 10 cm in back of the sample, then the optical quality is probably sufficient.

Safety

This experiment involves the use of high-intensity laser light. Precautions have been taken to reduce to a minimum the risk of exposure to dangerous levels of radiation. However, special **Laser Safety Glasses** must be worn whenever you are in the room with the laser in operation. Since the most intense radiation is in the infrared spectral region and therefore is invisible to the eye, eye damage cannot be avoided by simply avoiding visible light, so you must be careful to avoid visible laser beams striking your eyes directly. As a precaution, do not sit down with your eyes level with the laser beams.

The laser also includes a power supply with potentially lethal electrical current. You should never open the power supply, which would expose internal parts. Even when the power supply is off, charge accumulated on large capacitors can be hazardous.

The safety of the apparatus is also a matter of concern. This experiment makes use of optical elements including lenses, partial and high reflectors, polarizers, etc. They are very expensive! (Almost no elements cost less than $100, and most cost much more.) Please treat fine optics, and the precision mounts they are on, with respect. Be especially careful to avoid touching or damaging polished optical surfaces!

The chemicals used in this experiment include many dye solutions, including the Rhodamine 6G, which are suspected carcinogens.

Data Analysis

Overview Here we describe the background for the data analysis of the experiments on Rhodamine 6G solutions. Basically, two quantities change as the dye concentration is increased. First, the concentration of traps is increased since the traps are Rhodamine 6G dimers which are in equilibrium with monomers. Second, the rate of transport increases since an excited molecule has more neighbors in close proximity with which it can interact and exchange energy. Both of these effects increase the rate of trapping. In the limit of very low concentration, there are essentially no dimers and so the excited-state population decays with the monomer decay rate k_M. In the limit of very high concentration, rapid transport of energy leads to essentially immediate trapping on a dimer, and the excited-state population decays with the dimer decay rate k_D. At intermediate concentrations, some monomer excited states ultimately encounter traps and then relax to the ground state, and some relax to the ground state before encountering a trap. The overall decay rate depends on k_M, k_D, and the trapping rate k_T. The trapping rate, unlike the monomer and dimer decay rates, depends on the concentration: at low concentration it is essentially zero (excited states never find traps) and at very high concentration it is very rapid (all excited states are quickly trapped).

Rate Equations and Solutions The excited monomer and dimer concentrations, M^* and D^* respectively, are described by ordinary chemical rate equations:

$$dM^*/dt = -k_M M^* - k_T M^* \tag{2a}$$

$$dD^*/dt = -k_D D^* + k_T M^* \quad (2b)$$

The first equation describes the decay of excited monomers through relaxation to the monomer ground state (with rate constant k_M) or trapping on dimers (with rate constant k_T). Note that energy transfer from monomer to monomer does not enter into the equation since it does not change the monomer excited-state population. The second equation describes the formation of excited dimers through trapping (with rate constant k_T) and the decay of excited dimers through relaxation to the dimer ground state (with rate constant k_D).

The solutions to Eqs. (2) are

$$M^* = M_0^* \exp[-(k_M + k_T)t] \quad (3a)$$

$$D^* = [M_0^* k_T/(k_M + k_T - k_D)] \{\exp(-k_D t) - \exp[-(k_M + k_T)t]\} \quad (3b)$$

M_0^* is the initial concentration of monomer excited states produced by the excitation laser pulses. The initial dimer excited state population is assumed to be negligibly small.

In the transient grating experiment, production of excited states at the peaks leads to a change in the optical density (OD). The OD is therefore different at the grating peaks and nulls. The diffracted signal intensity, $S(t)$, is proportional to the square of the peak-null difference in OD:

$$S(t) = A(OD_p - OD_n)^2 . \quad (4)$$

The proportionality constant A involves many parameters such as laser spot sizes, photodetector efficiency, etc., and is not important to us. To calculate the OD values, we define the total dye concentration T as

$$T = M + D \quad (5)$$

where M and D are the monomer and dimer concentrations respectively. Since there are no excited states at the grating nulls,

$$OD_n = \epsilon l T, \quad (6)$$

where ϵ is the extinction coefficient at the probe wavelength and l is the sample length. The monomer and dimer absorption spectra are very nearly identical, and we have assumed that ϵ is the same for both species. At the grating peaks, the OD is given by

$$OD_p = \epsilon l(T - M_p^* - D_p^*) + \epsilon^* l(M_p^* + D_p^*) \quad (7)$$

where M_p^* and D_p^* are the monomer and dimer excited state concentrations at the peaks and ϵ^* is the extinction coefficient of the excited states (again assumed equal for monomers and dimers). The first term in Eq. (7) describes the decrease in probe pulse absorption by ground states due to the reduced ground state population. The second term describes the probe pulse absorption by the excited states. From Eq. (4), the diffracted signal intensity is

$$S(t) = A(\epsilon^* - \epsilon)^2 l^2 (M_p^* + D_p^*)^2. \tag{8}$$

The absolute diffraction intensity and the constant A are not determined experimentally. The important part of Eq. (8) is the time-dependence of the signal, which is given by the time-dependence of the excited monomer and dimer populations. Thus the main result for transient grating signal is

$$S(t) \propto (M^* + D^*)^2 = [T^*(t)]^2, \tag{9}$$

where $T^* = M^* + D^*$ is the total excited state concentration. This is the time-dependent quantity determined experimentally. From Eqs. (3),

$$T^*(t) = M_0^* \exp[-(k_M + k_T)t] + [M_0^* k_T/(k_M + k_T - k_D)]\{\exp(-k_D t) - \exp[-(k_M + k_T)t]\} \tag{10}$$

The limiting cases discussed earlier can now be reexamined. In the limit of very low concentration, k_T is very small, there is essentially no trapping, and $T^*(t)$ decays with the monomer decay constant k_M. In the limit of high concentration, $k_T >> k_M$ and k_D, all the excited states are trapped immediately on dimers which decay with the dimer rate constant k_D. At intermediate concentrations, the decay of excited state population can be more complicated. However, note that the dimer decay constant exceeds considerably that of the monomer, i.e., $k_D >> k_M$, and that at modest concentrations $k_D >> k_T$ as well since trapping is not fast. In this case, $T^*(t)$ decays with rate constant $(k_M + k_T)$. Basically, no significant D^* population ever builds up and the decay rate is that at which the monomer decays due to relaxation and trapping.

Data Analysis At each concentration, determine the decay kinetics of grating signal. The high- and low-concentration limits should give the monomer and dimer lifetimes, respectively. At intermediate concentrations the decays should be nearly single-exponential. As discussed above, for moderate concentrations you can assume exponential decays with rate constant $(k_M + k_T)$. At some concentrations, it may be necessary to fit the data with the functional form given in Equation (10). In all cases, remember that it is the square of $T^*(t)$ that is given by the TG signal.

A plot of k_T vs. dye concentration should be made. As discussed earlier, k_T increases sharply with dye concentration due to the presence of more dimer traps and due to a more rapid transport rate of excited-state energy. A theoretical treatment shows that k_T increases as M^3. This should be tested.

TG data from other samples should be analyzed. In particular, the optical generation of acoustic waves should be explained and the speed of sound in ethanol determined.

Acknowledgements

The picosecond laser system has been provided by National Science Foundation Instrumentation and Laboratory Improvements Grants CHE89-51738 and

CHE90-50584. Much of the design and all of the construction of the system were carried out by T.P. Dougherty and W. Wang.

Literature Cited

1. Fleming, G.R. *Chemical Applications of Ultrafast Spectroscopy*; Clarendon Press: Oxford, U.K., 1982.
2. Lutz, D.R.; Nelson, K.A.; Gochanour, D.R.; Fayer, M.D. *Chem. Phys.* **1981**, *58*, 325.
3. Fayer, M.D. *Ann. Rev. Phys. Chem.* **1982**, *33*, 63.
4. Hunter, C.N.; Van Grondelle, R.; Olsen, J.D. *Trends in Biochemical Sciences*, **1989**, *14*, 72.
5. Beddard, G.S.; Cogdell, R.J. in *Light Reaction Path of Photosynthesis*, Fong, F.K., Ed,; Springer-Verlag: Berlin, 1982, pp.46-79.
6. Van Grondelle, R. *Biochem. Biophys. Acta*, **1985**, *811*, 147.
7. Hänsch, T.W.; Permier, M.; Schawlow, A.L. *IEEE J. Quantum Electronics*, **1971**, *QE-8*, 45.

Laser system hardware

A pulsed, mode-locked frequency-doubled Nd:YAG-pumped dye laser system is used in this experiment. In this section, we describe the principles of operation of this system and provide a detailed schematic and parts list.

The Nd:YAG Laser In its simplest form, a laser consists of only the lasing medium and two end mirrors. An electronic excited-state population inversion (more higher-energy electronic states than lower-energy ones) is produced, usually by optical pumping of the lasing medium by a very bright arc lamp or similar light source. Stimulated emission then leads to the laser output. The output is a continuous beam of coherent light if the lasing medium is pumped continuously. The "continuous wave" or cw laser cavity is illustrated in Figure 12.

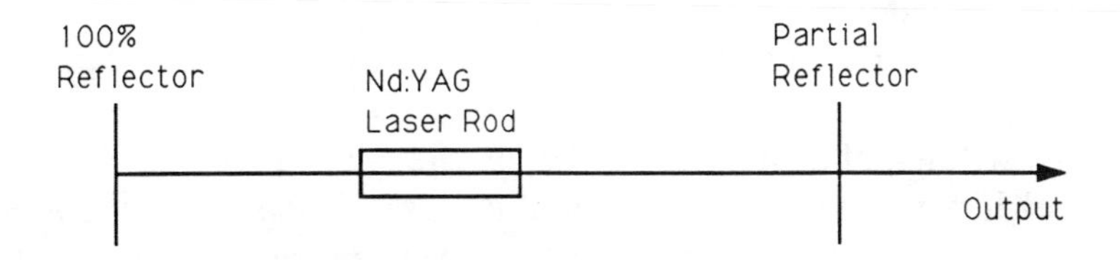

Figure 12. Nd:YAG laser.

In the laser system you will use, the lasing medium is a crystal of Yttrium Aluminum Garnet (YAG) doped with Neodymium ions. The Nd ions absorb light from an arc lamp and undergo stimulated emission and lasing in the near-IR, at a wavelength of 1.06 μm. The YAG crystal is essentially an inert host for the Nd ions.

Nd:YAG lasers are commonplace in scientific work and in many applications including laser surgery and laser machining.

There are several ways to get pulsed rather than continuous laser output. If the cavity is blocked, emission from the rod cannot reflect back and forth between the reflectors and there is no output. You could insert your hand to block the cavity, then remove it to open the cavity and quickly replace it to block the cavity again. If you have quick hands you may be able to produce pulses as short as about 0.1 sec. To produce nanosecond or picosecond pulses, some other means of closing and opening the cavity must be used.

Acoustooptic Q-switching and Mode-locking In part of experiment, ultrasonic waves are optically excited in a sample and detected by Bragg-diffraction of the probe pulse. This will demonstrate that ultrasonic waves can act as diffraction gratings. Now imagine that *inside the laser cavity* is a material with an ultrasonic wave running through it. In practice the material is a piece of optical quality glass with an acoustic transducer glued to the side of it. If the orientation is correct for Bragg diffraction, the laser beam inside the cavity will be diffracted by the acoustic wave. In this case the laser beam will be diverted away from the end mirror of the cavity in Figure 13.

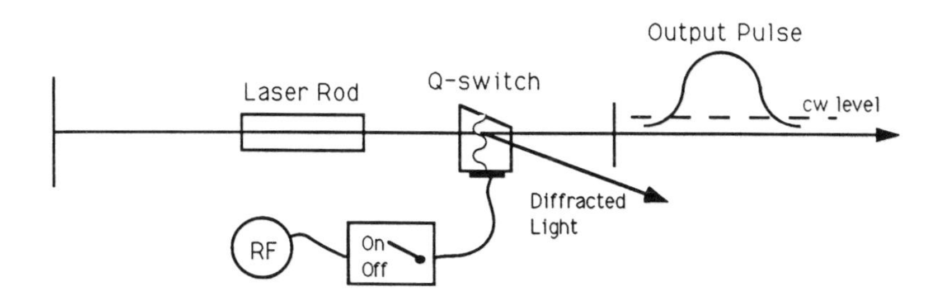

Figure 13. Q-switching.

When the radio frequency (RF) signal which generates the acoustic wave is on, light is diffracted out of the cavity and there is no laser output. During this time the population inversion in the rod builds up to a very high level since there is no energy given up as laser light. The RF signal can be turned off suddenly (in about 10 nanoseconds), at which time there is an intense burst of laser output. (Note that the output peak intensity is far higher than the cw level.) The RF then goes back on, the large inversion builds up again, and the process can be repeated. This method, called "Q-switching", can be used to produce pulses of about 100 ns in the laser you will use.

In Q-switching, the acoustic wave in the glass piece is launched by the transducer and hits the opposite face of the glass, which is beveled as shown in Figure 13. In this case any acoustic reflection off the beveled face is diverted away from the active area (the part that the laser beam passes through) of the glass piece. Within the active area there is an acoustic travelling acoustic wave moving down from the transducer, described in terms of density changes $\delta\rho$ by

$$\delta\rho(x, t) = A \cos(\omega t - kx) \tag{11}$$

where ω and $\mathbf{k}$ are the acoustic frequency and wave vector (assumed in the x direction) respectively. The acoustic wave diffracts light because the changes in density cause changes in refractive index and so there is a spatially periodic variation in refractive index, i.e., a grating. The ultrasonic wave diffracts light as long as it is present in the glass.

A variation on this theme, called "mode locking" can be used to produce considerably shorter pulses. In this case the top and bottom faces of the glass piece are parallel so the reflected acoustic wave returns along its original path. Then there are counterpropagating waves

$$\delta\rho(x, t) = A \cos(\omega t - \mathbf{k}x) + A \cos(\omega t + \mathbf{k}x) = 2A \cos \omega t \cos \mathbf{k}x \qquad (12)$$

The second equality shows that the two counterpropagating waves form a standing wave with fixed nodes (at positions where $\mathbf{k}x = 0, \pi, 2\pi$, etc.) and antinodes (at $\mathbf{k}x = \pi/2, 3\pi/2$, etc.). Unlike the travelling wave, which always diffracts light, the standing wave diffracts light periodically. When $\omega t = \pi/2, 3\pi/2, 5\pi/2$, etc., i.e., twice each acoustic vibrational period, $\delta\rho = 0$ at all points in space and there is no "grating" pattern to diffract light. When $\omega t = 0, 2\pi, 4\pi$, etc., there is a spatial variation of $\delta\rho$ which diffracts light. The acoustic frequency is typically about 50 MHz, so the acoustic diffraction grating "opens" and "blocks" the laser cavity with a period of about 10 ns. If the acoustic amplitude A is high so that the cavity is effectively blocked unless cos ωt is very nearly zero, then the "open" time can be as little as 50 picoseconds. The laser cavity is as illustrated in Figure 14.

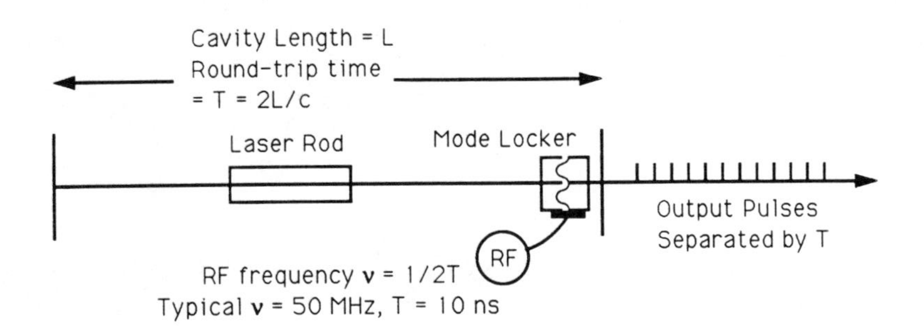

Figure 14. Mode locking.

The acoustic standing wave in the mode locker opens the cavity for about 100 ps at a frequency of 100 MHz (twice each acoustic cycle), or every 10 ns. Thus a 100-ps pulse of light is able to get through while the cavity is "open." This pulse reflects back to the laser rod and eventually returns to the mode locker in a time T which depends on the length L of the laser cavity. The cavity length and the acoustic frequency must be set correctly such that when the pulse returns to the mode locker, it finds it "opened" (i.e., the acoustic wave off) again. Partial transmission of the pulse through the end mirror of the laser yields the output, which is a series of pulses separated by time T.

Mode locking and Q-switching can be combined to produce 100-ps pulses of high peak intensity. This is illustrated in Figure 15.

This is a schematic illustration of the mode-locked, Q-switched Spectron Model 903 Nd:YAG laser that we use in this experiment. The output is a sequence

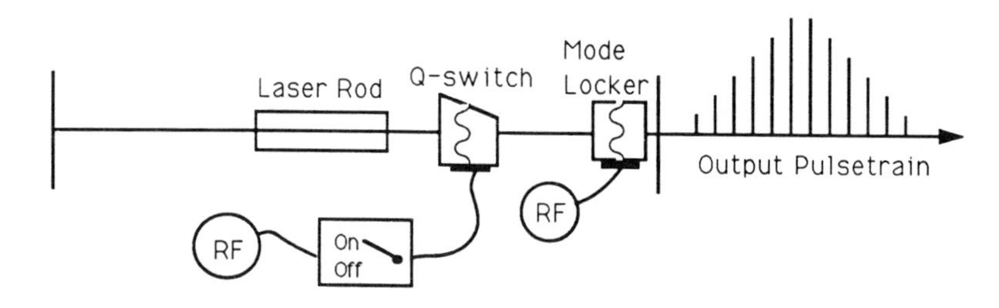

Figure 15. Mode-locked, Q-switched laser cavity.

of (about 30) pulses of 100-ps duration, separated by about 12 ns. The wavelength is 1.06 μm (near-IR). After frequency-doubling, the wavelength is 532 nm (green light). The biggest pulse in the sequence is isolated from the others and used for excitation of samples in your experiments.

The Synchronously Pumped Dye Laser The probe pulse in your experiments comes from a separate laser in which the lasing medium is an organic dye, Rhodamine 6G. This is the same dye that is studied in the transient grating experiments. While the Nd:YAG laser crystal is pumped by a bright arc lamp, the laser dye is pumped by the green pulses from the Nd:YAG laser. The pulses are absorbed by the dye and this leads to stimulated emission and lasing. The cavity lengths of the Nd:YAG and dye lasers are set to be equal, so that the dye laser pulse pumped by each green pulse returns to the dye cell to find many excited dye molecules from which it can increase its energy. The pulse in the dye laser builds up in intensity with each successive round trip through the cavity.

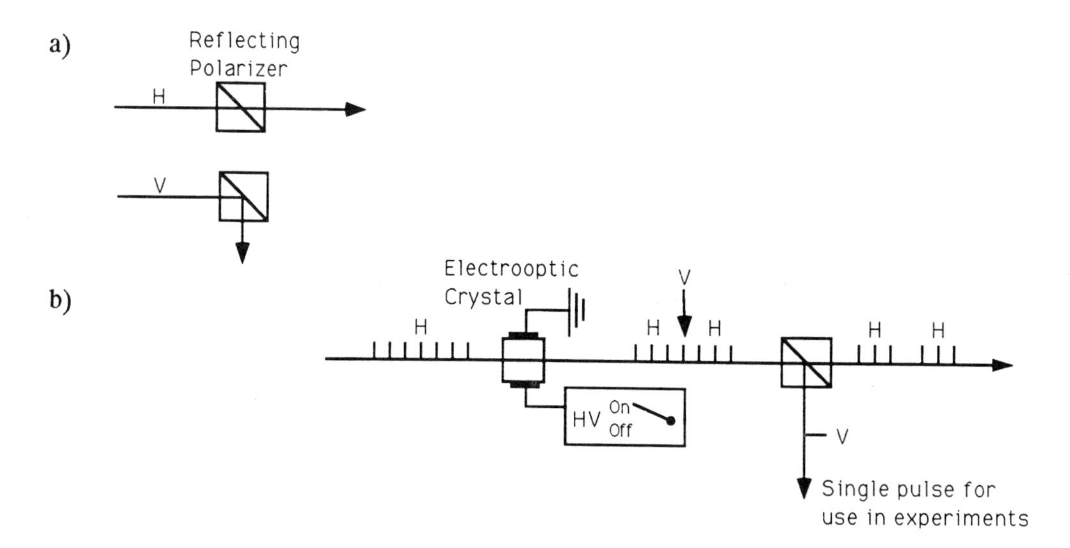

Figure 16. Pulse selection. (a) Horizontally (H) polarized light passes through the reflecting polarizer, vertically (V) polarized light is reflected. (b) H-polarized light is rotated to V by a birefringent crystal. An electrooptic crystal is birefringent only when voltage is applied. By applying voltage for only 12 ns, a single pulse can be selected.

Single Pulse Selection As indicated in Figure 15, the output of each laser is a sequence of pulses. A single pulse is used for excitation and probing of samples, so in some way it is necessary to isolate one pulse from each pulse sequence. This is done as follows. The pulses coming out of each laser are linearly polarized in the horizontal plane. If they are incident on a reflecting polarizer, they will pass through it. On the other hand, vertically polarized light is reflected as shown in Figure 16(a).

The polarization of light can be rotated by a birefringent crystal. To select a single pulse, an *electrooptic* crystal -- one which is birefringent only when voltage is applied -- is used. The voltage is applied for only about 12 ns, so only one pulse has its polarization rotated. This pulse is separated from the others and used for experiments. For the probe pulse, the single pulse is taken from inside the laser cavity itself into which the electrooptic crystal and polarizer are inserted.

The complete laser system is shown schematically in Figure 17.

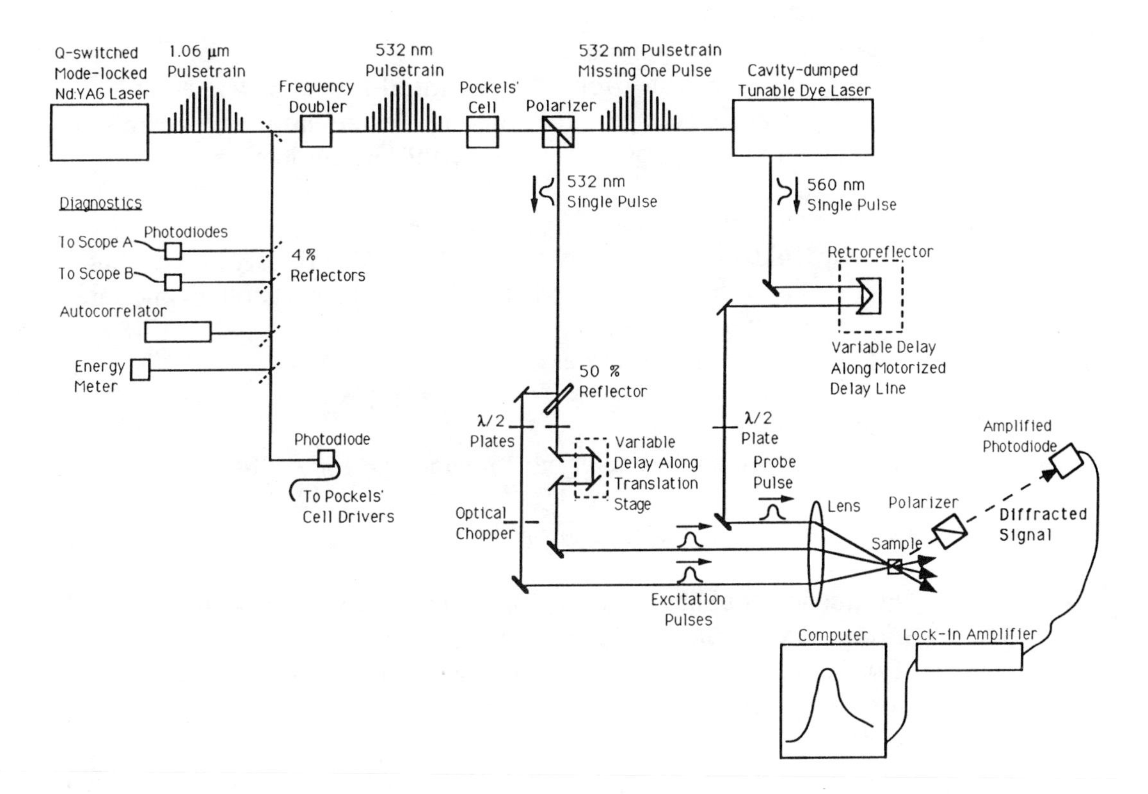

Figure 17. Schematic illustration of picosecond laser spectroscopy system and transient grating experiment. For pump-probe experiments, one excitation pulse is blocked and the transmitted probe light (not diffracted light) is monitored. Many lenses, dichroic reflectors, and other optical components are not shown. Notes: (1) Scope A and B refer to "fast" and "slow" oscilloscope channels. It is also useful to have fast diagnostics of the single pulse and the pulsetrain missing one pulse. (2) Four mirrors needing high-precision mounts are indicated with bold lines. 4% reflectors are indicated with dashed lines.

Parts List

1. Picosecond laser, laser table, and safety equipment

   A. Q-switched, mode-locked Nd:YAG laser
   Quantronix Model 4116MLQS

   B. Laser table and legs
      1. Laser table, Newport Model RS-412-8
      2. Four fixed legs for laser table, Newport Model NNH-4

   C. Laser safety glasses and goggles (10 total), Fred Reed Optical

2. Laser diagnostic equipment

   A. Autocorrelation equipment
      1. Autocorrelator for 1.06 μm, Inrad Model 514
      2. Pickoff (wedge) for aiming beam into autocorrelator, mounted on a mirror mount and magnetic base*

   B. Fast oscilloscope and auxiliary equipment
      1. Tektronix Model 2465A oscilloscope with cart and probes
      2. Photodiode detectors (4) with neutral density filters, mounted on magnetic bases*
      3. Pickoffs (4), same as A.2.

   C. Energy meter
      1. Molectron pyroelectric Joulemeter Model J3-05DW

3. Second and third harmonic generation equipment

   A. Frequency-doubling crystals (2), Aertron 3x3x5 mm KTP
   B. Frequency-tripling crystal, CSK Company, 4x4x6 mm BBO
   C. Rotation stages (3), translation stages (6), mirror mounts (3) and magnetic bases (3) for mounting frequency-doubling and tripling crystals*
   D. Dichroic mirrors for separation of 1.06μm, 532 nm, and 355 nm light (CVI)
   E. Mirror mounts (3) and magnetic bases (3) for dichroic mirrors*
   F. Polarization rotator for 1.06 μm and 532 nm for frequency tripling, Inrad Model 42-251
   G. AR-coated lenses (4) on lens mounts and magnetic bases*, for focussing into and collimating after frequency doublers and tripler

4. Single pulse selection equipment

   A. Pockels' cell and mount
      1. Pockels' cell, Inrad Model N3250
      2. Mount for Pockels' cell, Inrad Model 814-430

B. High-voltage electronics
   1. Power supply, Bertran Model 205A-05A
   2. Homebuilt Pockels' cell high-voltage switcher, including high-voltage VFETS and other components

5. Mounted optics for excitation and probe pulses and signal

   A. Polarization optics
      1. Half-wave plates for 1.06 μm (3), 532 nm (3), and 355 nm (1), mounted in rotation stages, Special Optics Model 8-R-3015
      2. Cube polarizers (3), mounted in rotation stages, Special Optics Model 7-1210-M-TM1-R
      3. Magnetic bases (9) for polarization optics*

   B. Partial and high reflectors
      1. 80% reflector for 1.06 μm, CVI
      2. 50% reflector for 1.06 μm and 532 nm, Virgo Optics
      3. High reflectors (20) for 1.06 μm and 532 nm, Newport 10D10ER.2
      4. High reflectors (5) for 355 nm, Virgo Optics
      5. Standard mirror mounts (18) for partial and high reflectors
      6. Precision mirror mounts (4), Klinger Model SL25.4
      7. Translational stage
      8. Magnetic bases for mirrors (18)*

   C. Focussing optics
      1. AR coated lenses (10)
      2. Mounts (10) and magnetic bases for lenses*

6. Probe pulse delay equipment

   A. Velmex Model B4000 stepping-motor delay line with Model 86MM controller
   B. Corner-cube retroreflector, Pyramid Optics

7. Signal detection and data storage equipment

   A. Signal detection equipment
      1. Optical chopper, Laser Precision Model CTX-534HD
      2. Lock-in amplifier, Stanford Research Systems SR510
      3. Amplified photodiode with power supply, EG&G

   B. Data storage equipment
      1. IBM PC-AT computer with hard disc and monitor
      2. A/D board, Data Translation
      3. GPIB interface card, CEC

8. Experimental alignment and sample apparatus

A. Experimental Alignment equipment
   1. Precision optical pinholes, 1 each 50, 100, 200 μm diameters
   2. 3-way translational stage for pinholes, Line Tool Model N

B. Sample mounting, flow, and temperature regulation equipment
   1. Flowing dye cells (2) for liquid samples, NSG Model 48-H
   2. Pumps (2) for sample and for T-regulated water in thermal contact with sample mount, Micropump Corp. pump and head
   3. Temperature-regulated bath, Techne Model TU-16A

C. Monochromator, PTI Model 01-001 1/4 - meter with adjustable slits and stepping-motor controller

* All translation stages are Line Tool Model N, unless otherwise specified. All rotation stages are Newport Model RSA-1T. Mirror mounts are Newport Model MFM-075 with mounting bracket and adaptor, or comparable unless specified otherwise. Lens mounts are Newport Model LM1-B2, Magnetic bases are Newport Model MB-2 with VPH-3 post, or comparable.

RECEIVED October 1, 1992

CHAPTER 13

Experiments in Laser Raman Spectroscopy for the Physical Chemistry Laboratory

Robert J. Moore, Jane F. Trinkle, Alpa J. Khandhar, and Marsha I. Lester

POTENTIAL HAZARDS: Lasers, High Voltage

Infrared and Raman spectroscopies are the predominant techniques used by molecular spectroscopists to study the vibrational structure of molecular samples *(1)*. The information obtained from the two techniques often is complementary. Although certain advantages of infrared spectroscopy (particularly FTIR) over Raman spectroscopy are familiar to most chemists, the converse is not true. For instance, by measurement of the polarization properties of the Raman scattered light, spectroscopists are able to determine which of the Raman bands observed in the spectrum result from excitation of the totally-symmetric vibrations of the molecule, i.e., those vibrations in which motion of the atoms along the normal coordinate do not cause a change in the symmetry of the molecule. Unless one uses special optics, infrared spectroscopy is limited to vibrations with frequencies greater than ca. 400 cm^{-1}; the Raman spectrometer described in this paper can be used to measure Raman bands to within 125 cm^{-1} of the Rayleigh line. It also is easier to study samples in aqueous solution because water is a weak Raman scatterer. The resonance Raman effect, which has no IR analog, can be used to selectively study the vibrational properties of individual chromophores in a complex biological system. For example, it is possible to study the prosthetic groups of proteins, such as the heme groups in hemoglobin and myoglobin, and retinal in rhodopsin.

Raman spectroscopy often is taught in the lecture portion of undergraduate courses in physical chemistry, but is rarely included in the laboratory portion of the course. This is in spite of the fact that several excellent experiments in Raman spectroscopy exist in the chemical education literature *(2-4)*. Raman signals are easily observed and measured, and Raman spectrometers are comparable in cost to analytical instruments that are commonly purchased by college and university chemistry departments. We believe that the absence of Raman experiments in the laboratory curriculum has resulted from the myths that Raman signals are exceedingly weak (Raman scattering is demonstrably not weak; see Figure 1), difficult to obtain, or require expensive high power lasers and monochromators.

Experiments in Raman spectroscopy that already have been published in the chemical education literature provide excellent examples of the versatility of this technique in molecular spectroscopy. Raman spectroscopy can be used to determine molecular force fields *(2)* and symmetries *(3)*, and to study dilute solutions of biologically significant molecules by resonance Raman spectroscopy *(4)*. In addition there exist "dry lab" experiments on pure rotational Raman spectroscopy *(5)*. (The apparatus described in this paper cannot be used to measure pure rotational Raman spectra. A spectrometer with better spectral resolution and a photomultiplier tube

2503–6/93/0217$06.00/0

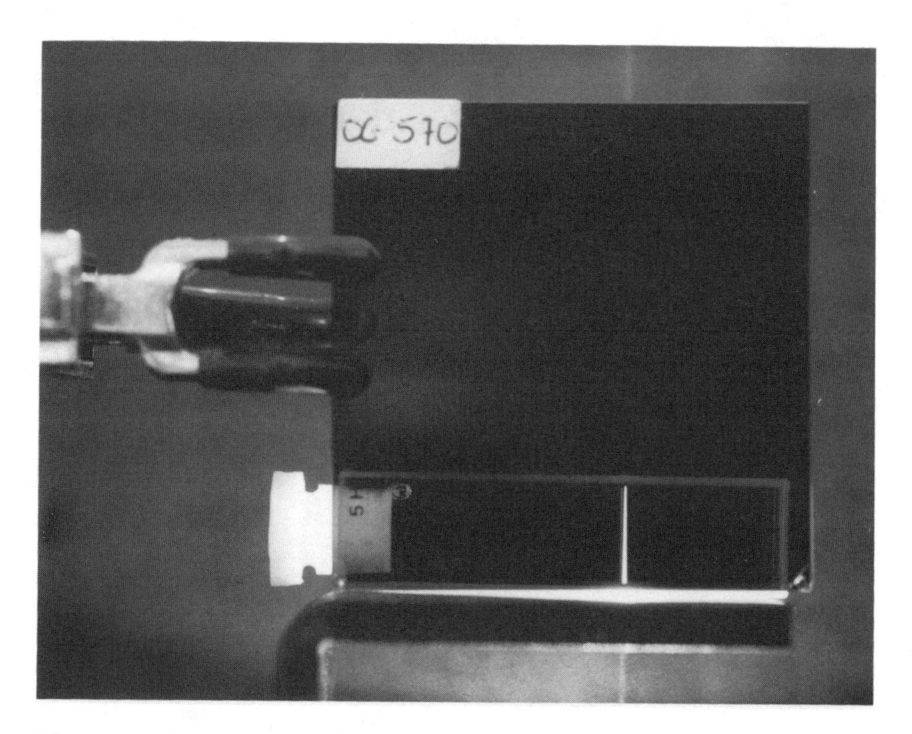

Figure 1. Raman scattering due to excitation of the C–H symmetric stretch in ethanol by a 500 mW Ar^+ laser at 514.5 nm as viewed through a colored glass filter.

with a lower dark count are required.) We will provide a review of the available experiments in Raman spectroscopy, some of which have been implemented in the Physical Chemistry Laboratory course at the University of Pennsylvania, and include one additional experiment of our own design. We also hope to explode some of the myths that currently are keeping Raman spectroscopy out of the laboratory curriculum.

Theory

An excellent and very readable discussion of the theory and experimental requirements for Raman and resonance Raman scattering is given by Carey *(6)*. Laser Raman scattering is due to inelastic collisions between photons and molecules. In these collisions energy is transferred from laser photons to sample molecules (Stokes scattering), resulting in excited molecules and scattered photons that have a lower energy than the laser photons. Alternatively, thermally excited molecules in the sample can lose energy in the collision, resulting in scattered photons that have higher energy than the laser photons (anti-Stokes scattering). The difference in the energy of the scattered photon from the energy of the laser photon is manifested as a change in the frequency, ν, of the scattered light from the laser frequency. A Raman spectrum is obtained by illuminating the sample with a laser and analyzing the frequencies of the scattered light with a monochromator. Because energy must be conserved and the energy levels of the molecules are quantized, energy transfer between photons and molecules occurs in discrete amounts. Thus, the spectrum will consist of a pattern of discrete Raman "lines" that are shifted in frequency from the laser excitation frequency, ν_L, by an amount proportional to the change in the vibrational energy of the molecule.

Light scattering occurs because the oscillating electric field associated with the laser beam interacts with the charged particles in the molecules to induce an oscillating dipole moment,

$$\pi(t) = \alpha \cdot \mathbf{E}(t) \tag{1}$$

where α is the electric polarizability and $\mathbf{E}(t)$ is the electric field vector (polarization) of the laser. For atoms, which are isotropic, $\pi(t)$ points in the same direction as $\mathbf{E}(t)$, but for molecules, which are anisotropic, $\pi(t)$ generally points in a different direction from $\mathbf{E}(t)$. This means that α is not a scalar quantity, but is a tensor, and that measurement of the polarization properties of the Raman scattered light can be used to obtain information about α. The general relationship of $\pi(t)$ to $\mathbf{E}(t)$ for molecules is:

$$\begin{vmatrix} \pi_x \\ \pi_y \\ \pi_z \end{vmatrix} = \begin{vmatrix} \alpha_{xx} & \alpha_{xy} & \alpha_{xz} \\ \alpha_{yx} & \alpha_{yy} & \alpha_{yz} \\ \alpha_{zx} & \alpha_{zy} & \alpha_{zz} \end{vmatrix} \cdot \begin{vmatrix} E_x \\ E_y \\ E_z \end{vmatrix} \tag{2}$$

When a molecule vibrates, the atomic nuclei are displaced from their equilibrium positions and the "electron cloud" is distorted. Therefore, each component α_{ij} of the polarizability tensor consists of an equilibrium contribution to the polarizability from the nonvibrating molecule as well as contributions from vibrational motions along each normal coordinate. The components of α can be written as a power series expansion in the normal coordinate displacements Q_k of the molecule. Retaining only the first two terms in the expansion, one obtains

$$\alpha(\mathrm{t}) = \alpha^{\mathrm{equil}} + \sum_{k}^{3N-6} (\partial\alpha/\partial Q_k)\, Q_k(t) \tag{3}$$

Each of the normal modes of vibration of a polyatomic molecule will make its own vibrational contribution to α.

The frequency shifts observed in the Raman spectrum of a liquid correspond to changes in the vibrational energy of the molecules in the sample. As in the case of infrared spectroscopy, there is a set of selection rules (derived from eq 3) that governs whether or not a certain vibration of the molecule is "Raman active"; that is, whether the vibration can be observed in the Raman spectrum **in principle**. For example, in both IR and Raman spectroscopy, the change in the vibrational quantum number of the kth normal mode, v_k, can only be ± 1. In IR spectroscopy, only those vibrations in which there is a change in the dipole moment during the vibration, $(\partial\mu/\partial Q_k \neq 0)$, are observable in the IR spectrum. Likewise, in Raman spectroscopy, only those vibrations in which there is a change in the polarizability of the molecule during the vibration, $(\partial\alpha/\partial Q_k \neq 0)$, are observable in the Raman spectrum.

The frequency distribution and polarization of the Raman scattered light contain information about molecular properties. In order to recover this information, spectroscopists measure a quantity called the depolarization ratio, ρ, which is the intensity of the Raman scattered light that is polarized perpendicular to the polarization of the laser, $I_{\perp}$, divided by the intensity of the scattered light that is polarized parallel to the polarization of the laser, $I_{\parallel}$. This is determined experimentally by placing a polarizer between the sample and the monochromator and measuring the spectrum twice, once for each orientation of the polarizer and evaluating $\rho = I_{\perp}/I_{\parallel}$ for each Raman band.

For crystalline samples, in which all the molecules have the same orientation, measurement of $I_{\parallel}$ and $I_{\perp}$ for different orientations of the crystal allow one to recover the individual components α_{ij} of the polarizability tensor. In liquids and gases, in which the molecules are oriented randomly relative to the polarization of the laser beam, some of this information is lost. Instead, $I_{\parallel}$ and $I_{\perp}$ are related to combinations of α_{ij} known as the tensor invariants. (The derivation of the tensor invariants and how they relate to $I_{\parallel}$ and $I_{\perp}$ are beyond the scope of this discussion. See reference 6 for a more complete discussion.) There are three tensor invariants: g^0, the isotropic component; g^s, the symmetric anisotropic component; and g^a, the antisymmetric anisotropic component. For a laser polarized perpendicular to the direction of observation, the depolarization ratio, ρ, is found to be

$$\rho = I_{\perp}/I_{\parallel} = (3g^s + 5g^a)/(10g^0 + 4g^s) \qquad (4)$$

The depolarization ratio for a given Raman line will depend on the relative magnitudes of the tensor invariants for that vibration. Except for special cases in resonance Raman spectroscopy, g^a is always zero. In addition, for vibrations that do not preserve molecular symmetry (non-totally-symmetric vibrations), g^0 is always zero. Thus, for non-totally-symmetric modes, $\rho = 3/4$, and for totally-symmetric modes (those which preserve molecular symmetry during the course of a vibration) ρ is between zero and 3/4. In the Raman literature, bands with depolarization ratios near zero are said to be strongly polarized.

The resonance Raman effect most often is used for the study of molecules that are of biological interest *(6,7)*. In resonance Raman spectroscopy, the wavelength of the laser is chosen to be coincident with or close to an electronic absorption band of the molecule. Under these conditions, the intensity of the Raman scattered light is greatly enhanced. For example, we have taken Raman spectra of 3×10^{-7} M β-carotene in acetone in which the β-carotene bands are more intense than the acetone bands, even though the concentration of acetone (13.6 M) is more than 7 orders of magnitude larger!

Although a detailed discussion of the theory of resonance Raman spectroscopy *(7)* is beyond the scope of this discussion, a few simple points can be made. First, in cases where the electronic transition that gives rise to the intensity enhancement is dipole-allowed, only totally-symmetric vibrations (with $0 \leq \rho \leq 3/4$) will appear in the resonance Raman spectrum. Furthermore, there is a Franck–Condon effect; for example, if the length of a certain bond in the molecule is substantially different in the ground and excited states, then totally symmetric vibrations in which the atomic displacements result in the stretching of that bond may be prominent in the spectrum. This so-called "A-term" scattering is in exact analogy to electronic absorption spectroscopy, in which only totally-symmetric modes that project onto large geometry changes upon electronic excitation are observed in the spectrum.

Second, in cases where a second excited electronic state is involved in the electronic transition (through vibronic coupling), both non-totally-symmetric and totally-symmetric modes may be observed in the spectrum, although the non-totally-symmetric modes tend to be more intense. This is called "B-term" scattering. In the case of heme proteins, some of the non-totally-symmetric vibrations observed in the resonance Raman spectrum have a depolarization ratio > 3/4. For these vibrations, the g^a tensor invariant is non-zero, and the Raman bands are said to be anomalously polarized. In fact, according to theory, these bands should be inversely polarized, with $\rho = \infty$. Thus, the remarkable signal enhancements of the resonance Raman effect can provide information about the electronic and vibrational structure of biologically significant molecules.

Safety

The experiments described herein use a continuous wave Ar^+ ion laser as a light source for Raman scattering. The laser beam is an extreme vision hazard and should be treated as such. First, require students to wear laser safety goggles at all times that the laser is on. Second, we suggest that when designing the apparatus, the aperture of the laser be placed in close proximity to the entrance port of the sample compartment to minimize the "length of beam" that is exposed to view.[1] Third, make sure all beams and reflections are stopped (that is, blocked by an object with a non-reflecting surface). Fourth, the controls for the apparatus should be placed as far as possible from the exposed beam to minimize the opportunity for students to come in contact with the beam. Fifth, use the minimum laser intensity that yields high quality spectra. Sixth, use the minimum laser power when making any adjustments to the laser beam. Seventh, require students to remove rings, bracelets and watches before doing the experiment to minimize the possibility of inadvertant specular reflections. Eighth, instruct students not to stare into the laser beam or to try to view it indirectly. Ninth, isolate the laser from the rest of the lab by putting it in a separate room or by encircling the apparatus with floor to ceiling blackout curtains.

Several of the samples, particularly $CHCl_3$ and CCl_4, used in the experiments are toxic. Toluene, acetone and petroleum ether are toxic and flammable. Load all cells or capillaries in a hood and properly dispose of wastes when finished.

Methods

An excellent overview of the experimental requirements for laser Raman spectroscopy is given by Tobin *(8)*. Our Raman spectrometer consists of a laser

---

[1]Please note that although placing the laser aperture near the entrance port of the sample compartment will reduce the safety hazard due to the laser beam, it is possible that plasma lines (spontaneous emission from excited atoms and ions in the laser plasma tube) may become prominent in the spectrum. Plasma lines can be eliminated by use of a commercially available plasma line filter.

light source, in this case a Spectra-Physics model 2020-05 Ar^+ laser, a Spex Industries model 1439 sample compartment, which contains focusing and collection lenses as well as mounts for polarizing optics, a Spex model 1680 0.22 m double monochromator controlled by a model 1673 Minidrive, a Hamamatsu R928P photomultiplier tube in a Products for Research model TE177RF cooled housing, and a Modern Instrument Technology model PRM-100E photon counter (see Figure 2). The operation of the spectrometer is as follows:

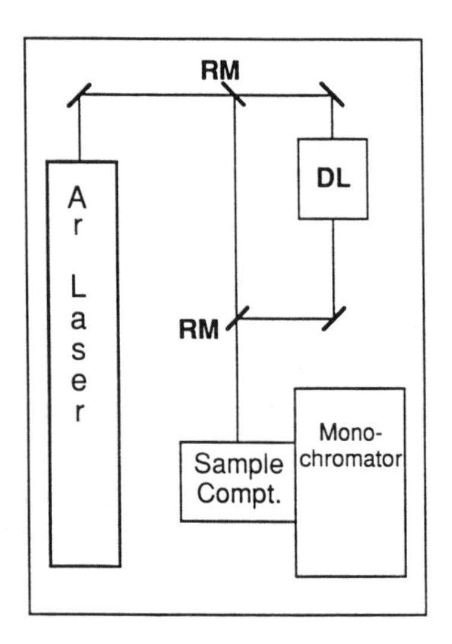

Figure 2. A schematic of the Raman spectrometer used in the undergraduate Physical Chemistry lab at the University of Pennsylvania. In this configuration, light from the Ar^+ laser (or dye laser) can be used as an excitation source in Raman or laser induced fluorescence spectroscopy. DL, dye laser; RM, removable mirror.

The laser beam is focused on the sample and a camera lens collects the scattered radiation from the sample and images it onto the entrance slit of the double monochromator (when setting up the sample compartment, one should make sure that the polarization of the laser is oriented perpendicular to the direction of observation). The double monochromator consists of two grating monochromators placed back to back and scanned in tandem. The function of the double monochromator is to filter out the strongly scattered Rayleigh light (which is at the same frequency as the laser light source) and to resolve the spectral components of the Raman scattered light. (Double monochromators are preferred over single monochromators because of their superior stray light rejection capability. This allows one to measure weaker signals and to work closer to the Rayleigh line. Some Raman spectroscopists use single monochromators equipped with holographically ruled gratings, which have fewer imperfections than mechanically ruled gratings and thus provide better stray light rejection. Although still inferior to double monochromators, these systems are substantially less expensive and allow a higher fraction of the light that enters the monochromator to strike the detector.) Because the gratings diffract horizontally and vertically polarized light with different efficiencies, a polarization scrambler is positioned between the sample and the entrance slit to convert the Raman scattered light into randomly polarized light which then enters the monochromator. When measuring polarization spectra, the polarizer is placed between the sample and the scrambler.

After being dispersed by the double monochromator, the scattered light strikes the photocathode of a cooled photomultiplier tube. The photocathode absorbs the scattered photons, and, by the photoelectric effect, emits electrons which subsequently are amplified in the photomultiplier tube. The output of the photomultiplier tube is a swarm of pulses of electrons resulting from the photoemission of single electrons that can be counted by a photon counting apparatus. The photomultiplier is cooled to reduce thermionic emission of electrons from the photocathode. A pre-amplifier/discriminator connects the photomultiplier tube to the photon counter. The discriminator is set so that only pulses of electrons that originate from the photocathode (generated by the emission of a photoelectron or by unavoidable thermionic emission ("dark counts")) are counted by the photon counter. The photon counter converts the photon counts into a voltage that is displayed on a chart recorder, or a digital signal that is stored in the memory of a computer. By scanning the monochromator and recording the number of counts, a spectrum of the scattered light is obtained.

We recommend four experiments that illustrate the utility and power of Raman spectroscopy. Two of these have been published previously in *The Journal of Chemical Education*, the third has been published in a laboratory text, and the fourth has been developed in our laboratory. For the previously published work, we refer the reader to the original literature, and will provide only a short discussion of the experiment and experimental requirements.

Raman Spectra of ZXY_3 Compounds: This experiment was originally published in 1974 as a "dry lab" experiment *(3)*. The purpose of the experiment is to use group theory and vibrational spectroscopy to elucidate the molecular geometry of a polyatomic molecule. The students are asked to consider the possible geometries of a molecule with the molecular formula ZXY_3. The students use group theory to determine the symmetries of the nine vibrations for each molecular geometry (Vincent's book, *Molecular Symmetry and Group Theory*, has an excellent introduction to this topic *(9)*). The results of their analysis will tell them which of these vibrations are allowed in the Raman spectrum, and which will give rise to polarized Raman bands.

The students should be required to work through this theoretical analysis before coming to lab. They should also be required to calculate the monchromator scanning range, given that they will find all of the Raman bands between 200 and 3100 cm^{-1} from the Rayleigh line. Two spectra are required; one taken with the polarizer oriented parallel to the polarization of the laser and the second with the polarizer perpendicular to the laser (100 mW @ 514.5 nm). $CHCl_3$ is a convenient sample, and only one three hour lab period is required for this experiment.

Molecular Force Fields: Carbon Tetrachloride: In this experiment students measure the Raman spectrum of CCl_4 *(2)*. Because all of the vibrations of CCl_4 are "Raman active", the frequencies of all the vibrational modes can be obtained from the Raman spectrum. The measured frequencies can be used to evaluate the force constants used in two different models of molecular force fields.

In previously published discussions of this experiment, the valence force field is calculated from the spectrum. In this case the forces are considered to be directed along the chemical bonds (these are forces that resist stretching and compression of the bonds from their equilibrium bond lengths) and "along" the angles between bonds (these are the forces that resist distortions of the molecule from its equilibrium shape, which for CCl_4 is a tetrahedron). In the case of CCl_4, since all four bonds are equivalent, only two force constants are required to specify this force field, k, the stretching force constant, and k_δ/l^2, the bending force constant, where l is the equilibrium C – Cl bond length. The equations relating the frequencies to the force constants can be found in Shoemaker, et al. *(2)* and will not be reproduced here.

A second force field model is the central force field *(1)*. In this case, we dispense with the concept of valence, and consider every atom to be "bonded" to every other atom in the molecule. At first glance, one might think that the central force model and the valence force model are nearly equivalent, in that a compression or stretching of a Cl – Cl "bond" is equivalent to the bending in the valence force model. However, this is not the case. The primary difference between the two models is that in the valence model, at the equilibrium molecular geometry (zero net force on each atom in the molecule), there is also zero force on each atom along the actual C – Cl chemical bonds and also no force along the bending coordinate. In the central force model although the net force on each atom in the molecule is zero at the equilibrium configuration, the individual forces between pairs of Cl atoms and between each Cl atom and the central C atom are not individually zero. For example, a repulsion between Cl atoms in the equilibrium position could be balanced by an attraction between C and Cl. Thus, in order to model the forces between individual atoms, an additional force constant, k', must be introduced in the central force model to account for the fact that all forces are not individually zero at equilibrium. The value of k' can be positive or negative. In our analysis, a negative value of k' is indicative of an additional attractive force between pairs of Cl atoms, whereas a positive value of k' represents a repulsion between pairs of Cl atoms (See Reference 1, pp. 165 - 168 for a more complete discussion of the application of the central force model to CCl_4).

In the central force field model, there are four equations that relate the four molecular frequencies to the three force constants. The force constant equations are given below, where k_1 is the force constant for the Cl – Cl "bonds" and k_2 is the force constant for the C – Cl bonds. In these equations, ν is in s^{-1}, m is in kg (per **atom**, not per mole of atoms!), and each k in kg/s^2 (N/m).

$$\lambda_1 = 4\pi^2\nu_1^2 = (k_2 + 4k_1)/m_{Cl} \tag{5a}$$

$$\lambda_2 = 4\pi^2\nu_2^2 = (k_1 - k')/m_{Cl} \tag{5b}$$

$$\lambda_3 + \lambda_4 = 4\pi^2(\nu_3^2 + \nu_4^2) = 2k_1/m_{Cl} + k_2(4m_{Cl} + 3m_C)/3m_{Cl}m_C - 2k'(16m_{Cl} + 3m_C)/3m_{Cl}m_C \tag{5c}$$

$$\lambda_3\lambda_4 = 16\pi^4\nu_3^2\nu_4^2 = 2\,(4m_{Cl} + m_C)\,(k_1k_2 - 8k_1k' - 5k_2k' - 8k'^2)\,/\,3m_Cm_{Cl}^2 \tag{5d}$$

The force constant k_1 is most easily determined by use of the following equation:

$$\lambda_3 + \lambda_4 - \lambda_1(4m_{Cl} + 3m_C)/3m_C - 2\lambda_2(16m_{Cl} + 3m_C)/3m_C$$
$$= -4k_1(4m_{Cl} + m_C)/m_{Cl}m_C \quad (6)$$

For this experiment we recommend that the students measure a survey spectrum to identify the Raman bands, followed by a high resolution spectrum with a Hg arc lamp placed near the entrance slit of the monochromator to serve as an internal calibration spectrum. Finally, polarization spectra are required to identify the totally-symmetric vibration ν_1. We use the 514.5 nm line of the laser at a power of ca. 100 mW. With our apparatus, the "splitting" of the vibrational bands due to the isotopes of Cl are not observable, except as an asymmetry in the shape of the ν_1 band. The students can determine the instrumental resolution from the arc lamp spectrum and predict the change in vibrational frequencies with isotopic substitution ($C^{35}Cl_4$ and $C^{37}Cl_4$) to show why these bands are not resolved. This experiment requires only one three hour lab period.

Resonance Raman Spectroscopy: β-Carotene: Resonance Raman spectroscopy of carotenoids was originally proposed as an undergraduate experiment by Hoskins in 1984 *(4)*. Students obtain the resonance Raman excitation profile for the 1155 cm^{-1} vibration of the carotenoids lycopene or β-carotene. Lycopene and β-carotene are conjugated polyenes that are pigments in tomatoes and carrots, respectively. This experiment can only be done with a line-tunable Ar^+ laser or dye laser. Average laser powers of 20 - 50 mW were found to be sufficient to acquire "good" spectra with our apparatus.

Strommen and Nakamoto have discussed the theory of resonance Raman spectroscopy *(7)*, and Hoskins has presented a simplified discussion of the dependence of the intensity of A-term Raman scattering in carotenoids in acetone solution as a function of laser frequency *(4)*. In this treatment, the intensity of a Raman band relative to a nearby solvent band, I, is related to the frequency of the laser:

$$1/I = (\nu_{00} - \nu_L)^2/C + \Gamma^2/C \quad (7)$$

In Hoskin's model, ν_{00} is the frequency of the absorption maximum of the electronic origin of the sample, ν_L is the laser frequency, Γ is the homogeneous broadening factor and C is a constant that is related to the intensity of the electronic transition. Γ is proportional to the width of the electronic absorption band, and, by the uncertainty principle, the inverse of the lifetime of the excited state. Plotting $1/I$ versus $(\nu_{00} - \nu_L)^2$ yields a straight line with slope $1/C$ and intercept Γ^2/C. If a nonlinear least-squares program is available to the students, then C, Γ and ν_{00} can be obtained as adjustable parameters in a fit of the intensities versus wavelength.

Although lycopene and β-carotene can be extracted from tomato paste and carrots, respectively, both are available commercially. β-Carotene is by far the cheaper compound, so we purchased 95% *trans*-β-carotene from Sigma and purified it by thin layer chromatography (TLC) with a 90% petroleum ether - 10% toluene developing solution. This procedure is necessary to remove fluorescent impurities that obscure the Raman spectrum at excitation wavelengths shorter than 514.5 nm. We prepared five TLC plates (6 cm wide) by spotting a saturated solution of β-carotene in acetone every three millimeters along the bottom of the TLC plate.

After development and drying at room temperature in a hood, we scraped the β-carotene spots (R_f ~ 0.8) from the plate and placed the scrapings in a 10-mL volumetric flask. We filled the flask to the mark with acetone and let the silica settle over night (the silica also can be removed by centrifugation). The resulting solution was decanted and found to have a concentration of 3 × 10^7 M β-carotene by measurement of the visible absorption spectrum, from which we obtained ν_{00}.

A Raman spectrum of neat acetone must be measured so that the pattern of solvent bands can be easily identified in the β-carotene spectra. If the data are to be fit to eq 7, then at least three different laser excitation wavelengths must be used, with at least one below 488 nm. If ν_{00}, Γ, and C are to obtained by fitting the data through a non-linear least squares regression, then a minimum of four laser excitation wavelengths must be used.

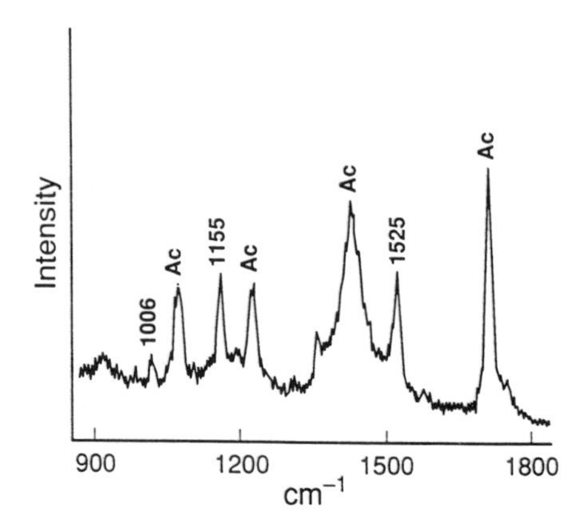

Figure 3. The resonance Raman spectrum of a 1.2 × 10^{-6} M solution of β-carotene in acetone, with the 1155 and 1525 cm^{-1} bands as indicated. The excitation wavelength was 514.5 nm and the laser power was 20 mW. The bands marked Ac are solvent bands.

In our work, the β-carotene solution was transferred to a capillary with a syringe and the Raman spectrum was measured for excitation wavelengths of 514.5 nm (Figure 3), 496.5 nm, 488.0 nm, and 457.9 nm (other Ar^+ laser lines at 501.7 nm and 476.5 nm also can be used if time permits). The intensity distribution that we measured was in excellent agreement with the literature *(10)*. The students should be asked to estimate the lifetime of the excited state of β-carotene based on the value of Γ obtained from their fit to equation 7. This experiment requires two three hour lab periods; one for sample preparation and the second for measurement of the Raman spectra.

Resonance Raman Spectroscopy of Ferrocytochrome c: In this experiment, the students measure the resonance Raman spectrum of ferrocytochrome *c*, and identify the symmetries of the Raman active vibrations from their polarization properties *(6, 11)*. Ferrocytochrome *c* is a heme protein that plays an important role in respiration. The chromophore and active site in ferrocytochrome *c* is an iron porphyrin with D_{4h} symmetry. The visible absorption spectrum of porphyrin containing proteins consists of three transitions, the α and β bands in the neighborhood of 530 nm, and the intense Soret transition at around 400 nm. The β band in the absorption spectrum is a vibronically induced transition. The vibrational modes that mix the dipole allowed α and Soret transitions are those that are enhanced in the resonance Raman spectrum.

The symmetry of the electronic states excited in the Soret and α transitions is E_u. Only those vibrations that are contained in the direct product of the group theoretical representations of the α and Soret bands will be found in the resonance Raman spectrum. These vibrations belong to the symmetry classes A_{1g}, A_{2g}, B_{1g} and B_{2g}. (A good student exercise is to have them obtain the direct product themselves.) The B_{1g} and B_{2g} modes give rise to the depolarized bands in the spectrum. The A_{1g} modes are totally symmetric modes, and hence are not vibronically active. The very weak polarized bands that are observed in the spectrum result from resonance enhancement of the A_{1g} bands from the Soret and α transitions. Finally, the scattering tensor for the A_{2g} modes is antisymmetric, so the tensor invariant g^a is non-zero for these vibrations. The result is that the A_{2g} modes give rise to inversely polarized bands.

A 0.5 mM aqueous solution of ferrocytochrome *c* was prepared by dissolving 62 mg of ferricytochrome *c* (horse heart cytochrome *c*, Sigma) in 10 mL of deionized water, followed by reduction to ferrocytochrome *c* by addition of an excess of sodium dithionate (the solution changes color upon reduction). The solution was diluted with deoxygenated deionized water to 0.05 mM and placed in a 1-cm pathlength fluorimetry cell. (The Raman spectrum of solution can be measured without further dilution by transferring the solution into a capillary with a syringe.) The excitation laser was tuned to 514.5 nm and the power set to 50 mW. Two spectra, with the polarizer oriented parallel and perpendicular to the laser polarization, were measured from 600 - 1700 cm^{-1} below the frequency of the laser. In this region there are four polarized Raman bands, four inversely polarized bands and six depolarized bands. These spectra are shown in Figure 4. From the polarization spectra the students can assign symmetry of the Raman bands to be A_{1g} (polarized), A_{2g} (inversely polarized) and either B_{1g} or B_{2g} (depolarized). One three hour lab period is required for this experiment.

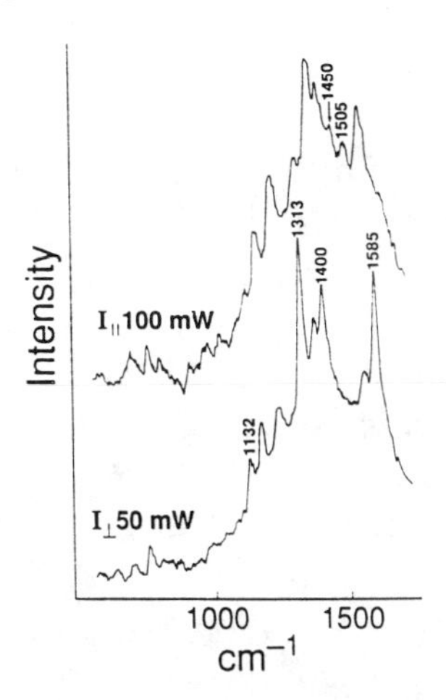

Figure 4. Resonance Raman spectrum of a 5×10^{-5} M solution of ferrocytochrome *c*, $I_{\parallel}$ and $I_{\perp}$. The excitation wavelength was 514.5 nm. Two of the four polarized Raman bands are labeled in the upper spectrum. Note that these bands are absent in the lower spectrum. The four inversely polarized Raman bands are labeled in the lower spectrum.

Summary

We have provided a review of four experiments in laser Raman spectroscopy that emphasize the power and utility of this technique in modern molecular spectroscopy. In all cases, the Raman signals are relatively large (on the order of 10^3–10^4 counts per second over a background of 2–3 counts per second), the samples are readily available and easy to prepare, and students are exposed to a modern apparatus that they find exciting to use. The cost of the Raman apparatus is about the same as a gas chromatograph equipped with a mass selective detector, and much less if an air-cooled Ar^+ laser is used. Undergraduate students (including AJK and JFT) helped to implement all of the experiments that we now use in the Physical Chemistry Laboratory at Penn, so the experiments are within the technical and scientific capabilities of undergraduates.

Acknowledgments

This work was supported by a grant from the Pew Mid-Atlantic Consortium of the Pew Charitable Trusts. We thank the Keck Foundation for an educational grant to purchase our laser. We also wish to thank Dr. David Van Dyke of the University of Pennsylvania for his help in the laboratory.

Literature Cited

1. Herzberg, G. *Molecular Spectra and Molecular Structure II. Infrared and Raman Spectroscopy of Polyatomic Molecules*; Van Nostrand Reinhold: New York, NY, 1945.
2. Shoemaker, D. P.; Garland, C. W.; Nibler, J. W.; *Experiments in Physical Chemistry*, 5th edition, McGraw-Hill: New York, NY, 1989; pp. 451-461.
3. DeHann, F. P.; Thibeault, J. C.; Ottesen, D. K. *J. Chem. Educ.* **1974**, vol. 51, pp. 263 - 265.
4. Hoskins, L. C. *J. Chem. Educ.* **1984**, vol. 61, pp. 460 - 462.
5. Hoskins, L. C. *J. Chem. Educ.* **1977**, vol. 54, pp. 642 - 643.
6. Carey, P. R. *Biochemical Applications of Raman and Resonance Raman Spectroscopies*, Academic Press: New York, NY, 1982.
7. Strommen, D. P.; Nakamoto, K *J. Chem. Educ.* **1977**, vol. 54, pp. 474 - 478.
8. Tobin, M. C. *Laser Raman Spectroscopy*; Chemical Analysis, vol. 35; Wiley-Interscience: New York, NY, 1971.
9. Vincent, A. *Molecular Symmetry and Group Theory*; John Wiley & Sons, Ltd.: London, UK, 1977.
10. Hoskins, L. C. *J. Chem. Phys.* **1980**, vol. 72, pp. 4487 - 4490.
11. Spiro, T. G.; Strekas, T. C. *Proc. Nat. Acad. Sci. U.S.A.* **1972**, vol. 69, pp. 2622 - 2626.

Hardware List

Laser: Spectra-Physics model 2020-05 Ar ion Laser

Monochromator: Spex Industries model 1439 sample compartment

Spex model 1680 0.22 m double monochromator (to mate the monochromator to the sample compartment we had to machine some legs to elevate the monochromator)

Spex model 1673 Minidrive scan controller (Spex now makes an interface so that a personal computer can be used to control the scan)

Spex also manufactures a complete turnkey Laser Raman Spectrometer, the model 1488I Analytical Raman System, for those who do not want to construct their own system from components.

Detection: Hamamatsu R928P photomultiplier tube

Products for Research model TE177RF cooled PMT housing

Modern Instrument Technology model PRM-100E precision rate meter for photon counting

Optics and Mounts: Newport Corporation provides a complete line of optical tables, mirror mounts and optics that are suitable for these experiments and other laser-based experiments described in this volume.

RECEIVED October 1, 1992

Laser Experiments in Thermodynamics

CHAPTER 14

Time-Resolved Thermal Lens Calorimetry with a Helium–Neon Laser

J. E. Salcido, J. S. Pilgrim, and M. A. Duncan

POTENTIAL HAZARDS: Lasers

The thermal lens effect results when a laser beam passing through a sample is absorbed, causing heating of the sample along the beam path. As shown below, the sample heating may in turn modify the transmittance of the same laser which caused the heating. Specifically, sample heating may cause expansion of the transmitted laser spot size in an effect known as "thermal blooming" or "thermal lensing." This relatively simple effect has been described in the literature as the basis for ultrasensitive analytical techniques. *(1,2)* At least one prior experiment has been described for undergraduate instrumental laboratory. *(3)* The utility of thermal lens calorimetry is derived from the brightness of laser light sources and the sensitivity with which laser beams may be detected. In the experiment described here, a simple experimental apparatus using inexpensive components is employed to observe the thermal lens effect and to derive useful experimental data from its magnitude and time dependence. Specifically, the heat capacity of different solvents and the extinction coefficients of blue dye solutions are determined. Because of its simple content, this experiment is recommended as the first in a curriculum which will eventually use more complex laser systems. Alternatively, it is inexpensive enough for even the most restrictive laboratory budgets where the purchase of larger systems is infeasible.

The origin of the thermal lens effect is easy to understand. Virtually any kind of laser may exhibit this effect in liquid, solid or gaseous samples. The first requirement is a sample which absorbs the laser radiation to be used. In the experiment described here, we use a red helium-neon (He-Ne) laser at 632.8 nm with various solutions which are blue in color. Blue solutions absorb a red laser beam passing through them, resulting in the desired heating along the laser path. The total amount of heating depends on how strongly the solution absorbs the laser and on the power of the laser. However, since the spatial intensity profile of the laser is non-uniform, the heating will not be evenly distributed throughout the sample. It is greatest at the center of the laser beam and less toward the outer edges. For the common case of a Gaussian beam profile, which applies to many continuous (CW) lasers including the helium-neon laser, the spatial variation in intensity is a well-characterized function. The heating caused by absorption of a Gaussian beam has radial symmetry along the laser path which creates a corresponding radial temperature gradient in the sample. The refractive index of most materials, including the solvents used here, decreases with increasing temperature. Therefore, corresponding to the temperature gradient in the sample there will be a gradient in the refractive index, with the refractive index lowest at the center of the beam path. Effectively, then, the laser path is shorter at beam center. This effect, with the radial symmetry, makes a diverging lens out of the sample. The same laser which heats the sample to cause lens formation can be observed to diverge, or "bloom," as it passes through the sample. Blooming can be detected

2503–6/93/0232$06.00/0

visually, or instrumentally with a photodiode positioned on the center of the laser beam axis. When blooming occurs, the diode detects a loss in laser intensity which is proportional to the absorption strength.

Another interesting aspect of thermal lensing is that the heating and subsequent lens formation is not instantaneous. It takes a finite time to develop depending on the laser power and the thermal properties of the sample. In solution, the thermal properties of the solvent (heat capacity and thermal conductivity) determine the time for blooming to occur (typically milliseconds). To best observe the thermal lens effect, laser light is focused with a lens at a precise position on the sample in the laser path. The light from the CW He-Ne laser is turned off and on with a rotating chopper. By correct choice of components, lens formation will occur during the "on" cycle of the laser through the chopper, and it will dissipate by cooling during the "off" cycle, so that the effect can be observed repetitively. Detailed equations predicting the magnitude and time behavior of thermal lensing are given below, which make it possible to select components and optimize the performance of the experimental system.

Theory

To derive useful data from the thermal lens waveform, it is important to consider briefly some more detailed concepts about the focussing properties of laser beams. *(4)* As shown in Figure 1, a laser focused with a lens does not achieve an arbitrarily

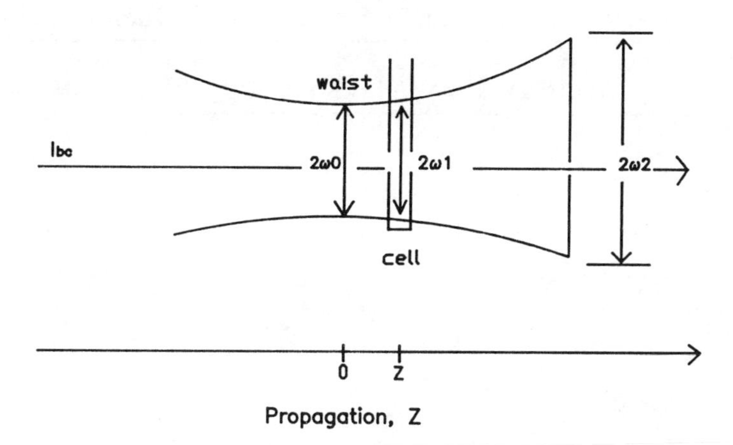

Figure 1. A diagram of the confocal waist of a focused laser beam (Gaussian profile). The sample is placed at the position Z downstram from the minimum waist to optimize the signal.

small spot size. In Gaussian laser beams, the actual spot size is limited by diffraction to a minimum beam waist, ω_0, defined as half the beam diameter. This minimum beam waist is related to the waist before focussing, ω_i, by

$$\omega_0 = \lambda f / \pi \omega_i$$

λ is the wavelength of the laser used and f is the focal length of the lens. Therefore, a shorter focal length lens results in a smaller spot size at the focus.

As shown in Figure 1, the focused beam achieves a minimum diameter, maintains nearly this same diameter for some distance, and then diverges again. A convenient measurement of the focussing is the "confocal length," z_c, which is the distance from the center of the minimum waist region to a point downstream where the beam size is $\sqrt{2}$ larger, i.e., $\omega_c = \sqrt{2}(\omega_0)$, where ω_c is the beam waist at the confocal length. It can be shown that the maximum thermal lens signal occurs when the sample is placed at the point z_c beyond the center of the focus. *(1)*

When the sample is positioned at z_c, a simple expression can be used to obtain the absorbance, A, of the solution,

$$\Delta I_{bc} / I_{bc} = -2.303\, P\, (dn/dT)\, A \,/\, \lambda\, \kappa$$

$$= 2.303\, E\, A$$

Here, I_{bc} is the initial intensity measured at the center of the laser beam and ΔI_{bc} is the loss in intensity after the blooming has reached its steady state value. P is the laser power in watts, dn/dT is the variation of the refractive index with temperature, λ is the wavelength, and κ is the thermal conductivity of the solvent. E represents an enhancement factor reflecting the improved sensitivity of the thermal lens effect over simple absorption experiments. It depends on the wavelength and the laser power used. E values for selected solvents are given in Table 1. As shown, the

Table 1. Thermo-optical Data for Solvents used in Thermal Lens Experiments.

Solvent	C_p(J/mol×K)	κ(mW/cm×K)	dn/dT×10^4/K	E/mW	ρ(g/cm^3)
Acetone	126.4	1.60	-5.0	2.16	0.7899
Benzene	135.6	1.44	-6.4	3.05	0.8765
Carbon Tetrachloride	132.6	1.02	-5.8	3.88	1.5867
Diethyl ether	172.0	1.30	-5.0		0.7138
Ethanol	113.0	1.67	-3.9		0.7893
Methanol	81.6	2.01	-3.9	1.33	0.7914
Toluene	156.1	1.33	-5.6		0.8669
Water	75.3	6.11	-0.8	0.09	1.000
Cyclohexane	152.3	1.24	-5.4	3.6	0.7785

enhancement is greatest for nonpolar solvents which usually have high dn/dT and low thermal conductivity. The equation above applies strictly to the case of a "weak" thermal lens. When the relative change in I_{bc} is greater than 0.1, a higher order quadratic term should be included in the analysis, i.e.,

$$\Delta I_{bc}/I_{bc} = 2.303\ E\,A + 0.5\ (2.303\ E\,A)^2$$

As mentioned above, the time variation in the thermal lens effect can also be used to determine the heat capacity, C_p, of the solvent. When the time constant for the decay of intensity, t_c, is determined as the $1/e$ lifetime of the exponential decay,

$$t_c = \omega_0^2\ \rho\ C_p\ /\ 4\ \kappa$$

where ω_0 and κ are defined above and ρ is the density of the solvent. *(1)*

Methodology

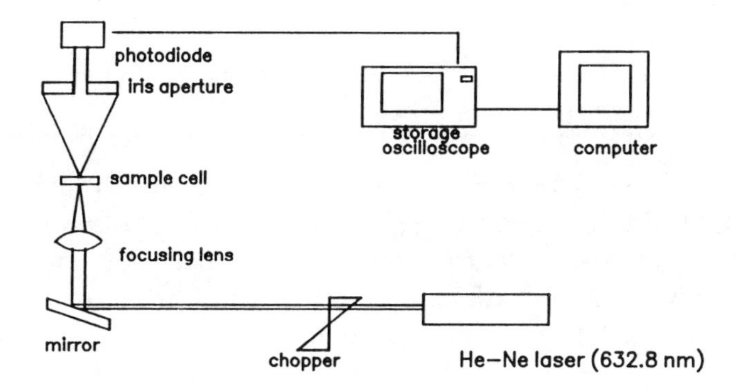

Figure 2. A schematic diagram of the components needed for a thermal lens experiment.

A schematic diagram of the thermal lens experiment in our laboratory is shown in Figure 2. The laser used is a He-Ne (Hughes series 3000) operating at 632.8 nm. This laser provides a nominal output of 5 milliwatts in a spot size of 0.83 mm dia. (at $1/e^2$ of the intensity at beam center) with a divergence of 1.0 mrad (another comparable laser is the Spectra Physics model SP-105-1). *(5)* The laser is passed through a 60 Hz chopper (homemade), bounced with a mirror on an *x,y* tilting mount (e.g., Newport model DM.4 mirror with MM-2 mirror mount), and then focused with a positive lens (80 mm focal length). Although alignment without the mirror mount can in principle be achieved by moving the He-Ne laser body, utilization of the mirror is highly recommended. At a critical distance beyond the focus of the laser beam, which must be adjusted to optimize the lens effect, the diverging laser passes through a 2 mm thick cuvette containing the sample at a carefully measured concentration. Adjustment of the sample-to-lens separation is accomplished with a basic translation stage (Newport model TSX-1A). At a point about 1 meter beyond the sample, an adjustable aperture (e.g., Newport model ID-1.0) is positioned on the laser beam. The light transmitted through this aperture is detected by a photodiode (Newport model 818-SL). The output of the photodiode is displayed on an oscilloscope, from which the experimental data of light intensity versus time is acquired. In our setup, a digital storage scope (LeCroy model 9410) interfaced to a PC (Zenith PC-286) is used. However, an equally effective and less expensive setup could use an analog oscilloscope (e.g., Tektronix model 2205), from which the waveform intensity versus time data are read visually.

We have observed the thermal lens effect in a variety of blue colored solutions using various solvents. Table 2 lists suggested compounds and typical

Table 2. Compounds that Exhibit the Thermal Lens Effect with a He-Ne Laser.

Compound	*Mol. Wt.*	*Solvent*	*Conc.* (M)	$\epsilon(M^{-1}cm^{-1})$
Acid Blue 25	416.4	MeOH	$2.6x10^{-4}$	4540
Indophenol Blue	276.3	MeOH	$2.3x10^{-4}$	3260
Bromophenol Blue	691.9	Acetone	$2.0x10^{-4}$	2475
Azulene	128.16	Acetone,MeOH	$2.4x10^{-3}$	254

concentrations for this experiment. These compounds are available from Aldrich and/or Sigma Chemical Companies. Table 1 provides a list of potential solvents for these experiments and their thermo-optical properties (see calculations below). Several solute/solvent combinations from these lists are possible. If alternate solutions are to be used, the concentrations may have to be adjusted for optimum performance.

Convenient operation of the experiment can be assured with a well-conceived layout for the optical components. As in all laser setups, system alignment and stability are critical. The various components must be mounted on a flat surface, such as an optical rail or breadboard optical table, to which optical elements can be attached securely. Commercial optical mounting hardware *(6)* is essential for items such the mirror mount and lens translator, but homemade brackets and clamps are acceptable for mounting the laser head, the focussing lens, the sample, and the aperture. However, unstable mounting of any component in the beam line will cause nothing but aggravation to the experimenter. It is most convenient if the laser and other optical elements are mounted in the same plane at a fixed height off the table. The laser should be mounted perpendicular to the beam axis containing the other components. In this position, its light can be bounced easily with the mirror to define the desired optical axis.

It is easiest to align the optical system initially without the sample or the focussing lens in place. Bounce the unchopped laser with a mirror mounted in the tilting holder through the aperture onto the photodiode. The output of the diode should be connected to one of the oscilloscope channels. With the scope triggering set on "auto" or "line," it should be possible to observe a DC voltage offset resulting from light detected by the photodiode. This level will be reduced if the aperture limits the laser spot size, and decreased to zero if the laser is blocked. These simple tests ensure that the photodiode and the oscilloscope are working properly. A 60 Hz level superimposed on the laser light level detected may be observed in rooms illuminated with fluorescent discharge lighting. This unwanted noise can be eliminated by turning the room lights out.

When the chopper is activated, and the scope triggering is set to "internal," the DC level should be replaced by a series of equally spaced square waves. The number of waves per time period and their separation can be used to verify the frequency and duty cycle of the chopper. A 60 Hz rate is convenient for the experimental components described here. However, it may be useful to vary this rate slightly to optimize performance.

When the sample is inserted into the optical path, the squarewave pattern on the scope will be diminished in intensity due to absorption of the laser beam by the sample. Adjustment of the voltage scale on the scope will restore the waveform to fullscale on the viewscreen. If the sample is too concentrated, the transmitted light will not be bright enough to detect easily with the diode and the scope may not trigger properly. Erratic or irregular scope traces and/or low square wave intensity can be corrected with a more dilute sample. However, if the sample is too dilute it will be difficult to locate the thermal lens signal. Some experimentation may be required to find an acceptable compromise in concentration.

The focussing lens should be the final optical element to be inserted into the beam path. If the lens is not positioned with its center on the laser beam, the laser will be deflected off the diode detector. The lens position should be adjusted to find the center of the optical axis already defined by the other components. Either the lens or the sample should be mounted on the translation stage to adjust the sample-to-lens spacing. Again, if the lens is not centered on the laser beam, or if the translation stage is not straight, adjustment of the lens-sample separation will cause the laser beam to "walk" off the detector. The desired thermal lens effect is extremely position sensitive. It is best to begin with the sample located near the focal length of the lens. By small displacements to either side of this position, the desired waveform shown in Figure 3 can be obtained on the oscilloscope. This is a

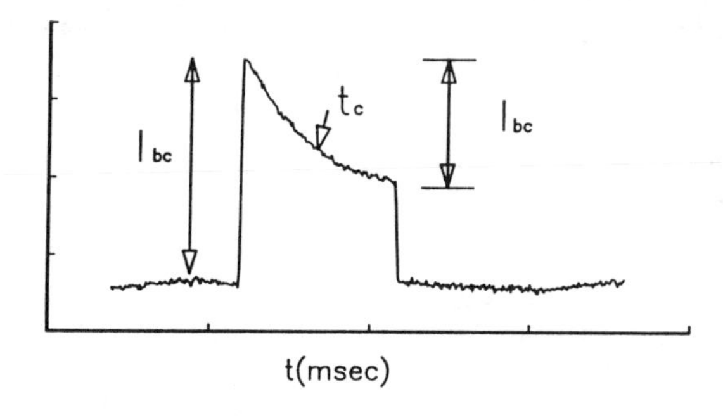

Figure 3. A typical thermal lens signal, showing the time-dependent output of a photodiode measured with an oscilloscope.

representative data set for a solution of indophenol blue in methanol (5.8×10^{-4} M). As shown, the detected voltage versus time waveform rises initially as it did with no sample, but it falls exponentially toward later time eventually converging to some final value. As shown below, the time dependence and the final value of this waveform are the desired quantities from which the final results will be calculated. At other lens positions nearby, an inverse lens effect may be observed which gives rise to a similar curved waveform, but one which rises in intensity with time. This is not the desired shape. With stable optical mounts and a little patience, the correct waveform can be achieved.

Once the thermal lens waveform has been optimized to obtain the correct shape and maximum depletion, two significant parameters are derived. The first is the ratio of the loss in beam intensity at beam center, ΔI_{bc}, to its initial value, I_{bc}, i.e. $\Delta I_{bc}/I_{bc}$. This ratio will be used to determine the extinction coefficient of the solute, as described below. The second parameter of interest is the time constant for the decay of intensity, t_c. Since the time decay is exponential (i.e., a "first order" decay process), the time constant is defined as the point where the I_{bc} value has fallen to a value $1/e$ times the total intensity change. In other words, at t_c the intensity will be I_c, where $I_c = I_{bc} - [\Delta I_{bc}\,(1 \quad e^{-1})]$. This value can be estimated visually from the oscilloscope. If the data is collected with a storage scope and can be plotted, t_c can be determined graphically. Perhaps best of all, if the waveform can be transferred to a computer, it can be analyzed and fit to an exponential decay via computer data analysis software. The final value of t_c will be used to calculate the heat capacity, as described below. We have found that $\Delta I_{bc}/I_{bc}$ and t_c values for several solutions in different solvents can be accumulated in a single lab period.

Safety

Please note that the toxicological effects of the organic dye compounds used in this experiment are not fully documented and they should therefore be handled with due caution. In particular, contact with skin or eyes should be avoided. A number of the chemicals listed in Tables 1 and 2 are toxic and/or flammable. In particular, benzene, carbon tetrachloride (CCl_4), methanol, and toluene are toxic to varying degrees, and cyclohexane is a skin and eye irritant. Most of the solvents are flammable.

Proper safety precautions are especially important when operating any laser equipment. The main concern from a He-Ne laser such as that recommended here is retinal damage which may result from direct on-beam viewing of the laser. Side-on viewing of the light scattered from dust particles in the air, viewing of the spot of the laser striking a darkened card, or laser impact on skin are not generally considered hazardous with low-powered He-Ne systems. To prevent accidental ocular impact, protective eyewear, or goggles, are highly recommended. *Laser-Gard* model LGS-HN, which are designed for use with He-Ne lasers, provide enough transmission (20%) to view and align the beam safely.

Data Analysis

The first step in analyzing the data for this experiment is to determine the laser spot size at the sample. For the laser used in our lab, focused with an 80 mm lens, the spot size at the minimum waist is

$$\omega_0 = (632.8 \times 10^{-6})\ 80\ /\ \pi\ (0.42) = 38\ \mu m$$

where mm units are used throughout. The maximum thermal lens effect is obtained at the point one confocal length beyond the minimum waist, where the beam has expanded to $\sqrt{2}$ times the minimum waist, i.e., $\omega_c = \sqrt{2}(38)$ or 54 μm.

Figure 2 shows a typical thermal lens time dependent waveform, with the parameters I_{bc} and ΔI_{bc} and the lifetime labelled. In this data, the exponential decay has not yet reached its final steady state value. The chopping speed used in the experiment could have been slowed slightly to observe the full decay. Alternatively, the data shown can be extrapolated to estimate the actual value of ΔI_{bc}. This extrapolation gives the ratio $\Delta I_{bc}/I_{bc} = 0.61$ for this waveform. Using a laser power of 1.8 mW (see below) and the *E* value of 1.33 per mW for the methanol solvent, we obtain,

$$0.61 = 2.303\ (1.33)\ (1.8)\ A$$

$$A = 0.11$$

The extinction coefficient, ϵ, is then found from Beer's law as,

$$\epsilon = A\ /\ cb = 0.11\ /\ (5.76 \times 10^{-4}\ \mathrm{M})\ (0.2\ \mathrm{cm}) = 960\ \mathrm{M}^{-1}\ \mathrm{cm}^{-1}$$

Table 2 shows that the value determined on a Cary UV-VIS spectrometer for indophenol blue at 632.8 nm is 3260 $\mathrm{M}^{-1}\ \mathrm{cm}^{-1}$. The value obtained here, therefore, has a 71% error.

The large error obtained here is typical of the results we usually obtain with our thermal lens setup. While we were at first concerned by this, it has turned out to provide an ideal opportunity for students to analyze the components of the apparatus, identify various sources of errors and determine how they contribute to the final results. In the present setup, we estimate that there is uncertainty in the laser power, in the solution concentrations, and in the true extinction coefficients. Our laser power is specified by the manufacturer to be 5 mW, but the laser is eight years old and is operating significantly below specifications. The Newport 818-SL photodiode we use is calibrated so that it can be used to determine the exact laser power. We measure 1.8 mWatt at the laser exit, but the exact power at the sample depends on the optical elements used and their transmission/reflectance properties. The effective power is therefore less than 1.8 mWatt, but not enough less to completely explain the large error. We have also been unsuccessful in locating literature values for dye solution extinction coefficients at 632.8 nm. We have therefore had to measure these values on a UV-VIS spectrometer, introducing another, albeit small, source of error. Perhaps our most significant problem is accurate determination of the concentration of the solutions. Dilute solution preparation requires either weighing small samples or diluting a stock solution, both of which must be done carefully. Inspection of the equations above shows that solutions must be less concentrated than expected to give a lower extinction coefficient than the actual value. These sources of error and their relative significance should be straightforward for students to identify, and simple enough problems to correct in further experiments.

The lifetime, t_c, obtained for the data in Figure 3 is 2.73 ms. To calculate the heat capacity C_p, we take the solvent data for methanol from Table 1, and use the calculated laser spot size of 54 μm, to obtain

$$C_p = (2.7 \times 10^{-3})\,(4)\,(2.01 \times 10^{-3}) \,/\, (0.0054)^2\,(0.792)$$

$$= 0.95 \text{ J} / \text{g·K}$$

$$= 30.4 \text{ J} / \text{mol·K}$$

This represents a 63% error relative to the literature value of 81.6 J/mol·K. Again, this magnitude of error provides an opportunity for the student to analyze the possible problems with the experiment. The only quantities which are not constants used in the calculation are the lifetime and the laser spot size. The lifetime determined cannot possibly be off by a factor of two. However, the laser spot size enters the calculation as a squared quantity, amplifying the effect of any uncertainties. To calculate the focused laser spot size, we assumed that the manufacturer's specified diameter at the laser exit could be used for the prefocused spot size just prior to the lens, ω_i. In other words, we ignored divergence which would make the actual prefocus spot larger at the point 1 m downfield where the lens was positioned. A larger beam diameter at this point would result in a smaller spot size at the laser focus, which would correct our result in the right direction. This problem with laser spot size measurement is pervasive in the thermal lens literature. The usual approach is to use a solution of known heat capacity as a standard against which new measurements are compared. By taking a ratio of the time constants measured for the "unknown" and the standard, the need to know the exact spot size is eliminated.

As shown here, an experiment in thermal lens calorimetry demonstrates several modern principles and experimental concepts without the need for an expensive apparatus. Lasers and laser optics should now form a component of every physical chemistry lab, and this experiment is a good introduction to this area. Time resolved detection of physical phenomena is pervasive throughout modern research in molecular spectroscopy and chemical dynamics, and this experiment also introduces relevant concepts in this area. The thermal lens experiment described here integrates these technologies with simple physical concepts (heat flow, heat capacity) which are familiar to all students. Perhaps most useful for the learning experience is that a so-called "high tech" experiment still gives experimental results with significant errors. Analyzing these errors, which is a major goal in any laboratory experience, is straightforward in this experiment. Perhaps the main advantage of the thermal lens experiment is that it is inexpensive to set up. As indicated in the Hardware List below, much of the equipment required is already available in most physical chemistry laboratories (i.e., oscilloscopes). If desired, however, it is easy to incorporate more sophisticated electronics (e.g., a storage scope and PC) to introduce computer data acquisition and data analysis. Overall, we have found this to be an enjoyable experiment providing a multifaceted learning experience.

Literature Cited

1. Harris, J.M.; Dovichi, N.J. *Anal. Chem.* **1980**, *52*, 695A.
2. Fang, H.L.; Swofford, R.L. in *Ultrasensitive Laser Spectroscopy*, Academic Press: New York, 1983, p.175.

3. Erskine, S.R.; Bobbitt, D.R. *J. Chem. Educ.* **1989**, *66*, 354.
4. O'Shea, D.; Callen, W.R.; Rhodes, W.T. *Introduction to Lasers and Their Applications*; Addison-Wesley: Reading, MA, 1977.
5. If a helium-neon laser is not available, another alternative is an argon ion laser like that used in Raman spectroscopy systems. However, solutions other than the ones described here are better for probing at the wavelengths available from this laser.
6. One recommended source for optical accessories is: Newport Corporation, P.O. Box 8020, 18235 Mt. Baldy Circle, Fountain Valley, CA, 92728-8020, Phone: (714) 963-9811.

Hardware List

Item	Vendor	Model
Required Equipment		
5 mW He-Ne laser	Spectra-Physics	SP105
Photodiode detector	Newport	818-SL
60 Hz chopper	homemade	
20 MHz oscilloscope	Tektronix	2205
Optics and mounts	Newport	
Optional Equipment		
Storage Oscilloscope	LeCroy	9310
PC Computer	Zenith, IBM, etc.	

RECEIVED October 1, 1992

CHAPTER 15

Determination of Thermodynamic Excess Functions by Combination of Several Techniques Including Laser Light Scattering

Gerald R. Van Hecke

POTENTIAL HAZARDS: Lasers, High Voltage

In the spirit of revitalizing the undergraduate physical chemistry laboratory, the experiment described here seeks to impart to the student some of the excitement of current science by introducing a variety of new, perhaps even exotic methods to perform what might be thought of as routine experiments. Common thermodynamic properties are measured, but done with modern equipment and techniques. The measurements required are laser light scattering intensities, densities, and refractive indices of the solutions of interest. This particular combination of measurements provides a means for detailed discussion of how the microscopic world of fluctuations can be connected to the macroscopic world of thermodynamic observables. The thermodynamic observables of interest here are excess volumes and Gibbs potentials of a non-ideal binary liquid mixture. The use of a photon to probe a binary liquid mixture for these values is an unusual feature of this experiment.

Rationale

Liquid mixtures are rarely ideal for their physical properties are rarely the composition weighted sum of the properties of the pure components comprising the mixture. At least two questions arise immediately from the observation of non-ideality. First, if a simple sum of pure component properties cannot adequately describe the experiment, what functional relationship can? Second, and more fundamental, what makes the simple sum concept not work? In other words, what causes the difference, the non-ideality? The answer to the first question, what functional relationship works was provided by Euler in his work on homogeneous functions by what are now called partial molar quantities. The answer to the second question is still largely unanswered but it is possible to at least quantify the differences by the concept of excess functions. Discussions of excess properties provide opportunities to try and understand from quite fundamental bases what factors are responsible for physical properties.

By how much does the simple sum differ from the observed value of the macroscopic, extensive property J? The answer defines what will be called J^{EXCESS} (J^{E}) and is calculated in general by

$$J^{\mathrm{E}} = J_{\mathrm{OBS}} - J_{\mathrm{IDEAL}} = J_{\mathrm{OBS}} - \Sigma N_i \overline{J}_i \qquad (1)$$

2503–6/93/0242$06.00/0

where $\overline{J}_i$ is the value of J for pure i per mole of i, the molar J_i and all the other J values refer to those for the solutions under study, and the sum is over all solution components i.

If J is an extensive thermodynamic observable, which we will take here to be any physical property of a system that depends on the size of the system such as mass, volume, or energy, then

$$J_{OBS} = \sum_{i=1}^{n} N_i \left(\frac{\partial J}{\partial N_i} \right)_{T,\, P,\, N_j} \tag{2}$$

where the derivatives are the partial molar values of the extensive property J. The sum of the partial molar J values via equation 2 now add to give the total value J. The whole point here is described by

$$J^E = \sum_{i=1}^{n} N_i \left[\left(\frac{\partial J}{\partial N_i} \right)_{T,\, P,\, N_j} - \overline{J}_i \right] \tag{3}$$

where for any non-ideal mixture the term in brackets is not zero.*(1)* The consequence of this equation is that the value of the property J per mole of pure substance, $\overline{J}_i$, is not equal to the partial molar property J_i evaluated under isothermal, isobaric, and all moles constant save N_i conditions. Equation 3 is a recipe to quantify J^E but what is needed is an explanation of the origin of J^E.

When studying a specific liquid mixture, two questions always arise. Is J^E zero? If J^E is not zero, why? A primary goal is to discover whether or not the mixture is ideal. This means some method is necessary to measure the excess properties. What excess properties are most often studied? The two most common are excess volume and excess heat which is just the heat of mixing. It should be no surprise that these two excess quantities are also the most easily measured. While excess volume and heat are useful thermodynamic descriptors of a system, neither are the proper criterion to predict the point of chemical or phase equilibrium. The Gibbs potential serves that purpose for systems under isobaric and isothermal conditions. If the system is non-ideal, then methods to obtain the excess Gibbs potential are useful. Several ways exist to estimate the excess Gibbs potential but none can do so directly - all require noting how certain other thermodynamic properties vary. The excess Gibbs potential can be estimated by vapor pressure measurements using activity coefficients, from heat of mixing data as a function of temperature, or excess volume measurements as a function of pressure.

The point of this discussion is not, however, to examine all the methods available to measure excess Gibbs potential but to point out there is good reason to make such measurements and then introduce a quite different measurement method.

Theory

Usually one does not think of measuring thermodynamic properties with photons. Measuring thermodynamic properties conjures up visions of calorimeters, vapor pressure measuring devices and the like, all used with meticulous care. The microscopic photon can be used, however, to probe macroscopic thermodynamic properties. The link between the two "worlds" depends on the ability of a photon to probe the microscopic fluctuations that are always associated with the macroscopic average of any thermodynamic property. To appreciate how laser-generated photons can be used, it is necessary to briefly discuss how photons can probe the microscopic fluctuations leading to macroscopic thermodynamic properties.

The interaction of light with matter is described to good approximation as the interaction of the oscillating electric field of the light beam with the charge cloud of the target molecules. If the frequency of the electric field of the light beam matches one of the natural frequencies of the target molecule, then the photons may be absorbed, that is, captured by the molecule, causing the latter to make a transition from one energy state to another. However, if the frequency of the electric field differs from the natural frequencies of the target molecule, then the photons are scattered, emerging from the target molecule at the same frequency but with a different direction and with their electric fields pointing in different directions. Polarizability is the term used to describe the ability of a molecule to form a temporary dipole moment under the action of an oscillating electric field. It is easy to realize that the light scattering strength of a molecule depends on its shape as well as on its constituent atoms because these factors determine the molecular polarizability.

In any phase, deviations from the macroscopic average value of any physical property in that phase are always occurring. These microscopic deviations are called fluctuations. Density and concentration fluctuations play a major role in scattering phenomena because they affect the bulk polarizability of the collection of target molecules. For a pure substance there are no concentration fluctuations, only density fluctuations. In a solution the number of molecules fluctuates, as does the number of each type of particle. These fluctuations are influenced by the nature of the intermolecular forces occurring in the mixture. Thus, studying fluctuations by light scattering provides a means of studying intermolecular forces.

The above comments are perfectly applicable to spherical molecules, that is, isotropic molecules. For nonspherical, anisotropic molecules the total scattered light intensity in any given direction can be thought of as arising from two sources: (1) the density and concentration fluctuations just mentioned above and (2) the inherent anisotropies of the molecules. The total scattered intensity at a viewing angle θ from incident is

$$I(\theta) = I_{is}(\theta) + I_{an}(\theta) \tag{4}$$

where the subscripts is and an refer to isotropic and anisotropic respectively. To avoid measuring absolute light intensities, scattering measurements are "normalized" using the Rayleigh ratio. Equation 5 defines three pertinent Rayleigh ratios: $R_{tot}(\theta)$ the total ratio which is what is actually measured, $R_{is}(\theta)$ the isotropic ratio, and $R_{an}(\theta)$ the anisotropic ratio *(2)*.

$$\begin{aligned} R_{tot}(\theta) &= (1 + \cos^2(\theta))\,(2r^2/I_0V)\,I(\theta) \\ &= (1 + \cos^2(\theta))\,(2r^2/I_0V)\,(I_{is}(\theta) + I_{an}(\theta)) \end{aligned}$$

or

$$R_{tot}(\theta) = R_{is}(\theta) + R_{an}(\theta) \tag{5}$$

In equation 5, r is the distance of the viewer from the scattering source, V is the volume of the scattering source and the angular factor depends on the viewing angle θ and whether or not the incident light is polarized. For light polarized perpendicular to the plane defined by the laser light source, sample, and detector, the angle factor is unity. For unpolarized light the angular factor is as given in equation 5 *(3)*. The incident light intensity is I_0. While scattered light is a function of θ, using an angle of 90° provides many conveniences. All further discussion will be based on 90° observations and for ease of notation, the $R(\theta = 90°)$'s will be just R.

To separate R_{is} and R_{an} from the measured R_{tot}, Cabannes suggested taking into account the polarizations of the scattered light *(4)*. Equation 6

$$R_{is} = R_{tot}\,(3\,I_{VV} - 4\,I_{VH})\;/\;(3\,I_{VV} - I_{VH}) \tag{6}$$

shows how R_{is} is obtained from R_{tot} by measuring the scattered light intensities defined in Figure 1 *(5,6)*.

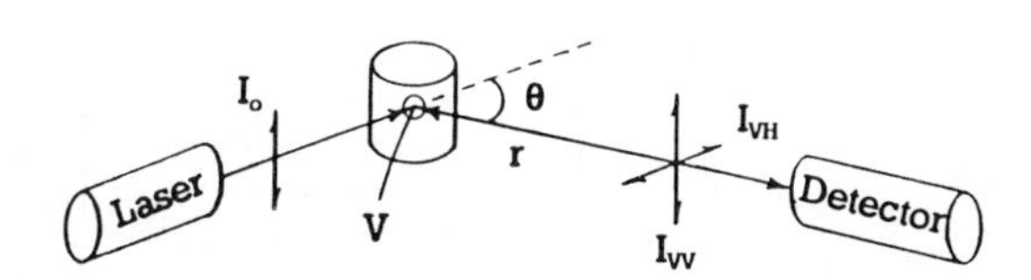

Figure 1. A typical block diagram of a light scattering apparatus suitable to carry out the measurements necessary for determination of excess Gibbs potentials that includes a schematic of the scattering cell to define the polarizations of the incident and scattered light.

In equations 5 and 6, R_{tot} refers to the total Rayleigh ratio measured using unpolarized light. Use of polarized laser light necessitates some extra experimental steps relative to older unpolarized procedures. Defining total Rayleigh ratios for vertically and horizontally polarized incident light, R_V and R_H respectively, the following is true *(7)*

$$R_{tot} = 0.5\,(R_v + 3\,R_H) \tag{7}$$

Using a substance whose Rayleigh ratios are known for calibration

$$R_{tot} = 0.5\,[R_{V,\,std}\,(I_{VV}/I_{VV,\,std}) + 3\,R_H\,(I_{VH}/I_{VH,\,std})] \tag{8}$$

Thus the necessary experimental measurements are just I_{VV} and I_{VH} for the samples and standard(s).

Though how R_{tot} and R_{is} are obtained has been discussed, R_{is} has not been connected to solution properties. As implied earlier, the polarizability of the collection of molecules measures the collection's response to the photon's oscillating electric field. The practical way to estimate polarizability, however, is the dielectric constant of the collection. Fluctuations of density and, if appropriate, concentration, affect the dielectric constant, and in turn, the scattered intensity. Thus, equation 9 expresses the Rayleigh ratio as a function of the fluctuations of the dielectric constant.

$$R_{\mathrm{is}} \propto \langle (\delta\epsilon)^2 \rangle \tag{9}$$

For solutions the isotropic scattering term R_{is} is further divided into contributions from density fluctuations R_{d}, concentration fluctuations R_{c}, and a coupling term R_{cd} that tries to account for how a density fluctuation might affect the concentration and vice-versa

$$R_{\mathrm{is}} = R_{\mathrm{d}} + R_{\mathrm{c}} + R_{\mathrm{cd}} \tag{10}$$

Noting how fluctuations in density and concentration might affect the fluctuations of the dielectric constant, yields *(4,8)*.

$$R_d = \left(\frac{\pi^2}{2\lambda^4}\right) kT\,\kappa_T \left[f\left(\frac{(n^2-1)(n^2+2)}{3}\right)\right]^2$$

$$R_{cd} = \left(\frac{\pi^2}{\lambda^4}\right) kT\,\kappa_T x_2 (1-x_2) \left[f\left(\frac{(n^2-1)(n^2+2)}{3}\right)\right]\left[2n\left(\frac{\partial n}{\partial x_2}\right)\right] \tag{11}$$

Here λ is the wavelength of incident light, κ_T is the solution isothermal compressibility, n is the refractive index of the solution, kT is the Boltzmann constant times temperature, and x_2 is the mole fraction of component 2, usually called the solute. The factor f corrects the gas phase refractive index expression for use in condensed phases. The correction is small and generally around 1.0 ± 0.15 *(9)*. Note that since κ_T and n are functions of composition measurable for each solution studied, R_{d} and R_{cd} are readily obtained for the T and λ of the experiment.

The concentration Rayleigh ratio R_{c} is obtained experimentally by subtracting R_{d} and R_{cd} from R_{is}. The theoretical expression for R_{c} is given by

$$R_c = \left(\frac{2\pi^2}{\lambda^4}\right) kT\,V_m (1-x_2) \left[\frac{n^2\left(\dfrac{\partial n}{\partial x_2}\right)^2}{\left(\dfrac{\partial \mu_2}{\partial x_2}\right)}\right] \tag{12}$$

where V_{m} is the molar volume of the mixture. It is the derivative of the partial molar Gibbs potential (the chemical potential μ_2) with composition, $\partial\mu_2/\partial x_2$, that contains the vital thermodynamic information.

For an ideal solution $\partial\mu_2/\partial x_2 = N_{\mathrm{A}}kT/x_2$ which means R_{c} reduces to what is called R_{id}, the ideal Rayleigh ratio given by

$$R_{\text{id}} = \left(\frac{2\pi^2}{\lambda^4}\right)\left(\frac{V_m(1-x_2)x_2}{N_A}\right)\left[n^2\left(\frac{\partial n}{\partial x_2}\right)^2\right] \tag{13}$$

where here N_A is Avogadro's number. For nonideal solutions $\partial\mu_2/\partial x_2$ becomes

$$\frac{\partial\mu_2}{\partial x_2} = \frac{N_A kT}{x_2} - (1-x_2)\left(\frac{\partial^2 G^E}{\partial x_2^2}\right) \tag{14}$$

Substituting equation 14 into equation 12, with some rearrangement, we obtain

$$\left[\frac{R_{\text{id}} - R_c}{R_c}\right] = \frac{x_2(1-x_2)}{N_A kT}\left(\frac{\partial^2 G^E}{\partial x_2^2}\right) \tag{15}$$

This equation connects thermodynamics, here the excess Gibbs potential (G^E), with laser-generated photon probes of the fluctuations. Since R_{id} and R_c are obtainable independently, equation 15 relates the isothermal composition dependence of the scattering (LHS) to a composition dependence of the excess Gibbs potential. Clearly any deviation from ideality appears immediately as a nonzero numerator of the LHS of equation 15 and this nonideality is reflected by the Gibbs potential.

Equation 15 allows the calculation of the second derivative of the Gibbs potential. To proceed further, some functional form is required for G^E. A variety of approaches are possible including Flory–Huggins theory.*(10,11)* Rather than recast the data in terms of volume fraction as required by the Flory–Huggins theory, the data was analyzed using the empirical Redlich–Kister polynomial expansion. This expansion provides thermodynamic consistency by satisfying the Gibbs–Duhem relation and is given below *(12)*.

$$G^E = x_2(1-x_2)\sum_{j=1}^{n} A_j(1-2x_2)^{(j-1)} \tag{16}$$

The A_j are parameters in principle dependent on temperature and pressure but independent of composition and are to be determined by data fitting. The calculation at this point proceeds by choosing a polynomial, taking its second derivative with respect to composition, substituting this into equation 15, and then solving for the unknown A_j coefficients by some least squares technique.

While it is not necessary to actually calculate the excess volumes of the solutions studied since only the molar volume is necessary, all the data necessary to make the calculations is available. The comparison of the excess volume and excess Gibbs potentials is instructive. To calculate the excess volumes the following expression is convenient

$$V^E(x_2) = \frac{(1-x_2)M_2 + x_2 M_2}{\rho_{\text{OBS}}(x_2)} - \frac{(1-x_2)M_1}{\rho_1} - \frac{x_2 M_2}{\rho_2} \tag{17}$$

The excess volume data can also be fit to a Redlich–Kister polynomial where the coefficients describe the temperature and pressure dependence of the excess volume. Thus

$$\bar{V}^E = x_2(1-x_2) \sum_{j=1}^{n} B_j(T, P)\,(1-2x_2)^{(j-1)} \tag{18}$$

Here also the B_j expansion coefficients are found by a least squares technique.

Experimental

Three types of measurements are made in this experiment: light scattering, density, and refractive index. Thermal compressibility data is taken from the literature.

Sample preparation.

The reagent grade *t*-pentyl alcohol was distilled over calcium hydride, filtered through 0.22 micron frit, and degassed by boiling. The HPLC grade toluene was distilled, filtered through 0.22 μm frit, and degassed by boiling. Solutions were made by weighing without buoyancy correction.

Laser light scattering.

A Brice–Phoenix light scattering photometer was modified for use with a laser by bringing the laser beam into the sample chamber through a hole drilled in side wall of the photometer case. In all cases a 90° scattering angle was used. The sample cell was made from a precision bore tube 1" in diameter and 1.5 " long. A flat piece of glass was fastened to one end of the tube using GE silicon cement. The cement is soluble in a variety of solvents so some care in systems studied has to be observed. About 20 mL of liquid was required to adequately fill the cell.

The detector was an air-cooled PMT whose output was fed directly into a Keithley picoammeter and the signal current then read off the analog meter. Dark currents were measured but generally proved to be insignificant. Signal currents generally ranged from micro to nano amps.

The laser used was a Coherent Nova 70 argon ion using the 488 nm line whose power incident on the sample was adjusted to be a constant 70 mW. The power was measured with a Coherent power meter. A 1.00 neutral density filter was placed in the beam to reduce the incident power with the final adjustment made to 70 mW with the laser power supply. The vertically polarized laser beam was aligned with the long (vertical) axis of the sample cell using a polarization rotator. A suitably sized piece of polaroid sheet was placed in front of the PMT to serve as the polarization analyzer. Actual scattering intensity measurements were made with the analyzer's polarization axis in a vertical position I_{VV}, then in a horizontal position I_{VH}. Since benzene is the only substance that has good published values for R_{VV} and R_{VH} at 488 nm, it was necessary to use it as the calibrating fluid. Temperature control was maintained with a thermostatted circulating bath circulating a 50/50 mixture of ethylene glycol/water through the cell jacket supplied with the Brice-Phoenix.

Density measurements.

An Anton Parr Model DMA40 density meter was used to determine the solution densities. The instrument was calibrated with water, methanol, and air. Temperature control was provided by a thermostatted circulating bath. Alternatively, densities could be determined using pycnometers.

Refractive Index measurement.

Refractive indices were measured at the sodium D line using a Bausch and Lomb Abbe' Model 3 refractometer. The dispersion correction to the laser wavelength 488 nm was made following the manufacturer's instructions and tables. The instrument was calibrated with a glass test piece. Temperature control was maintained by a thermostatted circulating bath.

Safety and chemical disposal

Clearly the greatest danger in this experiment is due to inappropriate exposure to the argon ion laser. Laser goggles for the argon lines must be routinely worn by all personnel working around the argon laser. In addition it is smart to remove jewelry, rings, especially watches whose crystal faces make excellent reflectors at potentially inopportune times.

The chemicals used are not known to be especially hazardous. The calibrant benzene is unfortunately a recognized carcinogen and its use requires precautions to reduce exposure to an absolute minimum. All liquids were transferred in well-vented hoods and always capped when removed from the hood area.

The studied chemicals were collected in waste bottles for future off-site disposal. The benzene was collected separately.

Data Analysis

Calculations.

The Rayleigh ratios R_{is}, R_{id}, R_d, R_c, R_{cd} were all calculated for each composition studied using the experimentally determined I_{VV}, I_{VH}, V_m and n data. The slope of refractive index versus composition graph was determined by fitting the n versus x data to a 2nd order polynomial and then differentiating. The necessary κ_T versus x data was obtained by assuming a linear dependence of κ_T on x and using literature values for the pure components. The data points calculated for each composition using equation (15) were set equal to the 4th order polynomial that resulted from differentiating a 6th order Redlich–Kister polynomial for G^E and the coefficients of the resulting 4th order polynomial found by a linear least-squares procedure. The calculated coefficients were in turn rearranged to yield the coefficients of the 6th order polynomial for G^E.

The excess volumes were calculated from the density data and fit to a 5th order Redlich-Kisler polynomial in composition.

All plotting and calculations were performed using the IBM PC based graphing program AXUM from Trimetrix of Seattle, Washington. Some type of graphing program and polynomial fitting routine are virtually a necessity to make the numerous, though straightforward, calculations tractable.

Results

All the results of the experiment are presented in Figures 2–7. Figure 2 presents the refractive index versus composition data at 30 °C and 488 nm. Figure 3 presents the excess volume data at 30 °C. Figure 4 shows the contributions of the various Rayleigh ratios as functions of composition at 30 °C. Figure 5 presents the fit of the second derivative of the Gibbs potential to the light scattering data calculated using equation (15). Figure 6 presents the excess Gibbs potential as a function of composition at room pressure and 30 °C using a 6th order Redlich–Kister polynomial. Figure 7 shows the results when the light scattering data is fit to only a 2nd order Redlich–Kister polynomial.

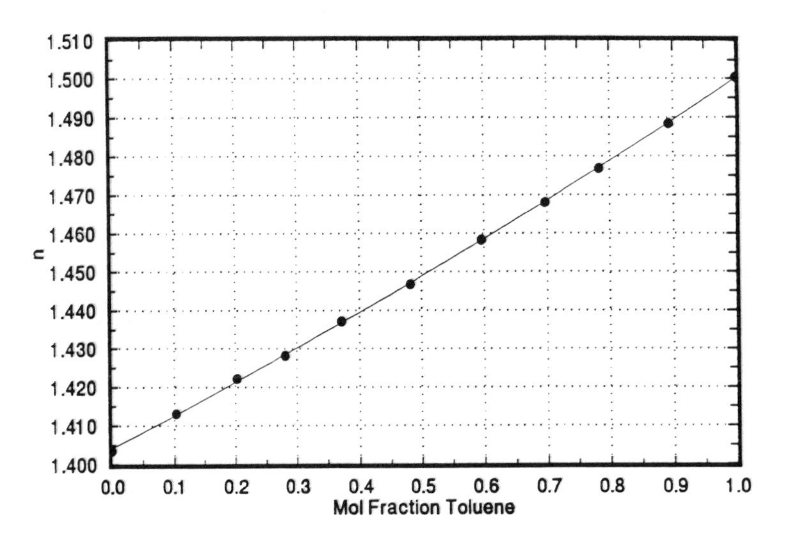

Figure 2. Refractive index at 488 nm, ambient pressure, and 30 °C for solutions of *t*-pentyl alcohol and toluene.

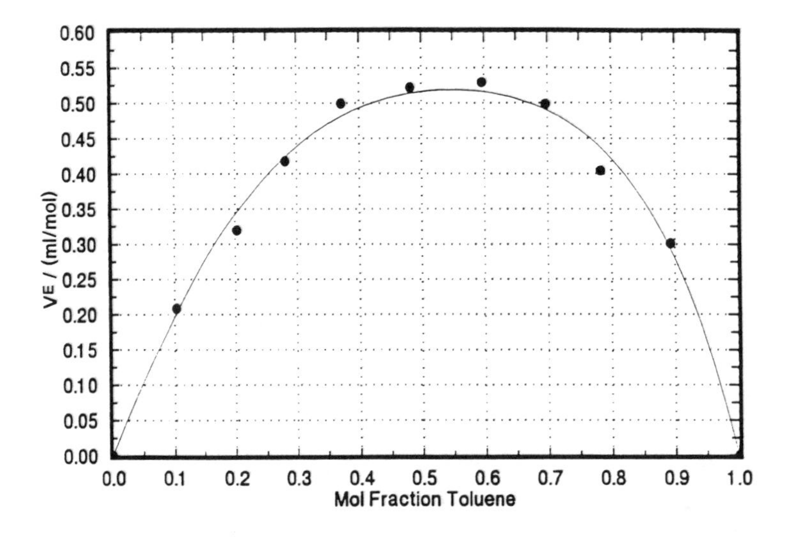

Figure 3. Excess volume at ambient pressure and 30 °C for solutions of *t*-pentyl alcohol and toluene.

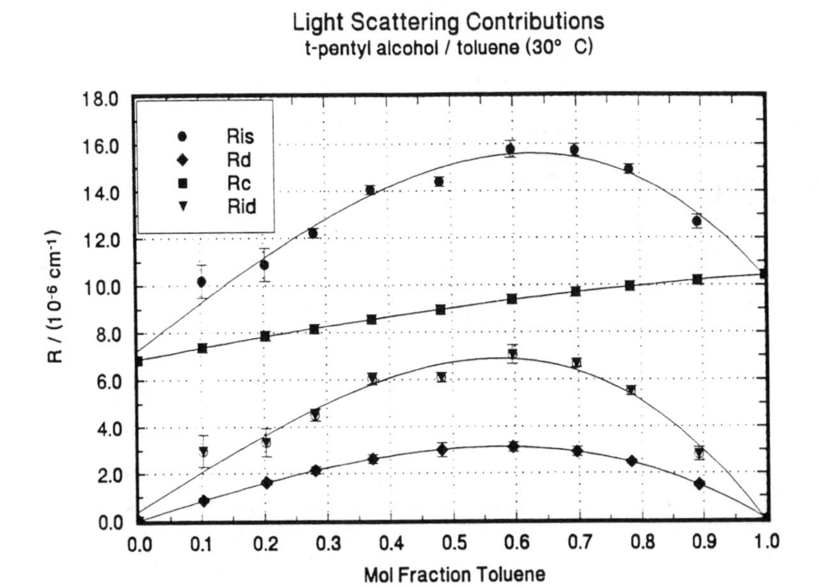

Figure 4. Rayleigh ratio contributions at 488 nm, ambient pressure, and 30 °C for solutions of *t*-pentyl alcohol and toluene.

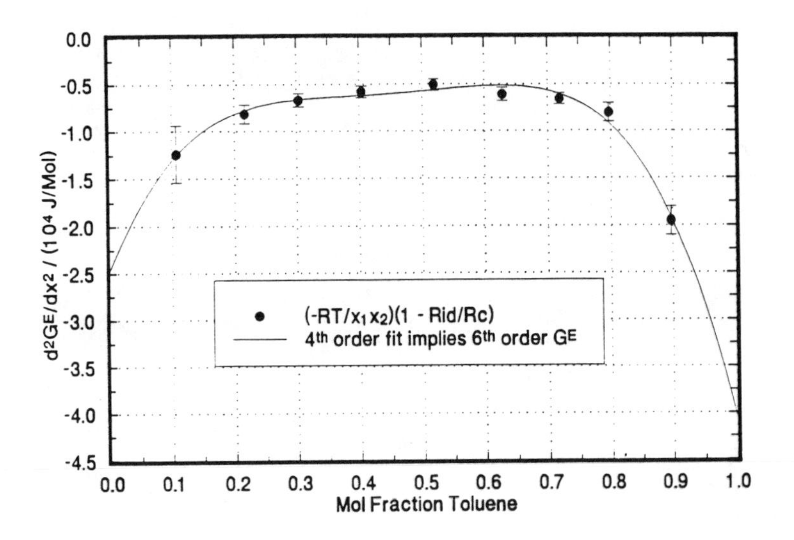

Figure 5. The 2nd derivative of a 6th order Redlich–Kister excess Gibbs potential fit to laser light scattering data obtained at 488 nm, ambient pressure, and 30 °C for solutions of *t*-pentyl alcohol and toluene.

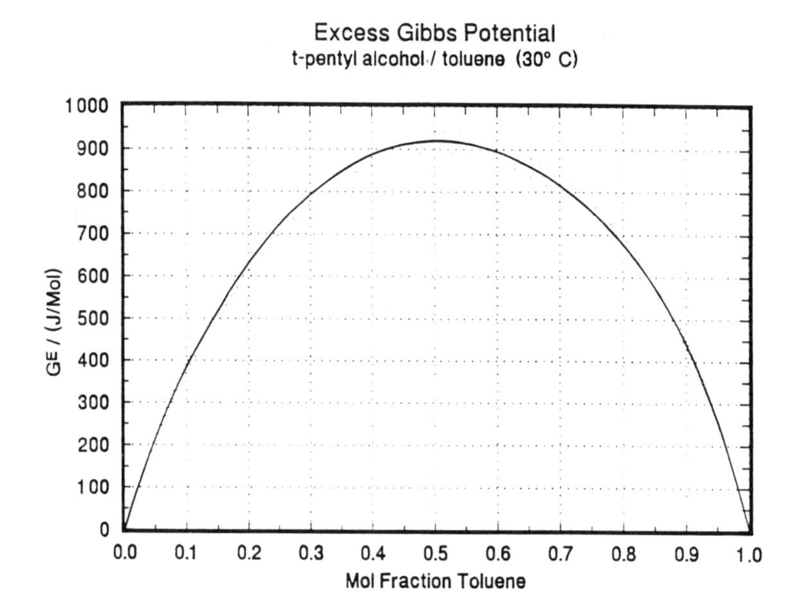

Figure 6. The 6th order Redlich–Kister excess Gibbs potential at ambient pressure and 30 °C for solutions of *t*-pentyl alcohol and toluene.

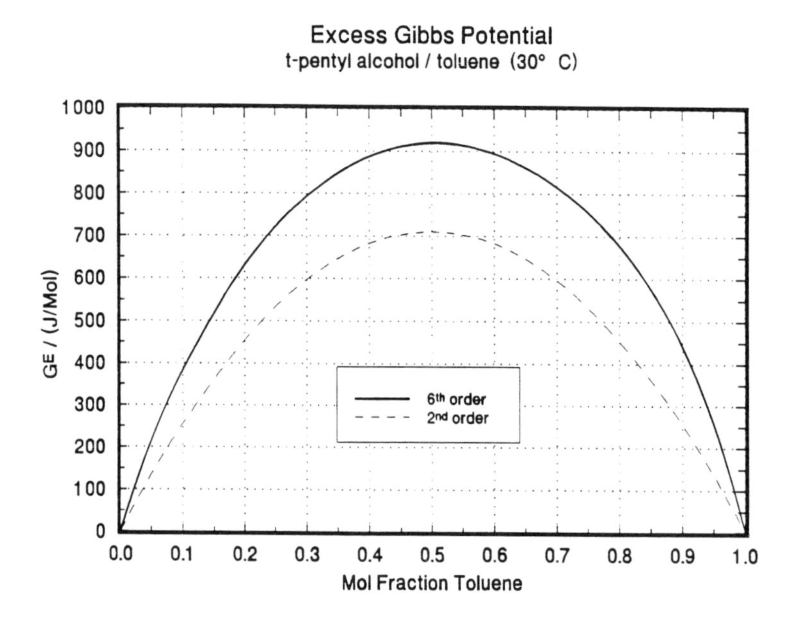

Figure 7. The excess Gibbs potential fit to 2nd and 6th order Redlich–Kister polynomials at ambient pressure and 30 °C for solutions of *t*-pentyl alcohol and toluene.

Discussion

It is not really the purpose here to discuss in any detail the meaning or origins of the excess volumes or Gibbs potentials but rather emphasize the ease with which the data was obtained and with available computer programs, analyzed. The experimental data, the scattering data, densities, and the refractive index measurements can be obtained in an afternoon, though it is better to reserve two periods for data collection. Also the extent of the experiment can be altered to do it all, or parts depending on equipment, or time constraints. What data is not obtained experimentally can be provided by the instructor. Data analysis programs can be left to the student or provided, again depending on time and equipment constraints.

There is no question that done completely this is an elaborate experiment, but one the student takes great pride in finishing, especially with the excellent data that is it possible to obtain.

Acknowledgments

Professors H.L. Clever, R.N. Zare, and G.D. Patterson are thanked here for their helpful comments and advice. The experimental assistance of Robert A. Westervelt is gratefully acknowledged. Financial support for the development of this experiment came from the National Science Foundation, the Pew Foundation, and the Mabel and Arnold Beckman Research Fund administered by Harvey Mudd College.

Literature Cited

1. In the cases of entropy and Gibbs potential, it should be noted that G_{mix} and S_{mix} exist whether the solution is ideal or not and in these cases, the "excess" values result from contributions to these properties beyond the simple statistical origins of G_{mix} and S_{mix}.
2. Coumou, D.J.; Mackor, E.L. *Trans. Faraday*, **1964**, *60*, 1726-1735.
3. Schmitz, K.S. *An Introduction to Dynamic Light Scattering by Macromolecules*, Academic Press: New York, 1990, p.19.
4. Cabannes, J. *La Diffusin Moleculaire la Lumiere,* Les Presses Universitaires de France: Paris, 1929, Chapter X.
5. Šegudović, N.; Deželić, G. *Croatia Chem. Acta,* **1973**, *45*, 385-406.
6. Kerker, M. *The Scattering of Light,* Academic Press: New York, 1969, p.580.
7. Kratohvil, J.P.; Smart, C. *J. Colloid Sci.,* **1965**, *20*, 875-892.
8. Schmidt, R.L.; Clever, H.L. *J. Phys. Chem.,* **1968**, *72*, 1529-1536.
9. Myers, R.S.; Clever, H.L. *J. Chem. Thermodynamics,* **1970**, *2*, 53-61.
10. Flory, J. *J. Chem. Phys.* **1941**, *9*, 660-661.
11. Huggins, M.L. *J. Chem. Phys.* **1941**, *9*, 440.
12. Redlich, O.; Kister, A.T. *Ind. Eng. Chem.,* **1948**, *40*, 345-348.

Bibliography

Coumou, D.J.; Mackor, E.L. "Isotropic Light Scattering in Binary Liquid Mixtures," *Trans. Faraday Soc.*, **1964**, *60*, 1726-1735.

Deželić, G. "Evaluation of Light-Scattering Data of Liquids from Physical Constants," *J. Chem.* **1966**, *45*, 185-191.

Schmidt, R.L.; Clever, H.L. "Thermodynamics of Binary Liquid Mixtures by Rayleigh Light Scattering," *J. Phys. Chem.*, **1968**, *72*, 1529-1536.

Myers, R.S.; Clever, H.L. "Excess Gibbs free energy of mixing in some hydrocarbon + alcohol solutions by Rayleigh light scattering," *J. Chem. Thermodynamics*, **1970**, *2*, 53-61.

Lewis, H.H.; Schmidt, R.L.; Clever, H.L. "Thermodynamics of Binary Liquid Mixtures by Total Intensity Rayleigh Light Scattering. II," *J. Phys. Chem.*, **1970**, *74*, 4377-4382.

Kerker, M. *The Scattering of Light*, Academic Press: New York, 1969.

Šegudović, N.; Deželić, G. "Light-Scattering in Binary Liquid Mixtures. I. Isotropic Scattering," *Croatia Chem. Acta*, **1973**, *45*, 385-406.

Deželić, G.; Šegudović, N. "Light-Scattering in Binary Liquid Mixtures. II. Anisotropic Scattering," *Croatia Chem. Acta*, **1973**, *45*, 407-418.

Schmitz, K.S. *An Introduction to Dynamic Light Scattering by Macromolecules*, Academic Press: New York, 1990.

Hardware

Light scattering. Because of the λ^{-4} dependence on scattering intensity, short wavelength lasers will perform better. This experiment was routinely performed with a Coherent Argon Ion laser. This laser actually had too much power which was reduced with a neutral density filter. The experiment can be done with a green He-Ne laser but the signal was generally only a factor of hundred above noise while with the argon the signal was usually 10^4 times the background. Virtually any PMT sensitive to visible light would be satisfactory. Of course a power supply for the PMT is necessary.

Densitometer. The Anton Parr Model DMA40 is a basic model of the resonant density meters. Any equivalent instrument would work. Of course simple pycnometers could always be used.

Refractometry. Virtually any type of Abbe refractometer could be used. In addition several types of critical angle experiments using lasers have been described in the literature.

Computing and graphics. Sophisticated graphing programs that incorporate polynomial fitting are readily available. Three competing programs are AXUM, Sigma-Plot, and Graftool. The plotting aspects of these programs are not in fact as important as their fitting aspects. Fitting programs are also available commercially or from various program compendia such as A.R. Miller, *Pascal Programs for Scientists and Engineers*, Sybex, 1981.

RECEIVED October 1, 1992

Fluorescence Probes

CHAPTER 16

Fluorescence Probes of β-Cyclodextrin Inclusion Complexes

Virginia M. Indivero and Thomas A. Stephenson

One of the most effective means of modulating the reactivity of species in solution is by altering the nature of the solute-solvent interaction. The weak, non-covalent forces that characterize this interaction are ubiquitous in all areas of chemistry, affecting phenomena as diverse as the tertiary structure of proteins *(1)* and collisions between gas phase molecules *(2)*. Few chemical laboratory experiments address the role of these intermolecular forces, however, and rarely is one required to come to an appreciation of the sometimes conflicting role that they play in chemical reactivity. In this chapter, we describe a multi-week investigation of the thermodynamics of the formation of cyclodextrin inclusion complexes. Our hope is that through this investigation, students will gain an appreciation for the both the importance and complexity of these issues, particularly in modulating chemical equilibria in solution. In the design of this experiment, we had several educational goals in mind. First, we sought to link an investigation of classical concepts of thermodynamics and intermolecular interactions with more modern spectroscopic instrumentation. Second, to facilitate integration of the seemingly abstract concepts contained in the physical chemistry curriculum with other sub-disciplines, we chose target species for our study that have widespread importance in biochemical applications. Third, we wished to introduce quantitative absorption and fluorescence spectroscopies into the physical chemistry laboratory curriculum, accompanied by non-linear least squares fitting of the data. Finally, we sought to develop an experiment that would foster collaboration among the students in the laboratory, by necessitating the sharing of experimental data and insights.

Cyclodextrins are cyclic oligosaccharide compounds *(3)* that have proved to play an increasingly important role as possible enzyme mimics *(4,5)* and in modifying often complicated photochemical processes. *(6)* Three different cyclodextrins are commonly commercially available, differing in the number of glucose subunits contained in each molecule. All cyclodextrins form the shape of a truncated cone, with the interior diameter varying with the number of glucose units. The most common, β-cyclodextrin, is composed of seven glucose monomers and forms a cavity with an interior diameter of approximately 7.5 Å. Much less widespread are α-cyclodextrin (six glucose units forming a 6.5 Å cavity) and γ-cyclodextrin (eight glucose units forming a 9.0 Å cavity). *(6)* In each case, the interior of the cyclodextrin "cone" is largely hydrophobic. The secondary and primary hydroxyl groups of the glucose subunits ring the openings of the cavity, thus providing a small degree of aqueous solubility (16 mM). *(3)* It is the ability to include compounds within the cavity on the basis of size and polarity, while operating in aqueous solution, that has resulted in the use of cyclodextrins as organic models for the active site of enzymes. *(4,5)* Cyclodextrins have also found a wide variety of other applications, including enhancing the aqueous solubility of pharmaceuticals, *(3)* providing conformational control over the substrates in photolysis reactions *(6)* and

2503–6/93/0258$06.00/0

acting as additives to the stationary phase in the chromatographic separation of chiral compounds. *(7)*

Numerous studies of peptides and proteins have made use of the ultraviolet spectroscopic properties of the amino acid tryptophan to probe a variety of phenomena, including, for example, the geometry of enzyme active sites *(8)* and conformational dynamics in solution. *(9)* The chromophore present in tryptophan is the indole moiety, which has a rich history of photochemical and photophysical investigations. A specific area of interest has been the role of the solvent in affecting the absorption and fluorescence spectra of indole. *(10,11)* This is an area of obvious crucial importance to studies of biochemical compounds, due to the dramatic changes in the local environment of proteins that can occur depending on the secondary and tertiary structure. *(1)* In this laboratory experiment, we investigate two facets of this problem, using the inclusion compounds formed by indole and indole derivatives with β-cyclodextrin as model systems. First, students investigate the thermodynamics of the inclusion process itself, providing a direct measure of the enthalpy and entropy effects that occur when indole (or derivatives) move from aqueous solution to the hydrophobic environment of the cyclodextrin cavity. Second, by examining a series of derivatized indoles, students explore the effect of steric bulk and changes in electronic structure on the thermodynamics of the formation of the inclusion complex. The ultimate goal of the investigation is to shed light on the microscopic solvent-solute and cyclodextrin-solute interactions that provide the driving force for this association reaction.

Theory

The methodology utilized to determine the association constant, K_a, as well as ΔH and ΔS for the formation of the indole–β-cyclodextrin inclusion complex is a modified form of that described by Örstan and Ross. *(12)* If indole forms a 1:1 complex with β-cyclodextrin, then we define the association constant according to

$$\mathrm{CD} + \mathrm{In} \rightleftharpoons \mathrm{CD{-}In},$$

$$K_a = [\mathrm{CD{-}In}]/[\mathrm{In}][\mathrm{CD}]. \quad (1)$$

(In these equations, we use the shorthand notation of CD for β-cyclodextrin, In for indole (or derivatives) and CD–In for the indole–β-cyclodextrin inclusion complex.) In this experiment, we use absorption and fluorescence spectroscopy to measure the changes in the indole spectra caused by the formation of the inclusion complex. Specifically, we measure the absorption and fluorescence spectra of solutions of constant indole concentration, while varying the concentration of β-cyclodextrin.

Absorption spectroscopy is a common experimental technique that many students encounter before reaching the physical chemistry curriculum. In brief, the absorbance of a sample at wavelength λ, A_λ, is given by the Lambert–Beer law, *(13)*

$$A_\lambda = \epsilon_\lambda\, b\, [\mathrm{X}],$$

where ϵ_λ is the molar extinction coefficient, b is the pathlength through the absorbing material, and [X] is the molar concentration of the absorbing species X.

For the indole derivatives used in this experiment, ϵ_{λ} is of the order 6×10^3 liter mole^{-1} cm^{-1} at the maximum of the near ultraviolet absorption band. Thus, a concentration of $\approx$ 170 μM is required to achieve an absorbance of 1.0 in a 1-cm pathlength absorption cell. When two species Y and Z are in solution, the absorbance of the combined sample will be simply the sum of the absorbances of the individual species,

$$A_{\lambda,\text{tot}} = A_{\lambda,\text{Y}} + A_{\lambda,\text{Z}} = \epsilon_{\lambda,\text{Y}}\, b\, [\text{Y}] + \epsilon_{\lambda,\text{Z}}\, b\, [\text{Z}]$$

For the experiment described here, species Y and Z are indole molecules that are free in solution and those that are complexed with cyclodextrin, so that

$$A_{\lambda,\text{tot}} = A_{\lambda,\text{In}} + A_{\lambda,\text{CD-In}} = \epsilon_{\lambda,\text{In}}\, b\, [\text{In}] + \epsilon_{\lambda,\text{CD-In}}\, b\, [\text{CD-In}]$$

Consider the case in which, at a particular wavelength λ', the extinction coefficients of the free and bound forms of indole are the same, i.e., $\epsilon_{\lambda',\text{In}} = \epsilon_{\lambda',\text{CD-In}} \equiv \epsilon_{\text{iso}}$. Then,

$$A_{\lambda',\text{tot}} = \epsilon_{\text{iso}}\, b\, (\, [\text{In}] + [\text{CD-In}]\,)$$

In other words, as long as the sum of the concentrations of the two species is a constant, the absorbance at the wavelength λ' will not be sensitive to the relative concentrations of the species. In the current experiment, the combined concentration of the indole and the indole—β-cyclodextrin complex is determined by the total indole concentration in solution, which we choose to keep a constant. This result commonly arises in situations, like this experiment, in which the chromophores in solution are in chemical equilibrium with one another and in the course of the investigation we act to alter only the position of the equilibrium. Under these conditions, the wavelength λ', at which the absorbance of the sample is invariant to the relative concentrations of the species, is known as an *isosbestic point*. In this experiment, we use absorption spectroscopy to identify not only the isosbestic points in the spectrum, but to also determine the wavelength of the maximum change in the absorption spectrum of indole upon inclusion into β-cyclodextrin.

An experimental tool that is complementary to absorption measurements is fluorescence spectroscopy, in which the intensity and wavelength of light emitted by chromophores following absorption is determined. In a typical experiment, light of a wavelength known to be absorbed (as determined from the absorption spectrum) is directed through the sample. A fraction of the molecules absorb this light and are elevated energetically to an excited electronic state. These excited species must dissipate this excess energy and a fraction of them will do so by re-emitting light to return to the lowest electronic state. The emitted wavelength differs in general from that absorbed, however, because of Franck–Condon effects, the solute–solvent interaction and rapid vibrational relaxation within the excited electronic state. *(14)* Under the condition that the absorbance of the sample is small ($\leq$ 0.05 at the wavelength of excitation), we can write the total (i.e., integrated over all emission wavelengths) fluorescence intensity, I_f, as *(15)*

$$I_f = I_0\, \epsilon_{\lambda}\, b\, [\text{X}]\, \phi_f,$$

where I_0 is the intensity of the incident light, ϵ_λ is the molar extinction coefficient at the absorbed wavelength λ, b is the pathlength of the sample and [X] is the molar concentration of the sample species X. The fluorescence quantum yield, ϕ_f, is defined as the fraction of molecules which absorb light that re-emit a photon. If one chooses to measure the fluorescence intensity at only one wavelength, λ'', then

$$I_{f,\lambda''} = I_0\, \epsilon_\lambda\, b\, [X]\, \phi_f\, \gamma_{\lambda''}$$

where $\gamma_{\lambda''}$ is the fraction of the total emission intensity that occurs at the wavelength λ''.

Consider the case that there are two species in solution that both absorb light and are capable of emitting light, indole and the indole—β-cyclodextrin complex. In this case, the total fluorescence intensity will be given by *(12)*

$$I_f = I_0\, \epsilon_{\lambda,In}\, b\, \phi_{f,In}\, [In] + I_0\, \epsilon_{\lambda,CD\text{-}In}\, b\, \phi_{f,CD\text{-}In}\, [CD\text{-}In]$$

In this equation, we take explicit account of the fact that both the quantum yields for fluorescence and the molar extinction coefficients for free indole and bound indole will, in general, be different. For notational simplicity we make the following definitions of constants,

$$\Omega_{In} \equiv I_0\, \epsilon_{\lambda,In}\, b\, \phi_{f,In}$$

$$\Omega_{CD\text{-}In} \equiv I_0\, \epsilon_{\lambda,CD\text{-}In}\, \mathrm{b}\, \phi_{f,CD\text{-}In}$$

so that

$$I_f = \Omega_{In}\, [In] + \Omega_{CD\text{-}In}\, [CD\text{-}In]. \quad (2)$$

(In the discussion that follows, we assume that the experimental measurements are of the total fluorescence intensity. Measurements of the fluorescence intensity at a single wavelength λ'' can also be used in this experiment by incorporating the factors $\gamma_{\lambda'',In}$ and $\gamma_{\lambda'',CD\text{-}In}$ into the definitions for Ω_{In} and $\Omega_{CD\text{-}In}$.)

To extract the association constant K_a from measurements of the fluorescence intensity I_f, we seek to relate these two quantities. To begin, we note that at all times,

$$[In]_0 = [In] + [CD\text{-}In]$$

$$[CD]_0 = [CD] + [CD\text{-}In].$$

$[In]_0$ and $[CD]_0$ are the total concentrations of indole and β-cyclodextrin present in the solution, while the quantities on the right hand side (which also appear in equations (1) and (2)) are the *equilibrium* concentrations of free indole, free β-cyclodextrin and the indole-cyclodextrin complex. Using these relations, we rewrite equations (1) and (2) as,

$$K_a = \frac{[CD\text{-}In]}{([In]_0 - [CD\text{-}In])([CD]_0 - [CD\text{-}In])} \quad (3)$$

$$I_f = \Omega_{In} ([In]_0 - [CD\text{-}In]) + \Omega_{CD\text{-}In} [CD\text{-}In] \tag{4}$$

We can rearrange (3) to find a quadratic equation in terms of [CD-In],

$$[CD\text{-}In]^2 - ([In]_0 + [CD]_0 + K_a^{-1}) [CD\text{-}In] + [In]_0[CD]_0 = 0$$

Solving for the roots of this equation (and eliminating the root which leads to $[CD\text{-}In] > [In]_0$), we obtain

$$[CD\text{-}In] = 0.5 \{ ([In]_0 + [CD]_0 + K_a^{-1}) - [([In]_0 + [CD]_0 + K_a^{-1})^2 - 4[In]_0 [CD]_0]^{0.5} \}$$

If we substitute this expression into equation (4), we find that upon rearrangement

$$I_f = \Omega_{In} [In]_0 + 0.5 (\Omega_{CD\text{-}In} - \Omega_{In}) \left\{ ([In]_0 + [CD]_0 + K_a^{-1}) - [([In]_0 + [CD]_0 + K_a^{-1})^2 - 4[In]_0 [CD]_0]^{0.5} \right\} \tag{5}$$

In this equation, we relate the experimentally measured fluorescence intensity I_f to the experimentally controlled total indole and cyclodextrin concentrations $[In]_0$ and $[CD]_0$, the collection of constants Ω_{In} and $\Omega_{CD\text{-}In}$, and the association constant K_a. For a given set of experiments, $[In]_0$ is a constant, known value. Ω_{In} is essentially the proportionality constant between the fluorescence intensity and indole concentration in a spectrum recorded *in the absence of cyclodextrin.* Ω_{In} can, therefore, be experimentally determined. If we measure I_f for a series of solutions with varying $[CD]_0$, a nonlinear least-squares fit of these data yields the values of the parameters K_a and $\Omega_{CD\text{-}In}$. *(12)*

Finally, we note that the relative values of the constants Ω_{In} and $\Omega_{CD\text{-}In}$ provides information on the sensitivity of the absorption and emission properties of indole on its solvent environment. Specifically,

$$\Omega_{CD\text{-}In} / \Omega_{In} = \epsilon_{\lambda,CD\text{-}In}\, \phi_{f,CD\text{-}In} / \epsilon_{\lambda,In}\, \phi_{f,In} \tag{6}$$

If the excitation wavelength λ is chosen to be an isosbestic point, then the ratio of Ω factors reduces to the ratio of quantum yields for fluorescence. In the case that fluroescence intensity measurements are made at only one wavelength λ'',

$$\Omega_{CD\text{-}In} / \Omega_{In} = \epsilon_{\lambda,CD\text{-}In}\, \phi_{f,CD\text{-}In}\, \gamma_{\lambda'',CD\text{-}In} / \epsilon_{\lambda,In}\, \phi_{f,In}\, \gamma_{\lambda'',In}$$

In this situation, if the excitation wavelength is an isosbestic point, then the ratio of Ω factors is the ratio of the quantum yields multiplied by a factor that takes into account any wavelength shifts or changes in shape of the fluorescence spectrum.

Methodology

In the first phase of the experiment, we measure absorbance difference spectra of solutions containing constant indole concentration and varying concentrations of β-cyclodextrin to locate appropriate wavelengths for excitation in the fluorescence measurements. Our solutions were prepared using indole and indole derivatives as obtained from Aldrich. β-cyclodextrin samples (Aldrich) were recrystallized twice from deionized water to remove the majority of a UV-absorbing impurity that was present in all of the commercial samples examined. Aqueous stock solutions of indole were made at up to 1 mM concentration, while all of the other indole derivatives were somewhat less soluble in water. We were able to make the following solutions, though with up to 24 h of stirring in some cases: 1-methylindole, 26 μM; 5-methylindole, 50 μM; 5-cyanoindole, 500 μM; 2,5-dimethylindole, 50 μM.

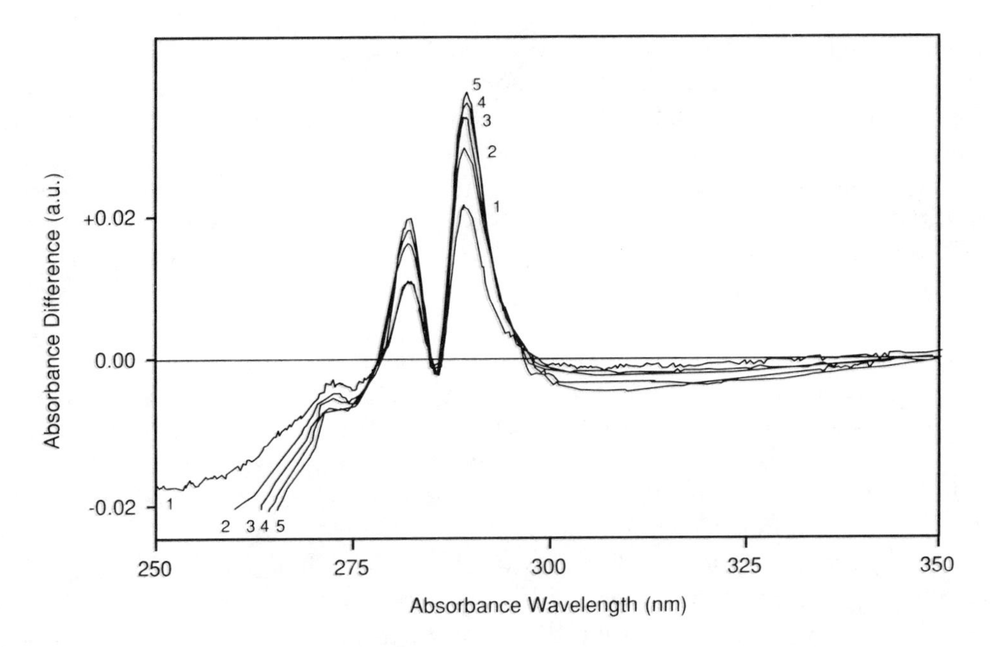

Figure 1. Difference spectra recorded by measuring the absorbance of solutions containing 50 μM indole with varying concentrations of β-cyclodextrin followed by subtraction of the spectrum of 50 μM indole. The β-cyclodextrin concentrations are: trace 1 : 2.9 mM; trace 2 : 4.8 mM; trace 3 : 6.2 mM; trace 4 : 7.3 mM; trace 5 : 8.1 mM.

In Figure 1, we display one set of absorption difference spectra. These particular spectra arise from solutions containing 50 μM indole and β-cyclodextrin concentrations that ranged from 0 to 8.1 mM. There exist two substantial complications in obtaining these data. First, the presence of a trace of a UV-absorbing impurity in all of our recrystallized β-cyclodextrin samples interferes with the absorption spectrum of indole. (Typically, following two recrystallizations, a 10 mM β-cyclodextrin solution has an absorbance of 0.01 at 280 nm. This absorbance is small compared to that of indole at the concentrations used, but is significant compared to the absorbance difference that this procedure is designed to measure.) Second, the small absorbance difference values observed requires that we carry out these measurements without removing the absorbance cuvette from the instrument, to avoid the normally small (but in these experiments, significant) errors associated with slight changes in sample geometry and/or variations in the cleanliness of the cuvette. To eliminate both of these considerations, we proceed as follows. We prepare an indole solution at the required concentration, and use a portion of this solution as solvent to dissolve β-cyclodextrin at a concentration, 15 mM, close to its aqueous solubility. We also prepare a cyclodextrin solution of the same concentration, but with no indole present. We begin our measurements by recording the spectrum of 50 μM indole, in the absence of cyclodextrin, with deionized water in the reference cuvette of the absorption spectrophotometer (Nicolet 9420 UV-Vis Spectrophotometer). We use identical volumes of solution in the sample and reference cuvettes, and take care to assure that we use the minimum volume required to obtain reliable measurements. A requirement of this procedure is that such spectra can be digitally stored for subtraction at a later time. After recording and storing the indole spectrum, we add equal volumes of solution to the sample and reference cuvettes. To the sample solution, we add an aliquot of the 50 μM indole solution in which 15 mM cyclodextrin is dissolved, thus maintaining the indole concentration while adding cyclodextrin. The cyclodextrin concentration in the cuvette can be calculated by considering the initial volume of solution in the cuvette along with that added. (We assume that for these relatively dilute solutions, additivity of volumes is an adequate approximation.) To the reference cuvette, we add an identical volume of aqueous 15 mM cyclodextrin solution, such that the cyclodextrin concentration is the same in both cuvettes. This procedure allows us to eliminate the effect of cyclodextrin impurities from the absorbance spectra through use of the double beam characteristic of most commercial UV-Vis spectrophotometers. The stored indole spectrum is then subtracted from the resulting spectrum, with the difference displayed as one of the traces in Figure 1. This procedure is repeated for each of several aliquots of indole/cyclodextrin solution (sample cuvette) and cyclodextrin solution (reference cuvette), until a family of absorbance difference spectra as shown in Figure 1 is obtained.

These absorption spectra provide the information needed to determine the excitation wavelength for fluorescence experiments. (One should note that it is possible to develop a relationship between the absorbance of a solution containing indole and β-cyclodextrin and the association constant K_a that is exactly analogous to that derived above for the fluorescence intensity. The small absorbance changes observed in these experiments suggests, however, that fluorescence measurements are the preferred method for examining the thermodynamics of the inclusion process. *(12)*) One choice of excitation is at an isosbestic wavelength, such as 278 nm for the case of indole. (See Figure 1; we note that in addition to this isosbestic point, in their original investigation, Örstan and Ross found that the minimum of the "dip" in the absorbance difference spectrum at 286 nm was also an isosbestic point. *(12)* The deviation from zero absorbance difference in our spectra is 0.003 absorbance units at this wavelength. We consider this discrepancy to be within our experimental error.) This choice has the advantage of providing an opportunity for directly measuring the ratio of fluorescence quantum yields for the bound and free indole using equation (6). Alternatively, one could choose to position the excitation wavelength such that the absorbance difference between indole and the indole–cyclodextrin complex is large. In this case, the interpretation of the variation in the Ω factors is more complicated, but the change in the fluorescence spectrum of indole upon addition of β-cyclodextrin may be enhanced. Anticipating that the latter choice leads to association constants that are more statistically significant, we have carried out our fluorescence studies with the excitation wavelength chosen as the maximum of the absorbance difference spectrum.

All fluorescence spectra were measured on a standand commercial spectrofluorometer (Spex F111A), equipped with a computer interface that allows digital integration of peak areas. A series of representative emission spectra are

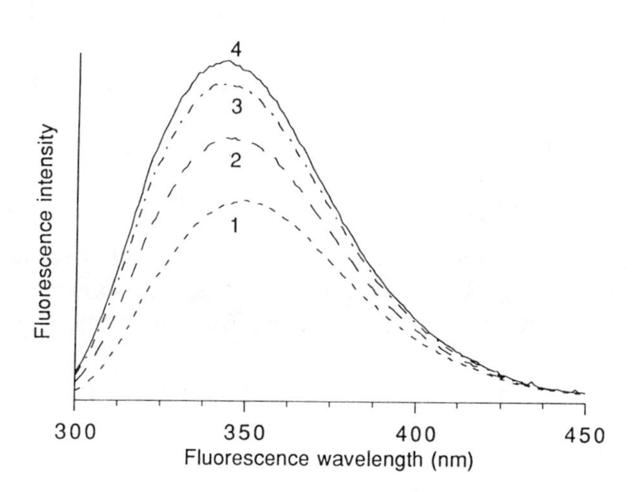

Figure 2. Fluorescence spectra from solutions containing 10 μM indole and various concentrations of β-cyclodextrin. The excitation wavelength is 289 nm. The β-cyclodextrin concentrations are: trace 1: 0 mM; trace 2: 1.96 mM; trace 3: 4.66 mM; trace 4: 6.43 mM.

shown in Figure 2. These spectra were recorded using a lower indole concentration (10 μM) than that used for the absorbance difference spectra to assure that the absorbance of the excitation wavelength remained ≤ 0.05. The initial spectrum was recorded on indole without added cyclodextrin, using the minimum volume of solution required to obtain reliable intensity data. This spectrum allows us to calculate the constant Ω_{In}. Subsequent spectra were recorded without removing the cuvette from the instrument by stepwise addition of small volumes of a solution containing 10 μM indole and 15 mM β-cyclodextrin. We again assume the additivity of volumes to calculate the β-cyclodextrin concentration. Thus we record a set of fluorescence spectra (typically a minimum of 5) with constant indole concentration, but with varying cyclodextrin content. All fluorescence measurements were carried out in a temperature controlled cuvette holder using solutions that were pre-equilibrated to the desired temperature. The solutions were stirred while in the cuvette to avoid photobleaching of the indole. We also found that indole is sufficiently photoreactive that high excitation light flux caused degradation of the sample over the course of an experiment. We minimize this problem by operating the excitation light monochromator with the smallest slit size available. The contribution to the fluorescence signal caused by scattered light and electronic noise was subtracted from each measurement by recording a baseline spectrum of deionized water. A spectrum of β-cyclodextrin in water without indole was also recorded to eliminate any fluorescence from the trace impurity in the sample. This small background signal is subtracted from each spectrum in proportion to the total cyclodextrin concentration.

The data of fluorescence intensity versus β-cyclodextrin concentration is analyzed using a nonlinear least squares program based on the grid-search method as described by Bevington. *(16)* In this program, we specify the functional form to be fit (equation (5)), along with initial estimates for the variable parameters K_a and Ω_{CD-In}. Should it be desirable, the data can be weighted by the experimental uncertainties, with the program calculating the associated error in the fitted parameters.

Finally, we carry out fluorescence measurements on this system at four different temperatures (typically 15, 25, 35 and 45 °C). Assuming that ΔH and ΔS do not vary significantly with temperature, we can use the relation,

$$\ln K_a = -(\Delta H/R)\, 1/T + \Delta S/R$$

to calculate the values of ΔH and ΔS from the temperature dependence of the association constant. In Table I we present a typical set of results obtained by a class of students at Swarthmore College for the association constant at 25 °C and the enthalpy and entropy of association for a series of indole derivatives.

Table I : Typical student data on the association of indole derivatives with β-cyclodextrin.

Compound	K_a *(M^{-1}; 25 °C)*	ΔH*(kJ/mol)*	ΔS*(J/mol K)*
Indole[a]	184	−17.2	−14.6
5-methylindole	311	−23	−31
2,5-dimethylindole	472	−19	−13
5-cyanoindole	326	−19	−15
1-methylindole	49	+41	+169

[a] Reference 12.

Safety

The only safety considerations in this experiment involve handling of the various indole derivatives. Indole is considered to be toxic, as well as a stench and irritant hazard. All of the other derivatives that we have investigated are either stench or irritant hazards. Preparation of all solutions should be carried out with adequate ventilation. Gloves and safety glasses should be worn while handling all solutions.

Data Analysis

A pair of students should be able to carry out absorption difference and fluorescence emission studies of one indole derivative at 25 °C and emission studies at three other temperatures in two four hour laboratory periods, assuming that a detailed experimental procedure is provided and all solutions are prepared in advance. Using this data and a nonlinear least-squares fitting routine, values of K_a for each temperature can be calculated. A linear regression of ln K_a versus $1/T$ then provides values of ΔH and ΔS. If considered desirable, a detailed error analysis to attach confidence limits to the thermodynamic quantities can be carried out.

The analysis of this experiment and the derived quantities can focus on some or all of the following issues:

1) Why is it true that fluorescence measurements provide a more reliable means (relative to absorbance difference spectra) to detect the association of indole with β-cyclodextrin?
2) At the highest cyclodextrin concentration used, calculate the fraction of the total indole that is included in β-cyclodextrin. Why is it not possible to record a spectrum of the "pure" indole–β-cyclodextrin complex?
3) How does one justify the sign of the quantities ΔH and ΔS for this reaction? What do their magnitudes reveal about the "local" environment of indole (a) in aqueous solution and (b) as part of an inclusion complex?
4) Assuming that a series of indole derivatives have been examined by different groups of students, compare the results. Comment on the importance of the following effects in modulating the formation of the inclusion complex:

    a) steric interactions.
    b) hydrogen bonding (involving the indole N–H and the cyclodextrin hydroxyl groups).
    c) the impact of electron withdrawing and/or electron donating groups on the hydrogen bonding interaction.
    d) the relative importance of electron withdrawing and/or electron donating groups on the stability of the indole compound free in solution and as part of the inclusion complex.

5) If an isosbestic point in the absorption spectrum is used as the excitation wavelength in the fluorescence studies, then the ratio of the fluorescence quantum yields for indole and the indole-cyclodextrin complex can be calculated (equation (6)). What might be the reason(s) that the fluorescence quantum yield changes when the complex is formed?
6) Finally, it might be useful to ask students to consider the importance of separating scientific observation from interpretations. Do the quantitative results derived from experiments such as this "prove" a given interpretation? How do we evaluate the amount of credence to lend to a given interpretation?

Acknowledgements

We are grateful to the Pew Charitable Trust and Swarthmore College for support of this work through the Mid-Atlantic Cluster of the Pew Science Program.

Literature Cited

1. Voet, D. and Voet, J.G. *Biochemistry*; Wiley: New York, NY, 1990; pp 169-181.
2. Atkins, P.W. *Physical Chemistry, 4th ed.*; W.H. Freeman: New York, NY, 1990; pp. 661-664.
3. Saenger, W. *Angew. Chem. Int. Ed. Engl.* **1980**, *19*, 344.
4. Tabushi, I. *Acc. Chem. Res.* **1982**, *15*, 66.
5. D'Souza, V.T. and Bender, M.L. *Acc. Chem. Res.* **1987**, *20*, 146.
6. Ramamurthy V. and Eaton, D.F. *Acc. Chem. Res.* **1988**, *21*, 300.
7. Alak, A. and Armstrong, D.W. *Anal. Chem.* **1986**, *58*, 582.
8. Mailer, K., Addetia, R. and Livesey, D.L. *J. Inorg. Biochem.* **1989**, *37*, 151.
9. Ferreira, S.T. *Biochemistry* **1989**, *28*, 10066.
10. Lami, H. and Glasser, N. *J. Chem. Phys.* **1986**, *84*, 597.
11. Smith, G.J. and Melhuish, W.H. *J. Phys. Chem.* **1991**, *95*, 4288.
12. Örstan, A. and Ross, J.B.A. *J. Phys. Chem.* **1987**, *91*, 2739.
13. Willard, H.H., Merritt, L.L., Dean, J.A., and Settle, F.A. *Instrumental Methods of Analysis, 7th ed.*; Wadsworth: Belmont, CA, 1988; pp. 159-162.
14. Atkins, P.W., ibid, p. 510.
15. Willard, H.H., et al., ibid, p. 201.
16. Bevington, P.R. *Data Reduction and Error Analysis for the Physical Sciences*; McGraw-Hill: New York, NY, 1969; pp. 208-215.

Hardware List

1. A scanning double beam UV-Vis spectrphotometer with spectral subtraction capabilities is preferred, but if such instrumentation is unavailable, the choice of excitation wavelength can be made by manually overlaying the unsubtracted spectra.
2. A spectrofluorometer, with temperature controlled cell holder for temperature dependent studies. A computer interfaced model that allows for digital integration of peak areas is preferred, but single wavelength analog fluorescence intensity measurements are also adequate.
3. Thermostated water bath for pre-equilibration of sample solutions.

RECEIVED October 14, 1992

CHAPTER 17

Measurements of Fluorescence Intensity and Lifetime

Lee K. Fraiji, David M. Hayes, and T. C. Werner

Fluorescence spectroscopy is a technique that is ideally suited for the undergraduate laboratory curriculum. Several workers have published experiments designed for the undergraduate lab employing fluorescence measurements*(1-5)*. In all of these cases, the measurement of the steady state fluorescence signal has been emphasized to extract information on analyte level or identity and on the efficiency of fluorescence quenching by an external quencher.

The time-dependent nature of fluorescence can also be exploited if the lifetime of fluorescence can be measured. This information is especially useful in evaluating the mechanism and efficiency of fluorescence quenching, especially when time-dependent data are combined with steady-state fluorescence measurements on the same fluorophore-quencher system. These data can often be used to determine whether the quenching mechanism is static (occurs because of a ground state complex between fluorophore and quencher), dynamic (occurs from diffusion of quencher to fluorophore while the latter is in its excited state) or if both mechanisms are occurring. Moreover, the binding constant for the ground-state quenching complex (K) and the rate constant for dynamic quenching (k_q) can often be calculated from these results. As a consequence, such measurements constitute an excellent experiment for the physical chemistry laboratory.

Theory

Intensity (Steady-State) Measurements and Fluorescence Quenching The intensity of fluorophore fluorescence can be quenched by ground-state quencher-fluorophore reactions (static quenching) and by excited-state quencher-fluorophore reactions (dynamic quenching). In the discussion that follows, static quenching is assumed to result from the formation of a non-fluorescent quencher-fluorophore complex in the ground state. Shifts of the fluorophore absorption spectrum with added quencher provide evidence of such complex formation. Another type of static quenching is often observed at high quencher concentrations due to the existence of increasing numbers of quencher-fluorophore pairs in which the quencher is close enough to the fluorophore to instantaneously quench its excited state*(6)*. Treatment of this type of quenching is less straightforward. Fortunately, it can be distinguished from quenching due to true ground-state complex formation because it does not produce changes in the fluorophore absorption spectrum.

The steady-state parameter which responds to added quencher is the fluorescence quantum yield ϕ_f, which is defined as the ratio of the rate of fluorescence emission to the rate of absorption (photons out / photons in). The fluorescence quantum yield in the absence of quencher (ϕ_f^0) is defined by the mechanism below, where A is the fluorophore ground state, A* is the fluorophore emitting state (lowest excited singlet), k_A is the rate constant for photon absorption, k_f is the rate constant for fluorescence emission and k_{nr} is the sum of first-order rate constants for non-radiative decay modes, such as internal conversion and intersystem crossing. n_A and n_{A*} are the number of fluorophores in the ground and excited state, respectively.

$$A + h\nu \rightarrow A^* \qquad \text{Rate} = k_A n_A \tag{1}$$

$$A^* \rightarrow A + h\nu \qquad \text{Rate} = k_f n_{A*} \tag{2}$$

$$A^* \rightarrow A \qquad \text{Rate} = k_{nr} n_{A*} \tag{3}$$

Since steady-state conditions exist, we can assume:

$$k_A n_A = (k_f + k_{nr})\, n_{A*} \qquad (dn_{A*}/dt) = 0 \tag{4}$$

Rearranging eq. (4) gives:

$$n_{A*} = k_A n_A / (k_f + k_{nr}) \tag{5}$$

The ϕ_f is formally defined as:

$$\phi_f = k_f n_{A*} / k_A n_A \tag{6}$$

Combining equations (5) and (6) gives the familiar form for the fluorescence quantum yield in the absence of quencher (ϕ_f^0):

$$\phi_f^0 = k_f / (k_f + k_{nr}) \tag{7}$$

When quencher (Q) is added, another A* decay mechanism is now possible, where k_q is the second-order rate constant for quenching and the product k_q[Q] is pseudo first-order.

$$A^* + Q \rightarrow A + Q \qquad \text{Rate} = k_q[Q] n_{A*} \tag{8}$$

Now equation (5) becomes:

$$n_{A*} = k_A n_A / (k_f + k_{nr} + k_q[Q]) \tag{9}$$

Combining equation 9 with equation 6 gives an expression for the fluorescence quantum yield in the presence of quencher, providing dynamic quenching (equation 8) is the only quenching mechanism.

$$\phi_f = k_f / (k_f + k_{nr} + k_q[Q]) \tag{10}$$

Dividing equation 7 by equation 10 produces the familiar Stern–Volmer equation*(7)*:

$$\phi_f^0 / \phi_f = 1 + k_q[Q] / (k_f + k_{nr}) \quad (11)$$

In the more familar form of this equation ϕ_f^0 and ϕ_f are replaced by F^0 and F, the intensities of fluorescence at a given wavelength, in the absence and presence of Q, respectively, and the term $[k_q / (k_f + k_{nr})]$ is set equal to K_{SV}, the Stern–Volmer quenching constant*(6)*.

$$F^0 / F = 1 + K_{SV}[Q] \quad (12)$$

If this quenching mechanism alone obtains, a plot of F^0 / F versus [Q] is linear with an intercept of 1 and a slope equal to K_{SV}.

When quenching can also occur by ground-state complex formation (static quenching) another reaction must be considered in the overall quenching mechanism*(8)*:

$$A + Q \overset{K}{\rightleftharpoons} AQ \overset{h\nu}{\rightarrow} AQ^* \rightarrow AQ + \text{heat} \quad (13)$$

K is the equilibrium constant for the formation of the "dark" complex, AQ. If both static and dynamic quenching are occurring, the intensity ratio (F / F^0) can be expressed as the fractional reduction due to quenching of A* (dynamic) times the fractional reduction due to complexation of A (f, static) *(6)*.

$$F / F^0 = (1 + K_{SV}[Q])^{-1} f \quad (14)$$

The fraction f can be expressed in the following way, given the definition of K in eq 13:

$$f = [A] / ([A] + [AQ]) = (1 + K[Q])^{-1} \quad (15)$$

Thus, it follows that

$$F / F^0 = (1 + K_{SV}[Q])^{-1} (1 + K[Q])^{-1} \quad (16)$$

This, in turn, can be rearranged to give a modified form of the Stern–Volmer equation:

$$F^0 / F = (1 + K_{SV}[Q]) (1 + K[Q]) \quad (17)$$

Equation 17 predicts upward curvature of the plot of F^0 / F versus [Q] in the event that both static and dynamic quenching are occurring.

If quenching **only** occurs by the static mechanism (k_q, $K_{SV} = 0$), equation 17 simplifies to:

$$F^0 / F = 1 + K[Q] \quad (18)$$

Note that equation 18 also predicts a linear relation between F^0 / F and [Q] with an intercept of 1, as does eq 12 derived for the case where dynamic quenching alone is occurring. In the former case the slope equals K, while in the latter it equals K_{SV}. Thus, one can not determine whether quenching is static or dynamic on the basis of a single linear Stern–Volmer plot. If such plots are obtained as a function of temperature, a determination might be made from the change in slope. Increased temperature often causes the slope (K_{SV}) to increase if quenching is dynamic (k_q increases with temperature), while the slope (K) should decrease with increasing temperature if the quenching is static *(6)*. Even so, there is no way to extract a value for k_q using steady-state measurements only.

Dynamic Measurements (Fluorescence Lifetime) and Fluorescence Quenching
When fluorescence is excited by a pulsed rather than a continous source the decrease of fluorophore fluorescence after the pulse is extinguished normally follows a single exponential decay in solution. The lifetime of fluorescence, defined as the time for the fluorescence signal to decay to $1/e$ of its original value, is given by eq 19 in the absence of quencher and eq 20 in the presence of quencher.

$$\tau^0 = (k_f + k_{nr})^{-1} \quad (19)$$

$$\tau = (k_f + k_{nr} + k_q[Q])^{-1} \quad (20)$$

Dividing equation 19 by equation 20 gives another form of the Stern–Volmer equation, **which applies only if quenching is dynamic.**

$$\tau^0 / \tau = 1 + k_q[Q] / (k_f + k_{nr}) = 1 + k_q \tau^0 [Q] \quad (21)$$

We now recognize that the Stern–Volmer constant (K_{SV}) obtained from steady–state measurements is equal to $k_q\tau^0$. Thus, if lifetime measurements are possible, the value of k_q can be extracted from plots of eq 21 or eq 12 and the value of the lifetime in the absence of quencher (τ^0).

It is important to note that lifetime measurements are not affected by the formation of a ground-state "dark" complex (AQ). Consequently, lifetime measurements can be used to separate dynamic quenching from static. Moreover, with both dynamic and steady-state measurements, it is often possible to extract k_q , if dynamic quenching only is occurring, K, if static quenching only is occurring, or k_q and K, if both mechanisms are occurring, from these measurements.

We report here three separate experiments which illustrate the usefulness of both types of quenching measurements for this end.

Methodology

Fluorescence spectra and intensity data were obtained on a Perkin-Elmer Lambda 5B Spectrofluorometer attached to a Perkin-Elmer R100 recorder. Excitation and emission slits were both set at either 5 or 10 nm for all measurements. The fluorescence source is a 8.3-W Xenon discharge lamp. Fluorescence lifetime data were obtained on a Photon Technology Incorporated Fluorescence Lifetime System, which employs a pulsed nitrogen discharge lamp for excitation and an optical boxcar detector. Lifetime values were extracted from the data by convolution of the lamp decay with exponential decays until the fit to the observed fluorescence decay was

Table I. Fluorescence lifetimes as a function of quencher concentration for the systems:

(1) N-MEAI/GMP, (2) PSA/Iodide, and (3) 2-AN/β-CD.

N-MEAI		PSA		2-AN	
[GMP]/10^{-3} M	τ/ns	[I^-]/10^{-3} M	τ/ns	[β-CD]/10^{-3} M	τ/ns
0.0	34.8	0.0	58.6	0.0	2.0
1.0	30.9	3.0	37.2	1.0	1.9
2.0	27.8	6.0	26.4		
3.0	25.0	9.0	20.7		
4.0	22.5	12.0	17.1		
6.0	18.8				
8.0	15.7				

deemed acceptable on the basis of the χ^2 statistic. The convolution was done on an IBM PC compatible computer. We include tabulated lifetime data in Table I for all of the fluorophore/quencher systems we studied so that others can perform these experiments without having access to lifetime measuring instrumentation. We are also willing to send copies of the lamp and fluorescence intensity profiles and fitted decay curves to anyone who would like to show these to students.

N-methylacridinium iodide(N-MEAI) and 1-pyrenesulfonic acid, sodium salt(PSA) were obtained from Molecular Probes (Eugene, OR). The GMP (guanosine-5′-monophosphate, disodium salt trihydrate), 2-acetylnaphthalene(2-AN) and β-cyclodextrin(β-CD) were obtained from Aldrich Chemical; 2-AN and β-CD were recrystallized from water before use. All other chemicals were reagent grade or better. Solutions were prepared using distilled water. Note: N-methylacridinum iodide is a potential skin and eye irritant. Solutions of GMP should be prepared within one day of usage. N-MEAI/GMP solutions were prepared in 0.1 M phosphate buffer at pH 7.0. The PSA/iodide solutions where prepared at a constant ionic strength of 0.0135 M with KCl. No attempt was made to control either pH or ionic strength in the 2-AN/β-CD system. Temperature should be controlled when performing the described experiments because k_q values are dependent on temperature and solvent viscosity. Moreover, the fluorophore used in the static quenching experiment (2-acetylnaphthalene) has a very temperature dependent fluorescence. Our experiments were performed at 22 °C.

Safety

Given the noted potential of N-methylacridinium iodide (N-MEAI) to act as a skin and eye irritant, we prepare the stock solution of N-MEAI in water for students to use in the preparation of the solutions containing N-MEAI in the presence of increasing concentrations of the quencher, GMP.

Since all of the solutions employed in these experiments are aqueous and very dilute, disposal of used solutions requires no special precautions.

Data Analysis

Quenching of N-methylacridinium iodide fluorescence by GMP Figure 1 shows typical results for the effect of quencher concentration on the fluorescence intensity and lifetime of N-MEAI. Lifetime data for this system are given in Table I. The

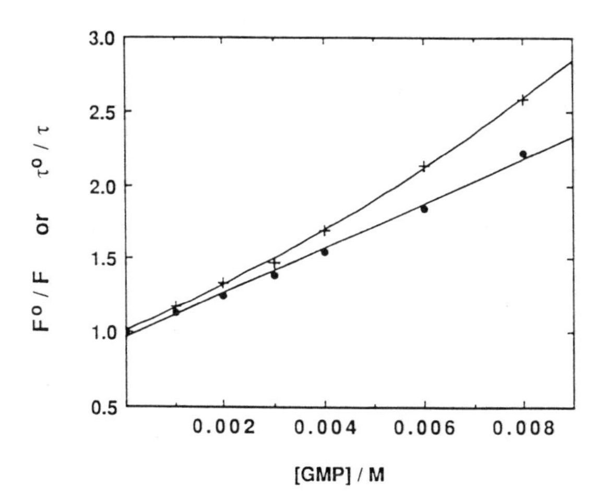

Figure 1. Fluorescence intensity and lifetime of N-MEAI plotted against the concentration of GMP quencher. The excitation wavelengths for the fluorescence intensity and lifetime determinations were 380 and 358 nm, respectively. In both experiments the fluorescence emission was monitored at 485 nm. The solid circles correspond to τ^0/τ and the crosses to F^0/F.

linear dependence of τ^0/τ and quadratic dependence of F^0/F on GMP concentration indicates the presence of both dynamic and static quenching in this system. From eq 21, the slope of the τ^0/τ plot and the independently measured value for τ^0 of 35 ns, one calculates a value for k_q of 4.4×10^9 M^{-1} s^{-1} in good agreement with 4.3×10^9 M^{-1} s^{-1} obtained by Kubota et al. *(9)*. Using eq 17 as a guide, one can calculate the N-MEAI/GMP association constant K from the coefficients obtained by fitting a second-order polynomial to the intensity data shown in Figure 1. Alternatively, rearranging eq 17 to give eq 22 shows that K can be obtained by fitting a first-order polynomial to a plot of $(F^0/F - 1)/[Q]$ versus [Q].

$$(F^0 / F - 1) / [Q] = (k_q\tau^0 + K) + k_q\tau^0K[Q] \tag{22}$$

This is the usual method employed to analyze these data *(6,9)* and is the procedure we have used. Figure 2 is a plot of $(F^0/F - 1)/[Q]$ versus [GMP]. From the slope and intercept one calculates K to be 48 M^{-1} in good agreement with Kubota et al.'s value of 44 M^{-1} *(9)*. Alternately, using just the slope and the previously determined k_q and τ^0, one calculates K to be 34 M^{-1}. That GMP is both a static and dynamic quencher is not surprising. Energy transfer between the

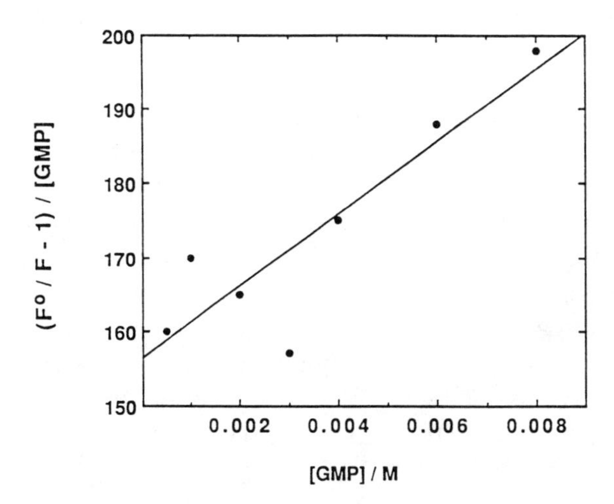

Figure 2. A plot of the N-MEAI fluorescence data of Figure 1 in a way suggested by eq. 22.

conjugated π-systems of the rings in N-MEAI and GMP should be efficient and the opposite charges on the fluorophore and quencher are certainly consistent with formation of a complex involving the ground state species.

Quenching of 1-Pyrenesulfonic Acid Fluorescence by Potassium Iodide As shown by the linearity and nearly identical slopes of the plots in Figure 3, quenching of PSA fluorescence by potassium iodide only occurs dynamically. Lifetime data for this

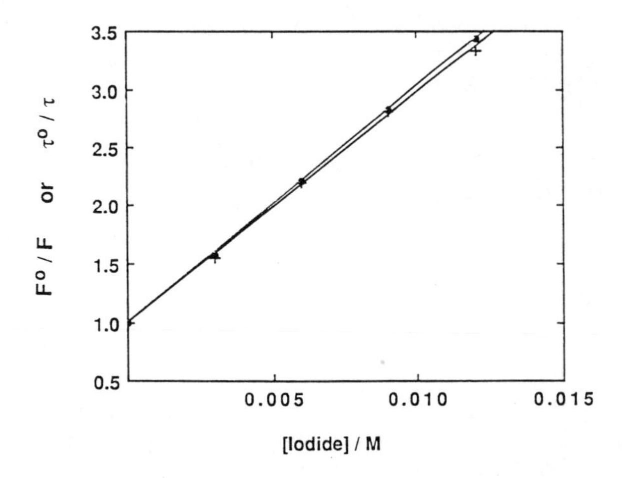

Figure 3. Fluorescence intensity and lifetime of PSA plotted against the concentration of iodide ion. The excitation wavelengths for the fluorescence intensity and lifetime measurements were 335 and 337 nm, respectively. In both experiments the fluorescence emission was monitored at 395 nm. The solid circles correspond to τ^0/τ and the crosses to F^0/F.

system are given in Table I. From eq 21, the slope of these plots, and τ^0 one calculates a value for the dynamic quenching rate constant of 3.4×10^9 M^{-1} s^{-1}.

The absence of static quenching is expected since both fluorophore and quencher exist as anions in aqueous solution.

Quenching of 2-acetylnaphthalene Fluorescence by β-Cyclodextrin During the course of an independent investigation by two of the authors, it was observed that the fluorescence of 2-acetylnaphthalene (2-AN) is quenched when this molecule binds in the internal cavity of β-cyclodextrin (β-CD) *(10)*. As the concentration of β-CD increases from 0 to 0.001 M, the intensity of 2-AN fluorescence decreases by 50%, while the lifetime of 2-AN fluorescence remains unchanged (2.0 ± 0.1 ns). Consequently, the 2-AN/β-CD system is an example of the third possibility mentioned above, where only static quenching is occurring. The slope of the fluorescence intensity plot, shown in Figure 4, is equal to the equilibrium constant for formation of the ground-state 2-AN/β-CD complex. The value we obtain for this equilibrium constant is 581 M^{-1} at 21 °C.

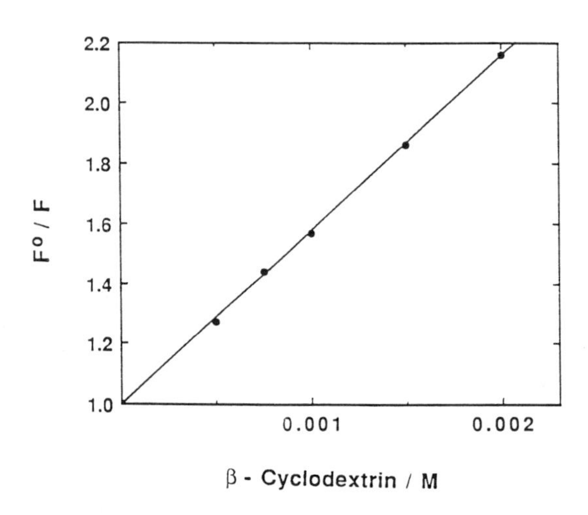

Figure 4. Fluorescence intensity of 2-AN plotted against the concentration of β-cyclodextrin quencher. The fluorescence excitation and emission wavelengths were 340 and 435 nm, respectively.

If students have sufficient time, they can repeat the fluorescence intensity measurements on this system at other temperatures and prepare a plot of ln K versus $1/T$ such as is shown in Figure 5. From the van't Hoff equation, the slope can be related to ΔH^0 of complex formation between 2-AN and β-CD and the intercept to ΔS^0. We obtain values of -11.9 ± 0.5 kJ/mol for ΔH^0 and 12.5 ± 1.6 J/K mol for ΔS^0. The negative sign for ΔH^0 is accounted for by the hydrophobic interactions between the naphthalene ring and the walls of the β-CD cavity *(11)*. The overall ΔS^0 of complex formation is the sum of several entropy changes with differing signs. Entropy loss accompanies restriction of fluorophore motion upon binding, while entropy gain occurs due to expulsion of water from the β-CD cavity and from disruption of the solvent shell around the fluorophore when it binds to β-CD *(12, 13)*. Apparently, the latter contribution dominates in this case.

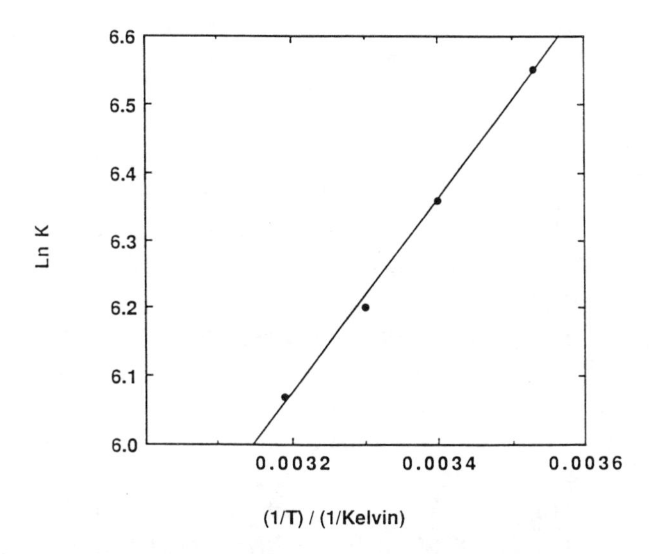

Figure 5. A van't Hoff plot of the equilibrium constant for formation of the 2-AN/β-CD complex vs. the inverse Kelvin temperature. ΔH^0 and ΔS^0 for complex formation are obtained from the slope and *y* intercept, respectively.

This quenching of 2-AN fluorescence by β-CD is in contrast to the normally observed fluorescence enhancement when fluorophores bind to β-CD *(13-15)*. The 2-AN molecule only shows significant fluorescence in strongly hydrogen-bonding solvents, such as water and fluorinated alcohols *(16)*, presumably due to the ability of these solvents to blue shift the low-lying n,π^* state to an energy where it can not interfere with fluorescence*(17)*. Such a strong hydrogen bond with 2-AN is not possible in the β-CD cavity and quenching therefore results.

Our students normally investigate two of the fluorophore/quencher systems during the course of two 4-h laboratory periods. Toward the end of this same course, students also undertake a more extensive 4-week (16 laboratory hours) kinetics project. The students assigned to do the project described here are able to investigate all three fluorophore/quencher systems as well as do the thermodynamic study of 2-AN/β-CD binding.

Acknowledgement

We wish to thank the National Science Foundation for a grant(#USE-9051249) through its Instrumentation and Laboratory Improvement Program which made possible the purchase of the fluorescence lifetime instrument used in this work.

Literature cited

1. Legenda, M.W.; Marzzacco, C.J. *J. Chem. Educ.* **1977**, *54*, 183-184.
2. Ebeid, E. *J. Chem. Educ.* **1985**, *62*, 165-166.
3. Marciniak, B. *J. Chem. Educ.* **1986**, *63*, 998-1000.

4. Sacksteder, L.; Ballew, R.M.; Brown, E.A.; Demas, J.N.; Nesselrodt, D.; DeGraff, B.A. *J. Chem. Educ.* **1990**, *67*, 1065-1067.
5. Byron, C.M.; Werner, T.C. *J. Chem. Educ.* **1991**, *68*, 433-436.
6. Lakowicz, J.R. *Principles of Fluorescence Spectroscopy*; Plenum: New York, 1983; Chapter 9.
7. Stern, O.; Volmer, M. *Phys. Z.* **1919**, *20*, 183.
8. Vaughan, W.M.; Weber, G. *Biochem.* **1970**, *9*, 464-473.
9. Kubota, Y.; Motoda, Y.; Shigemune, Y.; Fujisaki, Y. *Photochem. Photobiol.* **1979**, *29*, 1099-1106.
10. Fraiji, L.; Werner, T.C., unpublished results.
11. Bergeron, R.J. in *Inclusion Compounds*; Atwood, J.L.; Davies, J.E.D.; MacNichol, D.D., Eds; Academic: New York, 1984; pp 423-429.
12. Herkstroeter, W.G.; Martic, P.A.; Farid, S. *J. Am. Chem. Soc.* **1990**, *112*, 3583-3589.
13. Catena, G.C.; Bright, F.V. *Anal. Chem.* **1989**, *61*, 905-909.
14. Ueno, A.; Takahashi, K.; Osa, T. *J. Chem. Soc. Chem Comm.* **1980**, 921-922.
15. Hamai, S. *Bull. Chem. Soc. Jpn.* **1982**, *55*, 2721-2729.
16. Werner, T.C.; LoParo, K.A. unpublished results.
17. Kovi, P.J.; Capomacchia, A.C.; Schulman, S.G. *Spectrosc. Lett.* **1973**, *6*, 7.

Hardware List

Perkin-Elmer Lambda 5B Spectrofluorometer
Perkin-Elmer R100A Recorder
FTS Systems, Inc. Flexi Cool Unit for temperature control of fluorescence cell
Photon Technology International, Inc. Fluorescence Lifetime System with NEC PowerMate 286 Computer and Hewlett-Packard ColorPro Plotter

RECEIVED October 14, 1992

A New Look at Classical Topics

CHAPTER 18

Hückel Molecular Orbitals

Richard S. Moog

As the number of atoms in a molecule increases beyond two or three, detailed and accurate quantum mechanical calculations become increasingly more difficult to perform. Formulated in the early 1930s, the Hückel theory provides a method for examining fairly complicated molecular systems through a simple theoretical framework. Although this theory makes several crude and drastic approximations, quite often the results obtained are qualitatively (and often semiquantitatively) accurate. With the advent of inexpensive and easy to use programs for constructing and solving the corresponding secular determinants on personal computers, students may be readily exposed to the Hückel method and the types of predictions concerning molecular structure and reactivity which may be made from these approximate calculations. Of particular advantage over more sophisticated calculations of molecular structure is the ease with which calculations may be made on excited states involving $\pi \rightarrow \pi^*$ excitations. In this paper, a brief discussion of the Hückel method will be presented, followed by a description of two projects appropriate for undergraduates studying physical chemistry. The first project involves the calculation and analysis of Hückel molecular orbitals for the π systems of two related aromatic molecules. The relationship of quantities derived from the calculated orbitals to the molecular structure and reactivity are examined. The second experiment involves the analysis of a possible intramolecular excited state proton transfer reaction and the prediction of the relative energies of absorption and fluorescence assuming that the proton transfer does occur. These predictions may then be compared to the experimental absorption and fluorescence spectra.

Theory

Assumptions of Hückel Theory A brief discussion of Hückel Molecular Orbital (HMO) Theory is provided here. A more detailed description of the theory may be found in a number of texts (for example, see Streitweiser *(1)* or Lowe *(2)*). An excellent introduction for many undergraduate physical chemistry students is found in *Atoms and Molecules* by M. Karplus and R. N. Porter *(3)*. The most important assumptions of simple HMO Theory for planar hydrocarbons are listed below:

a) The wave function of a molecule may be factored into non-interacting portions representing the σ framework and the π system. The π portion of the wave function is made up of molecular orbitals (MOs) constructed by a linear combination of carbon atomic orbitals (LCAO). The component atomic orbitals are assumed to share a nodal plane, the plane of the molecule.

2503–6/93/0280$06.00/0

b) The resultant LCAO MOs are taken to be eigenfunctions of an effective one-electron Hamiltonian, $\hat{H}$, so that the variational principle may be applied.

c) The Coulomb integrals, $H_{ii} = \langle\phi_i \mid \hat{H} \mid \phi_i\rangle$, are set equal to a negative (but unspecified) energy quantity, α. The interpretation of α is that it represents the average energy of an electron in one of the atomic orbitals in a potential field due to the entire molecule. In this approximation, all sp^2 carbons are considered to experience identical fields, regardless of the actual environment in the molecule.

d) The resonance integrals (or bond integrals), $H_{ij} = \langle\phi_i |\hat{H}| \phi_j\rangle$, are taken as zero if i and j are nonbonded atoms and are set equal to β (also a negative energy quantity) if carbon atoms i and j are bonded. These integrals can be interpreted as the energy of the overlap charge between two atomic orbitals. Thus, all interactions between neighboring atoms are assumed to be identical, and atoms which are not bonded to one another are assumed to have no interaction.

e) The overlap integrals, $S_{ij} = \langle\phi_i|\phi_j\rangle$, are set equal to zero except when $i = j$. For normalized atomic orbitals, S_{ii} is taken as 1.

In simple HMO theory heteroatoms can be included through the introduction of two additional parameters, h and k. The parameter h is used to alter the Coulomb integrals for heteroatoms according to the formula $H_{ii} = \alpha + h\beta$, where α and β are the standard α and β for a carbon atom of benzene. The value of h will increase as the core potential increases, or roughly as the electronegativity of the atom increases. It also varies with the number of electrons contributed to the π system by the atom. The parameter k is used to modify the bond integrals. Its value depends on the type of bond and results in the formula $H_{ij} = k\beta$. Typical values of h and k are given in Table I.

Table I. Typical values of h and k for various atoms and bond types (from Ref. *1*).

Atom	Bond Type	Number of π electrons	h	k
C	C=C	1	0.0	1.0
N	C=N (pyridine)	1	0.5	1.0
N	C-N (pyrrole)	2	1.5	0.8
O	C=O (carbonyl)	1	1.0	1.0
O	C-O (furan or alcohol)	2	2.0	0.8
F	C-F	2	3.0	0.7
Cl	C-Cl	2	2.0	0.4
Br	C-Br	2	1.5	0.3
S	C-S (thiophene)	2	1.5	0.4

The general form of the HMO secular determinant is given by:

$$\begin{vmatrix} H_{11}-E_{11}S_{11} & H_{12}-E_{12}S_{12} & H_{13}-E_{13}S_{13} & \cdots & H_{1n}-E_{1n}S_{1n} \\ H_{21}-E_{21}S_{21} & H_{22}-E_{22}S_{22} & H_{23}-E_{23}S_{23} & \cdots & H_{2n}-E_{2n}S_{2n} \\ \cdots & \cdots & \cdots & \cdots & \cdots \\ H_{n1}-E_{n1}S_{n1} & H_{n2}-E_{n2}S_{n2} & H_{n3}-E_{n3}S_{n3} & \cdots & H_{nn}-E_{nn}S_{nn} \end{vmatrix} = 0$$

Note that this is a square determinant, where the dimension is determined by the total number of atoms considered to be part of the π system. Recall that in the simple HMO scheme that is considered here $S_{ij} = 0$ and $S_{ii} = 1$ so that the determinant becomes

$$\begin{vmatrix} H_{11}-E_{11} & H_{12} & H_{13} & \cdots & H_{1n} \\ H_{21} & H_{22}-E_{22} & H_{23} & \cdots & H_{2n} \\ \cdots & \cdots & \cdots & \cdots & \cdots \\ H_{n1} & H_{n2} & H_{n3} & \cdots & H_{nn}-E_{nn} \end{vmatrix} = 0$$

The elements of the determinant are then evaluated in terms of α, β, and E (and also h and k, if necessary) according to the prescriptions described above. Next, the entire determinant is divided by β, and the substitution $x = (\alpha - E)/\beta$ is made.

Expanding the resulting determinantal equation yields a polynomial equation in x of degree n. This equation will have n roots for x, and therefore n eigenvalues for E can be obtained (recall that $x = (\alpha - E)/\beta$.). The eigenfunctions corresponding to each of these eigenvalues is then calculated, providing the coefficients of each AO for each of the n HMO wavefunctions.

Calculations There are several quantities of interest which may be calculated given the complete set of MOs. For example, the total π-electron density q_i on atom i is given by

$$q_i = \sum_{j}^{all\ MO's} n_j\, c_{ij}^2$$

where c_{ij} is the coefficient for the AO on atom i in MO j and n_j is the number of electrons (0, 1, or 2) occupying MO j. The effective charge on an atom is obtained by subtracting the calculated charge density from the number of electrons contributed to the π system by that atom.

Similarly, the π-bond order p_{rs} between nearest neighbor atoms r and s may be calculated:

$$p_{rs} = \sum_{j}^{all\ MO's} n_j\, c_{rj}\, c_{sj}$$

As will be shown below, these calculated values can be used to make some qualitative predictions regarding the structure and properties of the molecule in the ground state and also in any $\pi \rightarrow \pi^*$ excited state.

The total π energy associated with a particular state of the molecule (a particular π electron configuration) may be calculated by simply summing the energies of the π electrons in each occupied orbital. In this way, an energy may be calculated for the ground state of the molecule, and also for any excited state. Thus, an energy level diagram for the *states* (S_0, S_1, ...) of the system may be constructed.

The resonance energy is the additional stabilization that the molecule has due to the delocalization of the π system. The resonance energy may be calculated according to the expression

$$E(\text{RESONANCE}) = E(\text{HMO}) - E(\text{LERS})$$

where E(HMO) is the total π energy described above and E(LERS) is the total energy of the electrons in isolated π bonds of a hypothetical molecule whose actual structure corresponds to that of the lowest energy resonance structure of the system. For example, for benzene, E(LERS) would be equal to the energy of 6 π electrons in isolated carbon-carbon double bonds. This is equivalent to the energy of 6 electrons occupying the π bonding orbitals in 3 isolated ethylene molecules. (For further discussion see pages 289 – 397 of reference 3.) Table II lists the energies of electrons in various isolated π bonds. Thus, for benzene, E(LERS) = $6\alpha + 6\beta$. For molecules with lone pairs of electrons involved in the π system, the energy of each such electron is given by H_{ii}. For pyrrole, which has 6 π electrons (2 C=C and one lone pair on N),

$$E(\text{LERS}) = 4(\alpha + \beta) + 2(\alpha + 1.5\beta) = 6\alpha + 7\beta.$$

Table II. Energy values[a] (in units of β) for some isolated π bonds of C, N, and O

Bond	$E(\pi)$	$E(\pi^*)$
C=C	$\alpha + \beta$	$\alpha - \beta$
C=N	$\alpha + 1.281\beta$	$\alpha - 0.781\beta$
C=O	$\alpha + 1.618\beta$	$\alpha - 0.618\beta$

[a]Calculated for the corresponding diatomic species using *Hückel Molecular Orbitals 2.4*, from Intellimation.

Methodology

Two projects involving HMO calculations and their interpretation in terms of molecular structure and reactivity are described below. Considering the complexity of the molecules involved in these projects, a computer program capable of calculating HMOs for relatively large molecules is essential. Many programs are currently available for both DOS and Macintosh platforms. (The Project Seraphim 1991 catalog lists several programs suitable for use on a DOS platform. The author is not familiar with these programs and thus cannot comment on their ability to perform calculations for all of the molecules suggested here.)

The author (and students) used the program Hückel Molecular Orbitals 2.4 by J. J. Farrell and H. H. Haddon, available from Intellimation, on Macintosh computers (Plus, Classic, SE, or higher). With this program, the user draws the lowest energy resonance structure and the computer constructs the appropriate secular matrix. On a Mac SE, the calculation of the HMOs for chlorobenzene (7 atoms) takes about 25 seconds and for tetramethylindigo (28 atoms) the calculation takes about 9 minutes. For the first project, some knowledge of group theory, the assignment of point groups, and the determination of the symmetry allowedness of electronic transitions is assumed. If the students have not been exposed to these topics, then those aspects of the project may be excluded.

Comparison of HMO Description of Two Similar Molecules Each student is assigned two similar molecules to examine and analyze using HMO theory. For example, two aromatic hydrocarbons, or a hydrocarbon and a simple derivative, such as a heterocycle or a halide may be used. Several possible pairs of molecules are listed in Table III. As an example, the calculated π charge densities for chlorobenzene are given in Table IV and bond orders for the ground states of benzene, naphthalene and chlorobenzene are given in Table V.

Table III. Some possible pairs of compounds for Experiment Part A

benzene	chlorobenzene
benzene	pyridine
benzene	toluene
naphthalene	1-chloronaphthalene
naphthalene	quinoxaline
naphthalene	quinoline
pyridine	pyrimidine
pyridine	4-chloropyridine
pyridine	3,5-dichloropyridine
1,4-dichloronaphthalene	quinoxaline
toluene	chlorobenzene
toluene	pyridine

Table IV. Calculated[a] π Charge Densities for Chlorobenzene.

Atom No.[b]	π Charge Density	Charge
1	0.9879	0.0121
2	1.0104	-0.0104
3	0.9995	0.0005
4	1.0075	-0.0075
5	0.9995	0.0005
6	1.0104	-0.0104
7	1.9848	0.0152

[a]Calculated with *Hückel Molecular Orbitals 2.4*, from Intellimation.
[b]See Figure 2 for atom numbering scheme.

For benzene and naphthalene, all of the π charge densities are equal (1.0000), as expected. Although all of the π bond orders are identical in benzene, this is not the case in naphthalene. The 1–2 π-bond order is greatest (0.7246), while the others are lower {2–3 (.6032), 1–9 (.5547), and 9–10 (.5182)}. This result is often a surprise to students, who frequently expect that all of the bonds in naphthalene should be identical, as they are for benzene. However, the fact that the 1–2 bond order is the largest in naphthalene is not unexpected based on the concept of resonance as a description of molecular structure. Figure 1 shows the three resonance structures which are commonly used to describe naphthalene. Based on concepts learned in a typical introductory organic chemistry course, each of these structures is expected to contribute equally to the overall molecular structure. Note that in two of them the 1–2 bond is designated as a double bond, whereas a double

Table V. Bond Orders[a] for Benzene, Chlorobenzene and Naphthalene.

Atom Pair[b]	C_6H_6	C_6H_5Cl	Atom Pair[c]	$C_{10}H_{10}$
1-2	0.6667	0.6616	1-2	0.7246
1-6	0.6667	0.6616	1-9	0.5547
2-3	0.6667	0.6676	2-3	0.6032
3-4	0.6667	0.6661	3-4	0.7246
4-5	0.6667	0.6661	4-10	0.5547
5-6	0.6667	0.6676	5-6	0.7246
1-7		0.1228	5-10	0.5547
			6-7	0.6032
			7-8	0.7246
			8-9	0.5547
			9-10	0.5182

[a]Calculated with *Hückel Molecular Orbitals 2.4*, from Intellimation.
[b]See Figure 2 for atom numbering scheme.
[c]See Figure 1 for atom numbering scheme.

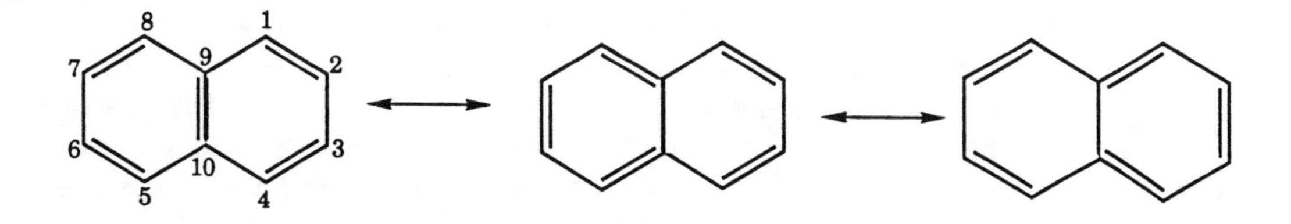

Figure 1. Major resonance structures for naphthalene. The numbering scheme used in Table V is also shown.

bond appears only once for all of the remaining bonds. Thus, the fact that the HMO treatment predicts that the 1–2 bond order is the largest in naphthalene is consistent with the predictions made using resonance structure arguments. Furthermore, HMO π-bond orders have been shown to be reasonably well correlated to experimentally determined bond lengths for unsaturated hydrocarbons *(1, 2, 4)*. For example, the 1–2 bond length in naphthalene is 0.1365 nm and the 2–3 bond length is 0.1404 nm *(5)*.

Another example of the relationship between HMO results and predictions based on resonance structures is provided by comparison of benzene and chlorobenzene. In the halide derivative, the aromatic ring bond orders are all very similar, but they are not identical. In addition, there is alternating charge on the carbon atoms in the ring, and the Cl atom is seen to have a net positive charge. Again, these results can be rationalized based on an examination of the resonance structures for chlorobenzene. There are several structures which can be considered to contribute to the overall species, but they are not all considered to contribute equally. The two major structures (I, II) are shown in the top line of Figure 2. In

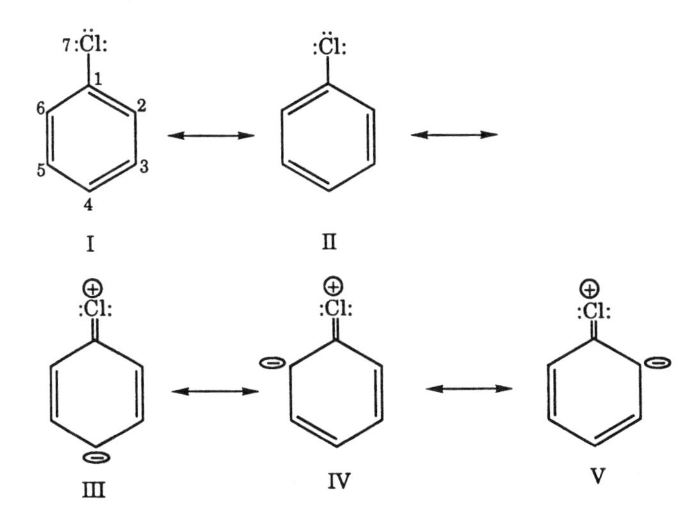

Figure 2. Major resonance structures for chlorobenzene. The numbering scheme used in Tables IV and V is also shown.

these resonance structures there is no formal charge present, and the combination of these two structures predicts that all carbon-carbon bonds are identical, as in benzene. The HMO results show that the bond orders are *nearly* identical, but there are slight differences. In addition, these two resonance structures do not suggest any π character to the C–Cl bond. Another inconsistency is that based on these two structures one would predict that all of the atoms would be neutral as in benzene. By invoking the greater electronegativity of Cl, one might predict that the halogen would have an excess of electron charge, and that the electron charge would decrease monotonically with distance from the electronegative center. (This is a frequently mentioned idea from students.) These results are not observed, and the rationale for the *positive* charge on Cl, the alternating charge of the C atoms around the ring, and the slight differences in C–C bond order can be found in the three additional resonance structures for this system (III, IV, V in Figure 2). These structures, in which there is a donation of electron density from Cl to the ring resulting in a formal separation of charge, are valid resonance structures for this molecule, although they are considered to contribute *less* to the overall structure than I and II. Still, these three are expected to contribute essentially equally. Based on the number of C–C double bonds present in these structures, one would expect that the 2–3 bond would have the greatest bond order, followed by 3–4, and then 1–2. Although the differences between the HMO bond orders for these species is slight, this is precisely the ordering found. These structures also provide a rationale for the small but nonnegligible C–Cl bond order (0.1228), and the net positive charge on the Cl atom (0.0152). Also worth noting is the prediction of excess charge density on atoms 2, 4, and 6, and neutrality for atoms 3 and 5. These predictions, which may be made from resonance structure analysis, are all observed in the HMO calculation. The excess of negative charge on atoms 2, 4, and 6 is consistent with the experimental result that electrophilic aromatic substitution reactions in chlorobenzene often result almost exclusively in substitution at the ortho and para positions.

As described above, concepts of resonance from organic chemistry are often useful in rationalizing the HMO results. This is not an *explanation* of *why* these results are obtained, but rather shows an interesting connection between material from previous courses and Physical Chemistry which is probably not apparent to students. In fact, it is not surprising that there is consistency between these two descriptions of molecular structure involving π systems. A resonance structure may be thought of as a pictorial representation of a particular valence bond description of electronic structure. In order to obtain the best description of a particular system with this approach, one takes a linear combination of many (or all possible) valence bond terms (resonance structures) and performs a variational calculation. The results of this calculation provide coefficients for the various terms which minimize the variational energy (give the best description). The magnitude of these coefficients is related to how much each term *contributes* (in organic chemistry language) to the actual molecule. This is the basis for indicating that a particular resonance structure is insignificant - what is meant is that the coefficient for the term corresponding to that structure in a valence bond description is very small. In the limit of including all possible terms (that is, all possible resonance structures), a molecular orbital description (including configuration interaction) and a valence bond description (including ionic terms) are mathematically identical. Thus, it is not surprising to find the relationships between simple HMO and resonance structure analysis described here.

Excited State Proton Transfer Reactions As a more advanced exercise, the students may perform some calculations related to a possible excited state proton transfer reaction. Such reactions are well-known and occur in a variety of molecular systems. *(6-15)* These excited state tautomerizations often occur in molecules in which there is a hydroxyl or amine group whose hydrogen is located in an appropriate position near a carbonyl group or other potential proton accepting site on an aromatic framework. In the $\pi\pi^*$ excited state, the tautomerized species may be of lower energy even though this is not the case in the ground state. Figure 3 shows an example of this reaction involving *o*-hydroxyacetophenone. Each student is assigned a molecule with appropriate structural characteristics so that the excited state proton

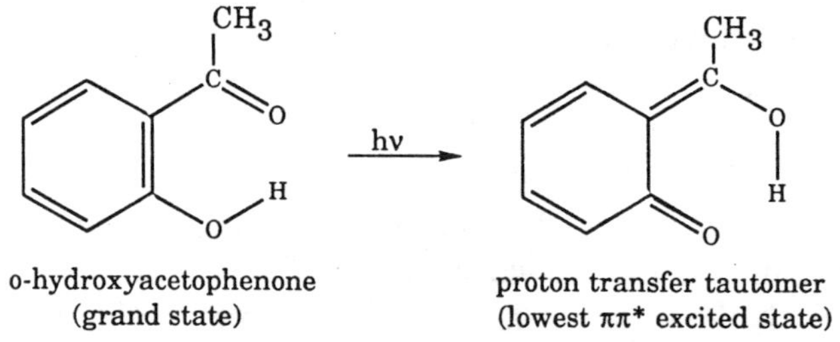

CH_3 C O H O — hν → — CH_3 C O H O

o-hydroxyacetophenone (grand state) — proton transfer tautomer (lowest ππ* excited state)

Figure 3. Photoinduced intramolecular excited state proton transfer reaction in *o*-hydroxyacetophenone.

transfer process is possible, and the student investigates whether or not the proposed tautomerization should be expected to occur in the excited state.

Table VI contains a list of some molecules which might be expected to possibly undergo an intramolecular excited state proton transfer reaction based on their structures. Also listed are literature references to experimental investigations of these reactions. The HMO calculation may be used to predict whether or not such a reaction should be expected on energetic grounds. The excited state proton transfer reaction may be considered to be energetically favorable if the total π energy of the first excited state of the original molecule (S_1^o) is greater than the energy of the corresponding excited state for the tautomer (S_1^t). This relationship is observed for HMO calculations of all of the species listed in Table V. In addition, it is generally the case that S_0^o is *lower* in energy than S_0^t, suggesting that a reverse proton transfer back to the original molecule is expected in the ground state. The

Table VI. Some molecules which may undergo intramolecular excited state proton transfer reactions

Molecule	Reference
2,5-Bis(2-benzothiazolyl)hydroquinone	*7*
1,5-Dihydroxyanthraquinone[a,b]	*8*
o-Hydroxyacetophenone[a,b]	*9*
2-Hydroxy-4,5-benzotropone	*10*
7-Hydroxy-1-indanone	*9*
2-Hydroxy-4,5-naphthotropone	*11*
2-(2'-Hydroxyphenyl)-benzothiazole[b]	*12, 13*
Methyl salicylate[a,b]	*14*
Salicylamide[a,b]	*9*
4,4',7,7'-Tetramethylindigo	*15*

[a] Available from Aldrich (1993).
[b] Available from Lancaster Synthesis (1993).

changes in bond order and π charge density upon excitation of the original molecule are also often consistent with the transfer of the proton in the excited state. For example, in o-hydroxyacetophenone, the carbonyl bond order in S_0 is decreased upon excitation to S_1 while the bond order of the C–O bond to the hydroxyl group increases. The changes in charge density of the two oxygens is also consistent with the expected transfer of the proton from the hydroxyl group to the carbonyl.

Several of the molecules listed in Table VI are commercially available. For these species, absorption and fluorescence spectra can be obtained to confirm whether or not the proton transfer has occurred, and to measure the relative energies of absorption and emission. The measurements provide an experimental determination of the relative energy separations of S_0^o and S_1^o and of S_0^t and S_1^t. For these experiments, anhydrous nonpolar solvents (heptane or methylcyclohexane, for example) are important to use because trace amounts of water or other hydrogen bonding species can interfere with the proton transfer process. Because for most of these species the absorption band of interest lies near or below about 330 nm, quartz cuvettes should be used. Solutions should be prepared immediately before use, and precautions should be taken to exclude water as much as possible. Concentrations of samples should be 10^{-5} to 10^{-4} M, and the maximum absorbance should be less than about 0.2 for the fluorescence experiments. Fluorescence excitation at the absorption maximum is preferred, since in many cases the fluorescence intensity is low. As an alternative to the experimental investigation, a literature search may be performed to find the appropriate spectra.

Safety

Alkane solvents such as heptane and methylcyclohexane are volatile and flammable. The commercially available solutes listed in Table VI are all considered irritants, and in some cases toxic. Only very small quantities of these materials need be used. Still, all sample preparation be done in a well-ventilated hood. Care must be taken in their use and preparation.

Data Analysis

Project A For the two molecules assigned in this part, HMO calculations should be made using a computer program. Then, the following exercises should be performed:

1) Determine the point group to which the molecule belongs.

2) Construct a MO energy level diagram, to scale, containing the appropriate number of π electrons. Determine the irreducible representation to which each MO belongs, and label the diagram accordingly. Remember that the convention is to use lower case letters for designating the representations for MOs.

3) Indicate the lowest energy $\pi \rightarrow \pi^*$ electronic transition on the MO diagram and label it according to its appropriate symmetry designation. Recall that the symmetry designation for the STATE of the molecule is determined by

the direct product of the representations of the orbitals that the electrons occupy. (By convention, this designation is written in capital letters.) Any state which consists of only fully occupied non-degenerate MOs must be totally symmetric. To determine the symmetry designation for a state with one or more partially filled orbitals, compute the direct product of the representations for those orbitals. Determine whether this transition is electric dipole allowed, and if so with what polarization. Recall that the transition is allowed only if the direct product of the ground and excited state representations is or contains the representation that transforms as either x, y, or z.

4) Use β = -20 kcal/mol to calculate the resonance energy for the molecule and the resonance energy per π electron.

5) Draw a stick diagram of the molecule with the π electron bond orders and the π electron charge densities indicated. Compare these values between molecules 1 and 2. Are these results consistent with the differences in chemical structure between these two species? Explain your reasoning clearly.

Project B In this project, HMO calculations are performed on the assigned molecule and its tautomer to investigate a possible excited state proton transfer reaction. Based on those calculations, do the following:

1) Construct a MO energy level diagram, to scale, for the original molecule and the tautomer. Indicate the ground state electronic configuration for the π electrons on your diagram.

2) Calculate the π energies of the ground and first excited states of the two molecules and construct an energy level diagram for the system (to scale). Label the ground and excited states of the original molecule S_0^o and S_1^o, respectively, and for the tautomer S_0^t and S_1^t, respectively. Based on the Hückel calculations, is the proton transfer process energetically favorable?

3) Determine the energy of the first $\pi \rightarrow \pi^*$ transition for the original molecule and the tautomer. Based on these calculations, what do you predict will be the relative energies of the origins of the absorption and fluorescence spectra if there is a proton transfer in the excited state?

4) Examine the effect of the excitation process on the distribution of electron density and the bond orders in the molecule (particularly those dealing with the heteroatoms involved in the possible proton transfer). Are these the type of changes that you would expect to see if an excited state proton transfer process were to take place?

5) Would you expect the tautomer to undergo a ground state proton transfer reaction to return to the original molecule? Why or why not?

6) Take absorption and fluorescence spectra of your molecule in a non-polar solvent to determine the energies of the lowest energy absorption and fluorescence for your system. (Alternatively, find absorption and fluorescence spectra for the molecule in the literature. The best spectra are those for which there is little or no interaction between the solute and the environment. Thus, spectra in nonpolar solution, or in rare-gas matrices or supersonic jets are preferable.) Compare the predicted relative energies of the absorption and fluorescence transitions (assuming proton transfer in the excited state) to the experimental values. Calculate a value for β from the experimental data, and compare it to the average value for a spectroscopic β, 55 - 60 kcal/mol.

Acknowledgments

This work was supported by a generous grant from the Pew Charitable Trusts through the Physical Chemistry Project of the Mid-Atlantic Consortium of the Pew Science Program, and by Franklin and Marshall College.

Literature Cited

1. Streitweiser, Jr., A. *Molecular Orbital Theory for Organic Chemists*; Wiley and Sons, Inc.: New York, NY, 1966.
2. Lowe, J. P. *Quantum Chemistry*; Academic Press: Orlando, FL, 1978.
3. Karplus, M.; Porter, R. N. *Atoms and Molecules - An Introduction for Students of Physical Chemistry*; W. A. Benjamin, Inc.: New York, NY, 1970.
4. Coulson, C. A. *Proc. Roy. Soc. (London)* **1939**, *A169*, 413.
5. Streitweiser, Jr., A.; Heathcock, C. H. *Introduction to Organic Chemistry, Third Edition*; Macmillan Publishing Company: New York, NY, 1985; p 976.
6. Kasha, M. *J. Chem. Soc., Faraday Trans. 2* **1986**, *82*, 2379-2392.
7. Ernsting, N. P.; Mordzinski, A.; Dick, B. *J. Phys. Chem.* **1987**, *91*, 1404-1407.
8. Van Benthem, M. H.; Gillispie, G. D. *J. Phys. Chem.* **1984**, *88*, 2954-2960.
9. Nishiya, T.; Yamauchi, S.; Hirota, N.; Fujiwara, Y.; Itoh, M. *J. Am. Chem. Soc.* **1986**, *108*, 3880-3884.
10. Jang, D.; Brucker, G. A.; Kelley, D. F. *J. Phys. Chem.* **1986**, *90*, 6808-6811.
11. Jang, D.; Kelley, D. F. *J. Phys. Chem.* **1985**, *89*, 209-211.
12. Elsaesser, T.; Kaiser, W. *Chem. Phys. Lett.* **1986**, *128*, 231-237.
13. Barbara, P. F.; Brus, L. E.; Rentzepis, P. M. *J. Am. Chem. Soc.* **1980**, *102*, 5631-5635.
14. Lopez-Delgado, R.; Lazare, S. *J. Phys. Chem.* **1981**, *85*, 763-768.
15. Elsaesser, T.; Kaiser, W.; Lüttke, W. *J. Phys. Chem.* **1986**, *90*, 2901-2905.

RECEIVED October 14, 1992

CHAPTER 19

Measurement of the Photoelectric Effect

Andrew Loomis and Richard W. Schwenz

The photoelectric effect remains one of the least often performed experiments in the undergraduate chemistry and physics curricula. The reasons for the lack of use are complex and involve expense, ease of performance, and how well the experiment "works", but not the desirability of the experience for the curriculum. Each of these items is a valid concern particularly when low numbers of students are using the apparatus. The "traditional" photoelectric effect apparatus consists of a light source, a photosensitive device, and a picoammeter to record the current passing between the photocathode and anode in a direct current manner. We will suggest several improvements in this scheme which we believe reduce the capital expenditures significantly, make the experiment easy to perform, and which continue to give adequate results (defined as within a factor of 2 of Planck's constant, h).

Theory

The general theory of photoemission of electrons from a metal surface was proposed by Einstein *(1)*. The quantitative experimentation measuring h with high accuracy was performed by Millikan *(2)*. Textbooks commonly treat the photoelectric effect as a lead-in to quantum chemistry or physics. Experimentally it is found that when light of frequency ν impinges on the emitter, photoemitted electrons result as long as ν is greater than some ν_0 (which depends on the type of metal used for the emitter and collector). The photoemitted electrons have an energy distribution, such that there exists a stopping potential V_s which corresponds exactly to the maximum energy electrons. The general form describing V_s as a function of ν is

$$eV_s = h\nu - \Phi \tag{1}$$

where Φ is the work function of the metal and is, of course, $h\nu_0$, and e is the electron charge. Millikan *(2)* also points out that the system is very sensitive to junction potentials and non-fresh surfaces, possible difficulties with most laboratory experiments. Rudnick and Tannhauser *(3)* have pointed out that the observed work function Φ is the work function of the collector, rather than the emitter. This counter-intuitive result arises from the definition of the work function as the energy required to escape the metallic surface with zero energy rather than the potential difference required to stop the electron, as most texts state. This definition implies that the energy associated with the stopping potential plus the work function of the collector, Φ_c, is then equal to the work function of the emitter, Φ_e, plus a contact potential difference, CPD, *vis.*,

$$eV_s + \Phi_c = \Phi_e + \text{CPD}$$

2503–6/93/0292$06.00/0

at the top of the energy barrier for an electron approaching the collector. This has been experimentally verified by McClellan et. al. *(4)* in the undergraduate laboratory, and much earlier by Millikan *(2)*. by using different emitter materials with the same collector. Textbooks in modern physics and chemistry do a reasonable job of treating the photoelectric effect within the error noted above.

Given the form of equation (1), a plot of eV_s versus ν should be linear with a slope of h and an x-intercept of Φ_c; or alternatively, a plot of V_s versus ν will have a slope of h/e and the same x-intercept. These graphs should be independent of the illumination intensity up to a saturation limit.

Experimental

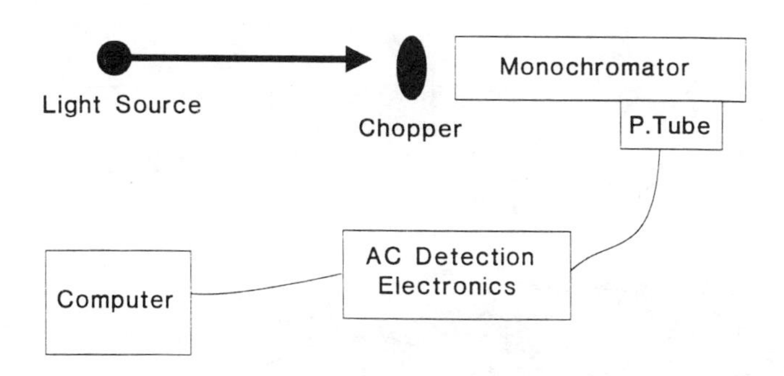

Figure 1. Block diagram of the photoelectric effect apparatus.

A block diagram of our experimental arrangement is given in Figure 1. Basically, light from a continuous visible light source (Kodak Ektagraphic III slide projector) passes through a surplus monochromator (Beckman DU optics and bench) after being chopped by a 5 bladed fan (surplus electronic cooling fan). The light exiting the monochromator has then been chopped at approximately 240 Hz into a square wave, and is approximately monochromatic, depending on the slit width of the monochromator. Alternatively, one could use chopped single frequency radiation (say from a mercury vapor lamp or narrow band filters following a continuum source). This chopped radiation beam impinges on the photocathode of a discarded phototube (A59RX from a Spectronic 20). Any photoemitted electrons then are produced at the chopping frequency. The resulting current pulses from the phototube are transformed into voltage pulses, frequency discriminated, rectified, and converted to an appropriate scale for analog to digital conversion by the circuit given in figure 2. This AC detection process is extremely important because the use of a chopped light source eliminates the electronic bias currents, random phototube noise, and interference from the room lights from the measurements. Measurements in the AC domain are customarily made for exactly these reasons, yet the photoelectric effect measurements have typically been performed, with great difficulty, in a DC fashion.

The circuit *(5)* shown in figure 2 has been optimized for the particular 5 bladed fan which we use. Some measure of tunability for differing fan frequencies around 240 Hz can be obtained using the potentiometer, R_8. Modification of the potentiometer value may be required for your particular fan. This circuit has also

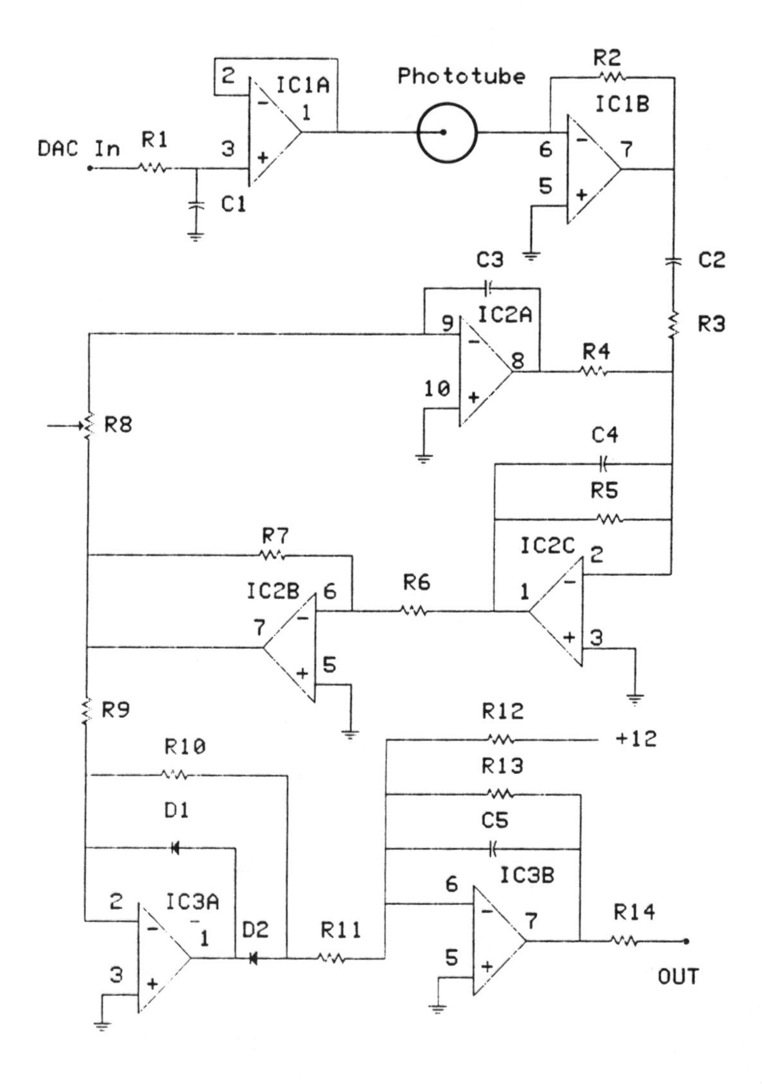

Figure 2. Electronic circuit for detection for photoelectrons and computer interface.

been adjusted so the output scales appropriately for the input to our analog-to-digital converter (± 10 V). In order to put the entire process under computer control, a digital-to-analog converter is needed to ramp the retarding potential to the point where zero current is observed (V_s).

A couple of notes are necessary at this point. First, the choice of a 5 bladed fan, and in fact, the chopping frequency, is arbitrary. The key is to use some frequency other than DC, and other than 60 Hz where fluorescent room lights interfere. Second, the computer control is optional, although better data results from the multitude of points which are collected, and from the lower level of noise in each point, particularly near zero current. Third, alternative types of phototubes should result in different values of Φ, and may require the use of a UV continuum source because of the higher value of Φ. Alternative phototubes must be of a single cathode-anode type (no photomultipliers), with preferable single element cathodes. Fourth, this experiment can also be performed using a commercial chopper and lockin amplifier should these be available.

Results

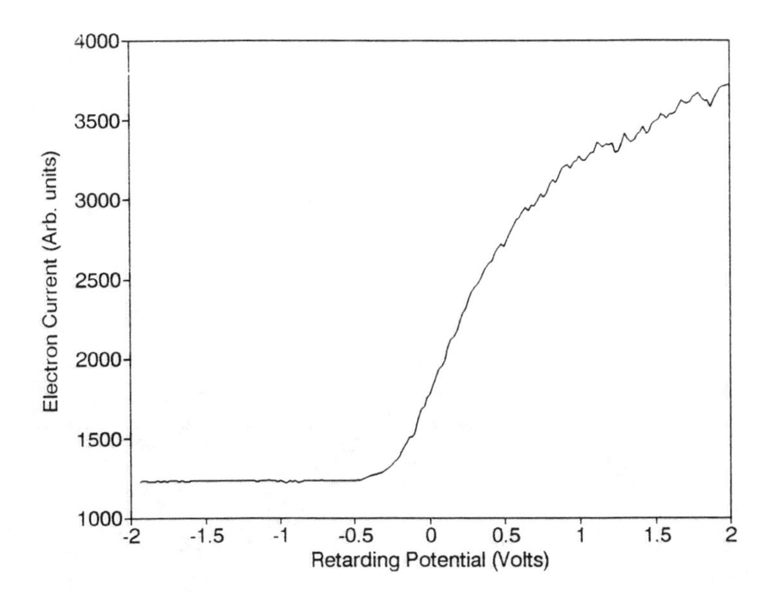

Figure 3. Photoelectron current as a function of retarding potential at 460 nm. The stopping potential is -0.44 V for this data set.

A representative sample data set at a single wavelength is shown in figure 3. The behavior of the current at large positive voltages represents saturation of the tube at that wavelength with respect to the intensity of the incoming light source, i.e., every possible photoelectron is produced. This observed saturation level is dramatically wavelength dependent. The actual level reached depends upon the quantum efficiency of the phototube and the photon flux at the wavelength used, and the gain in the electronics. The behavior near threshold probably results from the energy distribution of the emitted electrons at this wavelength (see figure 4 of reference 4). A technical question remains as to how to choose the stopping potential. The computer control program uses the first data point which is 2σ above the flat portion of the data at large negative voltages. An alternative method consists of constructing the straight line through the linear region in figure 3, followed by extrapolation to zero current. One technique which is useful in correcting for the observed changing wavelength sensitivity is to narrow the slit width so that the saturation voltage remains approximately constant. Doing this forces the instrumental sensitivity to remain approximately constant with changing wavelength. Figure 4 presents the data for multiple wavelengths, transformed to the graph appropriate for equation (1). The value of h obtained is $3.0 \pm 0.2 \times 10^{-34}$ J-sec and $\Phi = 1.74 \pm 0.11$ V, where the standard deviations are those from the fit to the data, and represent typical results from multiple experimental runs with a single phototube type. The value obtained for the work function is not unreasonable for a phototube, a limiting wavelength of 710-nm results, a quite acceptable value for an alkali phototube. The differences between this result and the accepted value of h probable arises from the uncontrolled, but reproducible junction voltages within the apparatus, and dirty, "oxidized" photocathode and anode surfaces. These conditions have

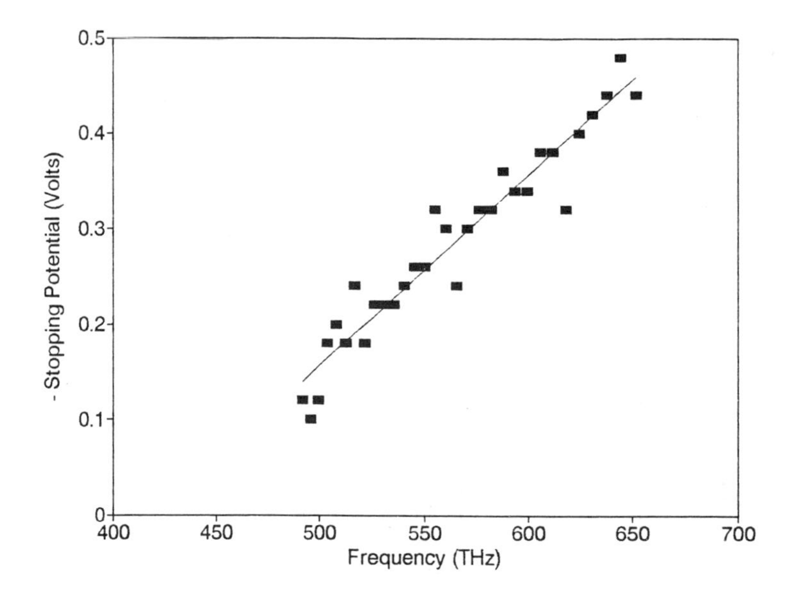

Figure 4. Stopping potential versus frequency. Results are presented in the text.

previously been observed to cause similar errors in photoelectric effect measurements *(2)*.

Acknowledgements

We would like to thank Drs. Ron Watras and Will Fadner for inspiring us in the development of this experiment and the experimental apparatus, and Douglas Brann for his help in programming the computer for data acquisition.

Conclusions

We have presented a new improved experimental apparatus for the measurement of the photoelectric effect. This apparatus allows the determination of h to reasonable accuracy, with inexpensive equipment (all surplus or easily found at your neighborhood electronic store), in a manner that is illustrative of how experiments are done in the AC domain, and that is easy to perform when compared with earlier designs for the photoelectric effect measurement.

Safety

Keep fingers out of fan blades.
Do not stare at UV light sources if used, as eye damage may result.
Take care with the power supplies used because of the shock potential.

Literature Cited

1. Einstein, A. *Ann. Phys. (Leipz.)* **1905**, *17*, 132; **1906**, *20*, 109.
2. Millikan, R. A. *Phys. Rev.* **1916**, *7*, 355; **1921**, *18*, 236.
3. Rudnick, J.; Tannhauser, D. S. *Amer. J. Phys.* **1976**, *44*, 796.
4. McClellan, G.; Didwall, E. M.; Rigby, C. J. *Amer. J. Phys.* **1978**, *46*, 832.
5. Horowitz, P.; Hill, W. *The Art of Electronics, 2nd edition*; Cambridge University Press: Cambridge, 1989.

Parts List

R_1 100 kΩ resistor
R_2 100 kΩ resistor
R_3 2.4 kΩ resistor
R_4 7.5 kΩ resistor
R_5 220 kΩ resistor
R_6 10 kΩ resistor
R_7 10 kΩ resistor
R_8 20 kΩ potentiometer, 15 turn
R_9 10 kΩ resistor
R_{10} 10 kΩ resistor
R_{11} 1 kΩ resistor
R_{12} 2.7 MΩ resistor
R_{13} 1 MΩ resistor
R_{14} 100 Ω resistor

C_1 0.1 μf capacitor
C_2 0.1 μf capacitor
C_3 0.1 μf capacitor
C_4 0.1 μf capacitor
C_5 0.1 μf capacitor

IC_1 LF 353 dual operational amplifier
IC_2 LM 324 quad operational amplifier
IC_3 LF 353 dual operational amplifier

D_1 1N914 diode
D_2 1N914 diode

Phototube AR59X phototube
± 12 V DC power supply (may be present from computer)

RECEIVED October 14, 1992

CHAPTER 20

High-Resolution Vibration—Rotation Spectra of Deuterated Acetylenes

J. I. Steinfeld and Keith A. Nelson

POTENTIAL HAZARDS: High-Pressure Systems, Cryogens, Vacuums

Spectroscopy is one of the most powerful methods available for determining the structure of molecules. Infrared (2-30 μm wavelength) spectroscopy probes the vibrational and rotational energy levels of molecules, and is widely used by chemists to study the structure, dynamics, and concentrations of chemical compounds. The general problem of analysis of spectra consists of obtaining assignments of observed lines to differences in energies of properly identified levels. It is carried out by matching a series of observed line positions with a series of expected level positions until all of the observed lines are accounted for in terms of the proposed level scheme. Intensity information can be useful in that expectations about relative intensities for different types of transitions or for different lines in a series are often available from theory. When the analysis is complete so that the observed spectrum can be predicted as the expected transitions among the levels described by a few simple equations, one is left with a set of vibrational and rotational parameters, which are the molecular constants. With a bit more analysis, these can usually be understood in terms of the molecular geometry, bond angles and distances, and the force constants for distortion of this geometry by bending and stretching.

In this experiment, students obtain the gas-phase infrared spectra of normal and deuterated acetylene. While the structure of this molecule can, and in fact has been deduced from its spectroscopy, we assume the linear symmetric structure, H-C≡C-H, and see how the appearance of the spectrum follows inexorably from that structure. Students analyze one band of each molecule in detail to obtain rotational constants, from which values for the carbon-carbon and carbon-hydrogen bond lengths can be calculated. The experiment includes the synthesis of C_2D_2, handling the gases on a vacuum system, recording the spectra on a computer-controlled Infrared Spectrometer, and computer assisted analysis of the spectra. The experiment is best done by students working together in pairs.

The infrared spectroscopy of acetylene has been suggested in the past *(1)* as a more substantial alternative to the more traditional study of diatomic species such as HCl and DCl *(2)*. The present experiment appears in the current edition of the Shoemaker et al., laboratory textbook *(3)*; this version takes advantage of the increasing availability of computer workstations in the undergraduate laboratory to facilitate data taking and analysis, thereby permitting the fundamental quantum mechanics underlying the experiment to be explored in much greater depth than is possible in the more traditional experiment.

Theory

Degrees of Freedom A structureless atom moving in three-dimensional space has three independent degrees of freedom (translational motion in the mutually perpendicular x, y, and z direction) When N atoms are hooked together to form a molecule, these degrees of freedom are not lost, but are transformed into other types of motion. The molecule as a whole possesses three translational degrees of freedom in (xyz)-space. In addition, it can rotate freely about three mutually perpendicular axes (linear molecules, however, cannot be said to rotate about their linear axis, and so have only 2 independent, but equivalent rotations). The remaining $3N - 6$ ($3N - 5$ for linear molecules) degrees of freedom appear as vibrations of the molecule, i.e., small-amplitude motions of the constituent atoms relative to each other.

Harmonic Vibrations In order to describe such vibrations, we customarily make use of the Born-Oppenheimer Approximation, which states that the atomic nuclei move on a potential surface which is set up by the much more rapid motion of the electrons in the molecule. Near the equilibrium positions for each atom, this potential can generally be approximated as that for a harmonic oscillator, i.e., $U(q) = (1/2)k\,(q-q_e)^2$, where k is the force constant for the vibration and q_e is the equilibrium value for the particular coordinate of the vibration. From quantum mechanics *(4-8)* the energy levels of a harmonic oscillator are given by

$$E(v) = (v + 1/2)h\nu \tag{1}$$

where v is the vibrational quantum number, h is Planck's constant, and ν is the frequency constant for the oscillator, which is related to the masses and force constants of a physical oscillator. For a polyatomic molecule, the total vibrational energy, G_v, is made up of $3N - 6$ ($3N - 5$ if linear) *normal modes*, each with its own characteristic frequency:

$$G_v = \sum_{i=1}^{n} (v_i + 1/2)h\nu_i + \sum_{i=1}^{n} \sum_{j=1}^{i} (v_i + 1/2)(v_j + 1/2)x_{ij} \tag{2}$$

where the anharmonicity constants x_{ij} correct for the fact that the potential surface is not exactly harmonic, and the sums are over the n ($3N - 5$ or 6) vibrational modes of the molecule.

The normal modes of a polyatomic molecule may be found by a straightforward but lengthy matrix-diagonalization procedure described in Refs *(5)* and *(7)*; for the case of acetylene, a linear 4-atomic molecule, the $(3 \times 4) - 5 = 7$ normal modes are shown in Figure 1.

The symmetry designation Σ refers to a vibration in which the displacements are along the molecular axis, and Π, refers to displacements perpendicular to the axis. (Note that the Π, or bending vibrations are doubly degenerate: the displacements may be along either of two directions perpendicular to each other and the linear axis. This gives the required total of 7 normal modes, of which two are degenerate pairs.) The subscripts g and u refer to displacements which either

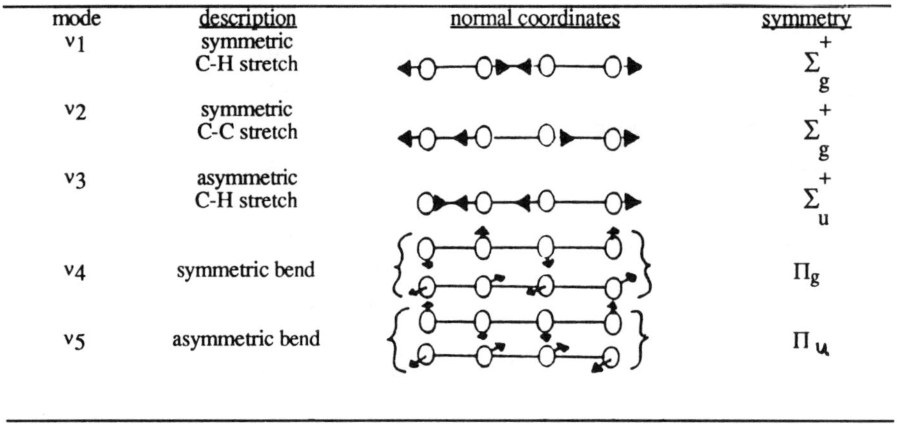

mode	description	normal coordinates	symmetry
ν_1	symmetric C-H stretch		Σ_g^+
ν_2	symmetric C-C stretch		Σ_g^+
ν_3	asymmetric C-H stretch		Σ_u^+
ν_4	symmetric bend		Π_g
ν_5	asymmetric bend		Π_u

Figure 1. Normal modes of acetylene (after *(7)*).

preserve the center of symmetry of the molecule, or are asymmetric with respect to that center, respectively.

Rigid Rotations Rotation of a linear molecule (diatomic molecule is a special case) can be described to a good approximation by the mathematics of a quantized rigid rotor, the energy levels of which are given by

$$E(J) = hcBJ(J + 1) \tag{3}$$

where B is the rotational constant (in wavenumber units, or cm^{-1}) and J is the rotational quantum number. The rotational constant B depends in turn upon the moment of inertia of the molecule, I_B:

$$B = h / 8 \pi^2 c I_B \tag{4}$$

where

$$I_B = \sum_{i=1}^{N} m_i r_i^2 \tag{5}$$

the sum is taken over all atoms in the molecule, having mass m_i and located a distance r_i from the center of mass of the molecule.

Second-order effects which may be observed in vibration-rotation spectra include (i) vibration–rotation interactions, in which the mean value of $\langle r_i^2 \rangle$, and thus the effective rotational constant B_v, is different for different vibrational levels; this is usually expressed as

$$B_v = B_e - \alpha_e(v + 1/2) \tag{6}$$

and (ii) centrifugal distortion, in which the internuclear distance stretches as the molecule rotates faster, which contributes a term $-D_e\,[J(J + 1)]^2$ to the rotational energy. Thus the rotation–vibration levels at this level of approximation can be expressed by the equation

$$\begin{aligned} E(v, j) / hc &= T(v, J) \\ &= G_v/hc + F_v(J) \\ &= G_v/hc + B_e J(J+1) - D_e J^2(J+1)^2 - \alpha_e(v+1/2)J(J+1) \quad (7) \end{aligned}$$

Absorption and Emission of Light ***Selection Rules*** Electromagnetic radiation provides a way of probing this manifold of levels by causing transitions of molecules from one level to another level. In general, this technique is referred to as spectroscopy. The observed line positions in a spectrum contain information about differences in energy levels; observed intensities tell us about populations of levels and the "allowedness" of transitions. We speak naturally of "lines" because only the transitions which can occur are those in which the photon energy exactly matches the energy difference between two available levels.

For a vibrational transition to be allowed in infrared absorption, there must be a net change in the dipole moment of the molecule. Inspection of the normal modes shown in Figure 1 should convince you that the **symmetric** (g) modes leave the dipole moment unchanged and in fact, precisely equal to zero for the symmetric C_2H_2 molecule. It is only the asymmetric modes ν_3 and ν_5 that induce a net dipole moment, so only these are "active," or observable in infrared absorption.

A similar line of reasoning, making use of the symmetry properties of the (spherical harmonic) wave functions for molecular rotation, leads to the following rotational selection rules:

for Σ vibrations, which are **parallel** to the molecular axis and therefore **perpendicular** to the rotational angular momentum:

$$\Delta J = -1 \text{ or } +1$$

for Π vibrations which are **perpendicular** to the molecular axis and therefore **parallel** to the rotational angular momentum:

$$\Delta J = -1, 0, \text{ or } +1$$

These "selection rules" that tell which modes absorb radiation are derived directly from the symmetry properties of a molecule. The study of molecular symmetry and its consequences is called "group theory" In group theory, each molecule is assigned to a "point group" based on its symmetry properties (Acetylene belongs to the $D_{\infty h}$ point group). Based on this assignment, selection rules and other properties can be derived *(9)*.

Transitions with $\Delta J = -1$, 0, and +1 are conventionally called P, Q, and R-branch lines, respectively.

Intensities, Populations, and Statistical Weights Before we can proceed to interpret the spectrum of acetylene, a few comments must be made about the relative intensities of the transitions. An infrared absorption intensity is proportional to the product of the square of the **transition moment** of the particular normal mode times the population of absorbing molecules. At room temperature, nearly all the molecules are in their vibrational ground state. If we make the further reasonable assumption that the transition moment is independent of J, then the intensity of lines in a particular band will be given by

$$I_j \propto N_j / N_{total} = g_I g_J \exp(-E_{rot}/kT)$$

$$= g_I (2J + 1) \exp[-hcBJ(J + 1)/kT] \qquad (8)$$

where k is the Boltzmann constant (1.38×10^{-23} J/K) and T is the absolute temperature.

The factor g_I takes account of nuclear-spin symmetry statistics, a purely quantum-mechanical effect. A detailed treatment is beyond the scope of this discussion, but may be found in Refs (*7*, pp. 16-18) and (*8*, pp. 117-120). Briefly, nuclei are found to obey certain symmetry rules with respect to exchange. Half-integral-spin nuclei, such as protons ($I = 1/2$), must be antisymmetric -- that is, the total wave function changes sign upon particle exchange -- and are said to follow Fermi-Dirac statistics. Integral-spin nuclei such as deuterium ($I = 1$) are symmetric, and are said to follow Bose-Einstein statistics. The total symmetry is the **product** of the rotational and spin wave function symmetries $[(+) \times (+) = (-) \times (-) = (+), (+) \times (-) = (-)]$. In the case of ordinary acetylene, or C_2H_2, the overall proton wave function must be antisymmetric. The combinations that satisfy this requirement are

$$\psi_{rot}(J_{odd}) \times \psi_{nucl\ spin}(I = 1)$$

and

$$\psi_{rot}(J_{even}) \times \psi_{nucl\ spin}(I = 0)$$

Since $I = 1$ is a triplet ($M_I = +1, 0, -1$) and $I = 0$ is a singlet, the nuclear-statistical weights are $g_I = 3$ for odd J, $g_I = 1$ for even J. For C_2D_2, the symmetric nuclear-spin functions are $I = 2$ ($g_I = 5$) and $I = 0$ ($g_I = 1$), while the antisymmetric nuclear-spin function is $I = 1$ ($g_I = 3$), so that

$$g_I(\text{odd } J) : g_I(\text{even } J)) = 3 : 6 = 1 : 2$$

This **intensity alternation** will be helpful in making the rotational assignments of the high-resolution spectra.

Isotope Effect In this experiment, spectra of normal and deuterium-substituted acetylene are recorded and analyzed. The Born-Oppenheimer separation of nuclear and electronic motion allows us to assert that the potential surface, and thus the equilibrium structure, remains unchanged upon isotopic substitution. For a simple diatomic molecule, AB, isotope shifts are easily calculated. Since the harmonic oscillator frequency is just

$$\nu = (1/2\pi)(k/\mu)^{1/2} \qquad (9)$$

where μ is the reduced mass $m_A m_B/(m_A + m_B)$, we have for the isotopically substituted molecule (designated by superscript i)

$$(\nu^i/\nu)_{harmonic} = (\mu/\mu^i)^{1/2} \quad (10)$$

From Eqs. (4) and (5), we have for the rotational constant

$$(B^i/B) = (I/I^i) = (\mu/\mu^i) \quad (11)$$

The relationships for a polyatomic species such as acetylene will be given at the appropriate points in the section on Calculations.

Experimental Methodology

Experimental work should be done by both lab partners working together. C_2D_2 and C_2HD should be synthesized, and the spectra run on the same day, since it is difficult to store the condensed gas overnight. C_2H_2 will be taken from a cylinder, using the same vacuum line setup. The experimental section will most likely take two days. C_2D_2 and C_2H_2 should be made and the spectra recorded during one lab period, and C_2HD should be run during another. All the samples can be run in one day if everything goes well and if both lab partners work efficiently.

Synthesis of C_2D_2 The vacuum line is set up in a hood (see Figure 2). The parts that must be supplied by the student are:

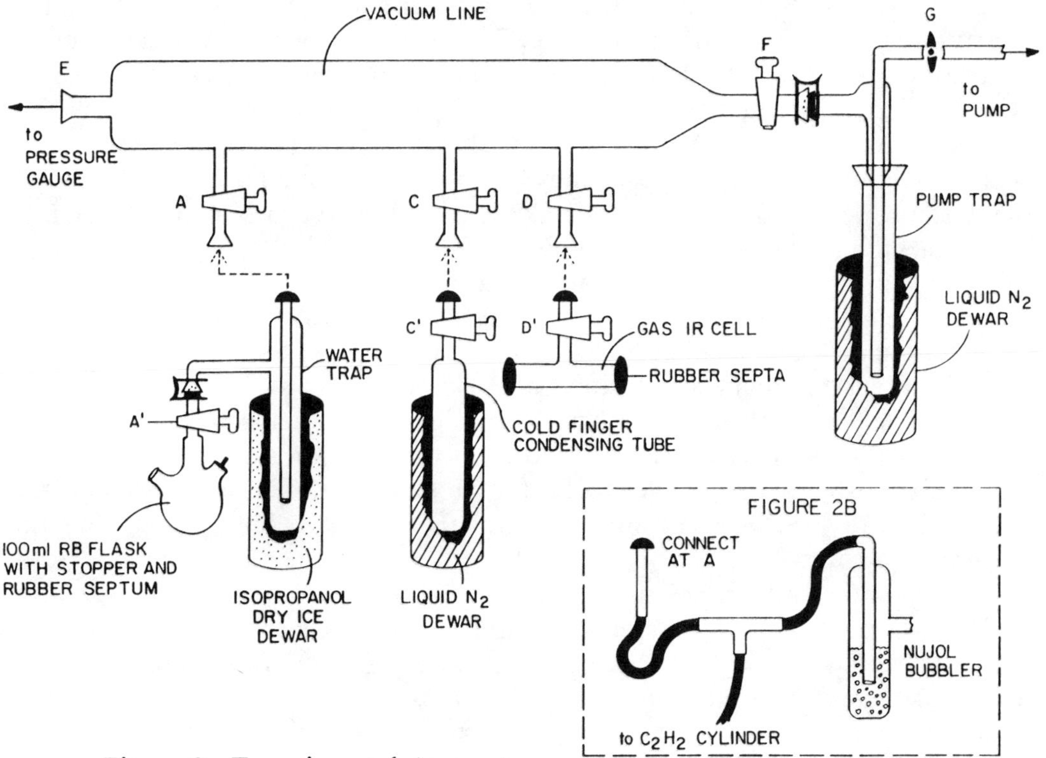

Figure 2. Experimental Apparatus.

1. one 3-neck round bottom flask, 100 mL size
2. one glass stopper
3. glass adapter (19/22 to 14/20)
4. rubber septa (to fit 14/20)

The gas IR cells should be kept in a glass desiccator. When using them on the vacuum line, keep the salt plate ends covered with rubber septa (also in a desiccator), to prevent water attack from etching the plates.

All glassware supplied should be cleaned and oven-dried. Assemble the glassware as indicated in Figure 2. To do this, use a light film of vacuum stopcock grease on the female parts of the o-ring joints, then clamp the o-ring joints with the clamps provided. Empty the pump if necessary. Pressure in the system will be read using a thermocouple pressure gauge.

Remove the stopper from the round bottom flask and add 2.5 g CaC_2 (kept in a desiccator in the hood). Do not try to crush the lumps of CaC_2; they are too hard, and the reaction will not be inhibited if the CaC_2 is in lumps. Replace the glass stopper, and check to make sure the rubber septum is tight.

Fill one of the Dewars with liquid N_2. Close stopcock F and turn on the vacuum pump. Close stopcocks A, A′, C, C′, D, D′ and E. Open stopcocks F and G. After about one minute, place the liquid N_2-filled Dewar around the pump trap and clamp it in place. (Raise the Dewar slowly to avoid splashing the liquid nitrogen, as it will cause skin burns.) The short delay before cooling the trap is to allow most of the air to be pumped out; otherwise, it would liquefy in the pump trap. Liquid air is very corrosive and reactive; it could cause organic compounds (such as acetylene) to burn when the trap is later warmed.

The pressure gauge should read less than 0.1 Torr (1 Torr = 1 mm Hg = 133.32 Pa) after a few minutes. Close stopcock F and wait about five minutes. The pressure should not rise more than 1 Torr. (This is called leak-testing the line.) If the pressure is stable, open F again, then one by one, open stopcocks A, A′, C, C′, D, D′. After each of these stopcocks is opened, the total system pressure will rise, but within a minute or two the pressure should fall back to its original level. If the pressure remains high after opening any one of these, then a leak has been found. If any stopcocks or o-ring joints are leaking, close the stopcock above the leak to keep a vacuum in the rest of the system, then repressurize the faulty part by taking apart the joint or stopcock, then cleaning and regreasing it.

When all stopcocks have been opened, and any leaks in the system have been repaired, the pressure should stabilize below 0.1 Torr, and this vacuum should be stable for at least 5 minutes with stopcock F closed.

While one partner is leak-testing the line, the other should prepare the remaining Dewars. One Dewar is to be filled with Liquid N_2; the other is to be filled with an isopropanol/dry ice bath; fill the Dewar 1/2 full of isopropanol and add crushed dry ice **in small amounts** until the violent bubbling stops and the mixture is viscous.

When the vacuum pressure has stabilized, slowly raise the isopropanol/dry ice Dewar around the water trap and clamp it in place. Close D and D′.

Since C_2D_2 is slightly acidic, it could exchange deuterium atoms with any stray proton sources in the system (such as excess grease). Therefore, the system must be "deuterated" before any C_2D_2 is collected. To do this, close stopcock F and inject 0.5 mL D_2O into the flask through the septum. Do not leave the cap off the D_2O bottle for long, as it will exchange protons with atmospheric water. The CaC_2

will bubble and the pressure will rise as C_2D_2 is produced. Do not let the pressure rise above atmospheric pressure, otherwise parts of the system might burst. (If too much D_2O is injected and the pressure rises too high, open stopcock F and let the pump withdraw the excess. It will be trapped in the pump trap.) Wait about 5 minutes with F closed to allow exchange with any available protons. Open F and pump out the gas. The vacuum will probably not be as good as before, since the CaC_2 will keep producing small amounts of C_2D_2 for a long time. However, the pressure should be less than 0.5 Torr. When this has been achieved, C_2D_2 may be produced and collected in the cold finger condenser.

Slowly raise the third Dewar (filled with liquid N_2) around the cold finger and clamp it in place. Be careful not to freeze C_2D_2 in the neck of the cold finger. Close stopcock F. Inject 0.5 mL D_2O into the flask. The pressure will rise and then fall slowly as the C_2D_2 is solidified in the trap at C′. When the pressure has fallen below ~10Torr, make another 0.5 ml injection. Repeat this procedure until a total of 4.0 ml D_2O has been injected.

Close C and evacuate the remaining gas by opening F. Close A and remove the flask and water trap; first remove the isopropanol/dry ice Dewar, then detach the water trap and reaction vessel from the system. The flask must be left open in the hood to allow C_2D_2 to escape. The water trap should also be left in the hood to warm up.

Now the IR cell can be filled by "boiling off" from the cold finger. Close F and open C, C′, D, and D′. Lower the liquid N_2-filled Dewar, allowing the cold finger to warm up slightly. When the pressure reads about 200 Torr, close stopcock C and quickly raise the Dewar around the cold finger. Wait a few minutes, then open F and evacuate the system to below 0.1 Torr, if possible. This procedure allows any air which has liquefied to boil off and also deuterates the IR cell.

Now close F, open C, and partially lower the liquid N_2 Dewar, allowing the upper half of the trap to warm up. Allow the pressure to rise slowly, using the height of the Dewar around the trap to control the rate. When the pressure reaches 50 Torr, close stopcocks D and D′, then close stopcock C, and immediately raise the Dewar around the cold finger. Open stopcock F, and pump out the remaining C_2D_2 in the system. The IR cell may now be removed and the spectra recorded. (See the following section.) Leave the Dewar around the cold finger in case another sample of C_2D_2 is needed.

Both high and low resolution spectra of C_2HD are required. The low resolution spectrum is necessary for the vibrational analysis program; the high resolution spectrum should be compared with the similar spectra for C_2H_2 and C_2D_2. Samples of C_2HD may be prepared by following the C_2D_2 directions with two modifications. A layer of alumina should cover the CaC_2. The alumina promotes the production of C_2HD over C_2H_2 and C_2D_2. Instead of injecting 4 mL of D_2O, 4 mL of a 50/50 mixture of D_2O and H_2O should be used.

Spectra of C_2H_2 must also be recorded. Samples may be prepared in a similar fashion as for C_2D_2, except the source of the acetylene will be a cylinder. Using a glass tube ending in an o-ring joint, connect the tank to inlet A. [See Figure 2B] Connect a clean, dry cold finger to C and the IR cell to D. Close A, C, D, and open F. Evacuate the system as before, then open D and C. Place a liquid N_2-filled Dewar around trap C. After leak-testing the line, close F. Open the valve on the C_2H_2 to solidify in trap C. If the pressure rises too quickly, slow the C_2H_2 flow by closing the tank valve slightly. After a few minutes, close A and disconnect the glass tube. As before, distill C_2H_2 from the trap into the IR cell. The pressure in the sample should again be 50 Torr.

After the spectra have been recorded, reattach the IR cell to the vacuum line and empty out the cell using the vacuum pump. If the sample pressure was too low or too high, the IR cell may be filled by adding more C_2D_2, or some of the gas may be pumped out. Alternatively if enough solidified C_2D_2 or C_2H_2 remains, the cell may be completely pumped out and a new sample made at the desired pressure.

When all spectra have been run, pump out the IR cell. Remove the cell, cold finger and cold trap; the IR cell must be placed in the desiccator with its rubber septa caps. The cold finger condenser and cold trap must be left in the hood to warm up **with the stopcock removed.**

Although variations on the above procedure are possible, **it is extremely important that no closed vessel containing solidified acetylene be allowed to warm up beyond the boiling point of the gas.** This includes the pump trap, which will contain some liquefied air as well. Any such rise in pressure above 760 Torr will cause glassware to burst and/or stopcocks to fly out of their bores.

Recording of Spectra After the gas IR cell has been filled with the desired pressure of C_2D_2 or C_2H_2 the complete IR spectrum (4000 – 600 cm^{-1}) should first be run. The C_2H_2 scan should be carried out from 4000 cm^{-1} to 600 cm^{-1} and a sample should then be run on an expanded scale between 1220 and 1420 cm^{-1}. The C_2D_2 should also be run; the expanded-scale scan between 960 and 1140 cm^{-1} can be run at this same pressure.

In any case, each sample should be run first under normal scan conditions to determine whether the pressure is in the right range (i.e., transmittance at the desired band between 0.20 and 0.50). Only after the sample has been checked in this manner should the expanded regions be run.

After recording all spectra, empty out the vacuum cell as previously described. Also record baseline spectra of the IR cell(s) used. Both high and low resolution baselines are necessary. Table 1 summarizes the experiments to be performed.

Safety

The acetylenes are very flammable gases with an irritating odor due to impurities. Therefore, all work is to be done in the hood, and **absolutely NO SMOKING or OPEN FLAMES** are allowed in the room. Approved safety glasses must be worn when carrying out any experimental work, particularly when using the glass vacuum system. However, safety glasses are not mandatory when working at the computer terminal (following section).

Data Analysis

This section describes the detailed analysis and assignment of the C_2H_2, C_2HD and C_2D_2 spectra. Students may benefit by attempting to work out parts of the analysis on their own, proceeding from the data and Equations (2) - (7), or as they progress with the analysis, proceeding from the results of previous steps. The analysis should

TABLE 1 - EXPERIMENTAL SUMMARY

Compound	Source	Pressure (Torr)	Low Resolution	High Resolution
C_2H_2	tank	~50	4000-600 cm^{-1}	1420-1220 cm^{-1}
C_2D_2	synthesis with 100% D_2O	~50	4000-600	1140-960
C_2HD	synthesis with 50/50 D_2O/H_2O	~250	4000-600	1260-1140
empty cell		0	4000-600	1420-1220 1140-960 1260-1140

1 cm^{-1} resolution has been found to be sufficient to give good-quality rotationally resolved spectra.

be carried out as soon as possible after the spectra are recorded, so that any measurements which may need to be repeated can be done so promptly. Quantitative results should be organized and presented in tables. Units should be given for **all** numerical quantities unless they are actually dimensionless.

Vibrational Analysis *C_2H_2.* First, tabulate the absorption bands observed in the survey spectrum. From the preceding discussion, recall that in C_2H_2 and C_2D_2 only the **asymmetric** vibrations (Σ_u^+, Π_u) can absorb infrared radiation. There are two such normal modes in acetylene – a C-H stretch and a bending vibration. From the observed frequencies, intensities, and band shapes the frequencies of these two **fundamental** vibrations may be assigned. For a perpendicular band (P, Q, and R–branches), take the pure vibrational frequency to be the peak of the central Q-branch; for a parallel band (P and R-branches only), take the frequency to be that of the minimum between the two branches.

In addition to these fundamentals, several **combination** bands should also be observed. In these bands, which two normal modes of vibration are simultaneously excited by one infrared photon. In particular, the prominent band in the 1300-cm^{-1} region (to be analyzed in detail shortly) is, in fact, a combination of the two bending modes ν_4 (π_g) and ν_5 (π_u). The symmetry of the combination state is the product of the symmetries of the fundamentals; since this includes Σ_u^+, the appearance of the band is that of a parallel band, with P and R-branches only. Identify as many other combination bands as you can find; by taking differences between the various bands, see if you can estimate the values of the three infrared-inactive normal modes.

C_2D_2. In order to eliminate potential confusion in the interpretation of the C_2D_2 spectrum, it is necessary to keep in mind that the sharp peak which almost certainly will be observed near 680 cm^{-1} is the ν_5 band (Q-branch peak) of C_2HD, present as an impurity in the C_2D_2 sample. From the intensity of that peak, estimate the percent C_2HD in the sample; assume that the transition moment is independent of isotopic substitution, and bear in mind that absorbance is proportional to the

concentration (i.e., pressure) of the absorbing species. The rest of the bands in the spectrum are the fundamental and combination bands of C_2D_2, shifted to lower frequencies from C_2H_2 by isotopic substitution; assign as many of these bands as possible, and obtain values for the normal-mode frequencies.

Isotope Effect The counterpart of Equation (10) for polyatomic molecules is the Teller-Redlich Product Rule. Its derivation is too technical to be developed here (see Ref. *(7)* pp. 231-235, for details). The result for the infrared-active modes of acetylene is quite simple, though; it is just

$$\frac{(\nu_3)C_2D_2}{(\nu_3)C_2H_2} = \frac{(\nu_5)C_2D_2}{(\nu_5)C_2H_2} = \sqrt{\frac{m_H(m_C+m_D)}{m_D(m_C+m_H)}} \tag{12}$$

Test the prediction of Equation (12) with the observed values of ν_3. Also use the formula to calculate the expected position of ν_5 for C_2D_2. Why can't this band be observed in the spectrum? What changes would be necessary in the experimental procedure to enable you to do so?

C_2HD Using the 250 Torr low resolution spectrum, tabulate the absorption bands. At this pressure, four of the five fundamental vibrations should be observable. The frequency of the other fundamental can be determined from a combination band. It may be helpful to use the frequencies of the bands you have determined for C_2H_2 and C_2D_2 as upper and lower limits for the corresponding C_2HD band. The C_2H_2 and C_2D_2 bands which appear in the spectrum should also be labeled.

Rotational Analysis *C_2H_2 and C_2D_2* The first step in dealing with the high-resolution scans of the $\nu_4 + \nu_5$ band in C_2H_2 and C_2D_2 is to **assign** the lines, that is, to identify the J-value of the molecule absorbing at each frequency. (*2*, p. 443). In this band, a weak Q-branch line will be observed at the band center. Assign the R(J'') and P(J'') branch lines on either side of the band center; the **intensity alternation** alluded to above will be a very useful guide in finding the correct J-values. (While the intensity alternation should be quite clear, the observed intensity ratios between adjacent lines will probably **not** be the 3:1 or 1:2 predicted by theory. Can you think of a reason why?) A computer program may be used to help locate the position of each line and to print out a table of line assignments and frequencies. Flow charts are given in an Appendix. By taking differences between the term values [Equation (7)] for the upper (v', J') and lower (v'', J'') states, we can find theoretical expressions for the frequencies of the lines:

$$\nu_R(J'') = \nu_o + (2B_e-3\alpha_e) + (2B_e-4\alpha_e)J'' - \alpha_e J''^2 \quad \text{for } J''=0,1,2,\ldots \tag{13a}$$

and

$$\nu_P(J'') = \nu_o + (2B_e-2\alpha_e)J'' - \alpha_e J''^2, \quad \text{for } J'' = 1, 2, 3, \ldots \tag{13b}$$

These two equations can be conveniently combined into a single one suitable for least-squares fitting,

$$\nu(m) = \nu_o + 2(B_e - \alpha_e)m - \alpha_e m^2 \tag{14}$$

with $m = J'' + 1$ in the R-branch and $m = J'$ in the P-branch. Taking differences between successive lines gives

$$\Delta\nu(m) = (2B_e - 3\alpha_e) - 2\alpha_e m \tag{15}$$

note that, if the weak Q-branch is ignored, there will be **no** value corresponding to $\Delta\nu(m = 0)$. The computer program finds the values of B_e and α_e by fitting the data to Equation (15).

Once the values of B_e for C_2H_2 and C_2D_2 are known, the interatomic distances, C≡C and C–H, in the molecule can be found. A simple way of doing this (Ref. 7, p. 397) is as follows. First, define the two distances $a = (1/2)\,r(\text{C≡C})$ and $b = (1/2)\,r(\text{C≡C}) + r(\text{C–H})$, as shown in Figure 3.

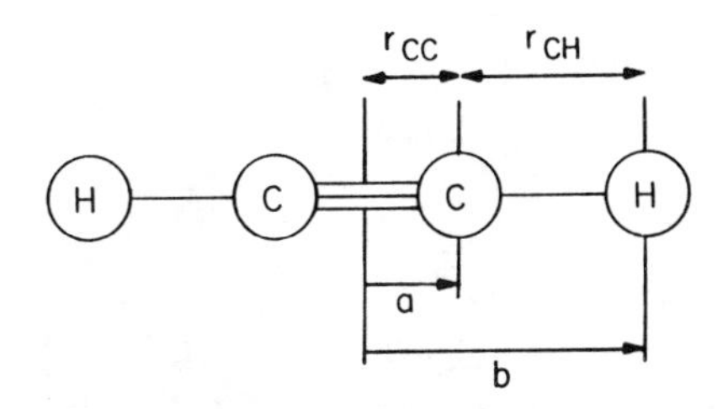

Figure 3. Bond distances in acetylene.

The moment of inertia [Equation (5)] is then just

$$I(C_2H_2) = 2m_C a^2 + 2m_H b^2 \tag{16a}$$
$$I(C_2D_2) = 2m_C a^2 + 2m_D b^2 \tag{16b}$$

Find $I(C_2H_2)$ and $I(C_2D_2)$ from the measured rotational constants. Solve the set of simultaneous equations (16a,b) to find a and b. A convenient set of scaled units would be masses in a.m.u. (m_i in grams = M_i (amu)/6.022 × 10^{23}) and distances in Ångstrom units (r_i in cm = r_i (Å)/1 × 10^8). Finally, evaluate the C≡C and C-H (C-D) bond distances. Be sure to give the units and an error estimate.

The intensities expected for the lines in the acetylene spectra are given as the product of the relative populations of the lower state levels

$$g_I\,(2J'' + 1)\exp\,[-BJ''(J'' + 1)hc/kT],$$

where $(2J'' + 1)$ is the degeneracy of the level and g_I is the appropriate nuclear-spin statistical-weight factor, multiplied by the transition probabilities $(J'' + 1)/(2J'' + 1)$ for $\Delta J = +1$ (R-branch) and $J''/(2J'' + 1)$ for $\Delta J = -1$ (P-branch). The relative intensities are thus given by

$$I(J'') \propto g_I(J'')\,(J' + J'' + 1)/2\ \exp\,[-BJ''\,(J'' + 1)hc/kT] \tag{17}$$

Calculate the expected distribution of intensities and make a bar graph to compare the observed and actual distribution. This may be done using the laboratory computer system; the printout should be included with the report.

C_2HD The C_2HD lines in this high resolution spectrum do not need to be assigned. Compare this spectrum with those measured for C_2D_2 and C_2H_2. How do they differ?

Calculations of Force Constants Once the vibrational and rotational spectra are correctly assigned, the force constants for C—H and C-C stretching and for bending the C-C-H linkage around the central C atom can be calculated.

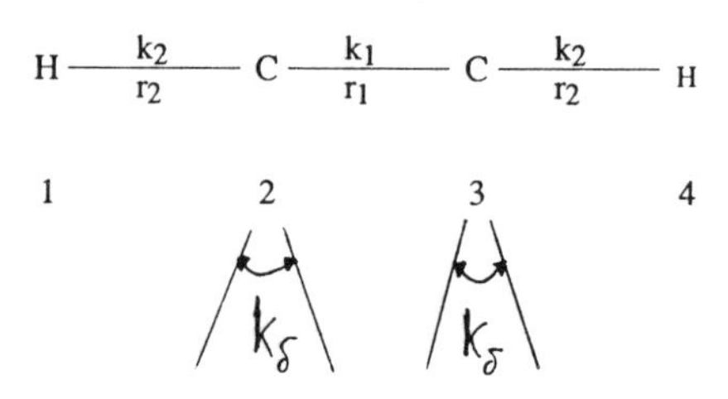

Figure 4. Force constants in acetylene.

Figure 4 serves to define terms. Atoms are labeled 1-4. Force constants are k_1, k_2, and k_δ. Equilibrium bond lengths are r_1, (C≡C bond) and r_2 (C-H bonds). Equilibrium bond angles are $\delta_{13} = \delta_{24} = 0$. We also define **deviations** from equilibrium bond lengths as x_{12}, x_{23}, and x_{34}. The vibrational potential energy of the molecule is

$$V = 1/2\ [k_1\, x_{23}^2 + k_2\,(x_{12}^2 + x_{34}^2) + k_\delta\,(\delta_{13}^2) + \delta_{24}^2)] \tag{18}$$

Note that the force constants, k_i, and bond lengths, r_i, are essentially the same for C_2H_2, and C_2HD.

This section should assist with calculating all three force constants from the C_2H_2, C_2D_2 and C_2HD spectra. Actually there are more than enough spectral information to do this, i.e., the k_i are overdetermined. Thus the internal consistency of the results may be verified.

It will be convenient to define λ_1 through λ_5 such that $\lambda_i = 4\pi^2\nu_i^2$, where λ_i is the frequency of the *i*th vibrational mode. Note that $\nu_i = (1/2\pi)\sqrt{\lambda_i}$, so that λ_i is a **normal mode** force constant divided by a normal mode reduced mass. By writing out the normal coordinates explicitly as linear combinations of the x_{ij} and δ_{ij} coordinates, it is easy to relate the normal mode force constants λ_i, to the interatomic force constants k_i. The results for C_2H_2 are (7, p. 181):

$$\lambda_1 + \lambda_2 = 2k_1 / m_C + (1 + (m_H/m_C))\,(k_2 / m_H) \tag{19}$$

$$\lambda_1 + \lambda_2 = 2k_1\, k_2 / m_C m_H \tag{20}$$

$$\lambda_3 = (1 + (m_H/m_C))\, k_2/m_H \tag{21}$$

$$\lambda_4 = (k_\delta / r_1^2\, r_2^2)\,(r_1^2 / m_H + (r_1 + 2r_2)^2/m_C) \tag{22}$$

$$\lambda_5 = (1 + (m_H/m_C))\ k_\delta/m_H\, r_2^2 \tag{23}$$

Results for C_2D_2 are identical except that m_D replaces m_H.

Some of the force constants can be calculated from just C_2H_2 and C_2D_2 spectra. Note that Equations (19) - (23) **do not apply** for C_2HD, since the normal modes of C_2HD do not look like those of C_2H_2 and C_2D_2. The form of the stretching modes of C_2H_2 and C_2D_2 may be found in Reference 7, p.292.

The C_2HD spectrum can be used to determine any remaining force constants with the help of the Teller Redlich relation (7, p. 289)

$$\frac{(\nu_1\ \nu_2\ \nu_3)\ C_2HD}{(\nu_1\ \nu_2\ \nu_3)\ C_2H_2} = \sqrt{\frac{m_H\ (2m_C+m_H+m_D)}{2m_D(m_C\ +\ m_H)}} \tag{24}$$

Having calculated the force constants, the internal consistency of your data can be checked using additional Teller-Redlich rules in Reference 7, p. 289-291. Noting that two of the normal modes in C_2HD look essentially like simple C-H and C-D stretches, one of the force constants can be estimated **directly** from the frequencies of these modes. How should the two frequencies be related?

Notebook and Discussion All procedures and observations should be recorded in the laboratory notebook as they occur in the lab. "Incorporate" all spectra and computer printouts into the notebook. Clearly explain who did what in the lab. The notebook of each lab partner should have all the data obtained from measurements of the spectra. **All calculations should be done independently in the lab notebook.** Be sure that the spectra are clearly labeled and turned in with the notebook pages.

Vibrational Spectrum Discuss the reliability of your assignment of the vibrational bands you have observed in the acetylene molecule. What additional experiments could be done to directly observe non-infrared-active fundamentals in C_2H_2 and C_2D_2? Discuss the implications of this work for quantum versus classical descriptions of molecular structure and behavior. In what respect(s) does this work show the failure of classical theory? To what extent is classical theory adequate?

Rotational Spectrum How pure was the C_2D_2? Discuss sources of C_2HD contamination. Discuss the accuracy of the results. Estimate the error in the measured line positions. Estimate the error in the values of the molecular parameters. Compare these parameters with literature values. Do they agree within the estimated error range? What are the largest sources of error in this experiment? Compare the value found for the C≡C triple bond with textbook values for the C=C double bond and C-C single bond (cite the source used!). How can these values be explained? Compare the C-H bond distance in acetylene with that in methane or ethane. What (if anything) does this imply about the electronic structure of acetylene?

Discussion and Extensions

While the overt purpose of introducing computerized data acquisition and analysis into the infrared spectroscopy laboratory was to relieve tedium and reduce the amount of time spent in analyzing data by hand, many students reported that the increased precision and efficiency available in the experiment enhanced both their enjoyment of the experience and their insight into the quantum - mechanical principles which the experiment was designed to elucidate *(10)*. This observation refutes the often-heard claim that students tend to use the computer as a "black box," with little or no comprehension of the principles underlying the operations being carried out. The analysis procedure can be readily extended to other molecular systems or to additional spectra of acetylene; for example, laser-Raman spectra *(11)* would be an excellent complement to the infrared study of C_2H_2 and C_2D_2, if the necessary equipment is available.

Acknowledgments

The data acquisition and analysis programs used in this experiment were written and tested by Mr. S. Garfinkel, with support from MIT Project Athena, which also provided the required computer equipment.

Literature Cited

1. Richards, L.W. *J. Chem. Educ.*, **1966**, *43*, 644.
2. Sime, R.J., *Physical Chemistry Methods, Techniques, and Experiments*, Saunders College Publishing: Philadelphia, PA, 1988, pp. 676-687.
3. Shoemaker, D.P.; Garland, C.W.; Nibler, J.W. *Experiments in Physical Chemistry*, 5th ed., McGraw-Hill Book Co.: NY, 1989, pp. 469-482.
4. Barrow, G.M. *Introduction to Molecular Spectroscopy*, McGraw-Hill, New York, 1962.
5. Wilson, Jr., E.G.; Decius, J.C.; Cross, P.C. *Molecular Vibrations*, McGraw-Hill, New York, 1955.
6. Herzberg, G., *Molecular Spectra and Molecular Structure. I. Spectra of Diatomic Molecules*, Van Nostrand, Princeton, NJ, 1950.
7. Herzberg, G. *Molecular Spectra and Molecular Structure. II. Infrared and Raman Spectra*, Van Nostrand, Princeton, NJ, 1945.
8. Steinfeld, J.I. *Molecules and Radiation*, 2nd Ed. MIT Press, Cambridge, MA, 1985.
9. Wigner, E., *Group Theory*, Academic Press: New York, 1959.
10. Turkle, S.; Schön, D.; Nielsen, B.; Orsini, M.S.; Overmeer, W. Project Athena at MIT, to be published.
11. Edwards, H.G.M. *Spectrochimica Acta*, **1990**, *46A*, 97.

Hardware and software requirements

Several different infrared spectrometers have been successfully used in this experiment. Among grating spectrometers, the Perkin-Elmer 781 and Perkin-Elmer 299 models provided the 1 cm^{-1} resolution required (see Table 1). The former may be interfaced directly to a computer via a IEEE-488 digital interface; the latter requires an A-to-D converter such as the Metrabyte DASCON–1. A Fourier Transform Infrared Spectrometer (IBM IR/32) supported by a Dell System 200 computer has also been used successfully for this experiment.

The data are recorded on 5 1/4-inch DOS-formatted diskettes and taken to a separate computer system for processing and analysis. The analysis programs are written in the "C" language, except for a few assembly language device drivers. IBM-XT personal computers have been used for both data acquisition and analysis; any current-generation DOS-based PC with a hard disk drive and graphics display should have the speed and memory capacity required for this application.

Appendix

The computer program for analyzing infrared spectroscopic programs offers two options, (A) vibrational analysis of low-resolution spectra and (B) rotational analysis of high-resolution spectra. Flow charts of each of these options are given here for convenience.

Option A - Vibrational Analysis

in this section of the program, you can subtract spectra from each other, and do a normal mode analysis to find the force constants in acetylene. Subtracting spectra allows you to remove artifacts in the spectra by the cell or by the reference beam. It also allows you to subtract C_2H_2 and C_2D_2 features from your C_2HD spectrum.

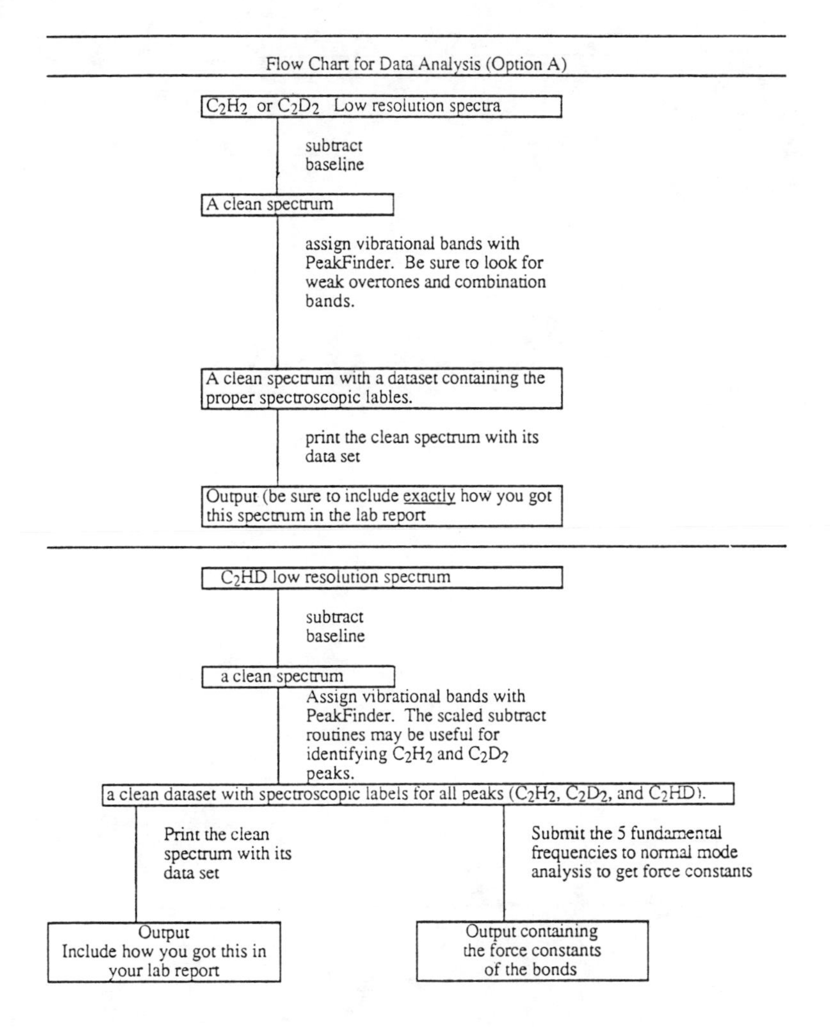

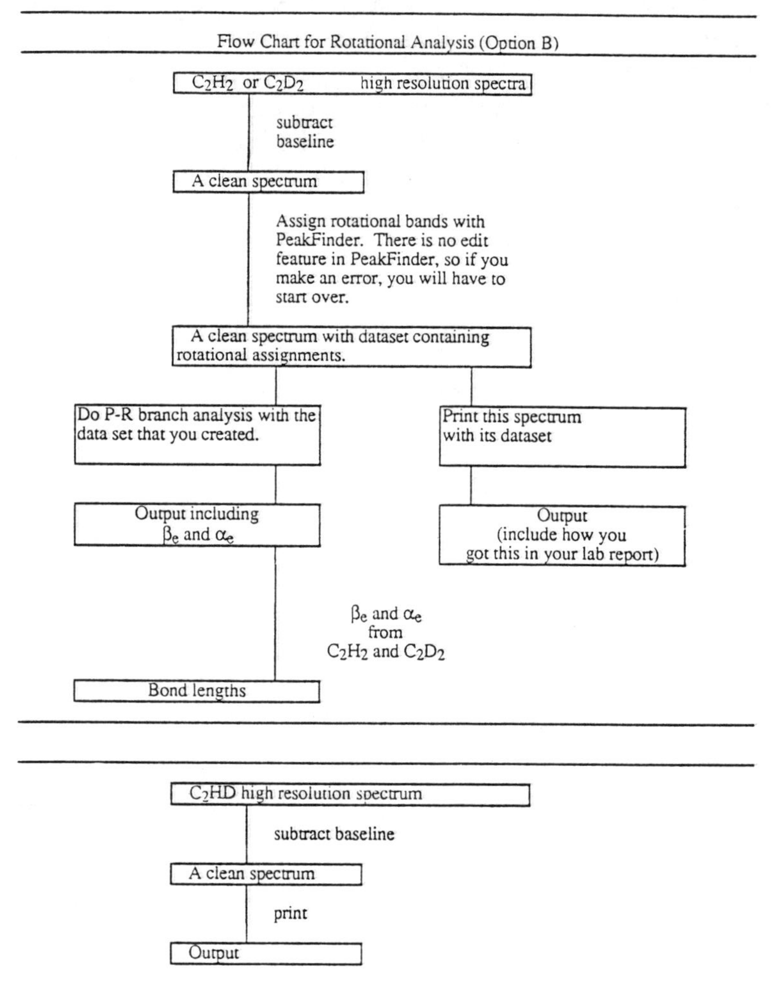

RECEIVED October 14, 1992

CHAPTER 21

NMR Relaxation Times

Kathryn R. Williams and Roy W. King

Carbon-13 spin–lattice relaxation times (T_1's) will be measured using the inversion-recovery method. Knowledge of relaxation processes is essential to the understanding and proper conduct of pulse Fourier-transform (FT) NMR. Under suitable conditions relaxation times can also be correlated with the environments of the carbon atoms and to the motional behavior of the molecules.

Students conducting this experiment are expected to have some experience in acquiring and integrating simple proton and ^{13}C spectra. The theory and use of FT-NMR is reviewed briefly in the following section, but the students will probably want to read references *(1-4)* for more in-depth discussions. For further background on the theory of relaxation mechanisms reference *(5)* gives a very readable presentation.

Theory

Fundamental Concepts of Magnetization A nucleus with nonzero spin, I, behaves as a small bar magnet with a magnetic moment, μ, whose magnitude is given by:

$$\mu = \gamma \hbar [I (I + 1)]^{1/2} \tag{1}$$

where $\hbar$ is Planck's constant divided by 2π, and γ is the magnetogyric ratio, which is an inherent property of a particular type of nucleus. In a collection of nuclei the magnetic moment vectors ordinarily have random directions and equal energies. However in a fixed magnetic field, B_0, the energies are split into $2I + 1$ states. Each of these states is specified by its magnetic quantum number, m_I, which has allowed values of $-I$, $(-I + 1)$, ..., $(I - 1)$, $+ I$. The corresponding energy is:

$$E = - m_I \gamma \hbar B_{\text{eff}} = -\mu_z B_{\text{eff}} \tag{2}$$

where the symbol B_{eff} means the effective magnetic field actually experienced by the nucleus. This depends on the degree to which the electronic environment shields the nucleus from the applied field (chemical shift) and the fields produced by the spin states of other $I > 0$ nuclei connected to the nucleus via chemical bonds (spin-spin splitting). In equation 2 μ_z, the observable value of μ, is the projection of μ onto the direction of B_0 (defined to be the z axis). The magnetic moment vectors themselves lie at an angle θ (where $\cos \theta = \mu_z/|\mu|$) to the magnetic field (Figure 1). Of particular interest in this experiment are nuclei with $I = 1/2$ (e.g., ^1H, ^{13}C, ^{31}P, ^{29}Si, ^{15}N), which have only two quantum states separated by

$$\Delta E = \gamma \hbar B_{\text{eff}} \tag{3}$$

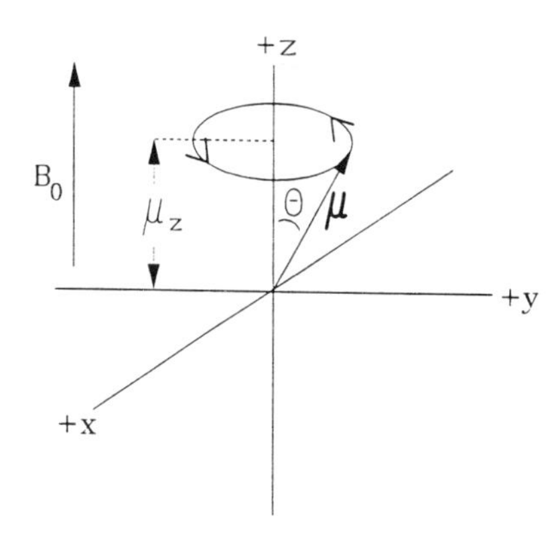

Figure 1. The magnetic moment vector, μ, and its projection on the axis (z) of the fixed magnetic field. According to classical mechanics the vector precesses about the z axis at the Larmor frquency. Reprinted with permission from ref. 1a. Copyright 1989 Journal of Chemical Education.

or

$$f = \gamma\ B_{\mathrm{eff}}\ /\ 2\pi\ (\mathrm{Hz}) \quad (4)$$

where f is the frequency of the transition. For nuclei with positive magnetogyric ratios (e.g., ^{1}H and ^{13}C with γ values of 26.7519 × 10^7 and 6.7263 x 10^7 rad T^{-1} s^{-1}, respectively) the lower energy state, often called the α state, has $m_I = +1/2$, with μ_z along the $+z$ axis. The higher energy state ($m_I = -1/2$; μ_z along $-z$) is called β. The value of ΔE given by equation 3 is very small compared to kT, where k is Boltzmann's constant. For example, the student can verify that for ^{1}H, which, except for its isotope, tritium, has the largest γ of all nuclei, the energy separation in a 7.05 T magnet is only 2.0 × 10^{-25} J (corresponding to a frequency of 300 MHz). According to the Boltzmann distribution, the ratio of the population of the α state to that in the β state is exp ($\Delta E/kT$). At room temperature this produces an excess in the lower state of less than 50 parts per million, and the imbalance is only one-fourth as large for ^{13}C.

The behavior of the nuclei can also be visualized in terms of a classical model (cf. Figure 1). The magnetic field exerts a torque on μ, and causes it to rotate, or **precess**, about the z axis. The frequency of this motion, called the Larmor frequency, is given by exactly the same expression as the quantum mechanical result shown in equation 4. For a positive γ, the two energy states correspond to precession about the $+z$ (α state) and $-z$ (β state) axes. When the spin system is in equilibrium in the magnetic field, the individual magnetic moment vectors are distributed evenly in a cone about $\mathbf{B_0}$. Because of the small excess population in the α state, the resultant is a net magnetization vector, **M**, along the $+z$ axis (Figure 2).

The Pulse NMR Experiment To detect the motions of the magnetic moments, there must be a component of the magnetization vector in the xy plane. This is accomplished by interaction with electromagnetic radiation at the Larmor frequency,

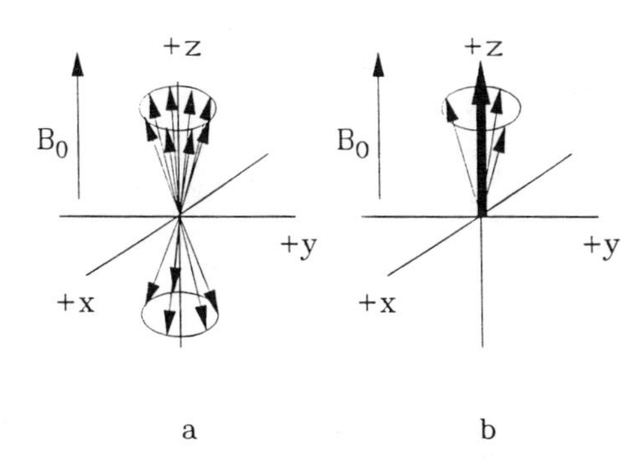

Figure 2. a) Schematic representation of a collection of nuclear spins, showing a greater number aligned with the magnetic field (lower energy state, if γ is positive). b) The excess spin population aligned with the field results in a net magnetization in the $+z$ direction.
Reprinted with permission from ref. 1a. Copyright 1989 Journal of Chemical Education.

which is in the radiofrequency region for all types of nuclei, directed along either the x or y axis. In a pulse NMR experiment the radiofrequency field (rf) is applied as a very short (5–20 μs) pulse of high intensity. The NMR transmitter produces a single frequency, called the spectrometer frequency, f_s, which in modern instruments is usually set approximately in the middle of the chemical shift range of the nucleus of interest. (For example, in a 7.05 T magnet the nominal spectrometer frequency would be 300 MHz for ^1H and 75 MHz for ^{13}C.) As a consequence of the Heisenberg principle, the short duration of the pulse necessitates an uncertainty in its energy (and hence its frequency). This means that the sample experiences a band of frequencies centered about f_s, which, if the pulse is short enough and has adequate intensity, usually has sufficient breadth to cover the Larmor frequencies of all the nuclei of interest.

The magnetic component of the electromagnetic radiation produces a magnetic field, called $\mathbf{B_1}$, which is oscillatory to the external observer (i.e., to a detector coil located in the xy plane). However, for the nuclei the $\mathbf{B_1}$ field is constant, because it contains a frequency which is in step (in resonance) with each of the precessing spins.[1] The magnetic moments for each chemical shift are brought into phase; i.e. they are moved together as a group by the pulse, which causes them to precess about $\mathbf{B_1}$ (Figure 3). If the pulse is applied very briefly, there is only partial revolution about $\mathbf{B_1}$, and the net magnetization vector is said to be "tipped" towards the xy plane. For a $\mathbf{B_1}$ field pointing in the positive x direction, rotation is towards the $-y$ axis. A pulse of sufficient length to locate $\mathbf{M}$ exactly on

[1]This statement is equivalent to use of the rotating frame convention, which places each chemical shift in a coordinate system rotating at the corresponding Larmor frequency. The axes of the rotating frame are designated x', y', and z, but it is customary to omit the primes, when discussing pulse directions, etc.

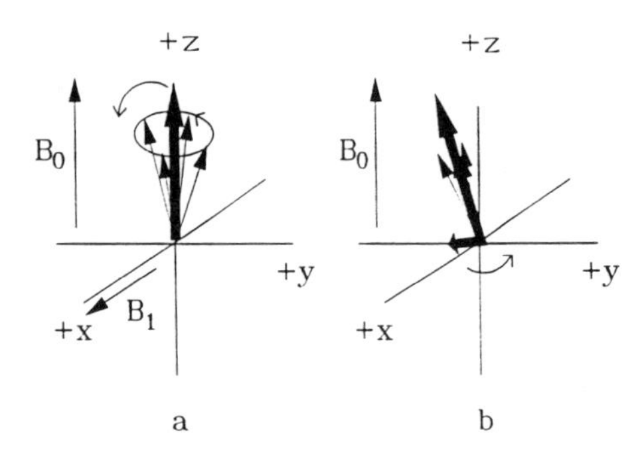

Figure 3. The effect of the rf pulse. a) The nuclei tend to precess about the $\mathbf{B_1}$ field. b) The pulse also causes the individual magnetic moment vectors to come into phase. After the pulse is removed the vectors precess as a group about the $\mathbf{B_0}$ field direction.

the $-y$ axis is called a 90°_x pulse. This tipping process involves absorption of energy from the rf field. The nuclear spins are induced to flip their spins, so that the populations in the two states do not have their equilibrium values.

After the pulse is terminated there is a component of the magnetization in the xy plane, and the receiver coil senses the magnetic moments as they precess. By electronic subtraction of the spectrometer frequency from the observed Larmor frequencies, offset frequencies in the kilohertz range, are produced, and these can be sampled adequately by available digitization equipment. If only one chemical shift is present, and the precession frequency is equal to f_s, the time domain signal, called the **free induction decay (FID)** shows a simple exponential decrease. If f_s does not exactly match the Larmor frequency, the FID is an exponentially decaying sinusoid with frequency equal to $f - f_s$. In a real sample with several chemical shifts a complicated interference pattern is observed, and this is converted into the frequency spectrum via a Fourier transformation (Figure 4).

Relaxation The question now arises as to how the nuclei rid themselves of the energy they have absorbed. In spectroscopies involving higher energy parts of the electromagnetic spectrum, for example electronic absorption in the UV/visible region, the return of an excited atom or molecule to the ground state occurs very rapidly. Although in liquids and solutions much or all of this energy may be dissipated by collisions with surrounding molecules (picosecond range), even a bare atom will spontaneously emit a photon within a small fraction of a second after excitation. The situation is much different in NMR spectroscopy, in which the very small energy difference between the nuclear spin states causes spontaneous emission to be very slow. (The lifetime of an unperturbed excited nucleus is estimated to be in the range of years!) As a result an excited nucleus must be induced by some external means to flip its spin and return to the ground state. As will be explained below, relaxation can cause experimental problems if it is either too fast or too slow. An understanding of the concepts of nuclear spin relaxation is, therefore, very important to anyone involved with NMR spectroscopy.

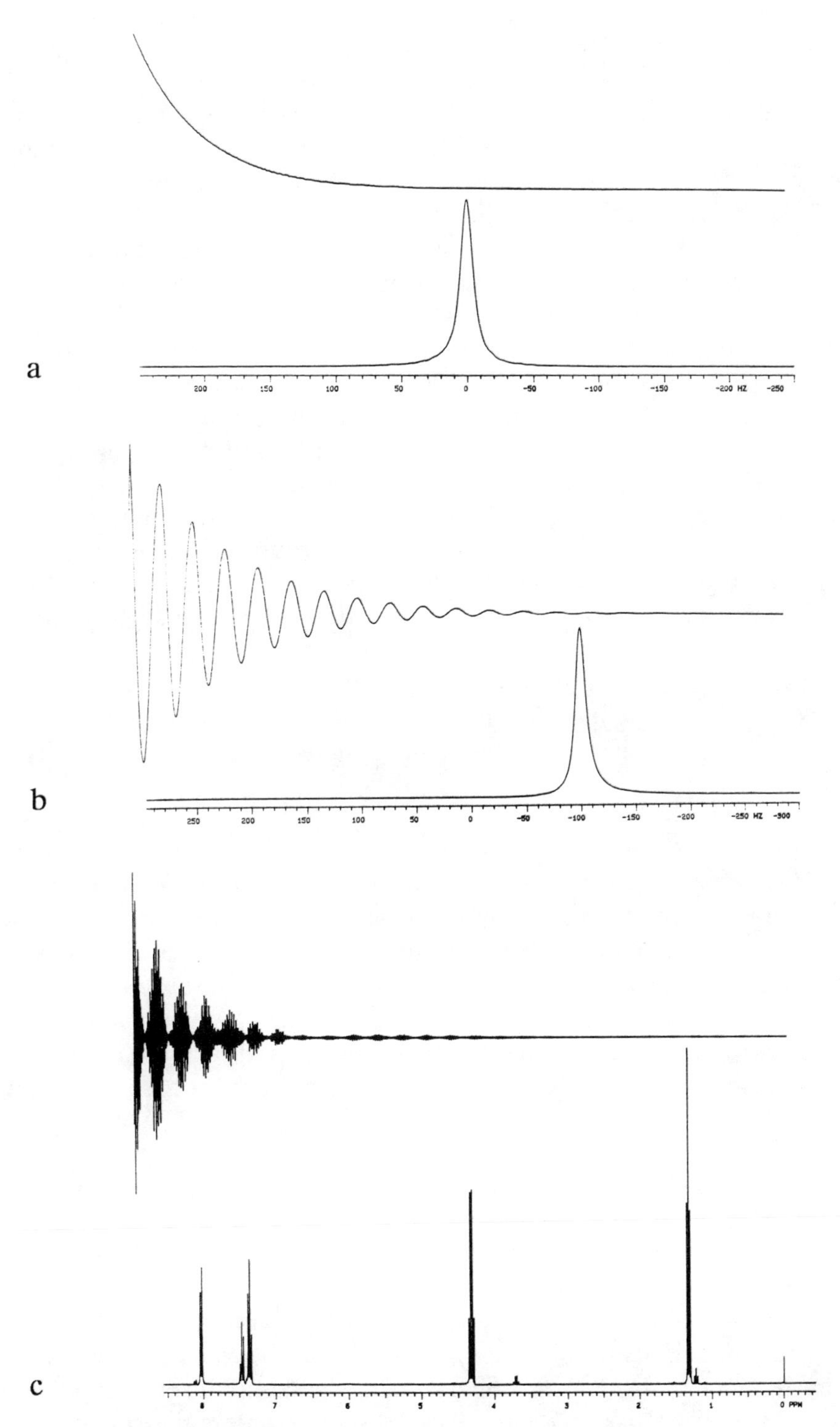

Figure 4. FID's and transformed proton spectra. a) Single-proton system (H_2O containing ca. 50 ppm Mn^{2+} to broaden the line for better presentation) with Larmor frequency equal to f_s. b) Single-proton system with Larmor frequency slightly less than f_s. c) Compound (ethyl benzoate at 300 MHz) with protons in several environments. Reprinted with permission from ref. 1a. Copyright 1989 Journal of Chemical Education.

There are two types of relaxation processes which cause the observed NMR signal to decrease following the pulse. **Spin–lattice** or **longitudinal** relaxation involves a net transfer of energy from the spin system to the surroundings (historically called the "lattice"), so that the populations in the two energy states can again return to the values dictated by the Boltzmann distribution, with **M** having its equilibrium length on the +*z* axis. The rate of return to equilibrium is equal to the reciprocal of T_1, **the spin–lattice relaxation time.**

There is also another class of relaxation processes which cause the spin vectors to become evenly distributed in the *xy* plane without transfer of energy to the surroundings. **Spin–spin** or **transverse** relaxation also leads to a loss in the detectable signal, because the vectors are no longer coherent in phase (no longer precessing as a group, as shown in Figure 3b). The rate of dephasing is given by $1/T_2^*$, in which T_2^* is called the **effective spin–spin relaxation time**. The word "effective" is included, because usually the main contributor to the dephasing is lack of homogeneity of the B_0 magnetic field. The sample is composed of many microscopic regions, each with a slightly different B_0. Equivalent nuclei within a region experience the same field and precess as a group, called an **isochromat.** Other isochromats experience slightly different fields and precess at faster or slower rates. Overall, the sample is a collection of isochromats rotating at slightly different frequencies. As the isochromats rotate out of step with each other, the phase coherence present during the rf pulse is lost.

The rate of dephasing includes the true spin–spin relaxation rate, $1/T_2$, plus a component, $1/T_{inhomo}$, which accounts for the magnetic field inhomogeneity:

$$R_{dephasing} = 1/T_2^* = 1/T_2 + 1/T_{inhomo}. \tag{5}$$

It is obvious from equation 5 that T_2^* must always be less than T_2, except when the effects of field inhomogeneity are negligible. It is also a fundamental requirement that T_2 be equal to or less than T_1. When T_1 relaxation is complete, there is no net component of **M** in the *xy* plane, and, therefore, T_2 relaxation must also be complete. On the other hand, it is very possible to have vectors which are randomly oriented in the *xy* plane, without an equilibrium resultant in the *z* direction (i.e, situation similar to Figure 2 but with a flattened cone). The overall order of relaxation times is, therefore: $T_2^* \leq T_2 \leq T_1$.

Relaxation times are very important in the conduct of the pulse NMR experiment and the appearance of the frequency spectrum. In the FT method the entire chemical shift range is observed in a time period much shorter than that required to obtain a spectrum with an older continous wave (scanning) instrument. The spectroscopist can take advantage of this time savings by acquiring several FID's, which are signal-averaged to improve the signal-to-noise ratio in the transformed spectrum (Felgett advantage). For less sensitive and/or low natural abundance nuclei, such as ^{13}C, signal enhancement is essential if spectral peaks are to be distinguishable from the baseline noise. However, unless the spin system is allowed to return to equilibrium after each successive pulse, the two spin states will soon have equal populations; i.e., the spin system will become **saturated**, and there will be no signal. Following a single pulse the fraction of the magnetization which has returned to equilibrium is given by $[1 - \exp(-\tau/T_1)]$, where τ is the elapsed time from the end of the pulse. The student can verify that a time period of about 5 times T_1 is required if relaxation is to be more than 99% complete. For ^{13}C

nuclei, which can in some environments have T_1's in the tens of seconds or minutes range, relaxation delays can lead to lengthy experiment times, if a large number of FIDs must be acquired, and saturation must be avoided.[2]

This discussion may lead one to believe that, within the limits of the instrumentation, short relaxation times are highly desirable. However, this initial conclusion ignores the requirements of the Heisenberg principle, which, as explained above in relation to the rf pulse, places a lower limit on the product of the uncertainties in the lifetime of a transition and its energy. Just as a short monochromatic pulse produces a wide band of frequencies, the width of a peak in the transformed NMR spectrum increases as the average lifetime of the observable signal becomes shorter. As described above, the rate of signal loss is given by $1/T_2{}^*$. For liquids the observed spectral peak theoretically has a Lorentzian lineshape, which falls to one-half its maximum value when the frequency is $\pm\ 1/(2\pi T_2{}^*)$ from the central value. Thus, the line-width (FWHM = full width at half maximum intensity) is equal to $1/(\pi T_2{}^*)$. For mobile liquids $T_2{}^*$ is usually dominated by the magnetic field inhomogeneity effect, but when T_1 is short, T_2 must also be short, and spectral resolution suffers as a result. The student may surmise that a worst-case scenario occurs if T_2 is inherently much shorter than T_1. The long time required for restoration of equilibrium necessitates long delays between pulses, and peaks may be excessively broadened by the short duration of the observable signal.

The Dipolar Mechanism In order for relaxation to occur, the individual magnetic vectors must interact with fluctuating magnetic fields within the sample. These fields arise as the molecules, and the nuclear dipoles contained in them, move about in space. The effectiveness of the relaxation process, and hence the rate of relaxation, depends on the strength of interaction of the spin system with the source of the microscopic field and also on the frequency of the fluctuation.

For diamagnetic compounds containing no quadrupolar ($I \geq 1$) nuclei, the principal means of both T_1 and T_2 relaxation is usually the **dipolar mechanism**. When two nuclear spins are close to each other in space, their magnetic moments interact. This produces a local magnetic field (in addition to the fixed $\mathbf{B_0}$ field) at each nucleus. As the molecule moves, the relative orientation of the dipoles changes (Figure 5), and this causes the fluctuation required for relaxation to occur. The strength of the dipole-dipole interaction is inversely proportional to r^6, where r is the distance between molecules. Thus, this mechanism is effective only for nuclei which are close to each other spatially. Although the atoms do not have to be linked by chemical bonds, in ^{13}C work very slow relaxation is observed for carbons with no bonded hydrogens. For example, carbons in carboxylate groups and at substituted positions in aromatic systems usually exhibit very long relaxation times (30–60 seconds compared to a few seconds for C–H carbons). This distance dependence has been very useful in the ^1H NMR study of macromolecules, especially biopolymers. Studies of the nuclear Overhauser effect, which occurs as a result of

---

[2] For quantitative ^{13}C spectra sufficient relaxation delays are essential to prevent partial or complete saturation and to allow decay of nuclear Overhauser enhancement (*1,2*), but the resulting experiment times may reach several hours. Often the needs of the chemist are satisfied by the chemical shift data (and possibly also the multiplicities). Users who do not need quantitative integration data often opt for shorter experiment times and/or an increased number of FID accumulations.

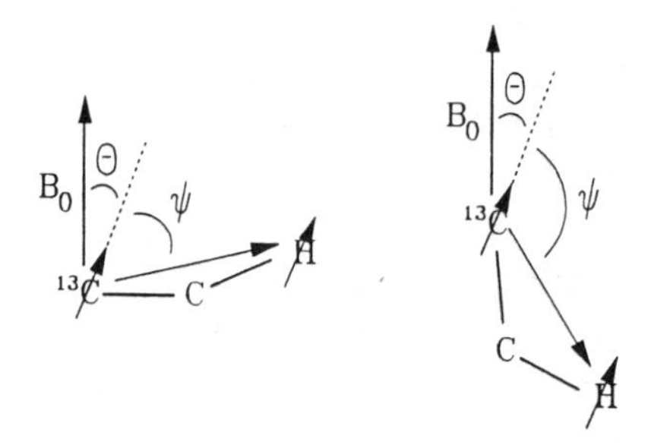

Figure 5. Representation of the dipolar interaction showing two orientations of the internuclear vector with respect to the magnetic field. The angle θ is defined by the direction of B_0, but ψ changes as the molecule tumbles. Reprinted with permission from ref. 1c. Copyright 1990 Journal of Chemical Education.

dipolar relaxation *(1b)*, can provide information on the proximity of amino acid groups in proteins, for example. Such work has helped biochemists to determine tertiary structures of these molecules.

The frequency of fluctuation of the local magnetic fields is often given in terms of a reciprocal quantity, called the **correlation time**, τ_c, which is usually defined as the average time that a molecule spends in a certain orientation. There are, of course, many possible orientations and an entire range of times between reorientations. The correlation time is an attempt to summarize these "waiting" times in a single parameter.

The BPP Equation and Its Consequences Spin–lattice relaxation by the dipolar mechanism was first treated in the late 1940's by Bloembergen, Purcell and Pound *(6)*, and their mathematical result is often called the BPP equation. For a ^{13}C atom interacting with a proton the rates of relaxation associated with T_1 (R_1) and T_2 (R_2) are:

$$R_1 = \frac{1}{T_1} = \left(\frac{\mu_0}{4\pi}\right)^2 \frac{\gamma_C^2 \gamma_H^2 h^2}{80\pi^2 r^6} \left[\frac{6\tau_c}{1+\omega_C^2\tau_c^2} + \frac{2\tau_c}{1+(\omega_C-\omega_H)^2\tau_c^2} + \frac{12\tau_c}{1+(\omega_C+\omega_H)^2\tau_c^2}\right] \quad (6)$$

and

$$R_2 = \frac{1}{T_2} =$$

$$\left(\frac{\mu_0}{4\mu}\right)^2 \frac{\gamma_C^2 \gamma_H^2 h^2}{80\pi^2 r^6} \left[4\tau_c + \frac{\tau_c}{1+(\omega_C-\omega_H)^2\tau_c^2} + \frac{3\tau_c}{1+\omega_C^2\tau_c^2} + \frac{6\tau_c}{1+\omega_H^2\tau_c^2} + \frac{6\tau_c}{1+(\omega_C+\omega_H)^2\tau_c^2}\right] \quad (7)$$

where ω_C and ω_H are the angular Larmor frequencies for the ^{13}C and ^1H nuclei, respectively, and μ_0 is the permeability of a vacuum.

It is now useful to consider three different ranges of correlation times:

1) For small molecules in the liquid state τ_c is very short, on the order of 10^{-12} second. In a 7.05 T magnet ω_C is 4.7×10^8 rad / s and ω_H is 1.9×10^9 rad / s ($2\pi \times 75$ MHz and $2\pi \times 300$ MHz, respectively). Thus, the denominators in equations 6 and 7 are all very nearly equal to 1 and the relaxation rates are given by:

$$R_1 = \frac{1}{T_1} = \left(\frac{\mu_0}{4\pi}\right)^2 \frac{\gamma_C^2 \gamma_H^2 h^2 \tau_c}{4\pi^2 r^6} \tag{8}$$

and

$$R_2 = \frac{1}{T_2} = \left(\frac{\mu_0}{4\pi}\right)^2 \frac{\gamma_C^2 \gamma_H^2 h^2 \tau_c}{4\pi^2 r^6} \tag{9}$$

This shows that for systems in rapid motion, which is the situation for nonviscous liquids containing small and moderately sized molecules, T_1 and T_2 are equal. Another important observation is the direct proportionality of R_1 and R_2 (inverse proportionality of T_1 and T_2) to the correlation time. Thus, if τ_c is short, the relaxation times are long. Both T_1 and T_2 are shorter for more slowly moving species, for example polymer molecules in solution.

2) For very viscous liquids (e.g. glycerol at low temperature) and large polymer molecules τ_c is of the order of ω. The student can verify that the spin–lattice relaxation rate reaches a maximum value (c.f., Data Analysis Section). T_2 relaxation is also rapid, but, as will be shown below, is not at a maximum. The very short T_2 results in broadened lines for such systems.

3) When τ_c becomes very long, as in glasses and solids, the denominators in equation 6 are dominated by the $\omega^2\tau_c^2$ terms. This leads to two important effects for spin-lattice relaxation. First, T_1 becomes dependent on ω, and increases as ω increases. The dependence of T_1 on τ_c is now opposite to that in case 1; T_1 increases with increasing τ_c, and this results in the T_1 minimum when $\tau_c \approx \omega$ (case 2).

The situation with T_2 is quite different, however. When τ_c is long, equation 7 is dominated by the first term, which does not contain an rf frequency component. The spin–spin relaxation rate remains independent of ω and is directly proportional to τ_c (T_2 inversely proportional to τ_c).

These results have major consequences for the NMR spectroscopy of solids. As described earlier, the high spin–spin relaxation rates cause very rapid loss of observable magnetization and severely broadened peaks. On the other hand, spin–lattice relaxation times are very long, and this necessitates long delay times between pulses when multiple FID's are acquired, unless special pulse sequences are used.

Although the expressions in brackets in equations 6 and 7 vary for different relaxation mechanisms, the forms show two common characteristics. For spin–lattice relaxation all terms contain denominators similar to those in equation 6, and, as observed for case 2 above, relaxation is most effective when $\tau_c \approx \omega$. On the other hand, equations for rates of spin–spin relaxation all contain an additional term directly proportional to τ_c. This means that T_2 is also sensitive to very low frequency changes in the local magnetic environment. For example, this situation can occur for substituents involved in chemical exchange reactions. In such cases there are effectively two correlation times: the τ_c for molecular tumbling, which is applicable to dipolar relaxation, and τ_{ex} for the exchange reaction. The latter mechanism dominates for spin–spin relaxation, leading to short T_2's and broadened lines. However, only the very high frequency motions are effective in promoting spin–lattice relaxation, which must still rely on the dipolar interaction.

Relaxation studies can give very useful information about motions of molecules and substituent groups. As described above, small molecules tumble rapidly, and their T_1's can be quite long. However, if the motion of a part of the molecule is restricted by steric effects or by intermolecular hydrogen-bonding, the longer τ_c results in a shortened T_1 value.

For a polymer molecule the motions are slower, and, as mentioned above, in general T_1's are smaller. Even so, T_1 values can still vary for different parts of the polymer molecule and can provide information about the mobilities of chain segments and pendant groups. Because of the low frequency component to spin–spin relaxation, T_2 values can serve as indicators of the motion of the entire polymer molecule. Tumbling is greatly slowed as the size of the molecule increases, and, as a result, T_2's can be used as probes for such important properties as molecular weight and the degree of cross-linking and network formation *(7)*.

Methodology

Inversion-Recovery Method for T_1 Spin–lattice relaxation times are commonly measured by the **inversion-recovery** method, which is one of the oldest multiple-pulse NMR experiments. The pulse sequence and vector description are shown in Figure 6. Initially the nuclei are at equilibrium in the $\mathbf{B_0}$ field, with the magnetization vector in the same direction as $\mathbf{B_0}$. The nuclei are subjected to the intense rf radiation for a time twice as long as that needed for a 90°_x pulse. The pulse causes **M** to rotate through 180° to the $-z$ axis. This corresponds to a reversal of all the spins in the sample, so that the β state has the population excess. Following the pulse there is a delay period, τ, whose length is varied as the pulse sequence is repeated. The experiment commonly involves eight to ten τ values, ranging from close to zero to four or five times T_1. During the τ interval the system attempts to return to equilibrium by the spin–lattice relaxation process. Spin–spin relaxation, which affects only magnetization components in the *xy* plane, is not a concern, because **M** lies on the *z* axis. As the nuclei gradually release excess energy in their attempt to return the population excess to the α state, the net magnetization vector shortens in the $-z$ direction. Depending on the length of the delay, **M** passes through zero and eventually recovers its full magnitude along the $+z$ axis.

To determine the extent of relaxation, **M** must be converted into observable magnetization in the *xy* plane. This is the purpose of the 90°_x pulse, which rotates

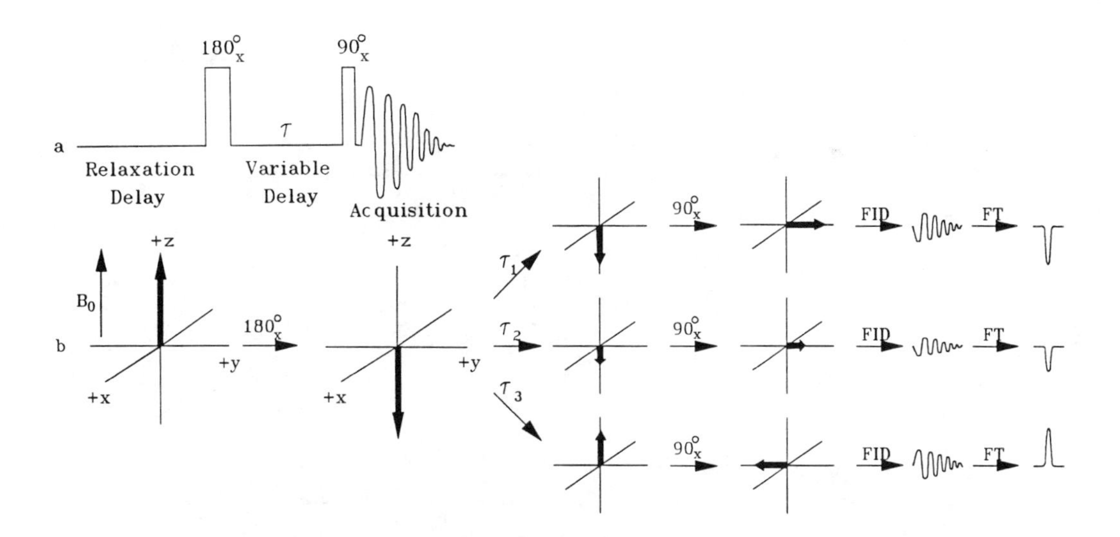

Figure 6. The measurement of T_1 by the inversion-recovery method. a) Pulse sequence. b) Vector description.

Reprinted with permission from ref. 1c. Copyright 1990 Journal of Chemical Education.

M to lie along the *y* axis. If **M** is still negative before the 90°_x pulse, the magnetization is rotated to the +*y* direction, but when relaxation is complete, the magnetization after the 90°_x pulse lies along the -*y* axis. If the transformed spectrum is plotted to give a positive peak for a 90°_x pulse applied to **M** at equilibrium, the areas of the transformed peaks vary with τ, from maximum negative for $\tau = 0$ to maximum positive for $\tau =$ infinity (practically four or five times T_1). The peak area, A, is related exponentially to T_1 by the equation:

$$A = A_\infty [1 - 2 \exp(-\tau/T_1)] \quad (10)$$

which may be rearranged to give:

$$\ln [(A_\infty - A) / A_\infty] = -\tau / T_1 + \ln 2 \quad (11)$$

For a plot of $\ln [(A_\infty - A) / A_\infty]$ versus τ, the slope is $-1/T_1$ and the intercept should equal ln 2.

Spin Echo Method for T_2 The molecular systems encountered by most chemists correspond to case 1 described above, for which T_2 (without the *) is equal to T_1. Therefore, measurements of T_2 values are rather uncommon, and only T_1's will be determined in this experiment. For the sake of completeness, the usual method of measuring T_2 will now be described. As mentioned above, the width of the NMR peak is given by $1/(\pi T_2^*)$, which is usually dominated by inhomogeneities in the $\mathbf{B_0}$ magnetic field. Because of this inhomogeneity contribution, T_2 usually cannot be evaluated from line-width measurements. Instead, the **spin echo** pulse sequence (Figure 7), which allows T_2 to be determined independently of the field inhomogeneity, is used. The initial 90°_x pulse turns the magnetization vector to the -*y* axis. In a coordinate system rotating at the Larmor frequency, the magnetization

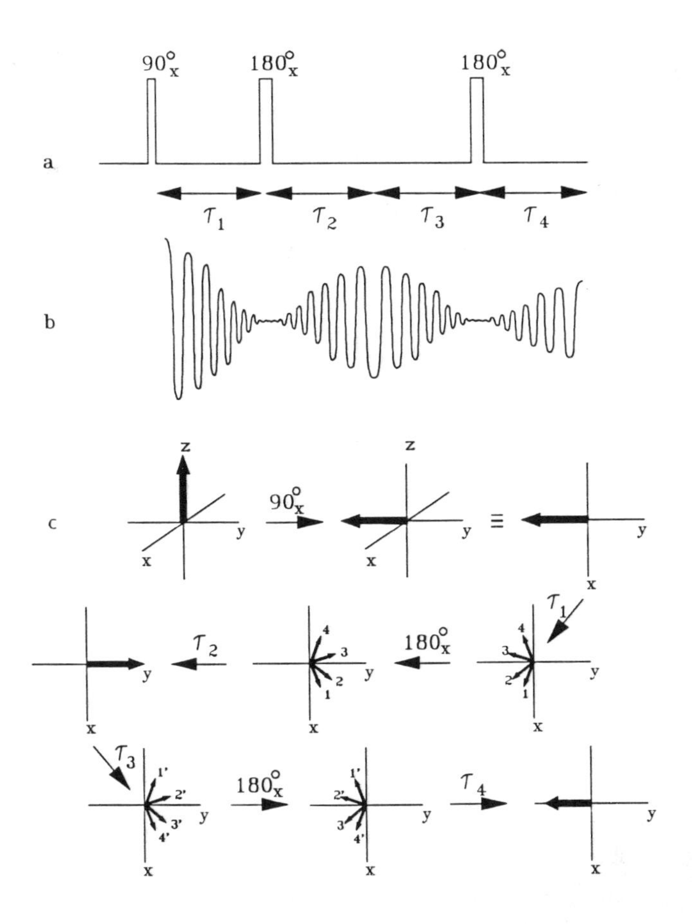

Figure 7. The spin echo experiment for the measurement of T_2. a) Pulse sequence (Carr–Purcell version). b) NMR signal showing "echoes" at the end of τ_2, τ_4, etc. c) Vector description. Reprinted with permission from ref. 1c. Copyright 1990 Journal of Chemical Education.

vector remains fixed in the -*y* direction, and the effect of T_2 is a gradual shortening of **M** along -*y* (no appearance of +*y* magnetization). However, inhomogeneities in $\mathbf{B_0}$ cause the nuclei in various microscopic regions to experience slightly different magnetic fields; some of the nuclei (groups 1 and 2 in the figure) precess faster than the rotating frame and appear to move in the counterclockwise direction, while others (groups 3 and 4) rotate clockwise.

At the end of the first τ delay a 180°_x pulse flips the spins about the *x* axis. After a second τ interval the slow and fast groups meet each other on the +*y* axis to give an inverted signal reduced only by the effects of the true spin–spin relaxation. If the signal is monitored throughout the $90^\circ_x - \tau - 180^\circ_x - \tau$ sequence, it dies away during the first τ interval, and then recovers to an inverted maximum when the spins are refocused at the end of τ_2. As shown in Figure 7, the 180°_x pulse is repeated at the end of τ_3, so that another echo is observed after τ_4. The time constant for the decrease in the amplitude of successive echoes is the true value of T_2.

Description of the Experiment

In this experiment the inversion-recovery method will be used to determine the ^{13}C T_1's in one or more molecules. The choice of chemical system is left to the discretion of the instructor, but a few suggestions may help to assure a meaningful class experience.

Aside from the data analysis itself, the students should appreciate the two major factors which affect the rate of ^{13}C relaxation by the dipolar mechanism: the need for nearby protons and the correlation time. Both effects can be observed by studying a substituted benzene molecule, such as ethyl benzoate or allylbenzene. Although the interacting protons need only be in spatial proximity, in most cases carbons which lack a directly bonded hydrogen relax very slowly compared to carbons with the requisite protons. This is shown quite dramatically by the carboxyl and substituted ring carbons in ethyl benzoate *(1c)*. It is also observed that the shortest T_1 corresponds to the *para* ring carbon, indicating that it experiences the longest correlation time. This provides evidence that a principal motion of the molecule is rotation of the benzene ring about the C_1–C_4 axis, a process which results in minimal reorientation of the C_4–H bond.

The effect of correlation time can also be demonstrated by studies of an alcohol, such as 1-octanol or 1-decanol. Intermolecular hydrogen-bonding restricts the motion of the substituted carbon, and causes it to have a comparatively long correlation time and a short T_1. The effect decreases with distance along the carbon chain, and the monotonic increase in T_1 can be used to assign the positions of the carbons in the chain. The effect of dilution on the degree of hydrogen-bonding can be observed if varying solution concentrations are investigated.

This experiment also provides an excellent opportunity to add concepts of polymers to the curriculum. Samples of poly(ethylene glycol)s of varying chain length, as well as the mono- and di-methyl ethers, are readily available commercially. A variety of motional effects can be demonstrated by suitable choice of molecular weight, end-group, and solution concentration.

Experimental Considerations

The pulse sequence for the inversion-recovery method has for a number of years been available in the standard software packages of most FT-NMR spectrometers. In order for it to set up a suitable range of delay intervals, a typical program requires as input a ballpark estimate of the minimum and maximum T_1's in the molecule. For carbons with directly bonded hydrogens suggested values are 1 – 3 s if motion is slow, and 4 – 6 s for a carbon in a less restricted environment. The T_1's of quaternary carbons can be 30 s or more. The program will include a relaxation delay between each 180°/90° pulse pair of four or five times the longest expected T_1. If NMR time is at a premium, it may be necessary to input a smaller maximum T_1 (e.g., 20 seconds for a quaternary carbon) to prevent excessive experiment times. If so, the longer T_1's will have a large error, but the effect of the lack of nearby hydrogens should be evident to the students. The temperature of the sample should be held constant, if the spectrometer is equipped with a control unit. A study of the temperature dependence of T_1 would make an interesting addition to the experiment or special project for a small group of students.

Although the students are to perform their own analysis of the inversion-recovery data, it is also instructive to compare results to those generated by the instrument software. The calculation routines in many spectrometers use peak heights, determined at the peak maxima in the fully relaxed spectrum, instead of areas. To reduce random error the use of areas is preferred, but the peak height should be proportional to the area, as long as the peak width for each carbon is invariant. For the inversion-recovery experiment this condition should hold, because there is no *xy* magnetization during the τ delay, and, therefore, T_2* effects should not cause peak widths to vary for different τ delays.

A possible cause of systematic error is the presence of dissolved oxygen, which is paramagnetic and can shorten long T_1's (cf. Data Analysis Section). Air may be removed by vacuum-degassing, followed by sealing the NMR tube. Somewhat less complete removal may be achieved by bubbling with nitrogen and secure capping. Normally this precaution is not necessary for samples with T_1's below about 10 s, but, of course, all other paramagnetic materials must also be excluded.

Another source of error, especially for long relaxation times, is molecular diffusion. The active volume of the spectrometer is a region between one and two centimeters long, corresponding to the size of the rf coils. To achieve adequate field homogeneity, the sample must extend above and below the coils, but for long τ delays (tens of seconds) it is possible that excited nuclei may diffuse out of the active volume, while molecules with unexcited nuclei diffuse in. To avoid this source of error the sample may be confined between long cylindrical plugs made of a material with the same magnetic susceptibility as the sample and placed at the edges of the active volume.

Analogous to the need for long relaxation delays described above, the effects of dissolved oxygen and diffusion will be worst for quaternary carbons in small molecules. Again, the instructor may find it necessary to omit the degassing step, and accept the resulting error in the T_1's, because the overall trend in the final results should still be instructive.

It is worth repeating a statement made earlier that students should already have had some practice with routine FT-NMR methods. If the students and/or laboratory instructor are unfamiliar with procedures such as sample preparation, specification of pulse lengths, phase adjustment, etc., an experienced instrument operator should be available to help with the experiment.

Safety

The most common solvent for routine NMR work with organic solutes is deuterated chloroform. This compound is known to be toxic to the liver and is a suspected carcinogen. Samples should be prepared in a hood, using gloves to prevent skin contact, and the tubes should be securely capped or sealed. The proper procedures for disposal of chlorinated solvent waste should be followed. Of the solutes suggested above, allylbenzene and ethyl benzoate are both irritants, but the polymers present no safety hazard.

Data Analysis

The students should obtain and analyze the data as follows:

1. For each carbon in the compound(s) studied obtain both the T_1 calculated by the spectrometer's software and the raw inversion-recovery data (τ intervals and corresponding peak areas or heights).

2. Use the raw data and a least squares fit to equation 11 to calculate some or all of the T_1's evaluated in step 1 and their 95% confidence intervals.

3. Compare the T_1's calculated in steps 1 and 2. The 95% confidence intervals should overlap, but the values may not be exactly equal if the software uses a non-linear regression method.

4. Discuss the T_1 values in terms of the dipole-dipole relaxation mechanism, including both the strength of the dipolar interaction and the motional characteristics of the molecule or substituent.

5. Using equations 6 and 7 prepare log-log plots of the ^{13}C relaxation times (log T_1 and log T_2) versus correlation time over the log τ_c range of −13 to −3 (τ_c in seconds). Use proton frequencies of 300 and 600 MHz (4 plots total). (Note: The purpose in making these plots is to observe the overall form of the τ_c dependence and not to calculate the actual values of log T_1 and log T_2. Therefore, the terms in front of the brackets in the two equations may be considered as arbitrarily equal to one.)

At the option of the instructor some or all of the following questions may also be assigned:

6. The dipole–dipole interaction is also the source of the nuclear Overhauser effect. Describe this effect and its importance in signal enhancement. Why is it sometimes a problem in ^{15}N and ^{29}Si spectra? How can NOE be used to aid in structure determination, especially for biological macromolecules?

7. Use the dipolar mechanism to explain why paramagnetic species (e.g., chromium(III) acetylacetonate) are sometimes added to NMR samples to decrease the need for long relaxation delays, when multiple FID's are acquired. (Hint: Consider the relative magnitudes of the magnetic moments of the electron and a ^{13}C nucleus.) What problem arises if an excessive concentration of paramagnetic species is added?

8. Describe other types of relaxation mechanisms. In what situations is each effective in promoting relaxation?

Acknowledgments

The authors thank Drs. Richard Briggs and Wallace S. Brey for their helpful comments on the text.

Literature Cited

1. King, R.W.; Williams, K.R. *J. Chem. Educ.* a) **1989**, *66*, A213, b) **1989**, *66*, A243, c) **1990**, *67*, A93, d) **1990**, *67*, A100, e) **1990**, *67*, A125.
2. Derome, A.E. *Modern NMR Techniques for Chemistry Research*; Pergamon Press: New York, 1987.
3. Atkins, P.W. *Physical Chemistry*, 4th Ed.; W.H. Freeman: New York, 1990, pp 536-557.
4. Harris, R.K. *Nuclear Magnetic Resonance Spectroscopy*; Longman: Harlow, Essex, U.K., 1986.
5. Farrar, T.; Becker, E.D. *Pulse and Fourier Transform NMR*; Academic Press: New York, 1971.
6. Bloembergen, E.; Purcell, E.M.; Pound, R.V. *Phys. Rev.* **1948**, *73*, 679.
7. Charlesby, A.; Folland, R. *Radiat. Phys. Chem.* **1980**, *15*, 393.

Hardware List

To conduct this experiment a pulse FT-NMR with inversion-recovery software is needed. At the present time instruments equipped with 7.01 T (300 MHz for ^1H) or 4.7 T (200 MHz) magnets are most common, but a 90 or 100 MHz instrument will provide adequate results, if the required software is available.

RECEIVED October 14, 1992

Polymer Experiments

CHAPTER 22

Polymers in the Physical Chemistry Laboratory: An Integrated Experimental Program

Donald A. Tarr, George L. Hardgrove, and Gary L. Miessler

POTENTIAL HAZARDS: High-Pressure Systems, Vacuums

Although a large fraction of chemists work with polymers in one way or another, most college chemistry programs do little to educate students in the fundamentals of polymer chemistry. After thinking about this deficit for some time, we have placed several polymer experiments in our physical chemistry laboratory to help draw our students' attention to the field and to give them some experience with polymers *(1)*.

There are many different options for such experiments *(2-5)*. We have chosen to study the kinetics of polymerization of styrene, prepare a copolymer of styrene and methylmethacrylate by a similar polymerization, measure the proportions of two monomers incorporated into the copolymer by NMR and FTIR, find the molecular weight distribution of the polystyrene product using gel permeation chromatography (GPC), and examine the melting and freezing behavior of several different polymers by differential scanning calorimetry (DSC). For most students, this provides their first experience using the NMR and FTIR instruments on their own and their only experience with GPC and DSC.

Since the polymers formed in the kinetics experiment are used in all the others, that experiment must be early in the lab sequence. The others are scheduled on the rotation basis used in many physical chemistry laboratories, depending on the size of the class and the apparatus available. The first experiment requires three separate laboratory times, first a short time to remove inhibitor from the monomers and dry them, second to distill them, and finally the kinetics experiment itself. The others are single-period experiments.

The experiments are described briefly below.

Kinetics of polymerization

Theory This free radical polymerization begins with formation of two free radicals from the initiator. These radicals react with the double bond of a monomer molecule, forming a larger radical. This process is repeated in propagation steps, terminated by combination of two radicals or transfer of a hydrogen atom to form two stable molecules. In the equations below, $Init_2$ is the initiator and M is the monomer:

Initiation

$$\mathrm{Init_2} \xrightarrow{k_i} 2\ \mathrm{Init}\bullet$$

$$\mathrm{Init}\bullet + \mathrm{M} \xrightarrow{k_p} \mathrm{InitM}\bullet$$

Propagation

$$\mathrm{InitM_n}\bullet + \mathrm{M} \xrightarrow{k_p} \mathrm{InitM_{n+1}}\bullet$$

Termination

$$\mathrm{InitM_m}\bullet + \mathrm{InitM_n}\bullet \xrightarrow{k_{tc}} \mathrm{InitM_{m+n}}$$

$$\xrightarrow{k_{td}} \mathrm{InitM_n} + \mathrm{InitM_m}$$

The overall termination rate constant is $k_t = k_{tc} + k_{td}$, the sum of the rate constants for combination and disproportionation.

Using a steady state approximation on the concentration of all radicals [R·], we can derive the final equation which shows a first order dependence on monomer and 1/2 order dependence on the initiator *(6)*:

$$[\mathrm{R}\bullet] = \left(\frac{k_i}{k_t}\right)^{1/2} [\mathrm{Init_2}]^{1/2}$$

$$\text{rate of propagation} = k_p[\mathrm{R}\bullet][\mathrm{M}]$$

$$= k_p\left(\frac{k_i}{k_t}\right)^{1/2} [\mathrm{Init_2}]^{1/2}[\mathrm{M}]$$

$$= k_{3/2}[\mathrm{Init_2}]^{1/2}[\mathrm{M}]$$

$$= k_1[\mathrm{M}]$$

Progress of the polymerization reaction is followed by precipitating, filtering, and weighing polymer from samples of the reaction solution. After calculation of the concentration of monomer remaining in the solution, plots of [M], ln [M], and 1/[M] are made to determine the order of the reaction and its rate constant. The concentration of initiator is assumed to be constant in these equations. One of the questions is designed to test this assumption, and further analysis of the kinetics with a declining concentration of initiator can be done after the molecular weight is determined by GPC, allowing calculation of the number of chains formed.

Experimental The monomers as usually supplied contain quinone-type inhibitors to prevent polymerization in storage *(7)*. These are first removed by washing with aqueous sodium hydroxide, which in turn is removed by washing with water. The monomer is then dried over calcium chloride and vacuum distilled before the polymerization reaction.

Wash 150 mL of monomer with 75 mL of 10% NaOH in a separatory funnel, repeated four times. In the same way, wash the monomer five times with 75 mL of water, and dry it over $CaCl_2$ at least overnight. Distill the monomer under reduced pressure (40–50 mm Hg is sufficient) with a nitrogen capillary bleed to avoid bumping and reduce oxidation. At this pressure, styrene boils at 50–55 °C and methylmethacrylate boils near 25 °C. The condenser water and receiver must be cooled with ice for the latter. Store the purified monomers in the dark until they are used, and avoid rubber stoppers, part of which may dissolve in the liquid.

Equilibrate 50 mL of styrene and a solution of 0.10 to 0.75 g 2,2′-azobis(2-methylpropionitrile) initiator in 50 mL of toluene in a water bath thermostated at 75–80 °C. The solutions may be deoxygenated by bubbling a stream of dry nitrogen through them, although the effect of oxygen in the presence of relatively large amounts of initiator is small. When they have had time to reach the reaction temperature, mix the two solutions in a flask equipped with a reflux condenser and remove an initial 10 mL sample. Remove subsequent 10 mL samples every 30 minutes for 3 to 3.5 hours and quench them in 200 mL methanol. This precipitates all polymers above MW about 300–400. Collect the precipitate in fritted glass filters, allow it to air dry, and weigh it to determine the amount of polymer formed. The later samples tend to yield rather sticky precipitates. This can be avoided by stirring rapidly while running the sample slowly into the methanol and by stirring and grinding the sticky mass against the wall of the beaker until it is broken up and more easily filtered.

To prepare the copolymer, a mixture of styrene and methylmethacrylate is polymerized and the polymer isolated. Equilibrate 10 mL of a specified mixture of styrene and methylmethacrylate (concentrations from 90% styrene to 70% styrene) and a solution of 0.020 g 2,2′-azobis(2-methylpropionitrile) initiator in 10 mL of toluene in a water bath thermostated at 75–80 °C. Larger concentrations of methylmethacrylate lead to sticky products that are hard to handle. When they have had time to reach the reaction temperature, mix the two solutions in a flask equipped with a reflux condenser. Allow the reaction to proceed for 2–3 h. Remove 10 mL of the reaction mixture and add it to 200 mL methanol with rapid stirring. Collect the resulting precipitate by suction filtration in a fritted glass filter, and allow it to air dry.

Safety As in all laboratory work, safety glasses or goggles should be worn. The main safety concern is with the chemicals used. All are moderately toxic. A number are skin and eye irritants. Sodium hydroxide (NaOH) is caustic. All the operations in which monomers or the reaction solution are handled should be done in the hood to reduce exposure to the volatile organic compounds. The remaining monomer, reaction solution, and methanol filtrate should be collected and disposed of according to regulations.

Data Analysis Calculate the amount of monomer remaining in solution at each of the times, and plot the data according to zero, first, and second order kinetics ([M], ln [M], and 1/[M] vs. time).

Assuming a molecular weight of 15,000 and that one initiator molecule initiates two polymer chains, calculate the concentration of initiator remaining at the end of your experiment. Is this enough change to have a significant effect on the rate of the reaction? If it does, explain the effect and how it would change the analysis of the data. In the gel permeation experiment, the molecular weights of your samples will be determined, which will allow a check on this factor without assuming a molecular weight. Try to predict whether the average molecular weight will increase, decrease, or stay essentially constant as the reaction proceeds. Remember that this is a free radical reaction and that free radical reactions in solution are likely to be very fast. Why are the purified monomers stored in the dark?

Polymerization of pure styrene results in first order plots that fit the data well and zero and second order plots that show distinct curvature at longer times. A similar analysis of data from the polymerization of mixtures results in an apparent order between 1 and 2 because of the different rate constants for the two monomers.

Typically, the kinetic data show curved lines for zero and second order plots, and a linear plot for first order. Unless the degree of polymerization is large enough to change the concentrations significantly, the choice between the orders is difficult. Increasing the initator concentration gives larger amounts of polymer and also reduces the molecular weight.

Gel permeation chromatography

Theory Gel permeation chromatography separates molecules according to size. Several different kinds of column packing can be used, all containing porous material that allows diffusion of smaller molecules into the solid packing. As a result, the small molecules are slowed in their progress and emerge from the column at later times. Many columns contain highly cross-linked polystyrene, which becomes a gel in the presence of the solvents used. A refractive index detector is most generally useful, although other detectors may be used for polymers with appropriate properties.

Separation in time depends on ln MW; a calibration curve can be prepared by using a solution of several commercial polystyrene molecular weight standards. Ideally, a plot of ln MW versus time (t) is linear, but a better fit is usually provided by an equation of the form $\ln MW = a + bt + ct^2$, where a, b, and c are parameters found by using nonlinear least squares. For comparison of polymers with different structures, ln (MW[η]), where [η] is the intrinsic viscosity, is a better function to use. The difficulty is avoided if the polymer used to calibrate the column is the same as that being analyzed; for this reason, use of polystyrene gives more accurate results than use of the copolymer would.

Average molecular weights of polymers are usually reported as number average or mass (or weight) average. The number average, M_n, is found in the usual way of taking averages. It is the total mass of the polymer sample, $\Sigma M_i n_i$ (where M_i is the molecular weight of species i, and n_i is the number of moles of species i) divided by the total number of moles, Σn_i. The mass average, M_w, is the sum of the products of the square of the molecular weight times the amount of material with that molecular weight divided by the sum of the product of the molecular times the amount of material with that weight.

$$M_n = \frac{\sum M_i n_i}{\sum n_i} \qquad M_w = \frac{\sum M_i^2 n_i}{\sum M_i n_i}$$

The number average weights each molecule equally, while the mass average weights the larger molecules according to their molecular weights. The difference in refractive index between the solvent and the solution when polymer is being eluted from the column is proportional to the mass, $M_i\, n_i$, of polymer in the sample. The molecular weight, M_i, can be obtained from the elution time using the calibration curve for the column. These data make possible the calculation of the two molecular weight averages. The molecular weights depend on the amount of initiator used:

$$\text{kinetic chain length} = \frac{\text{rate of propagation}}{\text{rate of initiation}}$$

$$\nu = \frac{k_p[\text{M}]}{2(k_i k_t)^{1/2}[\text{Init}_2]^{1/2}}$$

In the absence of chain transfer between growing polymer units and solvent or monomer and if the termination is by disproportionation, the molecular weight is νM_0, where M_0 is the molecular weight of the monomer. If termination is by combination of growing polymer units, the molecular weight is $2\nu M_0$.

Experimental Prepare solutions containing 8–12 mg of polymer in 10 mL of tetrahydrofuran, filter them using a membrane filter, and inject 0.5 mL of the solution into the gel permeation column, using tetrahydrofuran solvent at a 1.5 mL/min flow rate. After the sample has run through the instrument completely (about 20 min), inject a solution of four commercial polystyrene molecular weight standards (molecular weights from a few thousand to about 100,000 are appropriate) under the same conditions to determine the t versus ln MW curve. The refractive index is recorded, either on a strip-chart recorder or by computer. If time allows, samples from early and late in the kinetics run are used to find the time dependence of the molecular weights.

Safety Safety glasses or goggles should be worn during all laboratory operations. All of the chemicals used are moderately toxic, and a number are skin and eye irritants. Preparation of solutions should be carried out in a hood to minimize contact with vapors of tetrahydrofuran. Used solvents may be distilled and reused, or should be disposed of according to regulations.

Data Analysis If the data were recorded by a chart recorder, mark the time scale at 1 or 2 s intervals and measure the height of the curve at each interval. Using a plot of t versus ln MW, prepared from the standardization run, convert each time to a molecular weight. Since the height of the refractive index above the baseline is proportional to the mass of polymer in the solution at that time, it is proportional to $M_i\, n_i$. Divide the height by the molecular weight to obtain a value proportional to n_i, the number of molecules. Multiply the height by the molecular weight to find $M_i^2 n_i$. Sum these three columns and calculate M_n, M_w, and the ratio of M_w/M_n, using the equations in the Theory section. This ratio is a measure of the polydisperse nature of the polymer. If all the molecules were of the same size, the ratio would be 1. Explain why the ratio cannot be less than 1. Compare the molecular weight averages obtained from samples from early in the polymerization reaction to those from late in the reaction, and explain any difference found. Does this result agree with your prediction in the polymerization experiment? If not, rethink your prediction and explain what the error was.

FTIR Analysis of the Polymer *(8)*

Theory Although a strict treatment requires that the area under spectral peaks be used for quantitative use, if the total absorbance is kept small the height of the peaks can be used for this purpose. By choosing peaks characteristic of the phenyl group of styrene (698 cm^{-1}) and the carbonyl group of methylmethacrylate (1732 cm^{-1}), the relative amounts of the two monomers in the polymer can be found. Fourier transform infrared (FTIR) software also helps in this analysis by providing easy conversion of the data from the more common transmittance format to absorbance, which is proportional to concentration by the usual Beer–Lambert law:

$$I = I_0 e^{-\varepsilon lc}, \quad A = \log\left(\frac{I_0}{I}\right) = \varepsilon lc$$

where ε = molar absorptivity, characteristic of a specific compound and radiant energy
l = path length
c = concentration

For convenience, the spectra are measured as thin films deposited on salt plates. Because the peaks are likely to have different molar absorptivities, the path length is difficult to control in work with films, and the concentration is difficult to calculate in a solid, a different standard must be used. An equimolar mixture of polymethylmethacrylate and polystyrene is also measured as a thin film. The heights of the peaks of the equimolar mixture [A_{1732}(std) and A_{698}(std)] and of the experimental polymer [A_{1732}(exp) and A_{698}(exp)] can then be used to calculate the ratio of the mole fraction:

$$\frac{x_{MM}}{x_S} = \frac{\dfrac{A_{1732}(\mathrm{exp})}{A_{698}(\mathrm{exp})}}{\dfrac{A_{1732}(\mathrm{std})}{A_{698}(\mathrm{std})}}$$

Experimental Dissolve 0.03 to 0.05 g of polymer in 2 mL $CHCl_3$, and similar solutions containing polystyrene, polymethylmethacrylate, and equimolar amounts of polystyrene and polymethylmethacrylate as standards. Place two drops of solution on a sodium chloride disk, spread the solution evenly over the surface as it evaporates, and then record the spectrum of the film in absorbance units (where the concentration is proportional to peak intensity). Absorbances near 0.1 provide the most accurate results.

Test the reproducibility of the results several times, rotating the salt plate between the measurements. If any discrepancy is found, the results of all the measurements should be averaged for the final value.

Repeat the measurement on two or more samples, from different times during the polymerization reaction.

Safety Safety glasses or goggles should be worn during all laboratory operations. Preparation of solutions should be carried out in a hood to minimize contact with vapors of chloroform. All of the chemicals used are moderately toxic, and a number arre skin and eye irritants. Used solvents may be distilled and reused, or should be disposed of according to regulations.

Data analysis Using the equation above, calculate the mole fraction of styrene in your polymer samples. Using the repeat measurements of the same sample, estimate the precision of your measurements. Compare the results for polymers from different times in the polymerization and explain any change in the relative amounts of styrene and methylmethacrylate as the reaction proceeds. Figure 1 shows spectra of polystyrene, polymethylmethacrylate, a copolymer, and the mixture of the pure homopolymers.

NMR Analysis of the Polymer

Theory Nuclear magnetic resonance allows the measurement of the number of protons with specific electronic environments. In this case, the details of the spectrum are not needed, but the relative number of phenyl protons and nonphenyl protons is used. Polystyrene contains 5 phenyl protons and 3 nonphenyl protons; polymethylmethacrylate contains 8 nonphenyl protons. The area of the phenyl peak is proportional to $5x_S$ and the area of the nonphenyl peak is $3x_S + 8(1 - x_S)$, where x_S is the mole fraction of styrene in the polymer. The ratio of the areas of the phenyl to the nonphenyl peaks is then

$$R = \frac{5x_S}{3x_S + 8 - 8x_S}$$

and the mole fraction of styrene is

$$x_S = \frac{8R}{5 + 5R}$$

Experimental Dissolve 0.03 g of polymer in 0.7 mL $CDCl_3$ (for deuterium lock) or CCl_4. Obtain the proton NMR spectrum and integrate the phenyl region (6–8 ppm) and the nonphenyl region (0–4 ppm), as shown in Figure 2. Repeat the measurement with a second solution to check on reproducibility, then repeat with polymer from different times in the polymerization reaction.

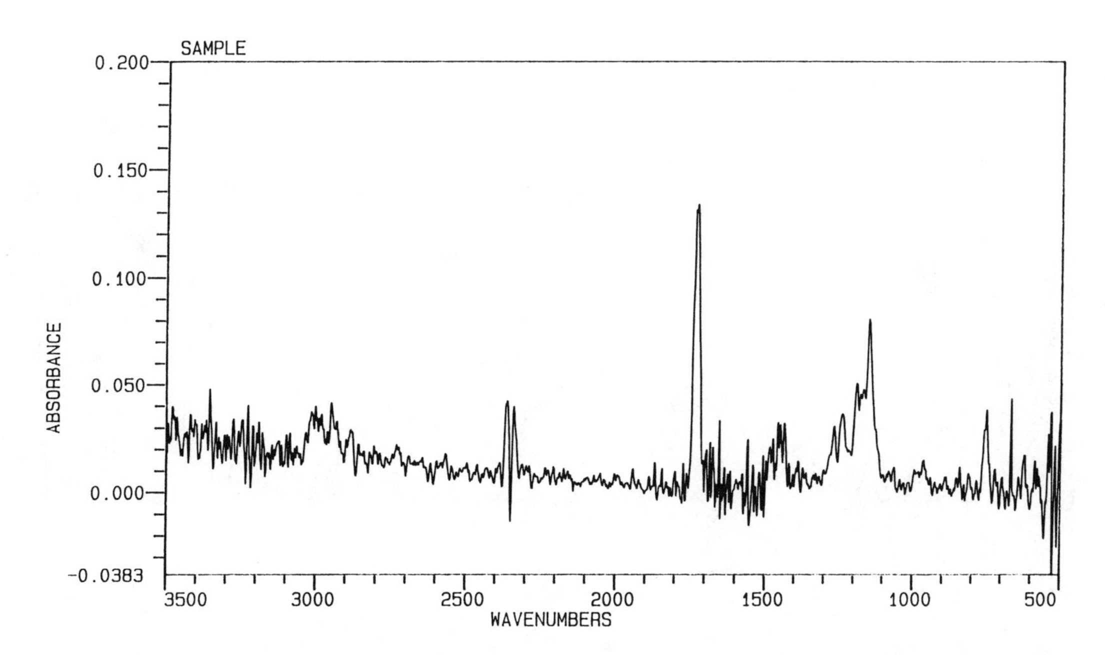

Figure 1. a. Infrared spectrum of polymethylmethacrylate (absorbance scale)

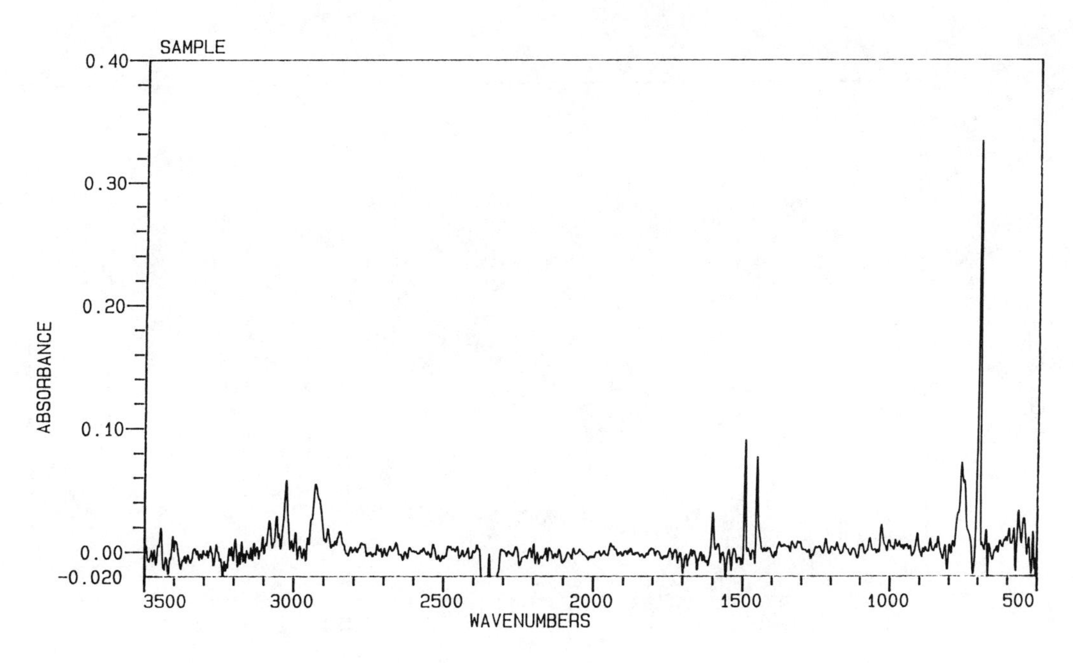

Figure 1. b. Infrared spectrum of polystyrene (absorbance scale).

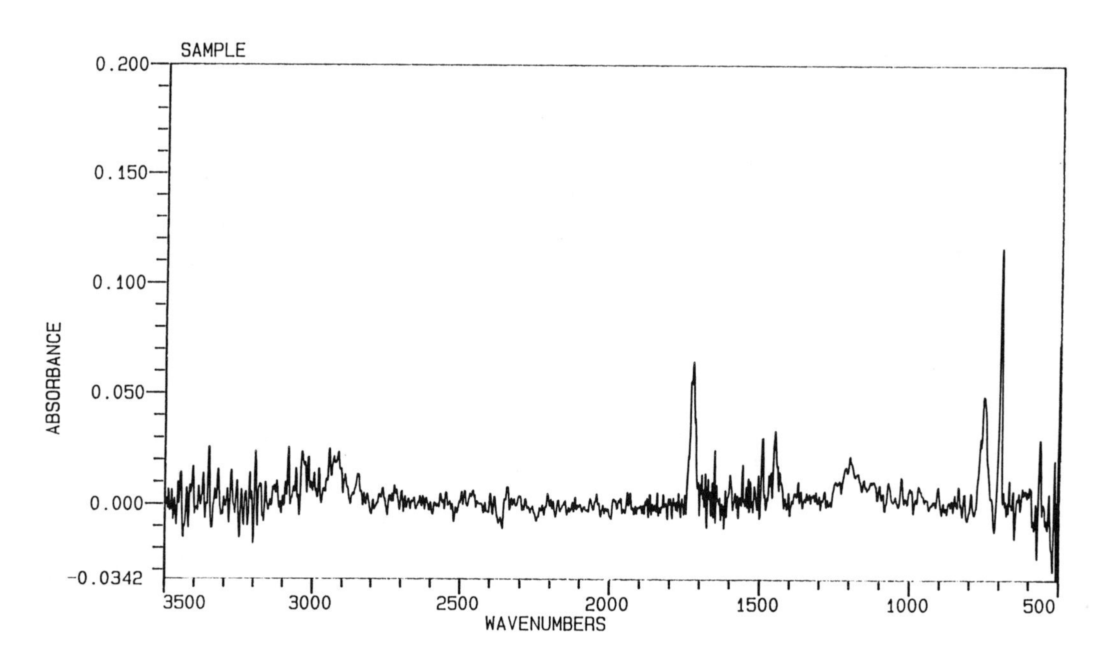

Figure 1. c. Infrared spectrum of a copolymer sample (absorbance scale).

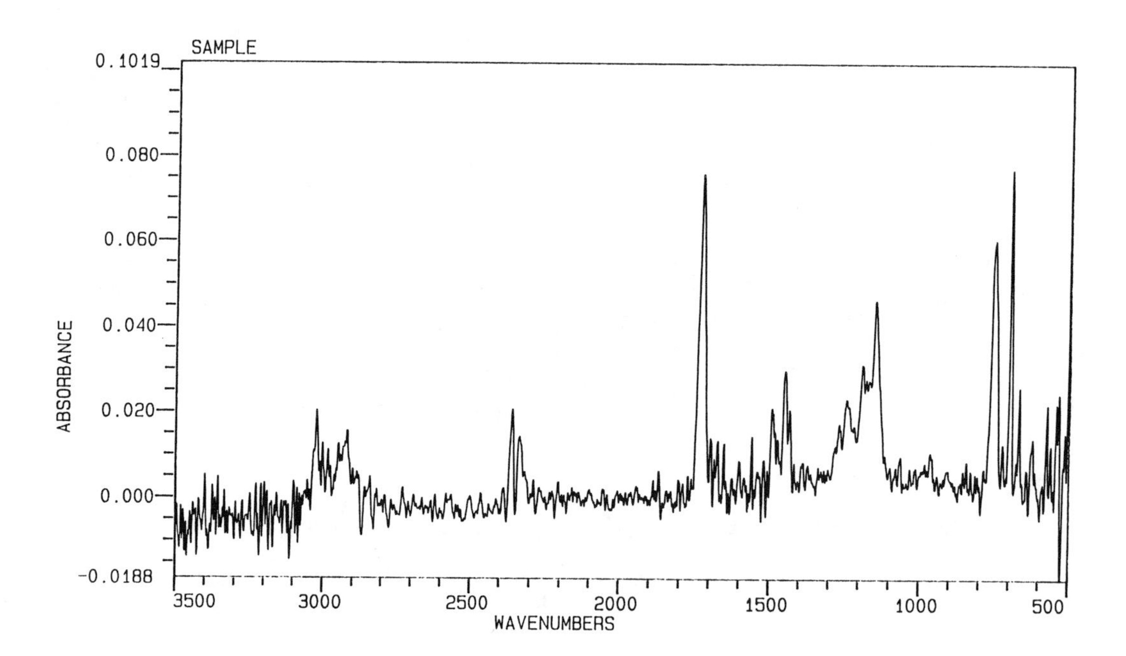

Figure 1. d. Infrared spectrum of a 1:1 mixture of polystyrene and polymethylmathacrylate (absorbance scale).

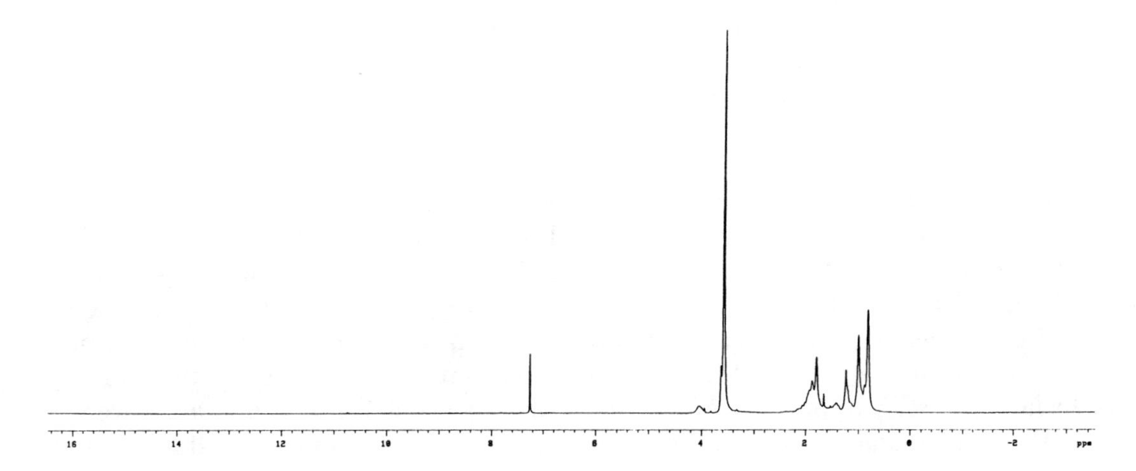

Figure 2. a. Proton NMR spectra of polymethylmethacrylate.

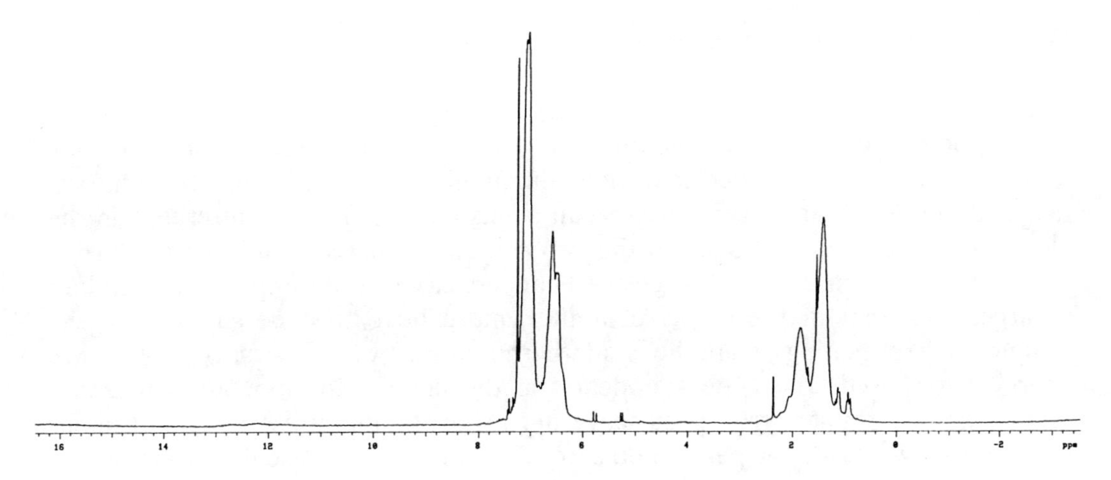

Figure 2. b. Proton NMR spectra of polystyrene.

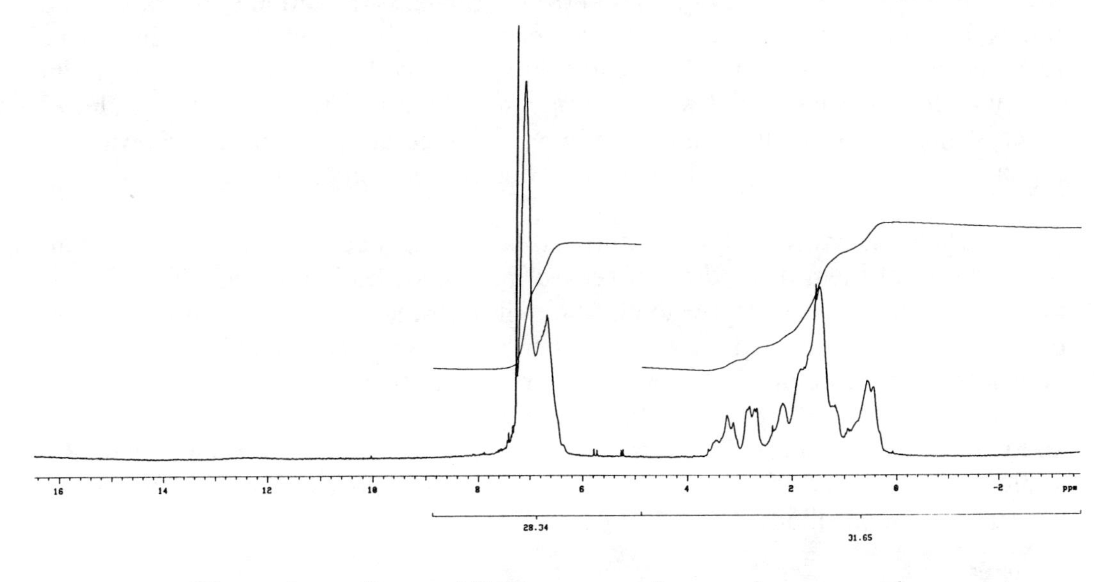

Figure 2. c. Proton NMR spectra of a copolymer sample.

Safety Safety glasses or goggles should be worn during solution preparation, and the work should be done in a hood. All of the chemicals used are moderately toxic, and a number are skin and eye irritants. Solutions should be disposed of according to regulations.

Data Analysis Using the spectra of styrene, polystyrene, methylmethacrylate, and polymethylmethacrylate provided and tables of chemical shifts, assign the major peaks to specific protons. You may not be able to assign all peaks precisely because of complicated splitting patterns, but you should be able to do most of them. Calculate the mole fractions of styrene and methylmethacrylate in the polymer samples, using the areas of the phenyl and nonphenyl peaks in the equation above. Include an estimate of the reproducibility of your results. If you find a difference in the relative amounts of the two monomers in samples from early and late in the polymerization, explain the reason for the difference. Compare the results of the NMR and IR analyses of the proportions of styrene and methylmethacrylate in the copolymer.

Differential Scanning Calorimetry

Theory Differential scanning calorimetry of polymers allows measurement of melting points, enthalpies of melting, and glass transition temperatures. Small samples (< 10 mg) are sealed in a small aluminum container, heated at a uniform rate and the temperature compared to an empty container. The difference in the electrical energy required to keep the two containers at the same temperature is plotted against temperature. Since the heat capacity of the sample and container is larger than that of the empty container, more heat must be supplied to the sample. When melting or another endothermic transition takes place, even more energy is required, and a peak appears on the curve. Compounds with sharp melting points exhibit narrow peaks covering a short temperature span; amorphous polymers show broad, low peaks, and show no return to the baseline. In this case, the transition is a more gradual conversion of the solid polymer through softening to a liquid, much as in the melting of glass. For this reason, the temperature at which the break in the curve occurs is known as the glass transition temperature, T_g, rather than the melting point, T_m. Polymers with a high degree of crystallinity have relatively narrow peaks in the DSC curve, approaching the shape of a melting point transition for a low molecular weight compound. Amorphous polymers have almost no crystallinity. Their thermal curves show a change in heat capacity, shown by a region with a slope similar to the lower temperature heating curve but at a different height.

The glass transition temperature has been described in a number of different ways, but all of them depend on increased molecular motion in the solid. Elastic materials such as rubber are above their glass transition temperature at room temperature. When cooled below T_g, rubber becomes rigid and brittle. Many other polymers are rigid below T_g, but have little or no elasticity above. Because there is increased motion in the chains of the softened polymer, the heat capacity is higher and gradually increasing. This causes a change in the baseline for such curves. An additional deviation due to enthalpy of melting may be superimposed if the polymer has sufficient crystallinity.

In cases where a reasonable baseline can be established, the area under the curve is proportional to the enthalpy change of the process. Many instruments have software that calculates the enthalpy change per gram automatically if the sample weight is provided. As in other methods of determination of melting points, the measurements are sensitive to the rate of heating. If heated slowly, the sample is more likely to melt over a small temperature range, and the glass transition temperature will be more distinct. Even when the polymer has a high degree of crystallinity, the transition will not be as sharp as that of an ordinary melting point. The polydispersity of the polymer, differences in the direction of bonding of monomer units such as styrene or methylmethacrylate, and the kinetic problems of arranging long polymer chains in crystalline patterns in a high-viscosity melt prevent perfectly crystalline materials.

Experimental Weigh 3–5 mg samples of benzoic acid, low density polyethylene, high density polyethylene, polystyrene from the kinetics experiment, and the copolymer synthesized at the same time, and seal them in the aluminum containers, using the instructions provided with the instrument. Handle the sample containers with tweezers only, not with the fingers. With an empty sample container in the reference position, record the thermal curves for heating and cooling over 110–130 °C for benzoic acid, and from 50–200 °C for the polymer samples, with heating rates of 10°/min or less. When possible, set the baseline and calculate (or have the computer calculate) the area under the curve for the enthalpy change. For glass transitions, have the computer plot the derivative curve to help in determining T_g. For the polymer samples prepared in experiment 1, more consistent results may be obtained by heating the sample rapidly to 150 °C and then returning to 50 °C to start the run. Any trapped monomer or solvent will be released, and mechanical stresses will be relieved *(9)*.

Data Analysis Compare the literature values for melting point and enthalpy of fusion of benzoic acid with your experimental results. Report the transitions observed for the polymer samples, and comment on the difference between the heating and cooling curves. If it is possible to determine an enthalpy of fusion for the polyethylene samples, use ΔH_f = 140 J/g as the enthalpy of fusion for pure crystalline polyethylene and calculate the fraction of crystalline polymer in the samples.

Figure 3 shows the melting and crystallization behavior of polyethylene; Figure 4 shows a glassy transition for a noncrystalline polymer sample.

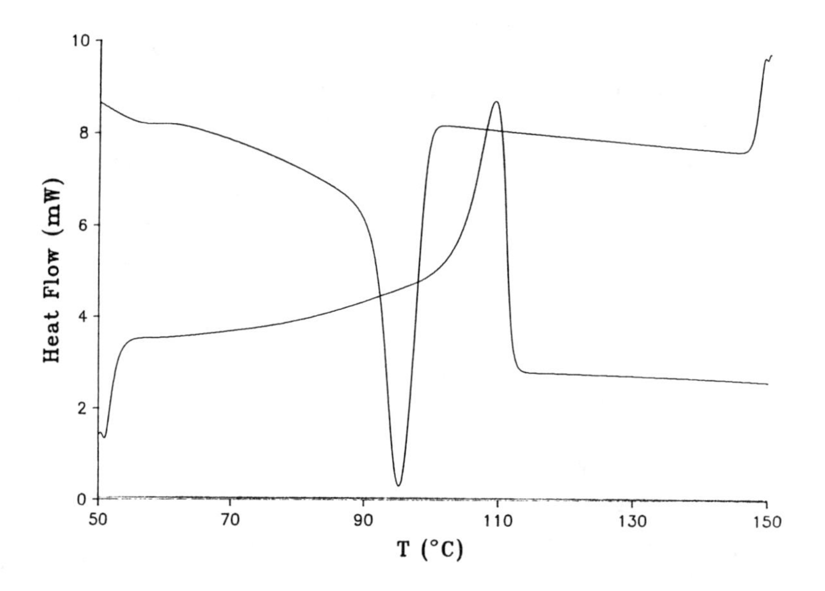

Figure 3. DSC curve of low-density polyethylene. Peak up, heating; peak down, cooling.

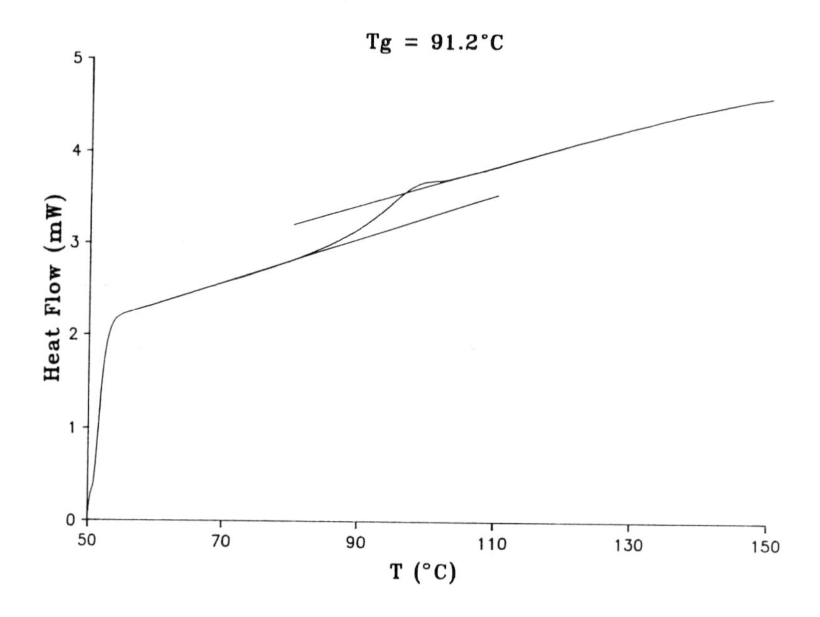

Figure 4. DSC curve of a copolymer sample. The energy absorption below 50 °C is an artifact.

Acknowledgment

We acknowledge with thanks NSF Grant CSI-8750342 for partial support for equipment purchases for this program. We also thank Wilmer Miller for many helpful discussions and for providing the laboratory manual for the polymer chemistry course at the University of Minnesota. The comments of the reviewers were very valuable in clarifying several sections of this paper.

Literature cited

1. Hardgrove, G. L.; Tarr, D. A.; Miessler, G. L. *J. Chem. Educ.*, **1990**, *67*,
2. Collins, E. A.; Bares, J; Billmeyer, F. W., Jr. *Experiments in Polymer Science*; Wiley, New York, 1973.
3. McCafferey, E. M. *Laboratory Preparation for Macromolecular Chemistry*; McGraw-Hill, New York, 1970.
4. Rabek, J. F. *Experimental Methods in Polymer Chemistry Principles and Applications*; Wiley, Chichester, 1980.
5. Mathias, L. J., *J. Chem. Educ.*, **1983**, *60*, 990-993.
6. Ferington, T. E., *J. Chem. Educ.*, **1959**, *36*, 174.
7. Collins, E. A.; Bares, J; Billmeyer, F. W., Jr. *Experiments in Polymer Science*; Wiley, New York, 1973, p. 40.
8. K. I. Ekpenyong; R. O. Okonkwo, *J. Chem. Educ.*, **1983**, *60*, 429.
9. Collins, E. A.; Bares, J; Billmeyer, F. W., Jr. *Experiments in Polymer Science*; Wiley, New York, 1973, p. 440.

Equipment list:

- Kinetics of polymerization
  - Glassware for extraction, vacuum distillation, kinetic run, and filtration of samples.
- Gel permeation chromatography
  - Waters M-6000 and M-45 pumps, Jordi-Gel 1000 Å gel permeation columns, and Spectra Physics SP8430 refractive index detector
  - Chart recorder or interfaced computer for recording data (we use a DataTranslation 8508 board for data input and A/D conversion).
  - Software for data collection and calculation (optional)
- FTIR analysis of the polymer
  - Nicolet PC/IR software controlling an IBM IR32 FTIR optical bench
- NMR analysis of the polymer
  - Varian VXR-300S NMR spectrometer
- Differential scanning calorimetry
  - Perkin-Elmer DSC-7 with PC software and control

RECEIVED October 14, 1992

CHAPTER 23

Determination of Polymer Molecular Weight in an Aqueous-Based Solvent System by Gel Permeation Chromatography

Susan Mathison, Dennis Huang, Arvind Rajpal, and Robert G. Kooser

POTENTIAL HAZARDS: High-Pressure Systems

Polymer chemistry is one of the largest branches of chemistry in terms of the number of individuals involved in it. Yet, in spite of years of tireless promotion by the Polymer Chemistry Division of the American Chemical Society, most undergraduate programs are still bereft of any polymer science. Recently, Hardgrove et.al. *(1)* have published a comprehensive polymer experience for the physical chemistry laboratory that includes styrene methylmethacrylate copolymer synthesis and characterization by NMR, infrared, gel permeation, and differential scanning calorimetry. One of the elegant features of the Hardgrove article is the use of gel permeation chromatography (GPC) to determine the number and mass average molecular weights of a polymer, both very important properties. The polystyrene system, while very common, must have an organic based solvent for GPC such as toluene or tetrahydrofuran *(1)* which has the disadvantage of the attendant waste disposal of large volumes of polymer solutions. Also, the system, as with almost all organic solvent based polymer systems, requires a refractive index detector for GPC that is not common in many laboratories.

Here we give a procedure for the determination of the polymer molecular weight in an aqueous based solvent system using the soluble polymer poly(sodium 4-styrene sulfonate). The polymer has commercial importance as a water soluble flocculating agent *(2)* and has the advantages that narrow molecular weight standards are readily available and that the phenyl ring chromophore allows for conventional ultraviolet absorbance detection.

Theory

Gel permeation or size exclusion chromatography is a type of liquid chromatography that uses a stationary phase consisting of porous particles whose pore dimensions approximate the size of polymer molecules *(3)*. Separation proceeds by size with the larger molecules being unable to penetrate the pores unlike the smaller sizes which permeate readily. The result is that molecules elute from the chromatographic column with the largest first and succeeding smaller sizes in order thereafter. The technique has been widely used in polymer science to estimate molecular weight distributions *(4,5)*; here the reasonable assumption is made that molecular size is proportional to molecular weight.

The problem of molecular weight determination in synthetic polymer systems is that the synthetic process itself almost always results in a distribution of polymer chain lengths; hence, a distribution of molecular weights. This results in what is

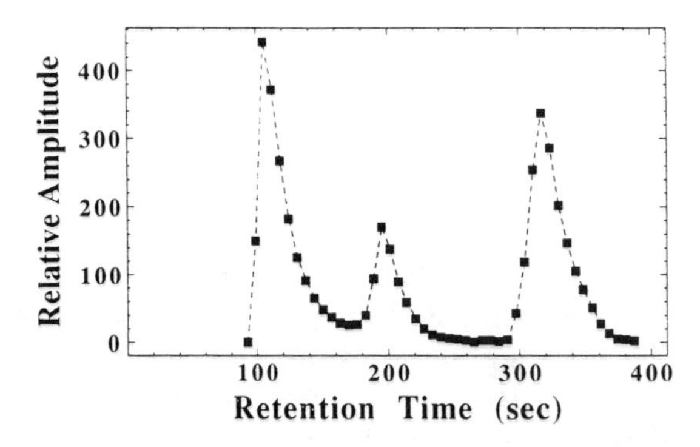

Figure 1. Discrete sample points in the GPC chromatogram of an unknown poly (sodium styrene sulfonate) sample in 0.1 M Na_2SO_4 aqueous solution.

called a *polydisperse* sample. Polymers whose molecular weights fall into a very narrow range of values are called *monodisperse*. When a polydisperse sample is subject to GPC what is observed is a broad band from a continuous distribution of molecular weights such as seen in Figure 1. It will be noted that the abscissa in liquid chromatography is usually in units of time and that there are no theoretical means to convert time into molecular weight *(4)*. Consequently, the gel permeation chromatograph must be calibrated using a series of samples of the known molecular weight in the same solvent system as the unknown sample or, failing that, fairly complex universal calibration procedures must be applied *(5)*. Once a series of monodisperse knowns are run, a calibration curve of the logarithm of the molecular weight versus retention time of the knowns can be constructed.

Since most polymer samples are polydisperse, it is not possible to characterize them with a single molecular weight; instead an average value must be quoted. Two common averages are used to characterize polymer molecular weights, the *number* and *mass (weight) average molecular weights (7)*. The number average molecular weight is defined as

$$\langle M_n \rangle = \Sigma N_i M_i / \Sigma N_i \quad (1)$$

where N_i is the number of polymer molecules having the precise molecular weight of M_i. This will be recognized as a simple number weighted average since $N_i / \Sigma N_i$ is just the fraction of molecules at the ith molecular weight. The mass average molecular weight is the mass weighted average defined as

$$\langle M_w \rangle = \Sigma (N_i M_i) M_i / \Sigma N_i M_i \quad (2)$$

The mass average value will be larger than the number average because higher molecular weights will contribute more to the average value. For a more extensive discussion of these concepts see Atkins *(8)*.

The transformation of the chromatographic data of detector response versus retention time requires a data analysis which demonstrates important principles of physical chemistry. First, equations (1) and (2) are based on a discrete sum, while the experimental data is in the form of a continuous distribution of molecular weights. To accommodate the calculation of averages, the chromatogram is marked off in equal intervals on the time axis as shown in Figure 1. At each interval the detector response and retention time is determined. From the calibration curve of log (Molecular Weight) versus retention time, each time can be assigned a molecular weight. For each retention time the detector response is proportional to the concentration of the polymer at that time in grams per liter. If h_i is the detector response at retention time t_i then

$$h_i \propto g_i / L = (N_i / N_A) M_i / V \quad (3)$$

where N_A is Avagodro's number and V is the solution volume. From Equation (3) it can be seen that the detector response is proportional to the product of N_iM_i and using the retention time, we can get the M_i at each point; thus the data of retention time versus detector response is proportional to M_i versus N_iM_i. An inspection of Eqs. (1) and (2) shows that the quantities that are needed for the calculation of the average molecular weights are N_i and M_i. To get a value proportional to N_i at each point, all that is needed is to divide the retention time ($\propto M_i$) into the detector response ($\propto N_iM_i$). The summations required in Equations (1) and (2) are readily done from the data of molecular weight, detector response, and detector response divided by molecular weight. The nature of the data and its transformations are ideally suited to the use of a computer spreadsheet.

Methodology

The equipment required for this experiment is an ordinary high performance liquid chromatograph (HPLC) system consisting of a pump, an injection valve, a GPC column, and an ultraviolet detector with an output device. The particular system used here consisted of a Beckman 110B pump connected to a Rheodyne 7010 injection valve with a 50 µL sample loop. The GPC column was an Alltech GPC 100 (catalog number 88133) and an ISCO UA-5 detector with an ISCO Type 6 optical unit operating at 254 nm was used. Data can be taken in analog form from the chart recorder in the ISCO unit, but in this case the analog output of the detector was fed into a Metrabyte Chrom-AT analog-to-digital board under the control of the software program LabCalc running on a 286 computer.

The molecular weight poly(sodium 4-styrene sulfonate) standards came in a kit of 10 knowns and were made by Polymer Laboratories and obtained from Alltech (catalog number 28303). The polymer standards were made up in 0.15% w/v solution per manufacturer's directions in 0.1 M Na_2SO_4 aqueous solutions in small quantities (2-5 mL). The unknown polymer was obtained from Aldrich (catalog number 24,305-1) and was dissolved in 0.1 M Na_2SO_4 aqueous solution at a concentration of 0.15% w/v. All polymer solutions were made up by gentle shaking or swirling to avoid mechanical degradation of the polymers *(9)*. Both the known and unknown solutions were doped with about 10 parts per thousand o-dichlorobenzene. This substance is not retained by the GPC column and gives the dead time for retention time measurements *(9)*. To remove particles that could clog the narrow bores of the chromatograph, the solvent was filtered through a nylon 0.45 µ filter (Alltech) before injection into the sample loop.

The solvent for all runs was 0.1 M Na_2SO_4 and all runs were done at a flow rate of 1.0 mL/minute as set on the pump (the samples must not be run at pump rates above 1 mL/min to avoid shear degradation of the polymers *(9)*). Each molecular weight standard was injected into the chromatograph a minimum of three times and the average retention time for each standard was determined. The unknown was injected and determined three times. The sampling rate of the software was set to give about 200 points through the detector response of the unknown. Alternatively, the data can be displayed on a chart recorder. If the data are obtained digitally, the unknown data file should be saved in a format that can be read by a spreadsheet in the form of retention time in one column and detector response in the other.

Safety

Since the pressures that exist in a typical HPLC experiment are high, eye protection must worn at all times. The UV light source should be shielded from view whenever it is on. Usual precautions in the handling of chemicals should be followed with the polymer solutions. *o*-dichlorobenzene is both volatile and toxic, appropriate safety precautions must be taken.

Analysis

The retention time of each standard molecular weight sample is gotten by measuring the time span from the dichlorobenzene peak to the peak maximum of the polymer sample *(10)* and a plot of the log of the molecular weight versus retention time is constructed. Ideally, a best fit to the data is done using a graphing computer program so as to get an empirical function that conforms to the data. In our case, the data are most easily fit by plotting the log of the molecular weight versus the reciprocal of the retention time and fitting the curve to a quadratic in inverse time (Figure 2). If the data work-up is by hand, a smooth curve should be drawn through the points so that interpolation can be done.

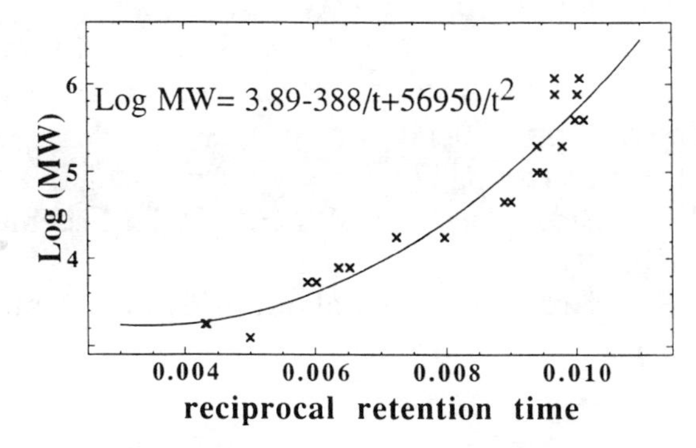

Figure 2. Calibration curve for the poly (sodium styrene sulfonate) standards in 0.1 M Na_2SO_4 aqueous solution. A better polynomial fit was achieved by plotting the inverse of the retention time rather than retention time. The quadratic fit is shown on the figure.

If the data are to be analyzed by hand, the chart tracing of the unknown should be marked off in equal intervals to generate 30 to 50 data points through the region where the unknown causes a detector response and for each point the retention time and detector response is recorded. Using the calibration curve, each retention time is assigned a molecular weight. A value proportional to the number of polymer molecules at each molecular weight (N_i) is determined by dividing the detector response by the molecular weight for each point. Also a column of data proportional to $N_iM_i^2$ is obtained by multiplying the detector response by the molecular weight assigned to each point. The columns of data representing N_i, N_iM_i, and $N_iM_i^2$ are summed and the number and mass average molecular weights are calculated as in Equations (1) and (2).

The preferable way to treat the data from this experiment is to acquire it digitally and import the unknown chromatograph into a spreadsheet program. The molecular weight can be assigned by the spreadsheet from the functional fit to the log (Molecular Weight) versus retention time data (see Figure 2). Data analysis follows as in the previous paragraph except that it is executed by spreadsheet commands.

Acknowledgments

Support of a National Science Foundation College Science Instrumentation Program grant number 8851423 is gratefully acknowledged, as is the Alltech Corporation for the donation of the GPC column.

Literature Cited

1. Hardgrove, G.L., Tarr, D.A., Miessler, G.L. *J. Chem. Educ.* **1990**, *67*, 979.
2. Billmeyer, F.W. *Textbook of Polymer Science*, 3rd Ed. ; Wiley-Interscience: New York, NY, 1984; p. 386.
3. Snyder, L.R.; Kirkland, J.J. *Introduction to Modern Liquid Chromatography*, 2nd Ed.; Wiley-Interscience: New York, NY, 1979; pp.483-540.
4. Billmeyer, F.W., *op.cit.* ; pp. 214-217.
5. Rudin, A., *The Elements of Polymer Science, An Introductory Text for Engineers and Chemists*; Academic Press: New York, NY, 1982; pp. 107-118.
6. Rudin, A., *op.cit.* ; pp. 110-116.
7. Billmeyer, F.W., *op.cit.*; p. 17.
8. Atkins, P.W. *Physical Chemistry*, 4th Ed.; W.H. Freeman: New York, NY, 1990; p. 683, p. 694.
9. Pl-Polymer Standards; Polymer Laboratories Inc.: Amherst, MA; pp. 3-5.
10. Rudin, A., *op.cit.* ; p. 111.

Hardware List

Any liquid chromatograph with ultraviolet detection can be used for this experiment by simple incorporation of the GPC column. The requirements for digitally acquiring the data by computer are not severe and several inexpensive solutions are possible including the use of the ubiquitous Apple IIe runnning data acquisition software and hardware from such vendors as Vernier Software (Portland, OR).

Unit	Manufacturer	Model
Liquid Chromatograph		
Pump	Beckman	110B
Injection valve	Rheodyne	7010
Column	Alltech	GPC 100 (88133)
UV detector	ISCO	UA-5 & #6 optical unit
Computer		
Computer	Austin	286
A/D board	Keithley-Metrabyte	Chrom-AT
Software	Galactic Enterprises	LabCalc
Filters		
	Alltech Nylon membrane	2047

RECEIVED October 14, 1992

CHAPTER 24

Thermodynamic Properties of Elastomers

Kathryn R. Williams

POTENTIAL HAZARDS: Cryogens

Rubber objects are common in daily life, and everyone is familiar with at least some of the physical properties of the class of polymers called elastomers:

1) They are capable of being stretched to several times their original length with relatively little applied force.

2) When the force is released they retract rapidly to the unstressed length (property of snap or rebound). The heat transfer on rebound is very close to zero.

3) They suffer no permanent deformation as a result of the extension process (property of resilience).

4) When they are fully elongated (or nearly so), they exhibit very high tensile strengths and stiffness (modulus).

The properties described above are all observable on the macroscopic level, which is the realm of classical thermodynamics. The classical treatment requires no knowledge or assumptions of molecular structure. However, in order to exhibit such behavior, the polymer must have certain molecular properties:

1) The polymer must have a large molecular weight, with, for the most part, very weak interactions between chains. For example, natural rubber, which is also called *Hevea* rubber after its source as the sap of the *Hevea brasiliensis* tree, has a molecular weight of about 350,000. Its chemical composition is poly(*cis*-isoprene) (Figure 1), which in the untreated material has only weak intermolecular forces.

2) The polymer must be amorphous (i.e., noncrystalline) and must be above its glass transition temperature, T_g. T_g is the temperature, or range of temperatures, over which the polymer exhibits a marked change in several physical properties, most notably specific volume, thermal coefficient of expansion, specific heat capacity, and refractive index. Below T_g there may be small local rotations (e.g., rotation around the C–C bond to a side-chain methyl group), but the polymer chains themselves are frozen into fixed positions (albeit not in a regular crystalline array), and the polymer is a hard,

2503–6/93/0352$06.00/0

brittle glass. Above the glass transition temperature the thermal energy is sufficient to allow rotations and limited translations of large segments of the polymer chain. On the macroscopic scale the polymer has the dimensional constancy of a solid, but on the molecular level the chain segments exhibit liquid-like properties.

```
a
        CH3      H     CH3      H
           \    /         \    /
            C=C            C=C
           /    \         /    \
   etc—CH2       CH2—CH2        CH2—etc

b
        CH3      H     CH3      H
           \    /         \    /
            C=C            C=C
           /    \         /    \
   etc—CH2       CH—CH2         CH2—etc
                 |
                 S
                 |
                 S
                 |
   etc—CH2       CH—CH2         CH2—etc
           \    /         \    /
            C=C            C=C
           /    \         /    \
        CH3      H     CH3      H
```

Figure 1. Structure of natural rubber a) before and b) after vulcanization to introduce disulfide cross-links.

A polymer with these properties can be envisaged as a disordered tangle of relatively compact random coils, as shown in Figure 2a. Because of their local mobility and lack of strong intermolecular attractions, the chains can be extended with essentially zero change in internal energy (Figure 2b). That they do not

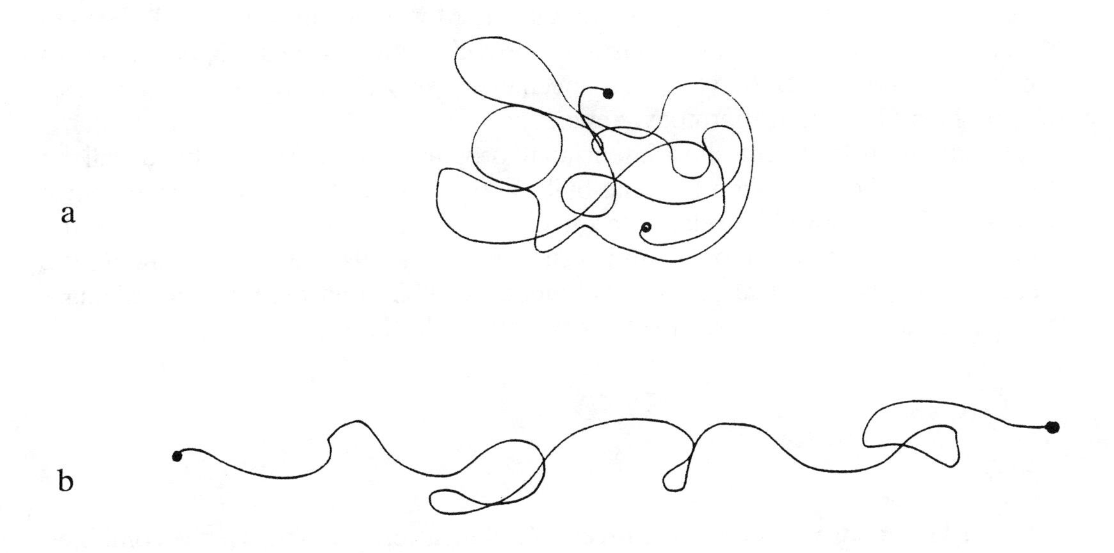

Figure 2. Representation of an a) unextended and b) extended random coil.

spontaneously revert to the stretched form is dictated by the entropy of the system. The elongated chains are more highly ordered than the random coils. In order to overcome the negative entropy effect, work must be performed to elongate the chains, and when the force is removed, the chains return to their disordered "spaghetti-like" state. The polymer will usually not regain its original dimensions, unless there is some overall network structure, and this leads to another requirement:

> 3) To prevent long-range movements, the polymer chains must be joined at a few points (about once in every one hundred C–C linkages) by chemical bonds, usually via a short segment called a cross-link. Figure 1 shows a disulfide group connecting two poly(*cis*-isoprene) molecules. The development of the process of vulcanization, which was discovered in 1839 independently by the Englishman Thomas Hancock and Charles Goodyear of the United States, made it possible to manufacture resilient rubber products, and led eventually to the large-scale usage of rubber in bicycle and automobile tires at the end of the nineteenth century.

The fourth characteristic listed above (high modulus at full extention) accrues from another molecular property:

> 4) The molecular order brought about by the stretching process induces formation of crystalline regions within the polymer at high elongations. The crystallites act as physical cross-links and the stiffness is increased as a result.

The commercial importance of rubber became evident soon after the discovery of vulcanization, and scientific investigations of the product began shortly thereafter. The classical thermodynamics were studied by such masters as Lord Kelvin and Joule. In the 1920's and '30's the molecular interpretation was developed by early polymer chemists, such as Staudinger, Meyer, Kuhn, Guth and Herman Mark. The study of rubber elasticity provides an excellent opportunity to demonstrate the utility of statistical mechanics, as well as to review the basic relationships of classical thermodynamics.

In this experiment the dimensional properties of a rubber band will be studied using both a simple home-built apparatus and a Thermomechanical Analyzer. The data will be related to the change in entropy with elongation by the equations of classical thermodynamics. The results may also be used to validate the predictions of the statistical mechanical model of rubber elasticity and to calculate the number of active chain segments at the molecular level.

Theory

Classical Thermodynamics For the reversible elongation of a polymer the combined first and second laws may be written in the form

$$dE = q - w = TdS - PdV + fdL \qquad (1)$$

In equation 1 there are two types of work: the usual PdV term, by which the system performs work on the surroundings when it expands, and fdL, which represents the work done on the system by an external force f when the polymer is stretched an increment dL. Elongation is essentially a constant volume process ($|PdV| < 0.001|fdL|$) *(1,2)*, and usually the PdV term can be ignored. Furthermore, it has been observed that the isothermal work input of elongation is almost exactly balanced by release of heat to the surroundings. (A qualitative observation of this evolved heat may be made by holding a rapidly stretched rubber band to the lips.) This means that the internal energy of isothermal elongation is effectively zero. This result requires that ΔE be zero for the reverse process as well. Indeed, as mentioned above, when the rubber rebounds (no work), the heat transfer is essentially zero. Recalling the condition that $(\partial E/\partial V)_T = 0$ for an ideal gas, the requirement that $(\partial E/\partial L)_{T,V} = 0$ is used to define an ideal rubber*(3)*.

To derive the important thermodynamic relationships for the elongation process, the Helmholtz free energy, A, which is applicable to constant volume processes, is written in differential form as follows:

$$dA = dE - TdS - SdT \tag{2}$$

and

$$dA = -PdV - SdT + fdL \tag{3}$$

by substitution for dE from equation 1. The following first derivatives may be obtained directly from equation 2.

$$(\partial A/\partial L)_{T,V} = f \quad \text{and} \quad (\partial A/\partial T)_{V,L} = -S \tag{4a and 4b}$$

Each of these may be further differentiated to give:

$$\left(\frac{\partial^2 A}{\partial L \partial T}\right)_V = \left(\frac{\partial f}{\partial T}\right)_{L,V} \quad \text{and} \quad \left(\frac{\partial^2 A}{\partial T \partial L}\right)_V = -\left(\frac{\partial S}{\partial L}\right)_{T,V} \tag{5a and b}$$

The equality of mixed partials shows that:

$$\left(\frac{\partial f}{\partial T}\right)_{L,V} = -\left(\frac{\partial S}{\partial L}\right)_{T,V} \tag{6}$$

By equation 4a, the force is equal to the derivative of A with respect to L at constant T and V. Thus, equation 2 may be differentiated to give:

$$f = \left(\frac{\partial A}{\partial L}\right)_{T,V} = \left(\frac{\partial E}{\partial L}\right)_{T,V} - T\left(\frac{\partial S}{\partial L}\right)_{T,V} \tag{7}$$

and because $(\partial E/\partial L)_{T,V} = 0$,

$$f = -T\,(\partial S/\partial L)_{T,V} \tag{8}$$

This exercise in differential calculus has led to two very important results given by equations 6 and 8. Considering equation 8 first, both the force and the absolute temperature must be positive. This means that the entropy of elongation at constant temperature, $(\partial S/\partial L)_{T,V}$, must be negative. Although this result is based strictly on macroscopic observations, it may be related directly to the molecular properties discussed above. When the elastomer is stretched, the randomly coiled molecules must be straightened, and this decreases the disorder, and hence the entropy, of the network.

According to equation 6, $(\partial S/\partial L)_{T,V}$ can be obtained by measuring the force as function of temperature with length and volume both held constant. However, as indicated by Flory (*2*, pp 440-444 and 489-491), the proposed experiment would not be simple, because normal thermal expansion would cause the volume to increase as the temperature is raised. Flory presents a detailed examination of this matter, and his treatment shows that equation 6 can be rewritten as follows:

$$\left(\frac{\partial f}{\partial T}\right)_{P,\lambda} = -\left(\frac{\partial S}{\partial L}\right)_{T,V} = -\frac{1}{L_0}\left(\frac{\partial S}{\partial \lambda}\right)_{T,V} \tag{9}$$

Equation 9 shows that the experiment can be carried out at constant pressure, which is easily achieved by working at the ambient atmospheric conditions. There is also a new variable, λ, which is the **relative elongation** of the polymer:

$$\lambda = \frac{L}{L_0} = \frac{L_0+\Delta L}{L_0} = 1 + \frac{\Delta L}{L_0} \tag{10}$$

where L_0 and L are, respectively, the lengths of the rubber without and with the applied force. The quantity $\Delta L/L_0$ is commonly called the **strain**, with the symbol ε. Another important quantity is the **stress**, often given the symbol σ, which is the force per unit cross-sectional area of the sample. Thus, both sides of equation 9 may be divided by the area to give:

$$\left(\frac{\partial \sigma}{\partial T}\right)_{P,\lambda} = -\frac{1}{V_0}\left(\frac{\partial S}{\partial \lambda}\right)_{T,V} \tag{11}$$

There is an obvious problem in the specification of the cross-sectional area, because it decreases as the rubber is stretched. To obtain equation 11 the unstressed area, A_0, was used. In later equations there will be occasion to use both the unstressed and stressed dimensions, and the symbol σ' will denote the value obtained with A, the area corrected for elongation (not to be confused with the symbol for Helmholtz Free Energy).

Summarizing these important results, equation 9 shows that the derivative of the entropy with respect to the elongation may be obtained by determining the force needed to maintain a constant relative elongation, λ, as the temperature is varied. Evaluation of this fundamental thermodynamic quantity is one of the goals of this experiment. Another property of interest, especially to engineers, is **Young's modulus, *Y*,** which is defined as the derivative of the stress with respect to the strain:

$$Y = \left(\frac{\partial \sigma'}{\partial \varepsilon}\right)_T = \left(\frac{\partial \sigma'}{\partial \lambda}\right)_T \tag{12}$$

in which the corrected cross-sectional area is used to calculate the stress. Young's modulus is also sometimes called the **stiffness** of the material; a large value of Y indicates that a large force per unit area must be applied to elongate the sample. Referring to equation 10, for a given force increment the change in ε must equal the change in λ. Thus, Y can be obtained from an isothermal experiment, by evaluating the slope of a plot of applied force versus λ and dividing by the cross-sectional area.

Statistical Mechanical Treatment Although molecular properties have been used to explain the equations derived so far, the thermodynamic relationships themselves have involved no assumptions about the microscopic behavior of the elastomer; all parameters (temperature, force, elongation, and volume) can be measured directly on the bulk sample. The molecular treatment of rubber elasticity involves the use of statistical mechanics, and the macroscopic observations can also be used to verify the predictions of the theoretical model. Several of the references *(1-4)* provide very readable derivations of the theory, which will be described in an abbreviated form here.

Consistent with the molecular description presented in the Introduction, the theory assumes that there is no internal energy contribution to the elongation process. The elongation work is needed solely to overcome the entropy effect, specifically the unfavorable change in the conformational entropy, S_{con}, which is related to the many ways that the chains can be arranged spatially by rotations about single bonds. The conformational entropy is given by the Boltzmann formula:

$$S_{con} = k \ln W \tag{13}$$

where k is Boltzmann's constant, and W is the total number of conformations available to the system. The goal, therefore, is to derive expressions for S_{con} in both the unstretched and the stretched rubber.

In its simplest form, the model treats a chain segment as a series of n uniform links (corresponding to the chemical bonds), which are assumed to be freely jointed; i.e. they are not restricted by bond angles or the volume that they occupy. In the polymer network a chain segment corresponds to the portion of the polymer extending from one cross-link to the next. Provided that a segment does not approach full extension, the probability that it extends randomly from an arbitrary point (x,y,z) to another point $(x+\Delta x_u, y+\Delta y_u, z+\Delta z_u)$ is given by a simple Gaussian distribution:

$$P(\Delta x_u, \Delta y_u, \Delta z_u) = \left(\frac{3}{2\pi <r^2>}\right)^{3/2} \exp\left[-\frac{3}{2}\left(\frac{\Delta x_u^{\,2} + \Delta y_u^{\,2} + \Delta z_u^{\,2}}{<r^2>}\right)\right] \tag{14}$$

in which the subscript u refers to the unstressed polymer. In the exponential the sum $\Delta x_u^2 + \Delta y_u^2 + \Delta z_u^2$ is equal to r_u^2, the squared distance between the two ends of the chain segment. This is divided by $<r^2>$, the mean square end-to-end distance, given by:

$$<r^2> = (1/N_0) \sum_{i=1}^{N_0} r_{u_i}^2 \tag{15}$$

where N_0 is the total number of possible conformations available to the chain segment. The total number of ways, W, that the segment ends can be separated by $(r_u^2)^{1/2}$ is the probability given by equation 14 multiplied by the total number of conformations:

$$W = N_0\ P(\Delta x_u, \Delta y_u, \Delta z_u) = N_0 C_0 \exp\left[-\frac{3}{2}\left(\frac{\Delta x_u^2 + \Delta y_u^2 + \Delta z_u^2}{<r^2>}\right)\right] \qquad (16)$$

where C_0 replaces the term $(3/2\pi <r^2>)^{3/2}$, which will remain constant throughout the derivation.

The mean square end-to-end distance, $<r^2>$ is shown in the complete treatment to equal nl^2, where l is the length of each link, so that the RMS distance itself is just $n^{1/2}l$. If the freely-jointed chain were to achieve full extension, its length would be nl. Because $n^{1/2}l$ is much less than nl for a large value of n (i.e. many links), this result justifies the assumption that the segment does not reach its fullest extension.

The polymer network is assumed to consist of ν identical segments. For this assembly the total number of conformations, W_u, in which the unstressed segment ends are separated by r_u^2 is

$$W_u = \prod_{j=1}^{\nu} W_i = \prod_{j=1}^{\nu} N_0 C_0 \exp\left[-\frac{3}{2}\left(\frac{\Delta x_{u_j}^2 + \Delta y_{u_j}^2 + \Delta z_{u_j}^2}{<r^2>}\right)\right] \qquad (17)$$

$$= N_0^{\nu} C_0^{\nu} \exp\left[-\sum_{j=1}^{\nu}\left(\frac{3}{2<r^2>}\right)\left(\Delta x_{u_j}^2 + \Delta y_{u_j}^2 + \Delta z_{u_j}^2\right)\right]$$

Thus, the conformational entropy of the unstressed rubber is

$$S_u = \nu k \ln (N_0 C_0) - \frac{3k}{2<r^2>}\sum_{j=1}^{\nu}\left(\Delta x_{u_j}^2 + \Delta y_{u_j}^2 + \Delta z_{u_j}^2\right) \qquad (18)$$

In equation 18 each of the $\Delta x_{u_j}^{\ 2}$, $\Delta y_{u_j}^{\ 2}$, $\Delta z_{u_j}^{\ 2}$, can be replaced by the mean-square average for the assembly to give:

$$S_u = \nu k \ln (N_0 C_0) - \frac{3k}{2<r^2>}\sum_{j=1}^{\nu}\left(<\Delta x_u^2> + <\Delta y_u^2> + <\Delta z_u^2>\right) \qquad (19)$$

$$= \nu k \ln (N_0 C_0) - \frac{3\nu k}{2<r^2>}\left(<\Delta x_u^2> + <\Delta y_u^2> + <\Delta z_u^2>\right) \qquad (20)$$

The expression for S, the conformation entropy for the stretched sample, is derived in the same manner, but with the additional assumption that polymer elongation is an affine deformation. This means that the changes in the x, y, and z components of each chain segment are proportional to the corresponding changes in the bulk material:

$$\Delta x = \lambda_x \Delta x_u;\ \Delta y = \lambda_y \Delta y_u;\ \Delta z = \lambda_z \Delta z_u$$

where λ_x, λ_y, and λ_z correspond to the respective relative changes in the bulk sample (i.e., $\lambda_x = L_x/L_{x0}$, etc.). Thus, the entropy of the stressed rubber is given by:

$$S = \nu k \ln (N_0 C_0) - \frac{3\nu k}{2<r^2>}\left(\lambda_x^2 <\Delta x_u^2> + \lambda_y^2 <\Delta y_u^2> + \lambda_z^2 <\Delta z_u^2>\right) \quad (21)$$

and the difference in entropies of the stressed and unstressed sample is:

$$S - S_u = -\frac{3\nu k}{2<r^2>}\left[\left(\lambda_x^2 <\Delta x_u^2> + \lambda_y^2 <\Delta y_u^2> + \lambda_z^2 <\Delta z_u^2>\right) - \left(<\Delta x_u^2> + <\Delta y_u^2> + <\Delta z_u^2>\right)\right] \quad (22)$$

For an isotropic network all directions are equally probable, so that on the average

$$<\Delta x_u^2> = <\Delta y_u^2> = <\Delta z_u^2> = <r^2> / 3. \quad (23)$$

With these substitutions the difference in entropies reduces to

$$S - S_u = -\frac{3\nu k}{2<r^2>}\left[\left(\lambda_x^2 \frac{<r^2>}{3} + \lambda_y^2 \frac{<r^2>}{3} + \lambda_z^2 \frac{<r^2>}{3}\right) - \left(\frac{<r^2>}{3} + \frac{<r^2>}{3} + \frac{<r^2>}{3}\right)\right]$$

$$= -\frac{\nu k}{2<r^2>}\left[<r^2>\left(\lambda_x^2 + \lambda_y^2 + \lambda_z^2\right) - 3<r^2>\right]$$

$$= -\frac{1}{2}\nu k\left(\lambda_x^2 + \lambda_y^2 + \lambda_z^2 - 3\right) \quad (24)$$

This equation simplifies further, because elongation is a constant volume process. If the x axis is specified as the direction of elongation, the following relations must hold:

$$\lambda_x \lambda_y \lambda_z = 1 \text{ and } \lambda_y^2 = \lambda_z^2 = \lambda_x^{-1} = \lambda^{-1} \quad (25)$$

Substitution of these conditions into equation 24 gives the desired expression for the entropy of the stretched polymer:

$$S = -\frac{\nu k}{2}\left(\lambda^2 + \frac{2}{\lambda} - 3\right) + S_u \quad (26)$$

Equation 26 may be differentiated with respect to λ to produce:

$$\left(\frac{\partial S}{\partial \lambda}\right)_{T,V} = -\nu k\left(\lambda - \frac{1}{\lambda^2}\right) \quad (27)$$

Combining this result with equation 8 and the definition of λ gives

$$f = \frac{\nu kT}{L_0}\left(\lambda - \frac{1}{\lambda^2}\right) \tag{28}$$

or

$$\sigma = \frac{\nu kT}{V_0}\left(\lambda - \frac{1}{\lambda^2}\right) = N_e kT\left(\lambda - \frac{1}{\lambda^2}\right) \tag{29}$$

where N_e is the number of elastically active chain segments per unit volume. Thus, a plot of the stress (using the uncorrected cross-sectional area) versus $(\lambda - \lambda^{-2})$ at constant temperature should be linear with slope equal to $N_e kT$. The relationship fails to hold for λ greater than about 4 (elongations of more than 300%), because the assumption that a polymer chain has an end-to-end distance considerably less than the length at full extension fails to hold at large elongations. Also, as discussed above, the elastomer may develop crystalline regions as the polymer chains are extended and adopt a more ordered arrangement.

The form of equation 29 is very similar to that of the ideal gas law, which may be written as

$$P = \frac{NkT}{V_0}\left(\frac{V_0}{V}\right) \tag{30}$$

where N is the number of molecules of gas, V_0 is the initial volume, and the term in parentheses is the expansion factor. The stress, which has units of force per unit area, is analogous to the pressure, and the number of elastically active chain segments replaces the number of gas molecules. The differences in equations 29 and 30 occur in the terms in parentheses. First of all, the quotients are inverted (L/L_0 versus V_0/V). For an ideal gas the pressure acts to decrease the volume, but the stress on a rubber band increases the length. Also, the strain factor in equation 29 is somewhat more complicated than the ideal gas expansion factor, due to the near incompressibility of the polymer *(3)*. The reasons for the analogy relate to the fundamental thermodynamics of the systems. For an ideal gas/ideal rubber there is no internal energy change for an isothermal expansion/elongation; the processes are entirely entropy-driven. The gas expands to maximize its entropy, whereas the molecules in the rubber retract to their disordered random-coil state, when the elongation stress is removed.

Equation 29 may be differentiated to give Young's modulus:

$$Y = \left(\frac{\partial \sigma'}{\partial \lambda}\right)_T = N_e kT\left(1 + \frac{2}{\lambda^3}\right)\left(\frac{A_0}{A}\right) \tag{31}$$

where the second term in parentheses is needed to correct the cross-sectional area to the value in the elongated sample. An average value for N_e may be obtained from plots of σ versus $(\lambda - \lambda^{-2})$. Then, at any specified temperature and elongation a theoretically predicted value of Y may be calculated from equation 31. The theoretically predicted Y should agree with the value obtained directly from the f versus λ plot at the same temperature.

Methodology

Isothermal stress/strain behavior may be studied using the apparatus shown in

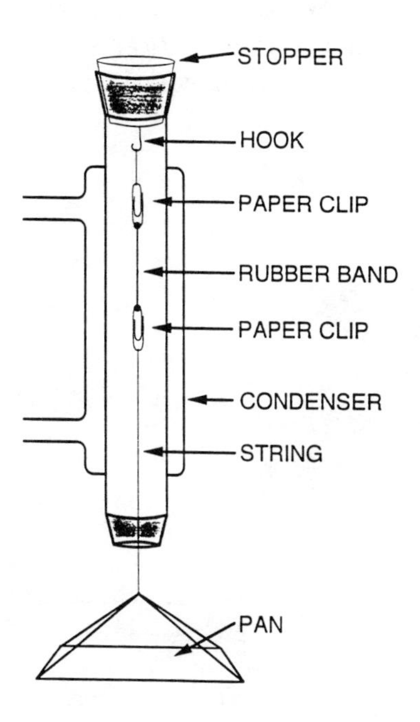

Figure 3. Diagram of the "weights and ruler" apparatus for isothermal stress/strain measurements.

Figure 3 (5). A series of weights is added to the pan, and the length of the band is measured with a ruler after each addition. The force is given by the total mass of the added weights plus the pan, clips, etc., all multiplied by the acceleration of gravity, 9.80665 m/s^2. To obtain the stress, the width and thickness of the unstressed band are measured with a calipers, and the force is divided by the calculated area. In the figure the band, string, etc. are surrounded by a condenser, which helps to keep air currents from disturbing the sample. If desired, the temperature may be controlled by means of the jacket and a circulating constant temperature bath.

This simple set-up is adequate for illustrative purposes, but in modern laboratories a **Thermomechanical Analyzer**, or **TMA**, is used for these types of investigations. This instrument measures the dimensional change as either the force on the sample or the temperature is varied. A schematic diagram of a TMA is shown in Figure 4a. The most important components are the furnace with associated temperature control circuitry and coolant reservoir for subambient work, the probe, and the linear variable differential transformer (LVDT). In the TMA shown, the lower part of the probe is connected to a metal rod, which fits into the force coil assembly at the bottom. When current is passed through the coil, the probe applies a force to the sample. The metal rod also passes through the core of the LVDT. The voltage output of the LVDT depends on the vertical position of the metal rod, and this is determined by the elongation of the sample.

For use in the study of elastomer elongation the TMA must be equipped with a tension probe (Figure 4b), which allows the rubber band to be stretched by a controlled force. A typical instrument is capable of three general modes of operation: **isothermal**, in which the dimensional change is measured as the force is changed; **isostress**, with measurements of dimensional change at constant force as the temperature is varied; and **isostrain**, in which temperature is varied and the force needed to maintain a constant elongation is measured.

The data obtained in an isostrain experiment may be used to prepare a plot

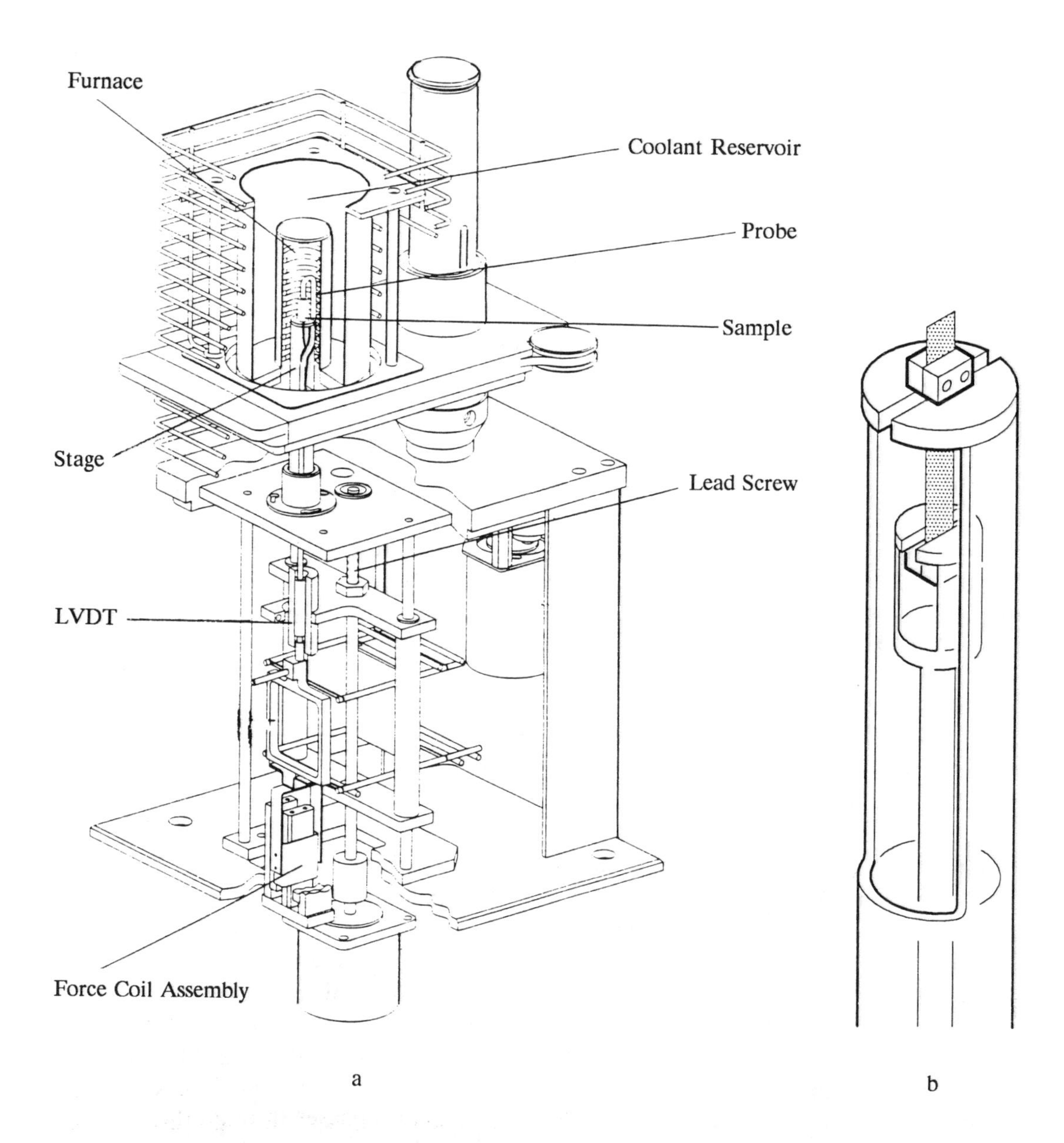

Figure 4. a) Schematic diagram of the TA Instruments Model 2940 Thermomechanical Analyzer. b) TMA probe used for elongation studies. Figures reproduced with permission of TA Instruments, Inc.

of force versus temperature, and the slope is equal to $(\partial f/\partial T)_{P,\lambda}$. Equation 9 is used to calculate $(\partial S/\partial\lambda)_{T,V}$ from the slope and the initial length of the sample between the probe clamps. The initial length may be measured with a calipers or a ruler, but this value is commonly obtained directly from the instrument. If so, the TMA's initial length measurement should be performed under minimal force load.

By suitable mathematical manipulations $(\partial f/\partial T)_{P,\lambda}$ may be evaluated using the other TMA operational modes. The total differential, d*f*, is given by

$$df = \left(\frac{\partial f}{\partial \lambda}\right)_T d\lambda + \left(\frac{\partial f}{\partial T}\right)_\lambda dT \quad (32)$$

For a constant value of *f*, d*f* is equal to zero, and $(\partial f/\partial T)_\lambda$ is given by:

$$\left(\frac{\partial f}{\partial T}\right)_\lambda = -\left(\frac{\partial f}{\partial \lambda}\right)_T \left(\frac{\partial \lambda}{\partial T}\right)_f \quad (33)$$

The first of the two contributing terms, the derivative of the force with respect to the relative elongation, may be obtained from the slope of a plot of *f* versus λ for an isothermal experiment. Because of the nonlinear relationship of *f* and λ, the slope must be evaluated for the line tangent to the curve at a specified elongation. To evaluate $(\partial f/\partial T)_\lambda$, $(\partial f/\partial\lambda)_T$ must be multiplied by the negative derivative of the relative elongation with respect to temperature at constant force. An isostress TMA experiment provides this derivative as the slope of a plot of relative elongation versus temperature. The value may also be evaluated from *f* versus λ plots at a series of constant temperatures. The λ corresponding to a specific force is obtained for each set of measurements; $(\partial\lambda/\partial T)_f$ is the slope of a plot of these λ values versus temperature.

In this experiment both the TMA and the "weights and ruler" apparatus (hence-forth called the manual method) will be used to evaluate $(\partial S/\partial\lambda)_{T,V}$ for a common rubber band. (A 1/16" or 1/8" band seems to work adequately.) The data obtained from the isothermal measurements will also be used to calculate Young's modulus and to verify the predictions of the statistical mechanical model.

For these measurements, the TMA offers the obvious advantages of automated data acquisition and an on-line computer for the computations, as well as the capability of performing the isostress and isostrain experiments described above. However, the instrumentation may tend to be a black-box approach, and the simple apparatus helps students appreciate the measurements being made. Use of the manual method also has scientific merits. TMA's are designed for highly sensitive measurements of dimensional changes, and as a result, they have small overall elongation limits. Thus, the thermodynamic relationships cannot be observed over a wide range of λ values. By measuring the length with a ruler the students should be able to test the validity of equation 29 for relative elongations up to about 4, with curvature observed at higher λ's due to the formation of crystallites in the polymer.

The materials for the manual set-up should be readily available in any laboratory. A bent wire gauze is an adequate pan, and a brass rod can be cut into short lengths to serve as weights (about 15–20 five gram pieces for the rubber bands specified above). Fluid from a circulating constant temperature bath can be used to provide several isothermal sample temperatures. However, if a TMA is available, manual measurements at the ambient temperature should be sufficient for students to understand the fundamental concepts, and the TMA can be used for the remainder of the experiment.

Considering now the TMA measurements, isothermal force/elongation data are collected for several accessible temperatures (e.g., 10, 25, and 50 °C). The range of forces to be applied depends on the extension limits of the TMA and the cross-sectional area of the rubber band. In general, the force is varied to produce relative elongations from about 1.1 (10% elongation) to the maximum allowed by the instrument and/or sample. The temperature ramps for the isostress and isostrain experiments should cover at least the range of temperatures used for the isothermal work (i.e., 10 to 50 °C for the above example). For the latter two modes of operation the band is initially strained to a relative elongation approximately in the middle of the elongation range observed in the isothermal sets. For the isostress work the force is fixed at the value needed to produce the desired strain at 25 °C or a mid-range temperature. Then the sample is equilibrated at one end of the desired temperature range, prior to starting the ramp itself. The isostrain experiment is conducted in the same manner, but the elongation is held constant and the force is measured as the temperature is ramped. (To the inexperienced user of Thermomechanical Analysis the design of the various acquisition cycles may seem formidable, especially since the operating manuals are commonly written for engineering and materials applications. The author has designed specific instructions for use with the TMA shown in Figure 4, and she will gladly provide them to interested faculty.)

The cross-sectional area of the rubber band may be calculated from the width and thickness of the unstressed sample measured with a calipers. The dimensions should be measured at several locations on the band, and care must be taken not to close the jaws of the calipers too tightly. (These values should theoretically be obtained at each of the specified temperatures, but the effects of thermal expansion/contraction are expected to be very small over this temperature range). The unstressed length of the sample should be obtained prior to and after each experiment, if possible. The values will show some random fluctuation, but should not increase systematically with time.

Although the elongations of the rubber band itself are sizable, the variations with temperature of the length (isostress mode) or the force (isostrain mode) are very small, especially over the temperature range suggested above. The TMA is capable of detecting these small effects, but there may still be considerable uncertainty in the final results. This is especially true of the isothermal data, for which a plot of λ versus T at a constant force must be prepared as part of the calculation of $(\partial S/\partial \lambda)_{T,V}$.

Probably the most important source of experimental error is failure to allow the system to reach equilibrium. Typically an experiment is designed to equilibrate the rubber band at the designated initial conditions, and then to start the temperature or force ramp. It is essential that the rubber band be allowed to sit at the initial conditions for a sufficient length of time, because some relaxation will occur. The ramps themselves should be as slow as student laboratory time will allow (suggested rates are 1 to 2°/minute for temperature; 0.05 newton/minute for force). If possible, runs should be repeated and/or data should be collected using both ascending and descending ramps. For temperature ramps, Flory has suggested that the sample be equilibrated initially at the highest temperature, with the descending changes first, followed by an ascending ramp back to the initial temperature. The author has also found that this procedure gives the most reproducible data.

Another major source of uncertainty is the rubber band itself. Students usually relate well to an experiment involving a consumer product, but, if possible, a crosslinked sample without fillers or other additives should be used.

Safety

Unless the TMA is equipped with a mechanical cooling device, either liquid nitrogen or dry ice will be needed for subambient work and/or controlled cooling ramps. Students should be advised of the hazards of cryogenic materials, and they should wear suitable gloves and eye protection.

Data Analysis

The students should collect and analyze the data according to the scheme described below. The software in the TMA may be capable of calculating several of the results. Alternatively, the raw data may be transferred to a spreadsheet or other analysis program. For each calculation suitable statistical analysis should be used to evaluate the uncertainty (as the 95% confidence interval) of the result.

1. Calculate the average width and thickness of the rubber band and its cross-sectional area. Assuming that no systematic change in the L_0 values was observed, calculate the average unstressed length, and the volume of the sample.

2. For each isothermal data set (manual and instrumental), plot the force versus relative elongation and construct a smooth curve through the points. On each plot locate the λ value used in the isostrain experiment (henceforth symbolized as λ_0). Evaluate $(\partial f/\partial \lambda)_T$ as the slope of the tangent line at that point. Use equation 25 to calculate the cross-sectional area at this elongation. Calculate Young's modulus at each temperature. If the manual apparatus and TMA were used at a common temperature, compare the results obtained by the two methods, and explain any significant difference.

3. For the 25 °C TMA data determine the force corresponding to λ_0. Determine the elongation produced by this force on each of the other isothermal plots. Prepare a plot of λ versus T; use its slope and $(\partial f/\partial \lambda)_T$ at 25 °C to calculate $(\partial S/\partial \lambda)_{T,V}$ at this temperature. (If the manual method is used at more than one temperature, a similar calculation can also be performed using that data.)

4. For the isostress TMA experiment plot λ versus T, and evaluate the slope at 25 °C. Use this result to calculate $(\partial S/\partial \lambda)_{T,V}$ in the same manner as in step 3.

5. For the isostrain experiment plot f versus T and evaluate $(\partial f/\partial T)_{P,\lambda}$ as the slope at 25 °C. Calculate $(\partial S/\partial \lambda)_{T,V}$ from this result and the length of the unstressed band.

6. Calculate the average value of $(\partial S/\partial \lambda)_{T,V}$, obtained in steps 3–5, and try to identify the cause(s) of significant variations in the results. Discuss the value of $(\partial S/\partial \lambda)_{T,V}$ (especially the sign) in terms of the molecular model for the elongation of an elastomer.

7. Using the data for each of the isothermal data sets (manual and instrumental) and the cross-sectional area of the unstretched band, prepare plots of σ versus $(\lambda - \lambda^{-2})$. Calculate N_e at each temperature and the average N_e. Use equation 31 and the corrected cross-sectional area to calculate Young's modulus at each temperature for λ equal to λ_0. Compare the Y values to those calculated in step 2.

8. Assume that the sample is composed of poly(*cis*-isoprene) molecules, each containing 5000 monomer units, with disulfide crosslinks at 100 bond intervals (25 isoprene units between cross-links). If the density of the polymer is 0.91 g/cm^3, calculate a value for the number of chain segments per unit volume, and compare it to the N_e obtained experimentally. Discuss the validity of this comparison *(3,4)*.

Acknowledgments

The author first learned of the "weights and ruler" stress/strain experiment during a workshop on polymer chemistry conducted by Dr. John P. Droske of the University of Wisconsin at Stevens Point. Dr. Droske continues to be active in promoting the incorporation of polymer concepts in the chemistry curriculum. Development of the present instrumental version of the elastomer experiment was made possible by a grant from the NSF-ILI program for the thermal analysis system. The author also acknowledges the assistance of Ensign Stephen E. Musson and Mr. Jeff Groh, who both helped with the initial experimental design, and Dr. Julianne Harmon for her helpful comments on the manuscript.

Literature Cited

1. Nash, L. K. *J. Chem. Educ.*, **1979**, *56*, 363.
2. Flory, Paul J. *Principles of Polymer Chemistry*; Cornell University Press: Ithaca, NY, 1953; Chapters X and XI.
3. Mark, J. E. *J. Chem. Educ.*, **1981**, *58*, 898.
4. Aklonis, J. J.; MacKnight, W. J. *Introduction to Polymer Viscoelasticity*, Second Edition; Wiley-Interscience: New York, 1983, Chapters 5 and 6.
5. Bader, M. *J. Chem. Educ.*, **1981**, *58*, 285.

Hardware

The instrumental version of this experiment requires a thermal analysis system containing a TMA module. The system should preferably be capable of operation by the three modes described in the text, but parts of the experiment can be performed, if one of the modes is not available. The TMA must be equipped with a probe similar to the one shown in Figure 4b. Instrument manufacturers usually sell probes of this design for the study of films and fibers. Before buying the TMA and film/fiber accessory, the purchaser should determine the extension capabilities of the instrument to make sure that the rubber band can be elongated sufficiently (preferably at least 100% to give a λ of 2.)

RECEIVED October 14, 1992

Incorporating Modern Instrumentation

CHAPTER 25

Oxygen Binding to Hemoglobin

Betty J. Gaffney and Paul J. Dagdigian

POTENTIAL HAZARDS: High-Pressure Systems, Cryogens, Vacuums

The co-authors have for a number of years at Johns Hopkins shared in the teaching of an undergraduate physical chemistry instrumentation laboratory, which is meant to be taken by junior-level chemistry majors after two years' laboratory experience in freshman and organic chemistry. The co-authors bring complimentary research expertise, in biophysical chemistry and chemical physics, to the design and implementation of experiments for this laboratory.

A large fraction of Chemistry majors at Johns Hopkins are planning careers in medicine or biomedical sciences. For this reason, we make an effort to find a number of biophysical applications in our choice of experiments for this laboratory. The first semester is taken concurrently with the first semester of the physical chemistry lecture course. As a result, students are learning thermodynamics while they are taking the lab. They have some background in organic chemical NMR and use of personal computers from previous courses. The choice of experiments reflects these facts. The Fall semester includes experiments on use of computers in data analysis with one experiment using a Fourier analysis of data from oscillating chemical reactions *(1)* and another on spectral simulation to analyse an EPR experiment on spin exchange *(2)*. Additional experiments explore the magnetic properties of molecules using NMR *(3)* and magnetic susceptibility. A calorimetry experiment is being developed to measure heats of ionization, with special emphasis on amino acids. A continuing concern has been to find appropriate experiments at this level to teach the principles of structure analysis by X-ray diffraction.

The experiments carried out in the Spring semester have a heavy emphasis on spectroscopy, to complement and illustrate the students' study of quantum mechanics in the concurrent lecture course. These include measurement of the absorption and emission spectrum of the sodium atom *(4)*, the infrared spectrum of the hydrogen halides or other simple molecules, and the electronic spectra of the nitrogen and iodine molecules *(5)*. The increasing use of lasers in chemical physics research is demonstrated by the infrared multiphoton dissociation of SF_6 *(6)*. The fluorescence spectrum of 2-naphthol is investigated to illustrate photophysics and excited-state reactions *(7)*. The kinetics of complexation of Fe^{3+} by SCN^- is also studied at various reagent and acid concentrations with the stopped flow method.

In this article, we describe an experiment in which the oxygen binding to hemes in hemoglobin is investigated. This reaction is followed by observing changes in the visible absorption spectrum of the heme group as a function of oxygen pressure in a closed sample chamber. The oxygen–hemoglobin equilibrium is complex because affinity of oxygen for one of the four hemes in a single hemoglobin

2503–6/93/0370$06.00/0

molecule is dependent on the state of oxygenation of the other three hemes. The data obtained in this experiment will be used to assess the extent of this "cooperativity" of the hemes.

Theory

Hemoglobin and myoglobin are two proteins which act as a pair in the transport of oxygen from the lungs, via the blood, to muscle. In the binding of oxygen, both proteins use heme groups, which are formed by the complexation of an Fe^{2+} ion with a porphyrin. Myoglobin molecules (molecular weight 17,800 daltons) contain one heme, while hemoglobin molecules (molecular weight 64,500 daltons) have four subunits, each containing a heme group.

These proteins differ, however, in the amount of oxygen they bind for an incremental change in the pressure of oxygen to which they are exposed *(8)*. The transfer of oxygen from hemoglobin to myoglobin in living systems is a dynamic process dependent upon oxygen gradients from capillaries to mitochondria. This process occurs because myoglobin has a higher affinity for oxygen than hemoglobin at low oxygen pressures. The binding of oxygen to myoglobin (Mb) shows the hyperbolic dependence of the fraction of oxygenated protein as a function of oxygen pressure that is expected for a simple equilibrium:

$$\mathrm{Mb} + \mathrm{O_2} \rightleftharpoons \mathrm{MbO_2} \quad (1)$$

The equilibrium constant for reaction (1) is given by

$$K_{\mathrm{Mb}} = [\mathrm{Mb}]\,[\mathrm{O_2}] \,/\, [\mathrm{MbO_2}] \quad (2)$$

Here the concentration of oxygen is measured by its pressure. The fraction y_{Mb} of Mb oxygenated at a given oxygen pressure is thus found to equal

$$y_{\mathrm{Mb}} = [\mathrm{Mb}] \,/\, \{ [\mathrm{Mb}] + [\mathrm{MbO_2}] \} \quad (3a)$$

$$= 1 \,/\, \{ 1 + K_{\mathrm{Mb}} \,/\, [\mathrm{O_2}] \} \quad (3b)$$

K_{Mb} equals $p_{50(\mathrm{Mb})}$, the pressure at which myoglobin is 50% saturated with oxygen.

The binding of oxygen to hemoglobin (Hb), however, is not hyperbolic but instead is S-shaped or sigmoidal, which means that the equilibrium constant must depend on greater than the first power of the oxygen pressure:

$$K_{\mathrm{Hb}} = [\mathrm{Hb}]\,[\mathrm{O_2}]^n \,/\, [\mathrm{HbO_2}] \quad (4)$$

Thus, the fraction y of Hb oxygenated equals

$$y = 1 \,/\, \{ 1 + K_{\mathrm{Hb}} \,/\, [\mathrm{O_2}]^n \} \quad (5)$$

K_{Hb} in equation (5) is equal to $(p_{50(\mathrm{Hb})})^n$, where $p_{50(\mathrm{Hb})}$ is the pressure at which hemoglobin is 50% saturated. The saturation curves for myoglobin and hemoglobin are compared in Figure 1.

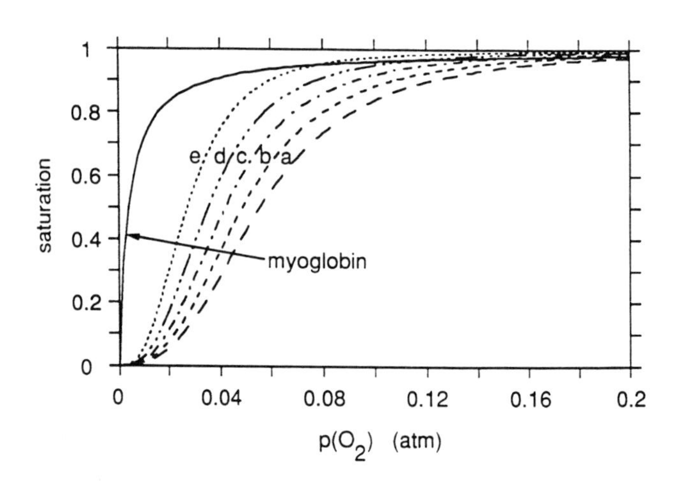

Figure 1. Oxygen binding curves for myoglobin and for hemoglobin at pH values of (a) 6.8, (b) 7.0, (c) 7.2, (d) 7.4, and (e) 7.6. The tendency to release oxygen at low pH is known as the Bohr effect. In the lungs, the oxygen partial pressure is approximately 0.13 atm, while it is 0.05 atm in the muscle.

The ways in which the four subunits of hemoglobin interact to give sigmoidal oxygen binding have been the subject of decades of experiment and theory. The simplest way to explain the dependence of the equilibrium constant on higher-than-first-power of the oxygen concentration is to assume that the only significantly populated oxygenated state of hemoglobin is the tetra-oxygenated one. In that case, equation (4) would become

$$K_{Hb} = [Hb]\,[O_2]^4 \,/\, [HbO_2] \tag{6}$$

and equation (5) could be rearranged to give

$$\log\,[y \,/\, (1 - y)] = 4 \log\,([O_2] \,/\, p_{50(Hb)}) \tag{7}$$

For hemoglobin it is found experimentally, however, that the dependence is not on the fourth, but on about the "2.8th" power of oxygen concentration. The power 2.8 is obtained from the slope of a plot of ($\log\,[y \,/\, (1 - y)]$ vs. $\log\,[O_2]$). This type of graphical presentation of data for oxygen uptake is called a *Hill plot*. The 2.8 power dependence in hemoglobin uptake of oxygen is now well described by the theory of Monod, Wyman, and Changeux *(9,10)*, to which the student can be referred for extra reading. In addition, the details of how distortions of the heme group by its protein surroundings alter the affinity of heme for oxygen have been examined by studies of chemical analogs of natural heme *(11)*.

Both direct plots, of the type shown in Figure 1, and Hill plots are used to compare the values of n and K_{Hb} for hemoglobin samples under a variety of conditions of pH, temperature, cofactor concentration, and medically abnormal states. It is the purpose of this experiment to demonstrate the techniques used to obtain these data.

The visible absorption spectrum of hemoglobin will be employed. In the visible spectral region, hemoglobin and oxyhemoglobin have two absorption bands: the very strong Soret (or B) band centered near 420 nm and the α and β (or Q) bands between 550 and 600 nm. The assignment of the optical transitions observed in porphyrins *(12)* and heme proteins *(13)* has evoked much interest. The α, β, and Soret bands are essentially prophyrin $\pi-\pi^*$ transitions.

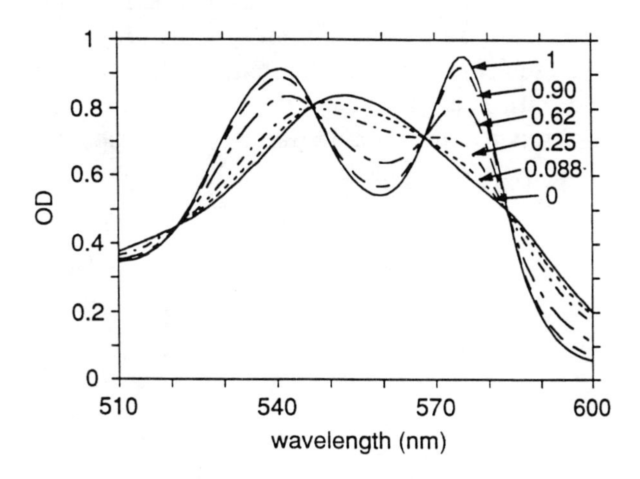

Figure 2. Absorption spectrum (510–600 nm) of horse hemoglobin at various degrees of saturation (indicated on the traces).

Figure 2 shows absorption spectra of hemoglobin over 510–600 nm at varying degrees of saturation. This region of the hemoglobin spectrum is useful for studies with 0.05 to 2% hemoglobin solutions and 1 cm optical path length. The Soret band can be used for even lower hemoglobin concentrations. In our work, the concentration of oxyhemglobin subunits will be estimated from the optical density at 576 nm, where the molar absorptivity is 1.65×10^4 M^{-1} cm^{-1}: *(14)*

$$\text{molarity of } HbO_2 \text{ (per subunit)} = OD_{576} / 1.65 \times 10^4 \quad (8)$$

Thus, the concentration of hemoglobin can be calculated using Beer's law. The two absorption bands observed for oxygenated hemoglobin in the 520–600 nm region represent different vibronic transitions. The deoxygenated form of hemoglobin exhibits one broad transition in this spectral range. The broad nature of the band and its lack of structure implies the presence of charge-transfer and Fe *d-d* transitions, with a consequent shortening of the excited state lifetime *(13)*

Methodology

Horse hemoglobin was used in the experiments reported here. It can be prepared from whole blood and purified by published procedures *(15)*. Briefly, red cells are lysed by adding 3 volumes of distilled water or 3 volumes of ice cold 1 mM Tris buffer (pH 8). Highly purified water should be used in all buffers because the rate of conversion of Fe^{2+} heme to Fe^{3+} heme (methemoglobin) is catalyzed by extraneous metal ions such as Cu^{2+}. After 1 hour at 4°C, 1 tenth volume of 1 M sodium chloride is added. Centrifugation at 4°C for 15 min at 28,000 g removes the cellular debris. The supernatant can be used as is or can be dialysed into a buffer of defined pH and organic phosphate concentration. The p_{50} of hemoglobin is very sensitive to pH (Bohr effect) and concentration of organic phosphates, and these can be varied in the dialysis step. (pH 7.2 is a convenient pH. See Figure 1.) Sheep hemoglobin might also be considered for this experiment because the p_{50} of sheep hemoglobin is quite insensitive to organic phosphates *(16)*. Hemoglobin solutions for several years' sets of experiments can be made at one time and frozen. In order to avoid thawing the entire sample to obtain an aliquot of solution, a convenient method of preparing frozen samples is to allow single drops of solution to fall into liquid nitrogen so that they form beads.

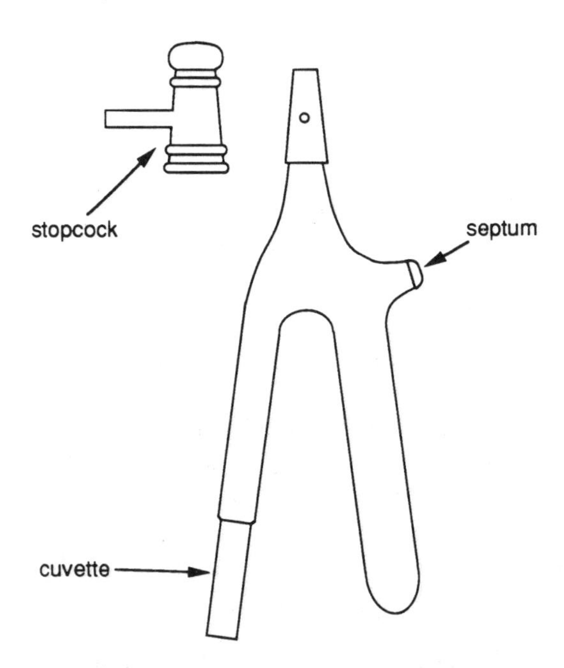

Figure 3. Schematic diagram of the tonometer (adapted from Reference 14). The overall length of the tonometer is approximately 20 cm.

In order to control the partial pressure of oxygen above the hemoglobin solution, the sample will be contained in a tonometer *(14)*, which is illustrated in Figure 3. Horse hemoglobin (approximately 0.1%) will be deoxygenated by slowly evacuating the space above the solution in the tonometer. After complete deoxygenation, the spectrum is recorded over the range 490–800 nm. Measured volumes of oxygen are then added, and after 5 minutes of equilibration, the spectrum is recorded again. Five aliquots of oxygen (or air) will be added to give five spectra corresponding to different degrees of saturation of hemoglobin. From these spectra, a plot of the saturation of hemoglobin vs. $\log_{10} p(O_2)$ (in atm) will be obtained.

The first step in the experiment is to remove oxygen from hemoglobin. This will take at least one hour where little attention to the sample is needed. During this hour, an estimate of the hemoglobin concentration must be made in order to make a crude calculation of the volumes of oxygen that will have to be used in the sequential oxygenation. To calculate these, assume a rough volume of the tonometer. (Later the volume will be measured more precisely.) Refer also to Figure 1 for the oxygen pressures that give various degrees of saturation of hemoglobin. It is important that these calculations be reasonably accurate, or there is a risk of getting 90% saturation after adding the first aliquot of oxygen.

During the period when the sample is being deoxygenated, use a normal cuvette and record the visible spectrum of an aliquot of the initial HbO_2 solution. The ratio of the peak at 576 nm to that at 540 nm should be noted. This ratio is sensitive to degradation of hemoglobin to methemoglobin, which is a Fe^{3+} derivative of Hb. The ratio will be measured again on the sample in the tonometer after step-wise oxygenation. Possible changes in the Hb concentration due to evaporation of water during evacuation should also be investigated by comparing spectra obtained at the beginning and end of the runs. The recorded range should be 490 to 800 nm. In all parts of this experiment, the reference is air, which means that no cuvette need be placed in the reference beam. Record a baseline with phosphate buffer in the sample cuvette.

The tonometer is filled by pipetting 5 mL of hemoglobin solution into the round arm of the tonometer. Place a greased septum in the side arm. The greased stopcock is placed on the tonometer and left in the open position. A narrow tube connected to a regulated supply of nitrogen is threaded through the stopcock opening to about 1 cm above the surface of the solution. The initial stage of deoxygenation is accomplished by blowing nitrogen gently over the Hb solution while the tonometer is rocked slowly back and forth. The apparatus is clamped on a rocking device with the arm containing Hb immersed in a Dewar of ice. Cooling the protein solution in ice minimizes formation of methemoglobin. After one hour, the nitrogen tube is withdrawn from the tonometer and the stopcock is closed. A vacuum pump is attached to the tonometer side arm by pressure tubing. (The vacuum pump should be protected with a Drierite trap and a trap cooled in dry ice-isopropanol.) After the tubing is evacuated, the stopcock to the tonometer is opened slowly. As soon as the Hb solution starts to bubble, the stopcock is closed. Opening and closing the stopcock is done repeatedly until there is no more evolution of oxygen. During deoxygenation, it is very important to minimize bubbling because proteins denature on the surface of bubbles. After bubbling ceases, rock the Hb solution (keeping it cooled) for 5 minutes so that dissolved oxygen will equilibrate with the space above the solution. Evacuate once more. Repeat these steps about 6 times with the sample in the ice bath and, finally, include two 3-minute periods of equilibration at room temperature, followed by evacuation, to remove the last traces of dissolved oxygen. Warming the sample to room temperature facilitates the removal of the last portion of bound oxygen since Hb binds oxygen less tightly at higher temperatures. After this period, the solution should be ready for recording the spectrum of deoxygenated hemoglobin. More evacuation-equilibration steps can be added if the optical spectrum indicates that deoxygenation is not complete at this point.

Most spectrophotometers require a special lid so that the sample compartment is extended sufficiently high to hold the entire tonometer. The lid can be contructed from masonite, with the seams sealed with an opaque material. Alternatively, a black cloth can be draped over the sample to prevent light leaks. Rotate the tonometer so that hemoglobin and condensed water mix and coat the whole tonometer. Tip the hemoglobin solution into the cuvette arm and place the cuvette in the holder. Place the cover on the sample chamber and record the spectrum of deoxy Hb from 490 to 800 nm. Remove and rotate the tonometer, replace the tonometer in the spectrophotometer and again record the spectrum. The two recorded spectra should be the same and should resemble the deoxyhemoglobin spectrum shown in Figure 2. The region from 770 to 800 nm will be used as the baseline.

Addition of air is made with a gas-tight syringe. Inject the air through the rubber septum and into the tonometer. After addition of air, the tonometer should be rotated so that the hemoglobin coats as much of the surface area as possible. Equilibration of the protein with oxygen will be much faster if the protein solution forms a thin film over a large surface area. However, it is best to avoid contact of the hemoglobin solution with the stopcock. Five minutes should be allowed for equilibration at room temperature. During this period, the tonometer should be rotated manually to keep the solution spread over the surface. *To minimize the risk of breaking the tonometer during manual rotation, hold it over a bench top with soft padding.* After equilibration, place the tonometer in the spectrophotometer and again record the spectrum between 490 and 800 nm.

Addition of air, equilibration, and recording of a spectrum are repeated four more times. It is best not to risk denaturation of the protein by trying to get more than 5 data points from one run. The tonometer is then opened to atmospheric pressure, and, after equilibration, the final spectrum of oxyhemoglobin is recorded. *It is important not to forget to take this last spectrum.*

The calculation of the exact oxygen pressures requires the internal volume of the tonometer. This can be measured at the end of the experiment. First, weigh the tonometer and final oxyhemoglobin solution. Then fill the tonometer completely with water. Care will have to be exercised to avoid bubbles and to fill the tonometer completely to the stopcock closing. The volume of water can be measured by weighing the full tonometer and subtracting the weight of the empty tonometer plus Hb solution.

Safety

If animal (e.g. horse, cow) hemoglobin is prepared from whole blood, the blood should be obtained from a licensed animal facility. Gloves should be worn during the experiment, and hemoglobin solutions should be pipetted using disposable plastic pipets and a pipet filler.

Data Analysis

The data will be plotted in the form of a Hill plot [$\log_{10} (y / (1 - y))$ vs. $\log_{10} p(O_2)$] and as the saturation vs. $\log_{10} p(O_2)$. [$p(O_2)$ is the oxygen pressure in atm.] The saturation is determined from the visible spectra using the optical density at a single wavelength for the deoxygenated solution as corresponding to zero saturation and the OD at the same wavelength for the solution exposed to atmospheric air as that of full saturation. Assume that the saturation at intermediate states of oxygenation is proportional to the fractional spectral change. Use the average saturation determined at 540, 560, and 576 nm wavelengths. The appropriate baseline should be subtracted.

The calculation of the oxygen pressure is based on the known volumes of oxygen added and the calculated amount of oxygen bound, i.e.

$$p(O_2) = [\,(\text{mL air added} \times 0.21) - \text{mL } O_2 \text{ bound}\,]\,(P' / V'') \qquad (9)$$

where P' = (atmospheric pressure of air) - (relative humidity multiplied by saturation water pressure at room temperature), and V' is total internal volume of the tonometer in mL less the volume of the solution. The mL O_2 bound can be calculated from

$$\text{mL } O_2 \text{ bound} = (\text{fraction oxy Hb})\,(\text{total moles Hb}) \times (22.4 \times 10^3 \text{ mL/mole})\,(T/273) \qquad (10)$$

where T is the room temperature in K. The total moles Hb can be obtained *(14)* from the OD at 576 nm at the beginning of the experiment:

$$\text{total moles Hb} = (OD_{576\text{ nm}} / 1.65 \times 10^4)\,(\text{liters Hb used}). \qquad (11)$$

Students should prepare a table of their data giving mL air added, $OD_{540,560,576\text{ nm}}$ and the calculated saturation. Plots should be made of $\log_{10} [y / (1 - y)]$ vs. $\log_{10} p(O_2)$ (Hill plot) and of the saturation vs. $p(O_2)$. The spectra of oxyhemoglobin at the beginning and end of the run should be compared with respect to the ratio of $OD_{576\text{ nm}} / OD_{540\text{ nm}}$. A least squares fit to the Hill plot should be made, and the Hill coefficient n and $p_{50(Hb)}$ should be determined. The uncertainty in the Hill coefficient should also be calculated. A curve can be calculated to fit to the plot of saturation vs. $p(O_2)$ using the values of n and $p_{50(Hb)}$ obtained from the Hill plot. As an example of the quality of the data which can be obtained, we show in Figure 4(a) a composite graph of oxygen uptake data from 9 student groups and, in Figure 4(b), a Hill plot based on the same data. The values of n and $p_{50(Hb)}$ obtained from a least squares fit *(17,18)* of these data are 2.61 ± 0.09 and 0.0274 ± 0.0004 atm, respectively. The quoted uncertainties represent 3 standard deviations.

If time permits, several variations on this experiment are possible. First, the optical spectra can be analyzed in more detail by calculating the amount of ferric heme formed during the experiment. The paper of Benesch *et al. (14)* gives molar absorptivites for deoxy-, oxy-, and methemoglobin at eight wavelengths. The oxygen affinity of myoglobin might also be examined although this experiment is very demanding because of the low value of $p_{50(Mb)}$ and the hyperbolic nature of the myoglobin saturation curve. An alternative approach to the study of the oxygenation of myoglobin and hemoglobin can be found in reference 16.

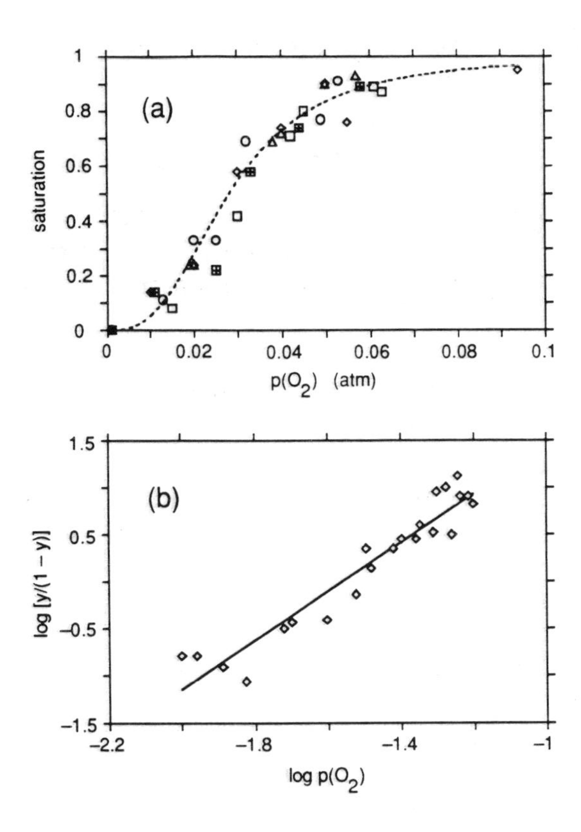

Figure 4. (a) Collected student data (9 separate groups) for the oxygen binding curve of hemoglobin. The curve through the data shows a fit with exponent n in eq. (5) set to 2.61 and $p_{50(Hb)}$ = 0.0274 atm. These values were determined from the least-squares fit to the data in the Hill plot shown in (b).

Acknowledgments

The authors gratefully acknowledge the Howard Hughes Medical Institute's Undergraduate Biological Sciences Grant to The Johns Hopkins University for funds used to acquire instrumentation for this laboratory.

Literature Cited

1. Eastman, M. P.; Kostal, G.; Mayhew, T. *J. Chem. Educ.* **1986**, *63*, 453.
2. Eastman, M. P. *J. Chem. Educ.* **1982**, *59*, 677.
3. Wooten, J. B.; Gurst, J. E.; Egan, W.; Rhodes, W. G.; Wagener, K. *J. Chem. Educ.* **1979**, *56*, 304.
4. Swiney, H. D.; Peters, D. W.; Griffith, W. B. Jr.; Mathews, C. W. *J. Chem. Educ.* **1989**, *66*, 857.
5. Shoemaker, D. P.; Garland, C. W.; Nibler, J. W. *Experiments in Physical Chemistry*, 5th ed.; McGraw-Hill: New York, 1989.
6. Quick, C. R. Jr.; Wittig, C. *J. Chem. Educ.* **1977**, *54*, 705.
7. van Stam, J.; Lofröth, J.-E. *J. Chem. Educ.* **1986**, *63*, 181.

8. Stryer, L. *Molecular Design of Life*; W.H. Freeman: New York, 1989.
9. Monod, J.;Wyman, J.; Changeux, J. P. *J. Mol. Biol.* **1965**, *12*, 88.
10. Lehninger, A. L. *Biochemistry: The Molecular Basis of Cell Structure and Function*, 2nd ed.; Worth Publishers, Inc.: New York, 1975.
11. Suslick, K. S.; Reinert, T. J. *J. Chem. Educ.* **1985**, *62*, 974.
12. Gouterman, M. In *The Porphyrins*; Dolphin, D., Ed; Academic Press: New York, 1978; Vol. III (Physical Chemistry, Part A), Chap. 1.
13. Adar, F. In *The Porphyrins*; Dolphin, D., Ed; Academic Press: New York, 1978; Vol. III (Physical Chemistry, Part A), Chap. 2.
14. Benesch, R.; Macduff, G.; Benesch, R. E. *Anal. Biochem.* **1965**, *11*, 81.
15. Riggs, A. *Meth. Enz.* **1981**, *76*, 5.
16. Barlow, C. H.; Kelly, K. A.; Kelly, J. J. *Appl. Spectrosc.* (in press).
17. Bevington, P. R. *Data Reduction and Error Analysis for the Physical Sciences*; McGraw-Hill: New York, 1969.
18. Taylor, J. R. *An Introduction to Error Analysis*; University Science Books: Mill Valley, CA, 1982.

Hardware List

A spectrophotometer capable of acquiring visible absorption spectra (in our laboratory, a Cary model 219) is required. A special lid for the sample compartment may need to be constructed to accommodate the large size of the tonometer. The tonometer was fabricated by a glass blower using a commercially available quartz cuvette with a graded quartz to pyrex seal (supplied, for instance, by Bodman Chemicals). The arm for the septum should be fabricated for a particular choice of septum. We have used 0.281" diameter, gas chromatography septa from Hamilton (part number 76010). Fixed-needle, gas-tight syringes with capacities in the 1-10 mL range are available from suppliers of Hamilton syringes. The design of the rocking device that we have used is shown in Figure 5. A commercial rocking device, for instance from Thermolyne, could also be adapted for use. For deoxygenating the Hb solutions, a tank of nitrogen is required, as well as a mechanical vacuum pump equipped with a Drierite trap and a trap cooled in dry-ice isopropanol.

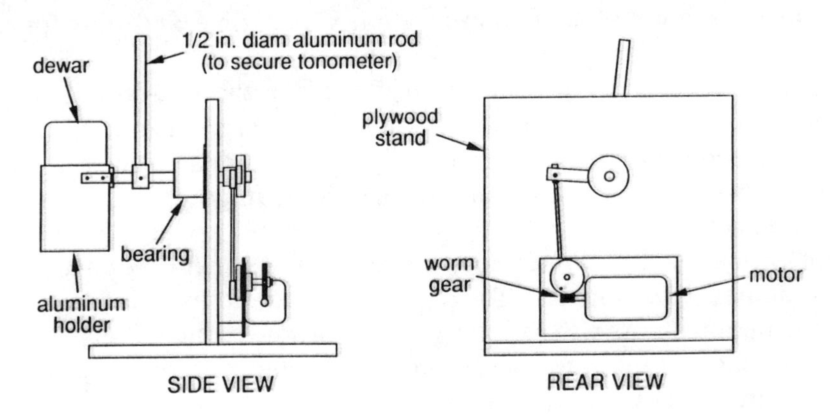

Figure 5. Schematic diagram of the rocking device used to equilibrate the Hb solutions. The dewar (filled with ice) containing the tonometer is tipped through an approximately 90° tilt once every ~30seconds.

RECEIVED October 14, 1992

CHAPTER 26

Mass Spectrometer or Mass Selective Detector Used To Study Gas-Phase Reactions

Colin F. MacKay

POTENTIAL HAZARDS: Cryogens, Vacuums

Much physical chemistry is done in the gas phase because intermolecular interactions can be minimized thus simplifying interpretation. The mass spectrometer, designed as it is to ionize and fragment molecules in the gas phase, is a natural and thus important analytical tool in many such studies. Its importance is enhanced by the fact that it is a universal detector. With the increasing availability of mass spectrometers linked to gas chromatographs in institutions both large and small, a series of experiments combining an introduction to simple gas phase techniques with the use of the mass spectrometer as an analytical tool seems of some value. Gas phase kinetics comes immediately to mind as one area in which such experiments might be developed.

The least expensive and therefore most widely available gas chromatograph–mass spectrometer combinations are tightly integrated and often make no provision for direct injection of gases. Such instruments could be used for gas phase kinetic studies by constructing an oven, withdrawing samples periodically with a gas syringe, quenching them if needed so that the reaction stops, and injecting them onto a suitable gas chromatographic column. By recording the time of withdrawal and calibrating the gas chromatograph so that peak areas can be converted into moles of significant reactants or products, an experimenter could generate the data for a plot of concentration versus time and thus for a kinetic analysis.

This work takes a different approach. Since the GC–MS already has an oven it choses to construct a vessel which will fit into that oven and which has a glass capillary connection for introduction of sample into the mass spectrometer source in the same way that the effluent from the gas chromatograph is introduced. Thus sample can be fed continuously from the reaction vessel to the mass spectrometer, and a plot of mass spectrometer counts versus time generated on the computer linked to that mass spectrometer. The partial automation that this represents is one aim of the experiment. As we shall see, with knowledge of the initial pressure of the reactant, extraction of the rate constant for a reaction from the data stored in the computer is straightforward.

As a demonstration reaction this work uses the dimerization of cyclopentadiene for several reasons. It is an example of an "evergreen" category of organic reactions, the Diels Alder reaction, one that has been continually under active investigation since its discovery by Diels and Alder in 1928.*(1)* Given its importance and its variety it is not surprising that the question of its mechanism has

2503–6/93/0380$06.00/0

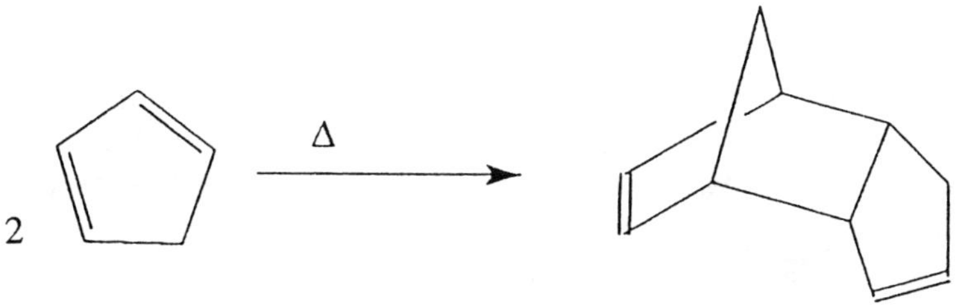

The dimerization of cyclopentadiene.

attracted the attention of many leading chemists and led to much discussion in the chemical literature. See, for example, the review by Sauer.*(2)* It provides opportunities for many variations, which is a desirable characteristic for the type of integrated laboratory at Haverford. Its kinetic parameters make it well suited for use under conditions easy to achieve with the the available instruments. There is a small problem in that dicyclopentadiene fragments easily into cyclopentadiene in the mass spectrometer, and so the major mass peaks of the two overlap. However, dicyclopentadiene does give a measurable parent peak and several significant fragmentation peaks above the cyclopentadiene mass range which are suitable for measurement.

The original kinetic measurements on the gas phase dimerization of cyclopentadiene came from two laboratories.*(3,4)* Neither experiment directly measured product as a function of time as does this work. Whether or not their Arrhenius parameters were in agreement within reasonable experimental error is not clear from the papers. Indeed, Harkness *et al.* expressed the opinion that the results of Benford, Khambata, and Wasserman *(3a)* were more reasonable than theirs. However, both groups were in general agreement that the activation energy for the reaction was low, and that there was a high negative entropy of activation. These results have been used as support for the position that the dimerization is synchronous, that is that the two bonds between the monomers are either formed simultaneously in one stage, or are formed so closely in time that the two events cannot be separated kinetically. Thus a single step encompasses two stages.

Background

You can find a discussion similar to this in the textbook by Berry, Rice, and Ross *(5)* as well as in other physical chemistry texts. The variation of the rate constant with temperature can be summarized in terms of two empirical parameters as expressed in the well known Arrhenius equation, $k = A \exp(-E_a/RT)$. The frequency factor is A, and E_a is the energy of activation. Various models have been used to interpret these parameters. One of the most widely used is that of Henry Eyring, which makes the simplifying assumption that the reaction occurs by a one dimensional motion along a reaction coordinate, that is along a minimum energy path between reactants and products. This path passes over an energy barrier, and the configuration of the system, that is the arrangement of atoms at the top of that barrier, is called the *activated complex*. It and quantities referring to it are identified by the symbol ‡ (double dagger). Thus the symbol for the energy difference between the activated complex and the reactants is $\Delta E^{o\ddagger}$, and the reaction forming the activated complex is represented as

$$\text{A} + \text{B} \rightleftharpoons (\text{AB})^{\ddagger} \rightarrow \text{Products} \quad (1)$$

Two further assumptions of the theory are that an equilibrium is established between the activated complex and the reactants and among the various degrees of freedom within the activated complex, and that the step from $(AB)^{\ddagger}$ to products occurs with a frequency k_BT/h, where k_B is the Boltzmann constant, h is Planck's constant, and T is the absolute temperature. This frequency corresponds to a time comparable to that for a bond vibration as is reasonable for a weak bond in a very unstable species. On the basis of the equilibrium assumption for the first step of (1) we formulate the equilibrium constant, K_V,

$$K_V = [(\text{AB})^{\ddagger}]/[\text{A}]\,[\text{B}] \quad (2)$$

and the free energy relations

$$\Delta G^{o\ddagger} = \Delta H^{o\ddagger} - T\Delta S^{o\ddagger} = -RT \ln K_V \quad (3)$$

where $\Delta G^{o\ddagger}$ is the free energy of activation, $\Delta H^{o\ddagger}$is the enthalpy of activation, $\Delta S^{o\ddagger}$ is the entropy of activation, R is the gas constant, T is the absolute temperature, and K_V is the equilibrium constant for reactants being converted to the activated complex. It is customary in this area to formulate the equilibrium constant in terms of concentrations rather than in terms of activities or pressures, in contrast to the common usage in thermodynamics.

On the basis of these assumptions we can write for the rate of the reaction

$$\text{d[Products]}/\text{d}t = -\text{d[A]}/\text{d}t = (k_BT/h)[(\text{AB})^{\ddagger}] \quad (4)$$

and since, from (2) $[(\text{AB})^{\ddagger}] = K_V[\text{A}][\text{B}]$ we arrive at the relation

$$-\text{d[A]}/\text{d}t = (k_BT/h)K_V[\text{A}][\text{B}] \quad (5)$$

This allows us to identify $(k_BT/h)K_V$ as the rate constant, k. From our thermodynamic relations we have $K_V = \exp(-\Delta G^{o\ddagger}/RT)$, which, by the use of (3) becomes

$$K_V = \exp[-(\Delta H^{o\ddagger} - T\Delta S^{o\ddagger})] \quad (6)$$

With this the equation for the rate constant becomes

$$k = (k_B\text{T}/h)K_V = (k_BT/h)\exp[-(\Delta H^{o\ddagger} - T\Delta S^{o\ddagger})/RT]$$

$$= A \exp(-E_a/RT) \quad (7)$$

We now take the natural logarithms of terms in (7) to give (8),

$$\ln k = \ln(k_BT/h) - (\Delta H^{o\ddagger}/RT) + (\Delta S^{o\ddagger}/R) = \ln A - (E_a/RT) \quad (8)$$

and differentiate (8) to give

$$\mathrm{d}\ln k/\mathrm{d}T = (1/T) + \Delta H^{o\ddagger}/RT^2 - (1/RT)(\mathrm{d}\Delta H^{o\ddagger}/\mathrm{d}T) = E_a/RT^2 \tag{9}$$

taking explicit recognition of the possible temperature dependence of $\Delta H^{o\ddagger}$. We follow through on this by beginning with the thermodynamic equation $\Delta H^{o\ddagger} = \Delta E^{o\ddagger} + \Delta(PV)^{\ddagger}$. Note that $\Delta E^{o\ddagger}$ is the energy of activation and differs from the activation energy, E_a. For a gas phase reaction in which n molecules are converted to $(AB)^{\ddagger}$ in the rate determining step, $\Delta(PV)^{\ddagger} = (1 - n)RT = -(n - 1)RT$, assuming ideal gas behavior. If we make the reasonable assumption that $\Delta E^{o\ddagger}$ is temperature independent, we can use the relation between $\Delta H^{o\ddagger}$ and $\Delta E^{o\ddagger}$to evaluate $(\mathrm{d}\Delta H^{o\ddagger}/\mathrm{d}T)$ as $-(n - 1)R$, and in turn use this in (9) to get our final expression

$$\mathrm{d}\ln k/\mathrm{d}T = (1/T) + \Delta H^{o\ddagger}/RT^2 + (n - 1)/T = E_a/RT^2 \tag{10}$$

Thus we have for the relationship between the $\Delta H^{o\ddagger}$ of our model and the experimental parameter E_a

$$\Delta H^{o\ddagger} = E_a - nRT \tag{11}$$

with $n = 2$ for this experiment. We can use this in (8) to derive an equation relating $\Delta S^{o\ddagger}$ to A giving relationships allowing evaluation of both of the parameters in the Eyring model.

In order to make an Arrhenius plot from which to extract the desired activation parameters the experimenter needs a procedure for extracting rate constants from the counts reported by the mass spectrometer. Such a procedure is provided by S. W. Tobey.*(6)* It requires only minor modifications for use in this work. $[A]_t$ is the concentration of cyclopentadiene at any time t, $[A]_0$ is the initial concentration of cyclopentadiene, λ_t is the sum of the counts in the mass peaks chosen to measure the product, and λ_∞ is the counts in those peaks at $t = \infty$. We can now write

$$[A]_t = 2\kappa(\lambda_\infty - \lambda_t) \tag{12}$$

where κ is a proportionality constant. At $t = 0$ we have

$$[A]_0 = 2\kappa\lambda_\infty \tag{13}$$

Putting $[A]_t$ and $[A]_0$ in the applicable form of the second order rate law equation we arrive at

$$\{[A]_0 - [A]_t\}/\{[A]_t[A]_0\} = kt \tag{14}$$

where k is the rate constant. After some straightforward manipulation we see that

$$\lambda_t = kt[A]_0(\lambda_\infty - \lambda_t) \tag{15}$$

Using (15) at times t and t' we write

$$\lambda = kt[A]_0(\lambda_\infty - \lambda) \tag{16}$$

and

$$\lambda' = kt'[A]_0(\lambda_\infty - \lambda') \tag{17}$$

Subtracting (16) from (17) and collecting terms we have

$$\lambda' - \lambda = k[A]_0\lambda_\infty(t' - t) + k[A]_0(t\lambda - t'\lambda') \tag{18}$$

Thus a plot of $(\lambda' - \lambda)$ versus $(t\lambda - t'\lambda')$ gives $k[A]_0$ as the slope, providing that $(t' - t)$ is held constant. Knowledge of $[A]_0$ allows calculation of k.

Methodology

The reaction vessel designed to fit in the gas chromatograph oven is shown in Figure 1. An experimenter with minimal glass blowing skills can easily fabricate it either with the parts in the figure caption or their equivalent. This 2-L reaction vessel will fit in any gas chromatographic oven of the size of that of the Hewlett-Packard 5890 gas chromatograph. A complete procedure follows.

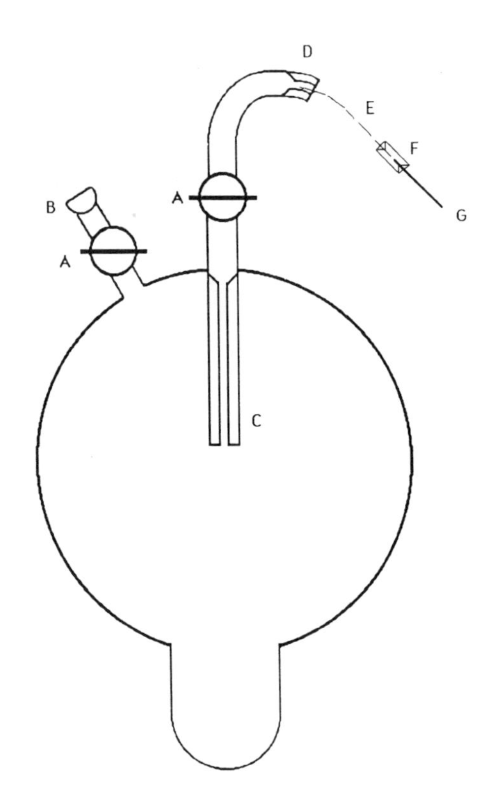

Figure 1. The Reaction Vessel

The bulb volume is nominally 2-L. The following is a key to the components attached to it.

A - Ace Glass high vacuum bakeable stopcock, 0-3mm orifice, Ace part number 8194-90. Can be baked at 230 °C.

B. - 12/5 mm "O" ring joint
C. - Glass capillary tubing, 8 mm OD, approx. 1.5 mm bore
D. - Glass capillary tubing, 7 mm OD, approx. 0.50 mm bore. Note the curve to relieve stress on the capillary.
E. - Uncoated deactivated fused silica tubing, 0.10 mm. ID, 0.30 mm OD, 60 cm long, Chrompack catalog number 004073. (Products of other companies may be used.) This serves as the pressure choke. A length of 60 cm gives a pressure in the Haverford MS of 2×10^{-5} torr. For lower pressures use a longer piece of tubing.
F. - 0.1 to 0.25 mm glass press-fit connector, Hewlett-Packard part # 5062-3553. The glass capillary tubing is sealed to this connector with epoxy. This work used gap filling slow curing epoxy, produced by Hardman, Inc., Belleville, N.J., #04003.
G - Uncoated deactivated fused silica tubing, 0.25 mm ID, 0.40 mm OD. Length should be such that it couples into the mass spectrometer source without strain or sharp bends. Chrompack catalog number 004074. (Products of other companies may be used.)

Preparation of Cyclopentadiene Crack dicyclopentadiene (tetrahydro-4,7-Methanoindene). For a procedure see *(7)* or *(8)*. The cracking may be carried out by students as part of the experiment or by the instructor beforehand. In the interests of saving time, at Haverford it will be carried out beforehand. Our dicyclopentadiene was procured from Aldrich Chemical Co., cat. no. 11,279-8. With a 2-L reaction vessel it takes about 3 mL of cyclopentadiene to achieve the recommended pressures. (See below.) **Clean-Up**. Acetone seems to be a suitable solvent. Placing the glassware containing the acetone in an ultrasonic cleaning apparatus containing some water facilitates this process. Again place all solutions into properly labelled containers.

Storage of the Cyclopentadiene Transfer the cyclopentadiene to a storage vessel equipped with a vacuum stopcock (teflon plug) and an "O" ring joint. This may be done after removing the stopcock by taking the cyclopentadiene up into a 5 mL syringe, inserting the syringe into the stopcock barrel, and ejecting the cyclopentadiene. Do this carefully. If cyclopentadiene holds up anywhere shake it down into the body of the storage vessel. If you have more than 5 mL of cyclopentadiene repeat the process as often as necessary. Insert and close the stopcock. Freeze the cyclopentadiene with liquid nitrogen, attach the storage vessel to the vacuum line using the "O" ring connector, and pump out the air. (Three styrofoam cups placed inside of each other make a suitable container for holding the liquid nitrogen.) Remove dissolved air from the cyclopentadiene using the following procedure. Close the stopcock and thaw the cyclopentadiene. This may be done by removing the liquid nitrogen and allowing the storage vessel to warm up. The process may be assisted by using a heat gun until liquid appears in the vessel, and then using a beaker filled with water as a heat source. Do not overheat. Refreeze in liquid nitrogen and repeat the pumping and thawing process at least one more time.

Filling the Reaction Vessel Attach the reaction vessel and the container with the thawed cyclopentadiene to a vacuum line equipped with a pressure measuring device.[1] (Note: The capillary tubing is quite sturdy unless it is kinked or scarred. Careful handling easily avoids this. Lay it out carefully when attaching it to the line.) Pump out the vacuum line and the reaction vessel. Close the stopcock leading to the vacuum pumping train. Transfer cyclopentadiene to the reaction vessel by opening the stopcock of the storage vessel. Close the stopcock of this vessel when the suggested pressure is reached (20 cm of mercury for most experiments). Close the stopcock of the reaction vessel and read the pressure of the sample. Record the temperature of the room. Place the styrofoam cups containing liquid nitrogen around the storage vessel and freeze the cyclopentadiene. Open the stopcock of the storage vessel. This draws the cyclopentadiene in the line back into the storage vessel. Close the storage vessel stopcock. If another experiment is to be done leave the storage vessel in place; if not remove it and store the cyclopentadiene in liquid nitrogen. (A spherical Dewar flask is ideal. Storage is not essential. The alternative is to crack the dicyclopentadiene as needed.) Close the stopcock to the pumping manifold and remove the reaction vessel. Support the glass capillary tubing when moving the vessel.

Running the Experiment Each student or pair of students should make a run at the temperature assigned by the instructor. Make sure that the mass spectrometer has been tuned.[2] If your gas chromatograph has a capillary column in the oven remove it as instructed. Place the reaction vessel in the oven of the gas chromatograph. Slide a nut of the kind used to attach capillary columns to the apparatus over the 0.25 mm capillary tubing. Slide an appropriate ferrule over the tubing. (Remember that the ferrule goes on backwards with a Hewlett-Packard instrument.) Thread the capillary tubing[3] into the channel leading to the source and adjust its placement so that it projects into the source but does not block the electron beam. Your instructor will show you how to do this. Tighten the nut to seal the capillary to the channel. Do not overtighten. Strong fingers will usually do the job. If not, careful use of a wrench is suggested. If you have a digital thermometer available put its probe through the injection port and place it at the site of the vessel. (The injection port can be easily dismantled to make this possible.) Allow the system to pump down until it reaches whatever is the usual pressure range for your mass

---

[1] At Haverford we used a simple mercury manometer. Cyclopentadiene complexes with mercury. This is not a serious problem in measuring pressure, but after several years of use the mercury may need to be changed. The vacuum line need not have a diffusion pump, but pumping is much faster with one.

[2] Note to the instructor: Use of the manual tune capability to tune for maximum counts on the131 mass peak of PFTBA (Perfluorotributylamine) is somewhat preferable to use of the Autotune capability, but is not essential.

[3] Note to the instructor: While the capillary tubing is sturdy, it is wise to keep a supply of 0.25 to 0.25 and 0.10 to 0.10 glass connectors handy for emergency repairs. These are available from Hewlett-Packard, Supelco, or any other source of gas chromatographic supplies.

spectrometer. (At Haverford we reach the 10^{-7} torr range.) Then open wide the stopcock in the line connecting the reaction vessel to the MS (A good guide is the lower "O" ring on the stopcock. It should be at least a quarter of the way above the beginning of the opening of the stopcock to avoid the stopcock being closed by expansion of components when the vessel is heated.) Read the pressure in the MS. Follow the instructions appropriate to your machine for calling up the method to be used and for initiating the experiment. The mass spectrometer method used for a run at 170° is shown in Table 1. Modify that method for your particular experiment according to Table 2 by changing time and final temperature. Make sure that you fill out and print out the scan acquisition form. Record the mass spectrometer parameters in your notebook. Make sure that you enter your assigned temperature and the time of the experiment in the proper places. When data has been collected for the time specified in the method the machine will automatically stop collecting. Use the gas chromatograph keypad to set the MS inlet temperature to 50 °C. When the oven has cooled off enough, open it, close the stopcock on the reaction vessel, unscrew the nut connecting the vessel to the MS (using a glove since it will be hot), and withdraw the capillary from the MS source. Install your installation's standard column in the gas chromatograph oven.
IMPORTANT. Remove the nut and ferrule on the capillary tubing immediately. This can be done by holding the capillary tubing with one hand, and using the other to sharply jerk the nut. The nut and ferrule can be removed most easily while the tubing is still a bit warm. Leaving the nut on the capillary tubing opens the possibility of a catapault effect caused by rapid movement of the nut at the end of the long lever arm provided by the tubing. This can stress the capillary tubing enough to break it.

Processing the Data Using the ion chromatogram capability of your data processing system, make and print out a plot of mass 132 versus time. Display a total ion chromatogram on the computer. Zoom in on the curve shown in 20-min segments. Inspect that curve to make sure that there are no discontinuous changes in the count rate. Data affected by discontinuities should be discarded during subsequent analysis. At the 12-min mark, use the average spectrum capability to average the data over a range of about 0.2-min. Print this out using the tab to printer capability. Repeat this at 2-min intervals for nominal temperatures below 190 °C, and at 1-min intervals for nominal temperatures of 190 °C and higher. (Nominal temperatures are those called for in the method used to govern the MS and the GC.) Remove the print-out of your data.

Setting up the Reaction Vessel for the Next Experiment Calculation of rate constants in conventional units requires knowledge of the initial cyclopentadiene pressure, and if this rather than relative rate constants is the goal, the vessel must be refilled for each experiment. Pump out the reaction vessel. If you use a vacuum line with a diffusion pump, bypass it so that the cyclopentadiene and dicyclopentadiene do not pass through the diffusion pump.[4] It is recommended that the outlet of the pump be connected to a hood using rubber tubing. We have had

---

[4] At Haverford we use a second simple line attached directly to a mechanical pump.

TABLE 1. THE MS METHOD USED

REAL TIME PLOT PARAMETERS

number of traces 2 initially ON time window 10.0

Plot #1	Spectrum	70.00	150.00	scale 240000
Plot #2	Total ion			scale 240000

TEMPERATURE PROGRAM AND HEATED ZONES

run time 106.00 equilibration time 0.00 splitless on time 0.00

level	initial temp	initial time	Rate (°C/Min)	final temp	final time	total time
1.	30	5.00	60.0	170	99.00	106.3
2.						
3.						
4.						
5.						

SCAN ACQUISITION

Solvent delay 0.00 eM volts 0 relative resulting volts 2550

start time	Low Mass	High Mass	Scan threshold	a/d samples (2^N)	scans per second
0.00	70.00	150.00	50	2	5.35

Comments: The 5 minute initial time is needed for the mass spectrometer to adjust to a steady target current. The oven begins heating at the 5 minute mark. The time that it takes to reach its final temperature increases as that final temperature increases. At 200 °C this takes slightly more than 5 minutes. Of course it takes longer for the gas in the bulb to reach this temperature. The quality of the fit to the second order equation is taken as evidence for constancy of temperature. As the temperature goes up the pressure in the MS source drops somewhat and then re-establishes itself. Students must adjust the run time and the final temperature to that assigned by the instructor as indicated in Table 2 below.

TABLE 2. SUGGESTED EXPERIMENTAL PARAMETERS

T(°C) (Nominal)	time of MS run(min)	frequency of averaging (min)	Tobey plot (t′-t) (min)
140	120	2	50
150	120	2	50
160	90	2	40
170	90	2	30
180	90	2	20
190	70	1	15
200	60	1	10

no problems with residual cyclopentadiene in the capillary line, but if there are such, place 20 torr of nitrogen in the vessel, attach it to the MS as described above, decouple the transfer line from the source, open the stopcock as described above, set the oven at 200 °C, and heat the apparatus for 30 min. Cool it, close the stopcock, and remove it from the GC oven as described above. At Haverford we intend that the instructor do this after the laboratory period if necessary. If there is a noticeable polymer deposit on the walls of the vessel it can be removed with acetone after a laboratory session.

Safety

Experimenters should wear safety glasses at all times. Avoid contact of dicyclopentadiene with the skin or eyes. Wear gloves and goggles while handling it. If contact is made wash off with water immediately. Avoid breathing it. Carry out the cracking in a fume hood. Dicyclopentadiene presents no unusual fire or explosion hazards. Residue of the distillation should be placed in a properly labeled container for proper disposal. The vacuum pump used in this experiment should evacuate into a hood.

Data Analysis

Calculate the quantities prescribed in the method of Tobey using Table 2 as a guide. At Haverford we use 12 min as a reference point, and at most temperatures calculate data for the Tobey plot using the appropriate $(t' - t)$ value, 2-min intervals below 190 °C and 1-min intervals at 190 °C and above. For the calculations which follow, use the sum of the counts in the mass 115, 116, 117, and 132 peaks. These are selected because they are major peaks in the dicyclopentadiene spectrum that are not found in the cyclopentadiene spectrum. Decide by looking at your ion chromatograph where you will start to analyze the data. For most of the experiments 20 min is a good place to start. Calculate λ by subtracting the counts at 12 min from those at 20 min. The time, t, for this would be 20 – 12 = 8 min. Add to 20 min the time interval recommended for the temperature of the experiment. Subtract the counts at 12 min from those at the time selected. This is λ', the net counts at time t'. (If the time interval is 20 min, this requires that you use the counts at 40 min. t' is then 8 + 20 = 28 min.) Make a Tobey plot of $(\lambda' - \lambda)$ versus $(t\lambda - t'\lambda')$. If available use a computer program which will do statistical analysis.[5] The slope of this plot is $k[A]_0$ where $[A]_0$ is the initial concentration of cyclopentadiene. This can be estimated using the ideal gas law since $n/V = P/RT$. Thus k can be calculated. Record your value for the rate constant k and your temperature on the sheet provided in the lab. When all students have recorded their values of k and T make an Arrhenius plot of the data. From your data calculate the activation energy, and the entropy and enthalpy of activation. Compare the activation energy to the energy of activation discussed

---

[5] Any spread sheet program is quite suitable. At Haverford we use Excel.

above. Comment on the significance of your results. In your discussion consider the relation between E_a and $\Delta H^{o\ddagger}$ for a reaction in solution.

Sample Results and Discussion

A sample Tobey plot is given in Figure 2. Figure 3 plots the Arrhenius data over the temperature range from 140 °C to 200 °C. As Table 3 shows, the agreement with the literature rate constants is not good at the higher temperatures. The Arrhenius parameters over the full temperature range reflect this. (Table 4). The rate constants at the lower temperatures are in better agreement with the literature.

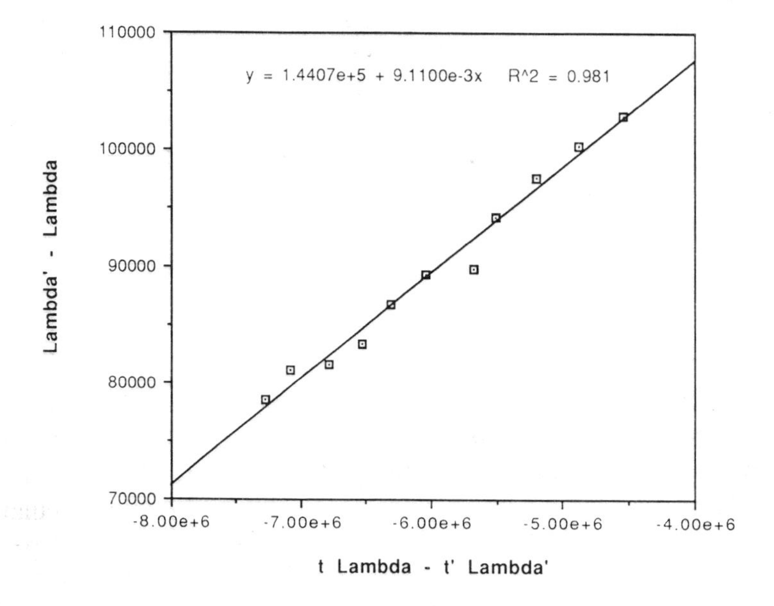

Figure 2. Sample Tobey Plot, T = 170 C

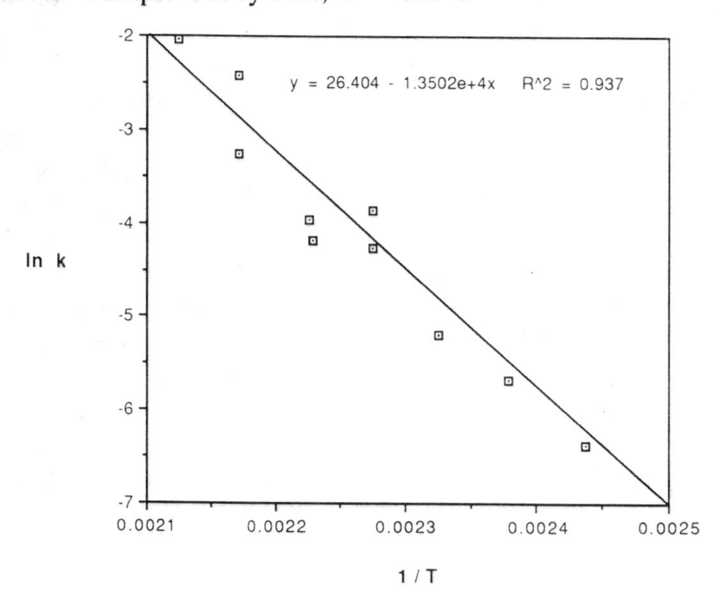

Figure 3. Arrhenius Plot 140 – 200 C

Thus a plot of the data from 140 °C to 160 °C (Fig. 4) gives more reasonable Arrhenius parameters. While these are in better agreement with the literature, that agreement still leaves something to be desired. The reasons for the increase in the discrepancy with the literature as the temperature increases are not established, but there is some evidence that there is a heterogeneous component to the reaction with that component being enhanced by the deposit of polymer on the walls of the reaction vessel. While the original hope was that the reaction could be run at the higher temperatures so that less mass spectrometer time would be used in data collection, the evidence now leads to the recommendation that it be run in the 140–160 °C range. In this range each experiment will take 120 min of mass spectrometer time.

The main focus of this work was on the development of a simple apparatus for injection of gases directly into the mass spectrometer part of a GC-MS. With such an apparatus other reactions may be run as well, the only limitation being that the stopcocks used are limited to a maximum temperature of 230 °C. Examples include the decomposition of *t*-butyl peroxide, the decomposition of acetic anhydride, and the conversion of dicyclopentadiene to cyclopentadiene. The gas phase

TABLE 3. COMPARISONS WITH LITERATURE

This work		Literature			
T(K)	k × 10^3	T(K)	k × 10^3(2)	T(K)	k × 10^3(3)
411	1.7±0.4	415	2.1±0.2	413	1.0
421	3.4±0.4	423	2.8±0.2	426	1.9
422	2.7±0.4				
431	5.6±0.6	431	4.3(4)	432	2.2
450	17±4(1)	450	9.7(4)	451	4.5
460	34±5	461	15(4)	463	8.1
459	34±5	459	14(4)		
471	110±15	471	22(4)		

(1)Average of 2
(2)Data from ref. (3).
(3)Data from ref. (4).
(4)These values are calculated from the values of the Arrhenius parameters given in ref (3a) which quotes an A value of 6.1±0.4 and an activation energy of 16.7±0.6 Kcal mole^{-1}. Given these generous error limits, the values found in this work, while consistently high, actually fall within them if both the A and E_a values are taken to their most favorable extremes. Despite this the agreement is not considered satisfactory.

TABLE 4. SUMMARY OF ARRHENIUS PARAMETERS

	$Log_{10}A$ (L mol^{-1}sec^{-1})	E_a (kJ mol^{-1})
This work (140-200 °C)	11±2	112±10
This work (140-160 °C)	8.3±1.8	87±14
Benford *et. al.* (3)	6.1±0.4	70±3
Harkness et. al.(4)	4.9	62

photolysis of acetone could be investigated with a similar apparatus equipped with a suitable window.

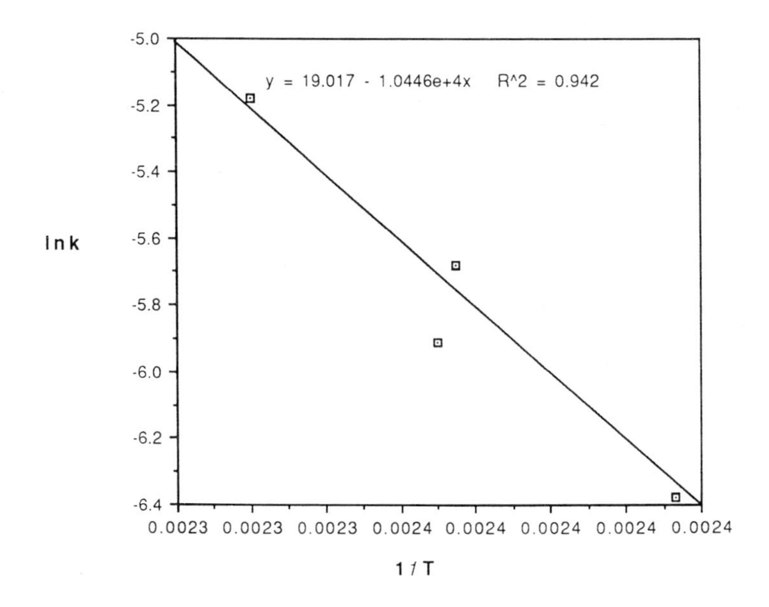

Figure 4. Arrhenius Plot 140 – 160 C

The experiment does introduce more material into the mass spectrometer than in its ordinary use as a gas chromatographic detector. One consequence of this is that the mass spectrometer will show a low target current after a run. Running it for 15–30 min with no sample present restores that current. A second consequence is that after 15–20 experiments the mass spectrometer source will need cleaning.[6] We have not had to clean either the quadrupole or the dectector. While the reaction vessel is quite sturdy it would be a good idea to have a second one available.

Acknowledgements

This work is part of the Physical Chemistry Project of the Mid-Atlantic Consortium of the Pew Science Program in Undergraduate Education sponsored by the Pew Foundation. The mass spectrometer used was purchased in part with funds provided by the National Science Foundation under the CSIP program. Additional funds were provided by the Keck Foundation and by the DuPont Company.

---

[6] Will Polik of Hope College has suggested that the use of an automated stopcock would greatly ameliorate this problem. It would, however, make the apparatus somewhat more complex.

Literature Cited

1. Diels, O., and Alder, K, *Justus Liebig's Ann. Chem.,* **1928**, *460*, 98-122.
2. Sauer, J., *Angew. Chem. Internat. Edit*, **1967**, *6*, 16-33.
3a. Benford, G.A., Khambata, B.S., and Wassermann, A., *Nature*, **1937**, *139*, 669-70.
3b. Benford, G. A., and Wasserman, A., *J. Chem. Soc.,* **1939**, 362-67.
4. Harkness, J. B., Kistiakowsky, G. B., and Mears, W. H., *J. Chem. Phys.*, **1937**, *6*, 682-694
5. Berry, R.S., Rice, S.A, and Ross, J., *Physical Chemistry*, John Wiley & Sons, New York, 1147-1157.
6. Tobey, S.W., *J. Chem. Educ.*, **1962**, *39*, 473-4.
7. Feiser, L.F., *Organic Experiments*, 2nd ed., Raytheon Education Co., Lexington, Mass. (1968), p. 85.
8. *Org. Synth.*, Rabjohn, N., Ed., John Wiley & Sons, New York, 1963: Coll. Vol. 4, 238.

Apparatus

The mass spectrometer at Haverford consists of a Hewlett-Packard 5988A mass spectrometer linked to a Hewlett-Packard 5890 gas chromatograph. While this mass spectrometer has more pumping power than the typical mass selective detector linked to a gas chromatograph, any gas-chromatograph mass spectrometer combination should do. A greater length of 0.10 mm capillary tubing will provide a suitably low pressure for a mass spectrometer with less pumping power and further will also lessen the degree of contamination of the mass spectrometer source. A Fluke model 51 digital thermometer was used to measure temperature in the oven. The components of the reaction vessel are detailed in the caption of Figure 1.

RECEIVED October 14, 1992

CHAPTER 27

An Introductory Experiment in Cyclic Voltammetry: The Oxidation of Ferrocene

David A. Van Dyke

Historically, electrochemistry has been a very important subdiscipline within physical chemistry. Much of what is known about solution-surface interfaces, ionic behavior in the solution phase and in molten salts, and the thermodynamics of oxidation-reduction (redox) reactions has been derived from electrochemical studies. Many of the greatest names in the history of physical chemistry made contributions in the area of electrochemistry: e.g., Faraday, Nernst, Gibbs, Helmholtz, Arrhenius, Tafel, Volmer, Stern, Planck, Gouy, Chapman, Cottrell, and Heyrovsky. Important advances have continued in the field, especially during the last quarter-century, due to the development of new techniques and continued advances in instrumentation. These advances have insured the indispensability of electrochemical methods for providing vital information about phenomena as diverse as adsorption, biological charge-transfer reactions, corrosion processes and other surface phenomena, and photoelectrochemical processes.

However, a perusal of the electrochemistry experiments included in the most recent editions of physical chemistry laboratory textbooks *(1-3)* would seem to indicate that advances in electrochemistry ended in the 1930's. In general, these experiments require only antiquated equipment, such as potentiometers, galvanometers and Wheatstone bridges in order to examine classical electrochemical concepts such as ion mobilities, the temperature dependence of standard potentials, and activity coefficients. (Happily, modern electrochemistry experiments are available in recent instrumental analysis texts, e.g. *(4)*).

The challenges to presenting a modern electrochemistry experiment in the physical chemistry lab are numerous. Modern experiments require modern equipment. So student-compatible equipment, preferably microprocessor- or computer-controlled, may need to be obtained. Another challenge is presented by the fact that electrochemistry represents a much smaller fraction of the total activity in the field of physical chemistry today than it once did. Hence, there is usually less time available for the study of electrochemistry; so the experiments and techniques that are chosen should concretely illustrate as many important electrochemical concepts as possible. Obviously it is desirable to select techniques that are versatile and broadly used.

Cyclic voltammetry is a technique that meets many of these criteria. Its prominence as an electrochemical method dates back to the groundbreaking theoretical work of Nicholson and Shain *(5)*. It can quickly provide insights into the redox behavior of a parent compound, but can also be used to probe the stability of the products of the initial electrochemical process. For reversible systems, it can be used to determine an estimate of $E^{0'}$, the formal potential for a redox couple; n, the number of electrons gained or lost by the molecule; D, the diffusion coefficient of the molecule; and whether adsorption of reactants or products is taking place. In systems where the initial redox process is followed by either chemical or

2503–6/93/0394$06.00/0

electrochemical reaction, cyclic voltammetry can sometimes yield valuable insights into the kinetics and mechanism of the process.

Several cyclic voltammetry experiments have appeared in the chemical education literature *(6-12)*. However, many of these are not suitable for an introductory physical chemistry laboratory, either because of their advanced level or because their focus is more appropriate to an inorganic chemistry laboratory. Most of these experiments involve the study of metal ions or complexes in aqueous solution. By contrast, cyclic voltammetry is used in this experiment to investigate the simple, reversible electron transfer reaction of the ferrocene/ferrocenium couple in an aprotic solvent *(13)*. The use of a nonaqueous solvent precludes the complex electrode/solvent interactions and pH dependencies that complicate many aqueous electrochemical systems. It also helps dispel the common misperception among students that electrochemistry deals only with the redox behavior of metals and metal ions in aqueous solution. Although an aprotic solvent (in this case, dichloromethane) allows the study of a simple electron transfer redox process, the solvent's high resistivity and low dielectric constant do introduce uncompensated resistance, which makes the results a bit more difficult to interpret.

Part of the emphasis in this experiment is on acquainting the student with some of the physical and electrochemical principles that underlie the observed voltammetric responses. Among the key electrochemical concepts that surface in this experiment are: formal potentials, thermodynamic and chemical reversibility, diffusion and the diffusion coefficient, faradaic current, charging current, and uncompensated solution resistance. This experiment is also meant to introduce the student to the basics of the cyclic voltammetry technique; it provides the necessary background for more advanced experiments, e.g. studies of structure/reduction potential correlations, or investigation of electrochemical redox processes that include coupled chemical reactions. For example, one such experiment used in our laboratory course involves a study of the kinetics and mechanism of the reaction of the cation radical of 9,10-diphenylanthracene with pyridine *(14,15)*. Other experiments of a similar nature are available in the literature *(6,7)*.

Theory

Basic Electrochemical Concepts In order to understand the cyclic voltammetry technique, students need a good grasp of the fundamentals of electrochemistry. An exceptionally lucid exposition of some of the distinctive concepts of electrochemistry *(16)* has appeared in a special issue of *The Journal of Chemical Education*, which also contained an overview of the concepts that should have been learned in general chemistry *(17)*. A more detailed treatment is found in *(18)*. Presented below is a sketch of basic electrochemical terms and concepts that are important in this experiment; for more details consult the references.

Potential and *current* are two fundamental concepts that are often confusing to students. From physics, the formal definition of potential is "the work required to bring a unit test charge [e.g., an electron] from infinite distance to a point of interest" [such as inside an electrode] *(16)*. Thus the potential of an electrode is a measure of the energy of the electrons on the electrode, which obviously depends on whether there is a net charge - an electron excess or deficiency - on the electrode. Current, in simple terms, is the net rate at which electrons flow into or out of an electrode. In general, one cannot simultaneously control both the potential and current in an electrochemical experiment. In amperometric techniques, of which cyclic voltammetry is an example, the potential is controlled and the resulting current is measured. By way of contrast, the familiar measurement of pH with a glass electrode and pH meter is an example of a potentiometric method, which involves keeping the current negligible and measuring the potential.

In most cyclic voltammetry experiments, the electrochemical cell contains three electrodes: a working electrode, a reference electrode, and an auxiliary or counter electrode. A potentiostat controls the potential "applied to" the working electrode and measures any current that flows into or out of the electrode as a result. In more exact terms, the potential that one wishes to control is the potential difference between the working electrode and the solution in the cell. Measuring this potential difference in an absolute sense is impossible because of the large, unmeasurable potential gradient that always exists at an electrode-solution interface and because a second electrode (with its own metal-solution interface) must be present for a complete circuit *(19)*. Therefore, a reference electrode is used to provide a stable, reproducible potential reference point in the solution being studied. The potentiostat keeps the working electrode at virtual ground and adjusts the potential applied to the auxiliary electrode so that the desired difference in potential between the working electrode and the reference electrode is maintained *(19)*. (This turns out to have important consequences for how the electrodes should be positioned in the cell.) If the potential of the working electrode is made positive enough with respect to species in the solution, those species may transfer electrons to the electrode and be *oxidized*, producing an *anodic* current. Conversely, if the working electrode potential is negative enough compared to the reduction potential of a species in solution, electrons may flow from the electrode to cause *reduction* of that species, producing a *cathodic* current. In keeping with an arbitrary convention, anodic currents will be considered negative and cathodic currents positive. Thus, a working electrode can be thought of as a "reagent" of adjustable oxidizing or reducing strength.

To be able to provide a fixed, stable potential, a reference electrode must contain both forms of a redox couple kept at a fixed composition. The potential of one such reference electrode, the normal hydrogen electrode (NHE), has been defined to be the origin of the electrochemical potential scale at all temperatures:

$$\text{NHE: } Pt/H_2(a=1)/H^+(a=1, \text{aq.}) \qquad E^0 = 0 \text{ V vs. NHE}$$

The NHE is rather inconvenient to use experimentally, so other reference electrodes are more commonly used, including the saturated calomel electrode (SCE), and the silver/silver chloride (Ag/AgCl) electrode, which are shown below:

$$\text{SCE: } Hg/Hg_2Cl_2/KCl \text{ (sat'd, aq.)} \qquad E = +0.2412 \text{ V vs. NHE}$$

$$\text{Ag/AgCl: } Ag/AgCl/KCl \text{ (sat'd, aq.)} \qquad E = +0.197 \text{ V vs. NHE}$$

where E is an electrode potential with respect to a reference and E^0 is a potential under standard conditions.

To have a complete circuit implies that any current flowing between the solution and the working electrode must be matched by an equal and opposite current at another electrode. Passing this current through the reference electrode is undesirable because the concomitant oxidation or reduction will alter the electrode's composition and thereby its potential. Instead, this function is performed by the auxiliary electrode. For example, if as is shown in figure 1, the working electrode oxidizes a species, such as ferrocyanide,

$$Fe(CN)_6^{-4} \rightarrow Fe(CN)_6^{-3} + e^- \text{ (on the electrode)}$$

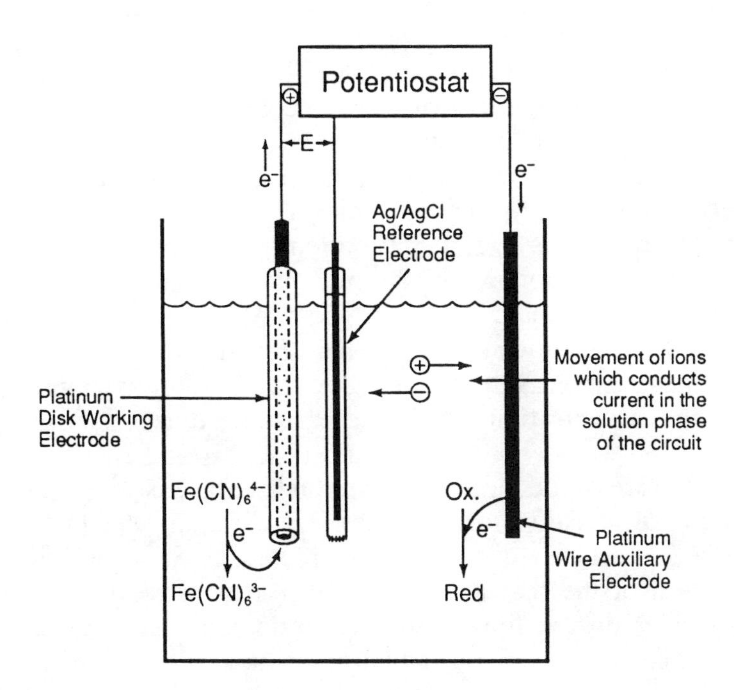

Figure 1. Diagram of an electrochemical cell in which ferrocyanide is being oxidized at the working electrode. The potentiostat controls the potential on the working electrode with respect to the reference electrode with an high impedance circuit so that negligible current flows between the two. The potentiostat measures the current flow between the working and auxiliary electrodes.

then there must be a reduction occurring at the auxiliary electrode. (The reaction that occurs there is usually not of interest.) To have a complete circuit also requires that charge pass through the solution between the working and auxiliary electrodes. For that reason, a fairly high concentration of supporting electrolyte is maintained in the solution; it is a salt or other ionic species that must have a reasonably high solubility and dissociation in the solvent being used. Thus in the solution phase the circuit is completed and charge buildup is neutralized by the electrostatically-induced movement of ions in the electric field, which is known as migration.

Because a potentiostat can precisely control the potential applied to the working electrode, in effect it becomes a reagent with variable oxidizing or reducing strength. Since extremely positive or negative potentials will respectively cause oxidation or reduction of the solvent and/or the electrode itself, there are limits to the useful potential range. By careful selection of relatively inert working electrode materials (e.g., Pt, Au, Hg), solvent and supporting electrolyte, electrochemists generally obtain a working range of 1 to 6 V in a given medium. Within this working range, as the potential is varied, one might observe one or more oxidative processes taking place at positive potentials, perhaps no electrochemical reactions at all at intermediate potentials, and then reduction processes occurring at more negative potentials.

It is crucial to recognize the heterogeneity of electrochemical processes; the critical electron transfer step occurs across the interface between two different phases, namely between the electrode and an electroactive species in the solution. Thus the "action" in electrochemical processes takes place at or near the electrode surface. The processes that bring molecules near the electrode surface where they can undergo an electrochemical reaction are collectively known as mass transport processes. These include migration, convection, and diffusion. Convection occurs when there is bulk movement of solution, as in a stirred solution. Diffusion is the net movement of randomly moving molecules in response to a concentration gradient; that is, molecules in random motion tend to move from regions of high concentration to regions of lower concentration. Diffusion is relatively easy to model theoretically, and is assumed to be the only significant mode of mass transport in cyclic voltammetry experiments. For this assumption to be valid, the solution must be quiescent and have a high concentration of supporting electrolyte relative to the concentration of the species of interest so that migration is minimized.

The structure of the electrode–solution interface is shown in Figure 2a. As an electrochemical oxidation or reduction occurs at an electrode, concentration gradients of both reactant and product are created, as shown in Figures 2b and 2c; the concentration of the reactant at the electrode surface is depleted and more reactant molecules diffuse toward the electrode, while the concentration of the product(s) builds near the electrode and product molecules diffuse away. The region of the solution thus affected is known as the diffusion layer. The diffusion layer typically extends about 10^{-4} – 10^{-2} cm out from the electrode surface, depending on how much time has elapsed since the onset of the electrochemical reaction. The portion of the solution too remote from the electrode to be affected is termed the bulk.

Diffusion processes are described by a pair of differential equations known as Fick's Laws that relate the flux of the diffusing species and its concentration as a function of time and position. (For the derivation of Fick's Laws see pp. 127–133 in *(18)*, pp. 9–19 in *(19)*). In the case of a planar electrode surface embedded in a non-conducting matrix, diffusion can be approximated as a linear (one-dimensional) process, for which Fick's first law is:

$$-J_{(x,t)} = D\,\partial C_{(x,t)}/\partial x$$

The flux $J_{(x,t)}$ is the net movement of the species per unit area at point x as a function of time t; its usual units are mol $cm^{-2}s^{-1}$. The flux is related to the concentration gradient $\partial C_{(x,t)}/\partial x$ (with units of mol $cm^{-3}cm^{-1}$) by a proportionality constant D, the diffusion coefficient (which has units of $cm^2\ s^{-1}$). The magnitude of the diffusion coefficient depends on factors such as the molecular size and shape

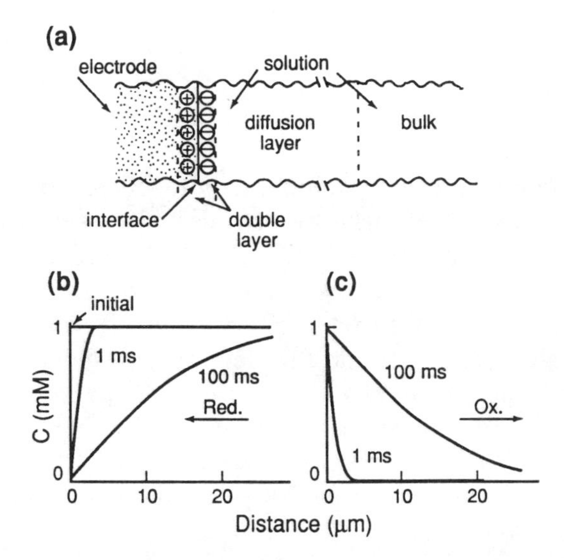

Figure 2. (a) Spatial structure of the electrode–solution interface. (b) Concentration profiles of a species being oxidized (at $E > E^{0'} + 120$ mV) showing the growth of the diffusion layer. The distances are from the electrode surface; times are from the initiation of electrolysis. (c) Concentration profiles of the oxidation product. Adapted with permission from *(16)*. Copyright 1983 Journal of Chemical Education.

of the species, as well as the temperature and viscosity of the solution. Fick's second law is

$$\frac{\partial C_{(x,t)}}{\partial t} = D \left(\frac{\partial^2 C_{(x,t)}}{\partial x^2}\right)$$

which describes the change in concentration of the diffusing species with time. Solutions of this equation provide concentration profiles, such as are shown in Figures 2b and 2c, while the current is obtained from the flux at the electrode surface.

Figure 2a also illustrates that a double layer exists at the immediate electrode–solution interface. Both the electrode surface (which usually has adsorbed ions) and the layer of solution next to it usually retain excess (and opposite) charges. As a result, the double layer behaves much like a capacitor, and changing electrode potential results in a change in the stored charge, which requires a charging current. In cyclic voltammetry experiments the charging current is responsible for a non-zero "baseline". This limits the sensitivity of cyclic voltammetry so that the concentration of the electroactive species must usually be at least 10^{-5} M.

Currents that arise from oxidation–reduction reactions are known as faradaic currents, which obviously include currents due to redox processes of the species being studied. One frequently has to contend with other, undesirable faradaic currents known as background currents, which are the result of oxidation or reduction of the solvent, supporting electrolyte, electrode material, or impurities therein. The reduction of oxygen (which has a fairly high solubility in many solvents) is a common source of significant background current at negative potentials. Obviously, background currents become an increasingly severe problem as the potential is made more extreme, until finally they become large enough to overwhelm any current from the species of interest at the end of the useful potential range.

The magnitude of the faradaic current reflects the rate of the overall electrochemical process at the working electrode. The overall rate is determined by the kinetics of the individual steps in the electrode process, including the rate of mass transfer of electroactive species to the electrode; the rate of adsorption and desorption on the electrode surface, if that occurs; the rate of the electron transfer(s) between the electrode and the electroactive species; and the rates of any reaction steps occurring in the solution that are part of the overall reaction. The step (or steps) that is the slowest kinetically will obviously be the chief determinant of the overall rate, and thus the current. Which step is rate-determining depends on the particular electrode reaction taking place, as well as the potential and other experimental conditions.

The applied potential is a critical factor in determining the net rate of the electron transfer step. Consider the electrochemical oxidation

$$\text{Red} \rightleftharpoons \text{Ox} + ne^-$$

where n is the number of electrons per molecule oxidized. The relationship between the potential and the relative concentrations of the oxidized and reduced forms **at equilibrium** is given by the *Nernst Equation*:

$$E = E^{o\prime} + \frac{RT}{nF} \ln \frac{[\text{Ox}]_{x=0}}{[\text{Red}]_{x=0}}$$

$$= E^{o\prime} + 2.303 \frac{RT}{nF} \log \frac{[\text{Ox}]_{x=0}}{[\text{Red}]_{x=0}}$$

where E is the applied potential, $E^{0\prime}$ is the formal potential of the redox couple, R is the gas constant (8.314 J $mol^{-1}K^{-1}$), T is the absolute temperature (K), and F is the Faraday constant (9.64846×10^4 C/equiv.). The subscript $x=0$ signifies that the concentrations under consideration are surface concentrations at the electrode-solution interface, *not* the concentration of the bulk solution. For $n = 1$ at 25 °C, $2.303\ RT/nF = 0.059$ V. So under those conditions and when the applied potential E is equal to $E^{0\prime}$, the surface concentrations of the oxidized and reduced forms at equilibrium will be equal; when E is 59 mV more positive than $E^{0\prime}$, $[\text{Ox}]/[\text{Red}]_{x=0}$ will be 10:1; or when E is 118 mV more negative than $E^{0\prime}$, then $[\text{Ox}]/[\text{Red}]_{x=0}$ will be 1:100. However, whether the Nernst Equation is applicable depends on the kinetics of the electrode reaction as well as the stability of the oxidized and reduced forms. For a thermodynamically reversible reaction, at applied potentials near $E^{0\prime}$, the charge transfer rates are rapid on the time scale of the experiment so that Nernstian concentration ratios of the oxidized and reduced forms are always present at the electrode surface. (This is also sometimes called charge-transfer reversibility.) Thermodynamically reversible reactions typically have simple electrode mechanisms, such as only a simple electron transfer. An example is the oxidation of ferrocyanide to ferricyanide:

$$Fe(CN)_6^{-4} \rightleftharpoons Fe(CN)_6^{-3} + e^-$$

In practice, reversibility is a relative concept which depends on the time frame of the experiment, since all electrochemical processes occur at finite rates. (See pp. 44-47 of *(18)* for a complete discussion.) For example, even though the oxidation of ferrocyanide is reversible in the time frame used for most cyclic voltammetry experiments, there are considerably faster, "more reversible" redox reactions.

Thermodynamically irreversible (or charge-transfer irreversible) electrode reactions, on the other hand, have infinitesimal electron transfer rates when the applied potential is set to $E^{0\prime}$, and an overpotential must be applied to force the reaction to occur. Irreversibility is typically seen in complex multistep electrode reactions. A classic example is the reduction of oxygen to water:

$$O_2 + 4H^+ + 4e^- \rightleftharpoons 2H_2O$$

Even at a platinum electrode, this reaction requires an overpotential of several hundred millivolts beyond $E^{0\prime}$ before it will proceed at a significant rate; the overpotentials at other metals are considerably greater. Reactions that fall between the reversible and irreversible cases are termed quasireversible.

Chemically reversible electrode reactions are those in which reversal of the cell current (by application of an appropriate potential) also reverses the electrode reaction. Chemically reversible reactions are not necessarily thermodynamically reversible. Chemical irreversibility occurs when the product of the initial charge-transfer reaction (denoted as an "*E*" step) is unstable or chemically reacts with other species in the solution (denoted as a "*C*" step), eg.:

$$E: \qquad Red \rightleftharpoons Ox + ne^-$$

$$C: \qquad Ox + X \rightarrow \textit{electro-inactive products}$$

Such a reaction is labelled an "*EC*" mechanism. Other more complicated mechanisms for chemically irreversible reactions (e.g. *ECEC*, *CEC*) are also encountered.

Cyclic Voltammetry Voltammetry techniques involve sweeping the potential applied to a working electrode and recording the resulting current. In cyclic voltammetry, the potential is varied at a constant rate from some starting potential out to a switching potential E_λ, then the scan direction is reversed back toward the starting potential. In this way the lifetime and fate of species generated by an electrochemical reaction in the forward scan can be examined by the response in the reverse scan. Besides Nicholson and Shain's pioneering paper *(5)*, a number of good introductions to the technique have appeared in the chemical education literature *(20-22)*) and texts, including *(18)*.

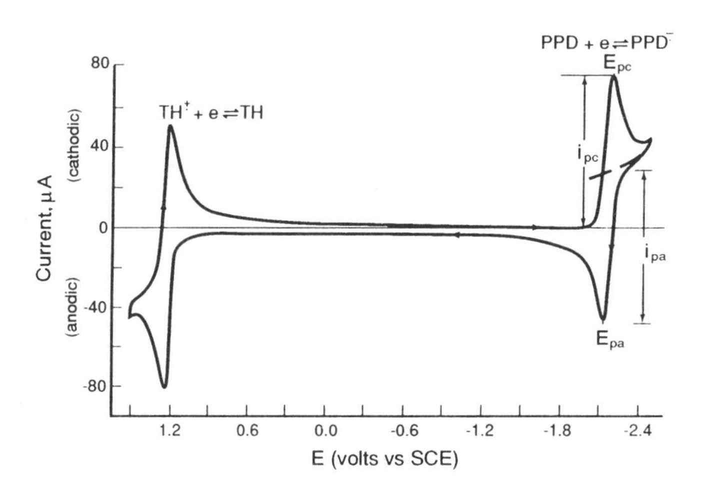

Figure 3. A cyclic voltammogram of a CH_3CN solution containing 1.0 mM thianthrene and 1.0 mM 2,5-diphenyl-1,3,4-oxadiazole with 0.1 M tetra-n-butyl ammonium perchlorate as supporting electrolyte. Adapted from *(23)* in *(13)*; used by permission.

Figure 3 is a cyclic voltammogram of a solution containing two electroactive compounds, thianthrene (TH) and 2,5-diphenyl-1,3,4-oxadiazole (PPD). The scan begins at -1.0 V vs. SCE, proceeds at 100 mV/s to –2.5 V versus SCE, then reverses back toward positive potentials. At +1.5 V the scan is reversed again; it continues toward negative potentials until it is stopped at –1.0 V. As the cyclic voltammogram shows, there is very little current at potentials between +1.0 V and –2.2 V which indicates that nothing in the solution is oxidizable or reducible in this range. At –2.2 V a cathodic current arises from the reduction of PPD to an anion radical, $PPD^{\overline{\cdot}}$. The current peaks at –2.3 V, then begins to drop as PPD is depleted from the diffusion layer, and fewer PPD molecules arrive at the electrode surface to be reduced. At –2.5 V the cathodic current begins to rise again, due to the onset of a background current from the reduction of the solvent or supporting electrolyte; the cathodic limit of the useful potential window in this medium has been reached. When the scan is reversed at –2.5 V, significant cathodic current continues to be generated until the threshold potential for PPD reduction is crossed. As more positive potentials are applied, an anodic current peak is observed at –2.2 V as the $PPD^{\overline{\cdot}}$ generated in the forward scan is oxidized back to PPD. As the $PPD^{\overline{\cdot}}$ in the diffusion layer is depleted, the anodic current declines back toward zero.

If the charge–transfer reaction is reversible and neither the reactant nor the product is adsorbed on the electrode surface, the height of the reverse current peak is an indication of the stability of the reduction product. If the ratio of the reverse and forward peak currents i_{pr}/i_{pf} = 1.0, then the product of the forward scan is stable, at least on the time scale of the experiment. Figure 3 shows how the cathodic peak current i_{pc} (= i_{pf}) and anodic peak current i_{pa} (= i_{pr}) are measured. In the case of PPD, the $PPD^{\overline{\cdot}}$ anion radical is quite stable in acetonitrile solution, as indicated by the equality of i_{pr} with i_{pf}.

Suppose instead that the $PPD^{\cdot-}$ reacted with another species in solution or decayed to an electroinactive product (an *EC* mechanism). The result would be a diminished concentration of $PPD^{\cdot-}$ in the diffusion layer and therefore a diminished reverse current peak. In fact, if the following reaction or decay were very fast, no reverse current would be seen at all and the i_{pr}/i_{pf} ratio would be zero. Even in that case, it might be possible to recapture a glimpse of the $PPD^{\cdot-}$ by using a very fast scan rate. This would reduce the time that the $PPD^{\cdot-}$ must survive before being detected by the anodic current peak from its oxidation in the reverse scan. If the scan rate were increased by several more orders of magnitude, the $PPD^{\cdot-}$ might appear to be essentially stable and the i_{pr}/i_{pf} ratio would approach unity. Thus by varying the scan rate in a cyclic voltammetry experiment one can obtain kinetic information about the reactions which electrochemically-generated species undergo. With appropriate instrumentation, useful scan rates range over about five orders of magnitude, from 10 mV/s up to about 2000 V/s. With an X–Y recorder, the maximum scan rate is limited to about 1 V/s.

In Figure 3, note that as the reverse scan proceeds beyond –2.0 V toward more positive potentials, no electrochemistry occurs until the oxidation potential of thianthrene (TH) is reached at about +1.2 V vs. SCE. The anodic current peak that is observed there is the result of the oxidation of TH to its cation radical, $TH^{\cdot+}$. This radical is also stable: the cathodic peak observed after the second scan reversal at E_λ = +1.50 V is equal in height to the anodic current peak. Finally, as the scan returns to the initial starting potential (–1.0 V), the current again is negligible and the composition of the diffusion layer has been returned to its initial state.

Although both the reduction of PPD and the oxidation of TH are reversible electrochemical reactions and therefore produce similar voltammetric responses, other types of electrode mechanisms will yield very different cyclic voltammograms. The various mechanisms and the corresponding voltammetric responses are presented in *(5)* and summarized in *(18,21)*. Only two cases will be discussed in detail here, namely thermodynamic reversibility and total charge-transfer irreversibility.

For a reversible reaction, where the electron transfer process is very facile and both the oxidized and reduced forms of the redox couple are stable, both the anodic peak potential E_{pa} and the cathodic peak potential E_{pc} are independent of scan rate and concentration. (But see the caveat on the effect of uncompensated solution resistance below.) In fact, the average of E_{pa} and E_{pc} is a good approximation to $E^{0'}$ for the redox reaction. The difference between E_{pa} and E_{pc}, denoted as ΔE_p, is a useful diagnostic quantity; it will be close to $57/n$ mV at all scan rates for a reversible reaction, but will be greater than $57/n$ mV and increase with scan rate in cases where the charge transfer is irreversible or quasireversible.

General expressions for currents in cyclic voltammetry have been derived by application of Fick's laws, the Nernst relationship, and the time dependence of the applied potential. (See pp. 213-231 of *(18)*). When it is assumed that 1) diffusion is the sole mode of mass transport; 2) the diffusion occurs in a linear fashion; and 3) the temperature of the solution is 298 K, then for a reversible system,

$$i_{pf} = (2.69 \times 10^5)\, n^{3/2} A\, D^{1/2}\, v^{1/2}\, C^*$$

where the current is in amperes, n is the number of electrons transferred per molecule, A is the electrode area (cm^2), D is the diffusion coefficient of the electroactive species (cm^2/s), v is the scan rate (V/s), and C^* is the bulk concentration of the electroactive species (mol/cm^3). For an irreversible system,

$$i_{pf} = (2.99 \times 10^5)\, n\, (\alpha n_a)^{1/2} A\, C^*\, D^{1/2}\, v^{1/2}$$

where all the units are the same as for the equation above. The α term is the transfer coefficient; n_a is the number of electrons in the rate-determining step of the electron transfer process. The transfer coefficient is a measure of the asymmetry in the energy barrier of the transition state (see pp. 86-100 of *(18)*); it typically has values between 0.3 and 0.7. The equation for the peak forward current for quasi-reversible systems is even more complex (pp. 224-227 of *(18)*), but it is essentially a hybrid of the reversible and irreversible cases.

Note that for both reversible and irreversible systems, the peak current is proportional to the electrode area, the concentration of the electroactive species, and the square root of the diffusion coefficient. In both cases, the linear dependence of the current on the square root of the scan rate is the result of mass transport of the electroactive species by diffusion alone.

The values of E_{pa}, E_{pc}, ΔE_p, i_{pr}/i_{pf}, and $i_{pf}/(v^{1/2}C^*)$ are often evaluated as functions of the square root of the scan rate in order to diagnose mechanisms. Consult references *(5)*, *(21)* and chs. 6,11 of *(18)* for details.

In obtaining peak currents and potentials for the diagnostic tests, two difficulties are frequently encountered. First, in solvents that have low conductivities (e.g., nonaqueous solvents), the solution between the working and reference electrodes may have a significant uncompensated resistance R_u. When a current i flows at the working electrode, the potential measured by the potentiostat between the working and reference electrodes will be in error by iR_u. The result is that the voltammetric peak is flattened and shifted toward more positive potentials for an oxidation, or toward more negative potentials for a reduction. See reference *(18)*, pp. 22–26 and 220–230. Uncompensated resistance can be minimized by proper electrode placement and by maintaining a high supporting electrolyte concentration. The reference electrode tip should be placed **between** the working and auxiliary electrodes, and as close to the tip of the working electrode as possible without shielding the electrode surface from diffusing species *(24)*). Some advanced potentiostats incorporate schemes to compensate for R_u. Second, making an accurate measurement of the reverse peak current can be problematic. Because of background currents, often the "baseline" from which i_{pr} must be measured is highly nonlinear, especially if the redox process occurs near the limit of the useful potential range. Figure 4, which is the cyclic voltammogram of 9,10-diphenylanthracene (DPA), illustrates this nonlinearity at the beginning of the reverse scan (between E_λ and the potential where the cathodic current peak begins). Sometimes a reasonably good estimate of the baseline can be obtained by careful extrapolation (see figure 4). Other more precise methods are discussed in *(21)* and pp. 228–229 of *(18)*.

Methodology

Instrumentation This experiment requires a potentiostat capable of performing cyclic voltammetry, a compatible X–Y recorder or plotter, a magnetic stirrer, a small glass electrochemical cell of 1 to 10 mL capacity, and suitable electrodes, including

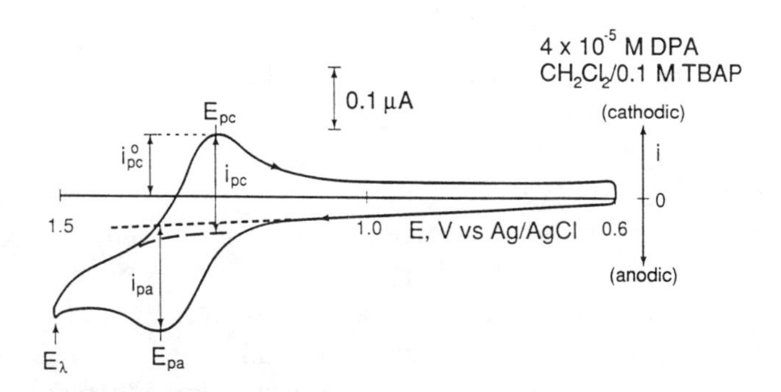

Figure 4. Cyclic voltammogram of 4×10^{-5} M 9,10-diphenylanthracene in CH_2Cl_2 solvent with 0.1 M TBAP supporting electrolyte. Adapted with permission from *(13)*.

a platinum disk working electrode, a Ag/AgCl or SCE reference electrode, and a platinum wire or helix auxiliary electrode. Although they are more expensive, computer-controlled or microprocessor-based potentiostats offer a number of advantages, including peak measuring capabilities which eliminate the tedium of measuring peak potentials and currents by hand. A partial list of suitable potentiostats is found in the hardware list below. Also listed are convenient and durable cells and electrodes that are commercially available. The purchase of an electrode polishing kit will also be necessary.

Chemicals It is important that the dichloromethane solvent be pure and uncontaminated by water. HPLC grade CH_2Cl_2 is usually acceptable as received. Ferrocene of adequate purity is available from a number of sources (Aldrich, Kodak, Æsar). The supporting electrolyte, tetra-*n*-butylammonium perchlorate (TBAP), can be obtained from GFS Chemicals or Bioanalytical Systems, and can also usually be used as received.

Procedure In 10 mL flasks, prepare solutions of 1.0, 2.0, and 4.0 mM ferrocene in dichloromethane; each solution should also contain approximately 0.1 M TBAP. (It is advisable to prepare a stock solution of 10.0 mM ferrocene in CH_2Cl_2 from which the other solutions can be made by dilution.) All glassware must be clean and dry, and the solvent and TBAP bottles must be kept tightly closed when not transferring materials. A small amount of water greatly degrades the working range in this solvent/supporting electrolyte system, so caution should be exercised to avoid contamination of the reagent supplies. Before use the electrodes should be thoroughly rinsed with methanol or ethanol, then with dichloromethane. Special care should be exercised to ensure that the reference electrode is thoroughly dried, since it is stored in an aqueous solution. (The rate of water leakage from the frits on commercial Ag/AgCl reference electrodes is usually too small to cause problems in this experiment. If necessary, the reference electrode can be isolated from the ferrocene solution entirely by the use of a non-aqueous salt bridge. Alternatively, a nonaqueous reference electrode, such as Ag/Ag^+ in CH_3CN, can be employed.) The bottom of the working electrode should be handled very gently to avoid scratches, which will hold water and result in poor or unusable results.

Transfer about 5 mL of a ferrocene solution to a clean, dry cell, add a micro stirring bar and then the cell cover. (Begin with the lowest ferrocene concentration.)

As discussed earlier, uncompensated solution resistance will be minimized if the reference electrode is positioned between the working and auxiliary electrodes and as close to the working electrode as possible (without interfering with diffusion to the electrode surface). Initially, the potentiostat can be set to scan from –2.00 V to +2.00 V, beginning at an intermediate potential (e.g., 0 V), using a moderate scan rate (e.g. 100 mV/s). Connect the proper potentiostat lead to each electrode, referring to the potentiostat manual if necessary. Before attempting to obtain a voltammogram from the solution, it is advisable to scan through the potential range with the potentiostat in the "Standby" mode (if it has one) or to leave the working electrode lead disconnected. This protects the working electrode from extreme potentials and allows one to make the proper range and zero settings if an X–Y recorder is being used. (The line that is traced out can be used as a zero-current axis on the chart.) Then a real voltammogram can be obtained from the solution. The results of this scan can be used to narrow down the potential range of the scan to exclude background current peaks, while retaining enough of the scan on either side of the ferrocene redox peaks for use as baselines for peak current measurements. (Retain at least 0.4 V on either side of the peaks if possible). Between scans the solution should be stirred briefly, then allowed to rest for at least 20 s. For each solution, obtain cyclic voltammograms at ν = 10, 20, 50, 100, 200, 300, and 500 mV/s. If an analog potentiostat is being used with an X–Y recorder, the potentiostat gain and the recorder *X* and *Y* spans will need to be adjusted so that the voltammograms are as large as possible without going off the page. By appropriate selection of the gain and recorder *Y*-span values, it should be possible to "nest" several voltammograms on the same page. It is critical that the gain and the recorder *Y*-span be recorded for each run.

The initial voltammogram should have fairly symmetrical, well-resolved peaks, similar to those shown in Figures 3 and 4. (Some spreading and flattening of the peaks may be seen at the higher scan rates and ferrocene concentrations.) There are three main causes for poor results. The first is poor solution preparation or improperly cleaned cells and electrodes, which result in water contamination. The second is a scratch or contamination on the working electrode surface. In this case the electrode will need to be repolished. The third is a bubble trapped on the working electrode surface. If visual inspection fails to show a bubble or an obvious scratch, the working electrode should be lightly repolished with alumina, then carefully rinsed and dried. If the results are still poor, the student should more carefully repeat the solution preparation process and start over.

Students will also need to measure the diameter of the platinum working electrode with a calipers in order to be able to calculate the electrode area.

When finished, all organic waste should be placed in the appropriate chemical waste containers. (Avoid disposing the micro stir bar at the same time!) All glassware and electrodes should be rinsed with CH_2Cl_2, then with methanol or ethanol. Glassware then should be washed with soap and water, then rinsed with deionized water. Use of a drying oven is helpful for insuring that the next group starts with dry glassware. Rinse the electrodes and stirring bar with deionized water as well; return the reference electrode to storage in 4 M KCl solution.

Safety

Dichloromethane (methylene chloride) is very irritating to the *eyes* and skin; it can cause corneal damage. **Safety glasses must be worn at all times.** In case of contact with the eyes or skin, flushing with water for 15 min is recommended. Inhalation of CH_2Cl_2 vapors should be minimized because of the toxic effects of high vapor concentrations on the nervous system and possible carcinogenicity. Tetra-*n*-butylammonium perchlorate (TBAP) is a strong oxidizer and should be kept away from flammable materials. Since CH_2Cl_2 is a halogenated solvent, all solutions must be placed in suitable chemical waste jugs for proper disposal, and **NOT** poured down the drain.

Data Analysis

The values of i_{pa}, i_{pc}, E_{pa}, and E_{pc} must be obtained from each cyclic voltammogram. Figure 4 shows how these quantities are measured on a typical voltammogram. These data are to be used to answer the following questions.

1) Plot i_{pf} versus $v^{1/2}$ for all three concentrations. Comment on the linearity or nonlinearity of the plots and discuss what it means.

2) From the plots of i_{pf} versus $v^{1/2}$ calculate the diffusion coefficient of ferrocene and compare the result to the diffusion coefficient of naphthalene in acetonitrile at 298 K, 2.7×10^{-5} cm^2/s *(25)*. In light of the fact that the viscosities of CH_2Cl_2 and CH_3CN are comparable, explain your results.

3) Plot i_{pf} versus C^* for v = 100 mV/s and v = 500 mV/s, and discuss.

4) Tabulate the values of $i_{pf}/(v^{1/2}C^*)$ for each scan and comment on the constancy of the values.

5) Calculate ΔE_p for each voltammogram for the 2 mM solution and plot ΔE_p versus $v^{1/2}$. Also plot ΔE_p versus C^* at 100 mV/s. Is ΔE_p a function of the scan rate? of ferrocene concentration? How do the ΔE_p values compare to what is expected for a reversible system? How about the ΔE_p values from extrapolations to zero scan rate and zero ferrocene concentration? Explain, remembering that CH_2Cl_2 is a highly resistive solvent. Account for all the factors that influence ΔE_p.

6) Compare $E_{1/2}$ (the average of E_{pa} and E_{pc}) for each voltammogram to the reported value of $E^{0\prime}$, +0.307 V versus SCE (in CH_3CN with 0.2 M $LiClO_4$ as supporting electrolyte; ref. *(18)*, p. 701). Are there any shifts in $E_{1/2}$ with $v^{1/2}$ or C^*?

7) Calculate i_{pr}/i_{pf} for each voltammogram. Do the values indicate any dependence on the scan rate or ferrocene concentration?

8) From the results above, give an overall summary of the extent to which the oxidation of ferrocene is a reversible process.

9) What effects, if any, do double layer capacitance and uncompensated resistance have on the results?

Additional Questions:

1) (Prelab) Give the electrode reaction that will be studied in this experiment, including the structure of ferrocene and its oxidation product, as well as the formal charges on the iron atom and the cyclopentadienyl rings in each case.

2) What is the difference between thermodynamic (charge-transfer) irreversibility and chemical irreversibility (due to following chemical reactions of the electrogenerated species)? What diagnostic tests could be applied to distinguish between the two in a particular case? Give an example of each.

3) Using your value of $E_{1/2}$ as an approximation to $E^{0'}$, what is the ratio of [ferrocenium]/[ferrocene] at 0 V versus Ag/AgCl? at 0.7 V versus Ag/AgCl? at $E_{1/2}$? (Assume that all activity coefficients are unity).

4) Calculate ΔG^0 at 298 K for the reaction

$$\text{Ferrocene} + H^+ \rightarrow \text{Ferrocenium} + \tfrac{1}{2} H_2$$

using the estimated value, +0.307 V versus SCE, as the standard electrode potential for the half-reaction

$$\text{Ferrocenium} + e^- \rightarrow \text{Ferrocene} \quad (CH_3CN, 0.2\text{ M } LiClO_4)$$

(Ignore liquid junction potentials).

5) Explain how the thermodynamic quantities ΔG^0, ΔS^0, ΔH^0, and K_{eq} can be obtained from electrochemical measurements.

Acknowledgments

The author initially encountered the use of the ferrocene for an introductory cyclic voltammetry experiment in a graduate-level electrochemistry course taught by Prof. Larry Faulkner at the University of Illinois-Urbana; he deserves credit for originally developing the experiment and for authoring an unpublished laboratory handout of which this paper is an expansion. The receipt of a matching funds grant for equipment from the Physical Chemistry Working Group of the Mid-Atlantic Pew Science Program in Undergraduate Education is gratefully acknowledged.

Literature Cited

1. Shoemaker, D. P.; Garland, C. W.; Nibler, J. W. *Experiments in Physical Chemistry*, 5th ed.; McGraw-Hill: New York, 1989, pp. 253-276.
2. Sime, R. J. *Physical Chemistry: Methods, Techniques, and Experiments*; Saunders College Publishing: Philadelphia, 1990, pp. 553-576.
3. Halpern, A. M.; Reeves, J. H. *Experimental Physical Chemistry*; Scott, Foresman and Company: Glenview, IL, 1988, pp. 119-155.
4. Sawyer, D. T.; Heineman, W. R.; Beebe; J. M. *Chemistry Experiments for Instrumental Methods*; John Wiley & Sons: New York, 1984, ch. 4.
5. Nicholson, R. S.; Shain, I. *Anal. Chem.* **1964**, *36*, 706.
6. Van Benschoten, J. J.; Lewis, J. Y.; Heineman, W. R.; Roston, D. A.; Kissinger, P. T. *J. Chem. Educ.* **1983**, *60*, 772.
7. Baldwin, R. P.; Ravichandran, K.; Johnson, R. K. *J. Chem. Educ.* **1984**, *61*, 820.
8. Brillas, E.; Garrido, J. A.; Rodriguez, R. M.; Domenech, J. *J. Chem. Educ.* **1987**, *64*, 189.
9. Piszczek, L.; Ignatowicz, A.; Kielbasa, J. *J. Chem. Educ.* **1988**, *65*, 171.
10. Ibanez, J. G.; Gonzalez, I.; Cardenas, M. A. *J. Chem. Educ.* **1988**, *65*, 173.
11. Carriedo, G. A. *J. Chem. Educ.* **1988**, *65*, 1020.
12. Geiger, D. K.; Pavlak, E. J.; Kass, L. T. *J. Chem. Educ.* **1991**, *68*, 337.
13. Faulkner, L. R. "*Cyclic Voltammetry*" (unpublished laboratory experiment handout for Chemistry 422), University of Illinois-Urbana, Urbana, IL, 1982.
14. Van Dyke, D. A. Manuscript in preparation for publication in *J. Chem. Educ*.
15. Evans, J. F.; Blount, H. N. *J. Phys. Chem.* **1979**, *83*, 1970, and references therein.
16. Faulkner, L. R. *J. Chem. Educ.* **1983**, *60*, 262.
17. Chambers, J. Q. *J. Chem. Educ.* **1983**, *60*, 259.
18. Bard, A. J.; Faulkner, L. R. *Electrochemical Methods*; John Wiley & Sons: New York, 1980. For an introduction to the fundamentals of electrochemistry, consult chs. 1-4 (pp. 1-135). Cyclic voltammetry is discussed in ch. 6 (pp. 213-248).
19. Kissinger, P. T. In *Laboratory Techniques in Electroanalytical Chemistry*; Kissinger, P. T.; Heineman, W. R., Eds.; Marcel Dekker, Inc.: New York, 1984, pp. 163-192.
20. Evans, D. H.; O'Connell, K. M.; Peterson, R. A.; Kelly, M. J. *J. Chem. Educ.* **1983**, *60*, 290.
21. Mabbott, G. A. *J. Chem. Educ.* **1983**, *60*, 697.
22. Kissinger, P. T.; Heineman, W. R. *J. Chem. Educ.* **1983**, *60*, 702.
23. Michael, P. R.; Ph.D. Thesis, University of Illinois - Urbana, Urbana, IL, 1976; p. 72.
24. Hawkridge, F. M. In *Laboratory Techniques in Electroanalytical Chemistry*; Kissinger, P. T.; Heineman, W. R., Eds.; Marcel Dekker, Inc.: New York, 1984, pp. 337-346.
25. Sawyer, D. T.; Roberts, J. L. *Experimental Electrochemistry for Chemists*; John Wiley & Sons: New York, 1974; p. 77.

Hardware List

Potentiostats, cells and electrodes for cyclic voltammetry are available from several sources, including Bioanalytical Systems (BAS, West Lafayette, IN), EG&G Princeton Applied Research (PAR, Princeton, NJ), and Cypress Systems (Lawrence, KS). Student-compatible analog potentiostats include the BAS CV-27, Cypress Systems Omni90 and PAR 264A-3. However, computer-controlled potentiostats are to be preferred, although they are more expensive. The Cypress Systems CS-1087 is a reasonably-priced computer-controlled potentiostat that appears to be suitable for educational purposes. More advanced and expensive research-grade instruments include the BAS 100-A, Cypress Systems CS-1090, and PAR 273.

The cells and electrodes available from BAS are convenient and fairly robust. To be specific, we use the VC-2 cell (P/N MF1065) which includes a teflon cell cover and platinum wire auxiliary electrode; a platinum voltammetry electrode (P/N MF2013); an RE-5 Ag/AgCl reference electrode (P/N MF2024); and a PK-3 polishing kit (P/N MF2056). Similar cells and electrodes are available from PAR, Cypress Systems and others.

RECEIVED October 14, 1992

Thermodynamics Experiments Without Lasers

CHAPTER 28

Critical Point and Equation of State Experiment

Ken Morton

POTENTIAL HAZARDS: High-Pressure Systems

A new apparatus has recently become commercially available that permits students to explore the properties of gases at pressures up to 50 atmospheres. This paper describes an experiment using this apparatus. The purposes of the experiment are to observe visually the behavior of a gas in the region of the critical point, and to apply theoretical models such as the van der Waals equation, Virial equation, and Clausius-Clapeyron equation in curve-fitting to experimental data. In a single 3-hour lab period, students can observe critical point behavior and record 5 isotherms (70 or more *P*, *V*, *T* data points) that bracket and include the critical isotherm. A second lab period is conducted at a computer to perform data analysis using a spreadsheet/graphics package that has been pre-programmed for this experiment.

The behavior and theory of gases is usually covered early in introductory physical chemistry courses as a foundation for subsequent development of several concepts--including ideality and non-ideality. Laboratory experiments with gases such as carbon dioxide *(1,2)* are available for illustrating behavior in the critical region. Several features of the present experiment recommend it for inclusion in the laboratory course. The apparatus is easy to use, allowing students to focus their attention on the behavior of the system under study. Direct visual observation of the fluid in the region of the critical point generates a strong interest within the students, and may be used to draw attention to recent applications of supercritical fluids in the areas of chromatography *(3, 4)* and solvent extraction *(5)*. Students generate data of sufficient quantity and quality to permit the introduction of data analysis and computerized curve-fitting techniques. Students are encouraged to ask "what if" questions in the data analysis stage. They quickly come to appreciate the power of spreadsheets in solving physical chemistry problems and to see the value of developing their own spreadsheets for subsequent applications. Finally, the apparatus need not be confined to the physical chemistry lab. It can be used for demonstrations in general chemistry classes--either directly (in small classes) or by an appropriate means of projection.

Theory

The theoretical background for this experiment is given in most physical chemistry texts *(6-8)* and only a brief review is presented here. Real gases may deviate significantly from the ideal gas law at high pressure and low temperature. Compression of a real gas at a constant temperature yields a set of *P*, *V* data that defines an isothermal line, or isotherm. For a given gas at a sufficiently low temperature, compression will eventually result in liquefaction, so that a plot of *P* vs. *V* exhibits a pressure plateau. At successively higher temperatures, the pressure plateau becomes narrower until it vanishes at a point when the critical temperature

2503–6/93/0412$06.00/0

(T_c) is reached. This point, the critical point, is unique for each gas and defines the critical constants for that gas, i.e., T_c, P_c (the critical pressure) and V_c (the critical volume, a molar quantity). Above T_c, liquefaction is impossible regardless of the pressure, but the gas will still deviate significantly from the ideal gas law if its temperature and pressure are too near T_c and/or P_c.

The van der Waals equation (Equation 1) introduces two simple modifications that extend the applicability of the ideal gas law to real gases. Intermolecular forces are represented by the inclusion of an "internal pressure" term (an^2/V^2 or a/V_m^2, where V_m is the molar volume); the actual volume occupied by the molecules of gas is represented by the "covolume" (nb).

$$(P + an^2/V^2)(V - nb) = nRT \text{ or } (P + a/V_m^2)(V_m - b) = RT \quad (1)$$

In Equation 1, a and b are the familar van der Waals coefficients and n is the number of moles of gas. A more useful form of the equation is obtained by solving algebraically for P (Equation 2). An alternative arrangement (Equation 3) illustrates that the equation is cubic in relation to V_m.

$$P = RT / (V_m - b) - a/V_m^2 \quad (2)$$

$$PV_m^3 - (bP + RT)V_m^2 + aV_m - ab = 0 \quad (3)$$

The van der Waals coefficients, a and b, are determined empirically to obtain the best description for a particular gas. They may be calculated from the critical constants using the fact that the critical van der Waals isotherm must exhibit a horizontal inflection at the critical point. At a horizontal inflection point, both the first and second derivative of a function must equal zero. Thus, the first and second derivatives of Equation 2 with respect to V_m may be set equal to zero (Equations 4 and 5).

$$dP/dV_m = -RT/(V_m - b)^2 + 2a/V_m^3 = 0 \quad (4)$$

$$d^2P/dV_m^2 = 2RT/(V_m - b)^3 - 6a/V_m^4 = 0 \quad (5)$$

Substituting T_c, P_c, and V_c, into Equations 2, 4, and 5 gives three simultaneous equations (6–8) that can be solved for the two unknowns, a and b.

$$P_c = RT_c/(V_c - b) - a/V_c^2 \quad (6)$$

$$0 = -RT_c/(V_c - b)^2 + 2a/V_c^3 \quad (7)$$

$$0 = 2RT_c/(V_c - b)^3 - 6a/V_c^4 \quad (8)$$

Since only two simultaneous equations are needed to solve for two unknowns, the values of a and b are overspecified. Because of this unusual situation, six different methods of computing a and b may be employed to obtain 5 unique a, b solutions (9). The method shown in Equation 9 is the only one of these that avoids the use of V_c, which is known with less precision than T_c and P_c in this experiment.

$$a = 27(RT_c)^2 / 64\, P_c \qquad (9a) \qquad b = RT_c / 8\, P_c \qquad (9b)$$

In order to predict the pressure of a gas in a given state using Equation 2, it is also necessary to know V, T, and n for that state. For non-ideal gases, n may be obtained from Equation 10 if the value of the dimensionless compressibility factor, Z, is known for the state.

$$PV_m/RT = Z \qquad (10)$$

For ideal gases, $Z = 1$. As the pressure is increased, Z usually decreases (due to attractive intermolecular forces) to a minimum of approximately 0.3 at the critical point and then increases (due to short-range repulsive forces) until it exceeds 1 at extremely high pressures. The value of Z may be obtained with sufficient precision for this experiment using the principle of corresponding states: all gases that are at the same reduced temperature (T_r) and reduced pressure (P_r), occupy the same reduced volume (V_r, a molar quantity), where the reduced variables are defined as $T_r = T/T_c$, $P_r = P/P_c$, and $V_r = V/V_c$. Because of this principle, it is possible to prepare a plot of Z as a function P_r and T_r *(10)*; such graphs are found in most physical chemistry texts and are useful for our purpose because they apply equally well to a large number of gases that differ greatly in molecular weight, geometry, and polarity. Thus, by calculating P_r and T_r for a given state, interpolation from the graph gives Z which can, in turn, be used to find n.

Instead of using the van der Waals equation for modeling the behavior of a real gas, the virial equation may be used. In this approach, a power series in $1/V_m$ is used to approximate Z

$$Z = PV_m/RT = 1 + B/V_m + C/V_m^2 + D/V_m^3 + \dots \qquad (11)$$

B, C, and D are temperature-dependent constants that are adjusted empirically to obtain the best fit to the data. As many terms as needed may be used, but a reasonably good fit is obtained by ignoring the last (and higher) terms in Equation 11; rearranging of the remaining terms gives Equation 12, which is in a form suitable for curve-fitting by the least-squares program described below.

$$P/RT - 1/V_m - B/V_m^2 - C/V_m^3 = 0 \qquad (12)$$

Although the virial coefficients lack the simple physical interpretation of the van der Waals coefficients, they can be related to physical reality through statistical mechanics. For example, the second virial coefficient (B) can be calculated from the intermolecular potential between pairs of molecules. In the simplest form, the calculation assumes the molecules to be hard spheres of equal size; the potential energy is assumed to be infinite within an excluded volume of radius σ, (where σ is twice the molecular radius) and is assumed to be zero outside the excluded volume. Under these conditions, the relationship between the virial coefficient B and σ is given by

$$B = 2\pi N_A \sigma^3/3 \qquad (13)$$

where N_A is Avogadro's number. Because of the great oversimplification involved in deriving Equation 13, the same relationship may be deduced without statistical mechanics by using the same simple mechanical model employed in developing van der Waal's coefficient b. However, in actual practice, the virial coefficient B and the van der Waal's coefficient b have different values.

Figure 1. Critical Point and Equation of State Apparatus

Methodology

Equation of state apparatus. This device (Figure 1) is sold in the United States by Cenco Scientific Company (Franklin Park, IL; catalog number 32019) and is manufactured by Phywe Systeme GMBH, P.O. Box 3062, D-3400 Gottingen/Federal Republic of Germany. It consists of a vertical, calibrated glass capillary that is sealed at the top and is mounted above a stainless steel pressure chamber containing mercury. Rotation of a 15-turn handwheel below the mercury chamber displaces mercury up into the capillary, thereby trapping and compressing the gas. The pitch of the threads on the handwheel mechanism permits good control of the volume of the chamber. The hydrostatic pressure of the mercury is read from a large gauge. The capillary is surrounded by a plexiglass water jacket fed from an external, thermostatted water bath. (To conserve student time in lab, it is helpful to use a calibrated water bath that can achieve equilibrium in about 10 minutes after a 10-degree jump.) The working ranges and readability of the measurements are as follows: pressure 0.30–5.00 ± 0.01 MPa, volume 0.20–4.00 ± 0.01 mL, temperature 0.00–55.00 ± 0.05 °C. Charging the apparatus with a gas is accomplished by means of separate vacuum (hose) and inlet (R 1/8-inch screw-type) ports. The instrument manual and advertising literature for the apparatus indicates that the following gases are suitable: sulfur hexafluoride, carbon dioxide, ethane, Freon 13 and Freon 23, or others with critical temperatures in the range 0–100 °C and critical pressures below 5 MPa. The supplier's catalog, however, only offers sulfur hexafluoride and ethane. Mixtures of gases could be employed for more advanced studies.

Computer and software requirements. Computerized data analysis is not required for obtaining educational value from the apparatus, but it adds an important dimension and illustrates concepts that are not readily approached in a more traditional format. A number of excellent spreadsheet/graphics packages are now available. The institutional context, interests of the instructor, class size, and computer literacy of the students will determine how these are used in analyzing the data. Our current practice (in a small liberal arts college where students vary widely in computer skills) utilizes a second lab period for instruction on use of a Macintosh IIcx computer (color monitor and 4 MB RAM hard disk), Wingz spreadsheet/graphics software (Informix, Inc., Lenexa, KS), and Wingz spreadsheets that have been pre-programmed for use in this experiment. Two printers are available; a Laserwriter IINTX and an Imagewriter; color printing of graphs is possible with the latter, but is time-consuming.

Spreadsheet programming. Two pre-programmed spreadsheets are used. The first one contains sections for data entry, displaying isotherms, fitting the data by the van der Waals and Clausius–Clapeyron equations, and reformatting the data for transfer to the second spreadsheet. The layout of the first spreadsheet is organized such that all graphs and calculations refer back to the data entry section to minimize the amount of data that must be entered. Students scroll downward from the data entry screen to reach the remaining sections, and may scroll to the right of a given graph to see the calculations that give rise to the graph. All graphs are pre-formatted and occupy one screen each in the spreadsheet. Programming this spreadsheet may be quite time-consuming if a visually polished product is desired, but the underlying operations are relatively straightforward for someone familiar with spreadsheets. To set up the van der Waals section, Equation 2 and the results of the students' preliminary calculations are used. This section also includes a standard deviation calculation (van der Waals vs. experimental pressures) that should be weighted if volume increments in the experimental data are not equal. The section on the

Clausius–Clapeyron equation requires students to enter the equilibrium pressure at four temperatures. This data is analyzed by the Wingz linear regression function to obtain the best least-squares fit. The last section of this spreadsheet converts the P, V, T data for the 50 °C isotherm to y vs. x data of the form P/RT vs. $1/V_{\mathrm{m}}$ (in units of mol/L), which is suitable for copying into the second spreadsheet.

The second spreadsheet is programmed for an iterative least squares curve-fitting procedure using the Virial equation. The spreadsheet version of this program was first obtained as a template in the Lotus 1-2-3/IBM format (see "LSGEN" program, ref. *11*); it was imported to the Wingz/Macintosh environment without error, except that it was necessary to format a new graph and a few modifications were necessary to control the iteration procedure. (In the Lotus version, the iteration is initiated by pressing the "Alt-C" key to activate a macro; in the current Wingz version, the spreadsheet instructs the student to select a value for two iteration parameters and then choose "Go Recalc" from the menu bar.) The Wingz version was then programmed with the desired function (Equation 12), its first derivatives with respect to x, y, and the two constants, and the experimental data. To avoid calculation errors, it is important for the data to be in units of mol/L rather than SI units, as the latter units cause the numerical values to range from 10^{-3} to 10^{3}. This range, in combination with the presence of many higher-order terms in the least-squares calculation, is presumed to be the source of the errors. To insure convergence of the iterative procedure to the correct values, the user must enter appropriate initial values for the two constants obtained by the simultaneous solution of two equations. Tedium and errors in calculating the initial values have been avoided by programming the spreadsheet with the general solution of the Virial equation as follows:

$$B = (\alpha - \beta)/(x_1 - x_2) \qquad C = (x_1\beta - x_2\alpha)/(x_1 - x_2)$$

where α and β are defined as

$$\alpha = x_1^2(x_1y_1 - 1) \qquad \beta = x_2^2(x_2y_2 - 1)$$

Student laboratory procedure. Students work in pairs. The gas selected for study is placed in the apparatus by the instructor before lab and its identity is not known to the students. The observations and data in the remainder of this report are based on sulfur hexafluoride at an initial charging pressure of 0.5 MPa. Students set the water bath at the lowest temperature to be used (ca. 20 °C), compress the gas in specified increments ranging from 0.50 mL (initially) to 0.1 mL, record 16 pressure-volume data points along this isotherm, and observe and record the behavior of the gas. If the data analysis software imposes constraints on the number of data points or the spacing intervals between measurements, the instructions to the students must indicate the number of measurements and intervals that are required. The behavior of the fluid is clearly visible at all times. The point at which the first liquid appears is noted. Shortly before reaching this point, both the pressure and volume of the gas respond differently than expected for a given angular displacement of the handwheel. Pressure equilibration is almost instantaneous at lower pressures, and is usually complete within 10-30 seconds at any pressure. By the time the minimum calibrated volume has been reached, most or all of the sample volume consists of liquid; the pressure has reached a plateau at ca. 2 MPa--well below the safety limit for the apparatus. Students are then instructed to return the mercury to the reservoir. To their surprise, a sudden decompression of the chamber causes a transient boiling of the liquid and partial condensation of any vapor that is present. This provides an opportunity for discussing non-equilibrium conditions and the temperature effects of expansion.

Additional isotherms are recorded at assigned intervals (ca. 30, 40, and 50 °C) that are known to bracket the critical temperature. The width of the pressure plateau decreases as the temperature is raised and the plateau disappears above the critical point. By comparing their P,V,T results to a data table that gives the critical temperature and pressure for a few gases (carbon dioxide, ethane, Freon 13, Freon 23, and sulfur hexafluoride), students are immediately able to identify the gas as sulfur hexafluoride and learn its critical temperature (ca. 45 °C) and pressure (ca. 3.6 MPa). Then they are asked to observe the behavior of the fluid in the region of the critical point. This is accomplished by setting the temperature approximately one-half degree below the critical temperature, adjusting the volume to approximately the middle of the pressure plateau, increasing the water bath thermostat setting by one or two degrees, and carefully observing the fluid in the capillary for several minutes. The meniscus gradually disappears, a cloud then appears in the region formerly occupied by the liquid, and eventually the fluid appears homogeneous. However, if a white card ruled with a few diagonal lines is then placed behind the capillary, the lines will appear to be bent in the region of the former meniscus since the density and refractive index are not yet uniform throughout the sample chamber. Color changes in the critical region have been noted in the literature *(12)*. The final laboratory exercise is to record one additional isotherm at or slightly below the critical temperature. Thus, in a three hour lab period, students have thoroughly explored the behavior of a real gas by visual means and have collected enough P, V, T data to define five isotherms that bracket and include the critical isotherm.

Safety

Eye protection must be worn in the laboratory. The equation of state manual, which is printed in English, French, and Spanish, gives clear safety instructions and states that each glass working capillary is tested at 7.5 MPa and 60 °C before shipment. No accidents have occurred in the first year of using the device; students seem to readily understand the safety precautions and follow them. Several factors help prevent overpressuring the apparatus: the overpressure zone is clearly marked in red on the pressure gauge, charging the apparatus with a gas is done by the instructor, and the initial run (where students are less familiar with the device) is conducted at a temperature sufficiently low that the pressure limit is not exceeded. Should mechanical failure of the pressure chamber occur during normal usage, it appears that the worst outcome would be release of mercury into the plexiglass-walled water bath or onto the handwheel. In the latter case, it would seem prudent to have the apparatus located on a tray to facilitate cleanup. When evacuating the apparatus to change gases, the vacuum pump should not be vented directly into the lab, as the exhaust may contain mercury vapor. After approximately 18 months of continual presence of sulfur hexafluoride in the apparatus, there is no apparent evidence of corrosion.

Data Analysis

Preliminary calculations. In the experimental design currently employed at Carson-Newman College, data analysis occurs in two distinct stages. First, students are required to perform a number of preliminary calculations in SI units using a hand-held calculator. They are required to complete these before coming to lab the following week, where they perform the computerized data analysis exercise. The preliminary calculations are conducted for a single data point and serve two purposes: familiarizing students with the calculations and providing numerical values needed for the computer exercise. Instructions for the preliminary calculations are given below.

Select the data point at which the gas is behaving most ideally (i.e., largest volume, highest temperature) and calculate the reduced temperature (T_r) and reduced pressure (P_r) for this data point. Conduct the remainder of these preliminary calculations for this data point. Interpolate from a graph of the principle of corresponding states to obtain the compressibility factor (Z). Calculate the molar volume (V_m); use this and the experimentally measured volume to find the number of moles of the gas (n). Calculate the van der Waals coefficients (a and b). Calculate the pressure predicted by the van der Waals equation for the gas at this point.

The results for sulfur hexafluoride at the most ideal point (V = 4.00 mL, T = 50.0 °C, P = 0.98 MPa) in one experiment were: Z = 0.90, n = 1.6 mmol, V_m = 2.5 L/mol, van der Waals a = 0.81 Pa m^6/mol^2, van der Waals b = 9.1×10^{-5} m^3/mol, P(vdW) = 1.00 MPa.

Computerized data analysis. The first screen of the first spreadsheet provides space for entering the student names, date, experimental P, V, T data and the results of the preliminary calculations. Data for the lowest-temperature isotherm is entered as run 1; the highest is run 4, and the critical isotherm is entered as run 5. Students then scroll or page downward through the remaining screens that develop various aspects of the analysis.

The first graph (Figure 2) is a plot of the experimental isotherms. A second graph (not shown) compares the pressure to the compressibility factor for the 50 °C isotherm, i.e., the pressure (referenced to the left ordinate) and the compressibility factor Z (referenced to the right ordinate) are plotted versus the volume in milliliters. Students readily observe that Z varies inversely with P in the region of this experiment. The third graph focuses on the 50 °C isotherm and occupies the majority of a split screen. One part of the screen displays spreadsheet cells for entering up to 10 sets of trial values for the van der Waals coefficients. The corresponding isotherms predicted by the van der Waals equation are displayed as continuous lines for comparison with the experimental P, V data points. This visual comparison of the goodness of fit is accompanied by a weighted standard deviation (predicted minus experimental) that appears alongside each a, b trial set in the spreadsheet. For trial 1, students enter the values of a and b from their preliminary calculations. For trials 2–4, a and b are set equal to zero (independently, then simultaneously) to see the effects of ignoring intermolecular attractions and molecular volumes. Following this, they make incremental changes in a and b to try to improve on the fit obtained in trial 1. Working with this screen is generally enjoyable and highly educational for the students. Subsequent screens duplicate this

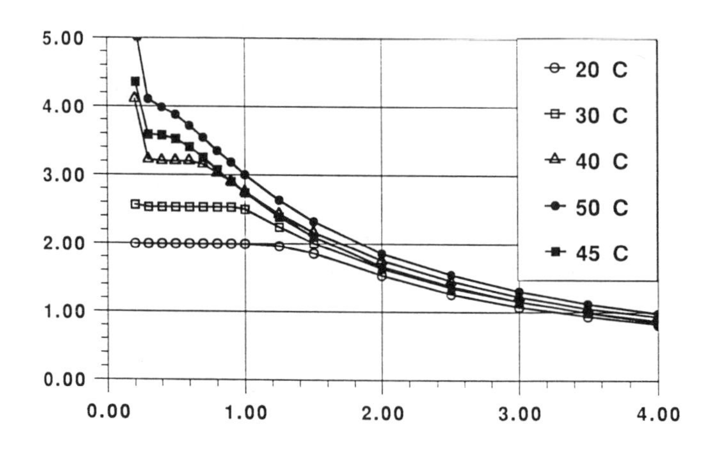

Figure 2. Experimental Isotherms for Sulfur Hexafluoride

analysis for lower temperatures. When assisting the students in interpreting their data it is helpful to point out that a high-precision study of the critical behavior of sulfur hexafluoride *(12)* has shown that the classical van der Waals description fails for this substance because the critical isotherm includes a pressure plateau rather than an inflection point.

To fit the data with the virial equation, students copy the appropriate section of the first spreadsheet into the second one and make a few other entries as prompted by the second spreadsheet. The best fit values of the second and third virial coefficients are calculated automatically and a pre-programmed graph depicts the experimental values as compared to the calculated best fit line. Students can observe the iterative process in "real time" by choosing values of the iteration parameters that cause the convergence to occur in a series of small steps and by viewing the calculated line that is displayed automatically after each step. In one experiment, the calculated value for the second virial coefficient (−0.223 L/mol) was within about 5% of a value interpolated from the literature *(13)*.

Students file a lab report one week after the data analysis exercise. The report is to contain all of the preliminary calculations, plus a calculation of the excluded radius and volume for a single sulfur hexafluoride molecule. Narrative responses to the following are also expected: Describe the behavior of the gas in the region of the critical point. Explain the behavior of the liquid-vapor system upon sudden decompression. For each van der Waal's coefficient, briefly explain on a physical basis what effect an increase in the magnitude of this property would have on the pressure of the gas, and explain mathematically by reference to the van der Waal's equation how an increase in the magnitude of this property affects P(vdw).

The experiment as described also yields data that can be used later in the semester to illustrate the effect of temperature on equilibrium vapor pressure of a liquid. When the pressures on the plateaus of the isotherms in Fig. 2 were plotted versus temperature, a curved line was obtained. Plotting in the form of the Clausius–Clapeyron equation (ln P vs. $1/T$) yielded 4 points in an apparently straight line with a standard deviation of 0.00595. From the slope of this line, the enthalpy of vaporization was found to be 18.5 kJ/mol. However, a closer look at the residuals revealed a slight curvature to the data points (ln P vs. $1/T$), which served as a reminder to students of the simplifying assumptions made in deriving the Clausius–Clapeyron equation.

Acknowledgements

This material is based upon work supported by the National Science Foundation under Grant No. USE-8951744; additional support for equipment purchases was provided by the Pew Charitable Trust and private donations. Helpful comments were provided by the reviewers and by attendees at meetings where this experiment was described (The Fourth Chemical Congress of North America, New York, August, 1991, and The Southeastern Regional Meeting of the American Chemical Society, Richmond, VA, November, 1991).

Literature Cited

1. Halpern, A.M. and Lin, M.F. *J. Chem. Educ.* **1986**, *63*, 38-39.
2. Shoemaker, D.P., Garland, C.W., and Nibler, J.W. *Experiments in Physical Chemistry, 5th ed.*; McGraw-Hill Book Company: New York, NY, 1989; pp. 246-252.
3. Chester, T.L. and Pinkston, J.D. *Anal. Chem.* **1990**, *62*, 394R-402R.
4. Palmieri, M.D. *J. Chem. Educ.* **1988**, *65*, A254-A259.
5. Hawthorne, S.B. *Anal. Chem.* **1990**, *62*, 633A-642A.
6. Alberty, R.A. *Physical Chemistry, 7th ed.*; John Wiley and Sons: New York, NY, 1987; pp. 8-20.
7. Atkins, P.W. *Physical Chemistry, 4th ed.*; W.H. Freeman and Company: New York, NY, 1990; pp. 14-21.
8. Noggle, J.H. *Physical Chemistry, 2nd ed.*; Scott, Foresman, and Company; Glenview, IL, 1989; pp. 5-27, 43-46.
9. Eberhart, J.G. *J. Chem. Educ.* **1989**, *66*, 906-909.
10. Su, G-J. *Ind. Eng. Chem.* **1946**, *38*, 803.
11. Whisnant, D.M. *J. Chem. Educ.: Software* **1989**, *IIB*, 65-67.
12. Wentorf, R.H., Jr. *J. Chem. Phys.* **1956**, *24*, 607-615.
13. *TRC Thermodynamic Tables, Non-hydrocarbons.* (1972 update) Thermodynamics Research Center, Texas A&M University System, College Station, TX, p. h-310.

Hardware

Equation of state apparatus. Cenco Scientific Company, Franklin Park, Il; catalog number 32019

Vacuum pump.

Water bath. Preferably calibrated and having a response time of about 10 minutes for a 10-degree increase in the dial setting.

RECEIVED October 14, 1992

CHAPTER 29

Joule–Thomson Refrigerator and Heat Capacity Experiment

Richard A. Butera

POTENTIAL HAZARDS: High-Pressure Systems, Cryogens, Vacuums

The ability to produce a controlled low temperature environment has permitted the investigation of the temperature dependence of material and thermodynamic properties. This has advanced our understanding of phenomena such as magnetic ordering, superconductivity, and vibrational excitation in solids. It has been standard practice to include an experiment which determines the Joule–Thomson (J–T) coefficient in the physical chemistry laboratory. A practical use of the J–T effect is to produce stable reduced temperature environments. This is accomplished by the use of a heat exchanger which cools the incoming high pressure gas before expansion by the J–T cooled low pressure exhaust gas. This configuration can be used to produce a variable temperature environment by adding a heater to that portion of the system containing the J–T expansion port and inputing enough heat to balance the cooling power of the system at the desired temperature. This is the configuration that is used in the following experiment. If the system is well designed and operating conditions are controlled, the cooling power of the refrigerator will be a reproducible function of temperature. The rate at which the refrigerator cools as a function of time under constant inlet gas pressure will depend on the magnitude of the heat capacity of the system and the heat leak from the external environment. The former will depend on the properties of the material from which the refrigerator is constructed (which for a given refrigerator are constant) and the properties of any sample attached to the cold end of the refrigerator. The latter will depend on the quality of the insulating vacuum separating the refrigerator from the external environment and also the thermal conductivity of any additional electrical leads connecting the low temperature end of the refrigerator to the external environment.

Theory

When a real gas passes through a very small orifice (or porous plug) from a high pressure on one side to a low pressure on the other, where the resistance of the orifice (or porous plug) is great enough to ensure a nearly constant pressure in the incoming and also in the outgoing gas, a temperature difference, ΔT_{JT}, is produced. The magnitude of ΔT_{JT} is dependent on the pressure difference across the orifice (or porous plug). The apparatus must be constructed of such poor thermal conductors that no appreciable amount of heat can pass into or out of the system. At ordinary temperatures and pressures, all gases, except hydrogen and helium, show a cooling effect in such a free expansion. This amounts in the case of nitrogen to $\approx$ 0.25 K/atm. This effect is large enough to be of practical importance, and the most common apparatus for the production of liquid nitrogen makes use of this phenomenon. Thus, if a certain portion of the compressed gas undergoes free

2503–6/93/0422$06.00/0

expansion and the cooling effect is used to precool another portion, that portion upon expansion will fall to a still lower temperature. By continuing this process a certain fraction of the original compressed gas can be liquefied. The experiments of Joule and Thomson and of others who have used the same method show that the cooling produced by a given pressure drop is nearly independent of the pressure (although it is noticeably smaller at the highest pressures which have been studied). On the other hand, it increases rapidly with diminishing temperatures.

In order to clearly understand the theory of the Joule–Thomson effect consider the following schematic representation (Figure 1).

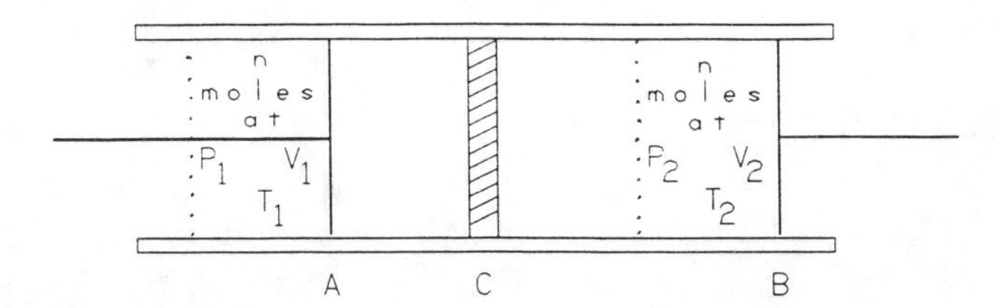

Figure 1. Joule-Thomson Expansion Schematic.

The gas is passing through the porous plug C from left to right. The pistons A and B are moved at such a rate as to keep each of the pressures P_1 and P_2 constant. Assuming now that the apparatus is constructed of such good thermal insulators that the process is adiabatic, then $\delta q = 0$ and $dU = \delta w = -PdV$, where q is the heat, w is the work, U is the internal energy, and V the volume. Thus for this system, when one mole of gas has passed through the plug, the work $P_1(0 - V_1)$ will have been done upon the system and the work $P_2(V_2 - 0)$ will have been done by the system. The net work is

$$w = -P_1V_1 + P_2V_2 \quad (1)$$

and

$$U_1 - U_2 = -P_1V_1 + P_2V_2 \quad (2)$$

thus

$$U_1 + P_1V_1 = U_2 + P_2V_2 \quad (3)$$

$$H_1 = H_2 \quad (4)$$

The process is one that occurs at constant enthalpy, H, and the Joule-Thomson coefficient, μ, is defined as

$$\mu = (\partial T / \partial P)_H \quad (5)$$

In order to understand the use of the J–T effect to produce a low temperature environment let us consider the schematic of the refrigerator shown in Figure 2.

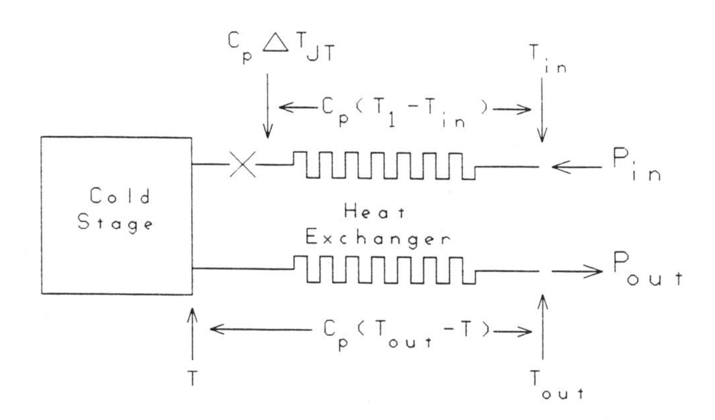

Figure 2. Refrigerator Schematic. X - JT expansion orifice.

The equations for the process going from P_{in} at T_{in} to P_{out} at T_{out} can be written as

$$N_2(g,T_{\text{in}},P_{\text{in}}) \rightarrow N_2(g,T_1,P_{\text{in}}) \qquad Q_1 = C_p(T_1 - T_{\text{in}}) \tag{6}$$

$$N_2(g,T_1,P_{\text{in}}) \rightarrow N_2(g,T_2,P_{\text{out}}) \qquad Q_2 = C_p\Delta T_{\text{JT}} \tag{7}$$

$$N_2(g,T_2,P_{\text{out}}) \rightarrow N_2(g,T_{\text{out}},P_{\text{out}}) \qquad Q_3 = C_p(T_{\text{out}} - T_2) \tag{8}$$

where

C_p = the heat capacity at constant pressure per mole for nitrogen gas
ΔT_{JT} = the J-T cooling of the nitrogen gas
T_{in} = temperature of the incoming high pressure nitrogen gas
T_{out} = temperature of the outgoing low pressure nitrogen gas
T_1 = temperature of the high pressure nitrogen gas prior to expansion
T_2 = temperature of the low pressure nitrogen gas just after expansion
Q_1 = heat removed per mole from the gas in going from T_{in} to T_1
Q_2 = J-T cooling heat equivalent per mole of nitrogen gas
Q_3 = heat gained per mole by the gas in going from T to T_{out}

Thus, if $T_{\text{in}} = T_{\text{out}} = T_{\text{RT}}$ (room temperature), the overall process is

$$N_2(g,T_{\text{RT}},P_{\text{in}}) \rightarrow N_2(g,T_{\text{RT}},P_{\text{out}}) \tag{9}$$

and

$$Q_1 + Q_2 + Q_3 = 0 \tag{10}$$

The cooling power of the refrigerator depends on the flow rate of the gas passing through the J-T expansion. The rate at which the cold stage will cool depends on the total heat capacity of the apparatus plus the heat capacity of any sample which is in thermal contact with the cold stage. The rate of cooling will also depend on the amount of heat which can reach the cold stage and heat exchanger from the external environment. This heat leak will depend on the quality of the insulating vacuum separating the cold stage and heat exchanger from the external environment and also on the thermal conductivity of the glass support material and any electrical leads brought in from the outside.

We now consider the determination of heat capacity by measurement of refrigerator cooling rates. Let

Q = the amount of heat transferred to a material,
T = the absolute temperature of the material

and

C = the heat capacity of the material as a function of T.

From thermodynamics we can write

$$dQ = C\, dT \tag{11}$$

The time dependence of this equation is given by

$$Q' = (dQ/dt) = C(dT/dt) \tag{12}$$

This is the basic equation that we will use to describe the experiment. We assume that the cooling power of the refrigerator is a reproducible function of the cold stage temperature for a given gas flow condition. For the empty refrigerator we can write

$$Q' = C_0(dT/dt)_0 \tag{13}$$

where

Q' = the cooling power of the refrigerator as a function of T

C_0 = the total effective heat capacity of the refrigerator including the thermal load due to any heat conducted to the cold stage through the support material as a function of T

$(dT/dt)_0$ = the measured cooling rate of the empty refrigerator as a function of T

We can write similar equations for the refrigerator with the standard copper sample mounted on the cold stage and also for any other sample mounted on the cold stage. These are given below.

$Q' = [C_0 + C_{std}](dT/dt)_{std}$ for the standard copper sample and
$Q' = [C_0 + C_{samp}](dT/dt)_{samp}$ for any other sample.

In the above equations, C_{std} is the heat capacity of the standard copper sample and C_{samp} is the heat capacity of whatever sample is mounted on the cold stage. Thus if Q' is the same in all cases, by determining (dT/dt) as a function of T and by knowing the heat capacity of the standard copper sample as a function of T, the equations can be solved for the heat capacity of the sample of interest. We now do this by eliminating C_0 from the equations. Since Q' is the same for all three equations we write

$$[C_0 + C_{std}](dT/dt)_{std} = C_0(dT/dt)_0 \tag{14}$$

rearrangement of this equation yields

$$C_0 = C_{std}(dT/dt)_{std}/[(dT/dt)_0 - (dT/dt)_{std}] \tag{15}$$

we can also write

$$[C_0 + C_{samp}](dT/dt)_{samp} = C_0(dT/dt)_0 \tag{16}$$

which upon rearrangement yields

$$C_{samp} = C_0\,[(dT/dt)_0 - (dT/dt)_{samp}]\;/\;(dT/dt)_{samp} \tag{17}$$

We now substitute the equation for C_0 into this last equation and rearrange to obtain the following equation

$$C_{samp}/C_{std} = \{[(dT/dt)_0/(dT/dt)_{samp}]-1\}\;/\;\{[(dT/dt)_0/(dT/dt)_{std}]-1\} \tag{18}$$

Since

$$C_{std} = \text{(standard Cu sample mass) (molar heat capacity of Cu)} \,/\text{(Atomic Weight of Cu)}$$

we can use the NBS tabulated values for the molar heat capacity of Cu and by determining the three (dT/dt) values we can, in principle, determine the heat capacity of any other material using the refrigerator. **Note: This is only valid to the extent that the cooling power of the refrigerator and the thermal contact between the samples and the refrigerator are reproducible.**

Methodology

The experiment has the following goals:

1. To demonstrate the use of the J–T effect to produce useful cooling and a stable low temperature environment.

2. To demonstrate the effect of an increase in the cold end heat capacity on the cooling rate of the refrigerator.

3. Under the assumption that the cooling power of the refrigerator is a reproducible function of the temperature of the J–T region of the refrigerator, use measured values of (dT/dt) for:
   a) the empty refrigerator,
   b) the refrigerator with an added standard copper sample,
   c) the refrigerator with a sample whose heat capacity is to be determined;

   to determine the heat capacity of the last sample as a function of temperature.

4. To evaluate the accuracy and precision of this method of determining the heat capacity using different data acquisition and reduction methods.

5. To demonstrate the usefulness of the Debye Theory in representing and explaining the low temperature behavior of the heat capacity of a solid.

The experiment will be conducted using a MMR Technologies Model K-2205 Cryogenic Microminiature Refrigeration System IIB combined with a MMR Model K-77 Temperature Indicator/Controller (cost ≈ $ 1,800). The apparatus is interfaced to a computer controlled data acquisition system (PC-XT and a Data Translation DT-2805 board) which will log the data. Programs to accomplish the data acquisition and reduction are provided to the student. Turbo Pascal source code and listings for these programs can be obtained by writing to the author.

Before beginning the experiment, the lab instructor is to go over the various parts of the equipment with the students. **IT IS IMPORTANT THAT THE STUDENTS TAKE CARE HANDLING THE REFRIGERATOR WHEN MOUNTING SAMPLES AS THE HEAT EXCHANGER AND SUPPORT ARE MADE OF GLASS PLATES AND CAN BE EASILY BROKEN.**

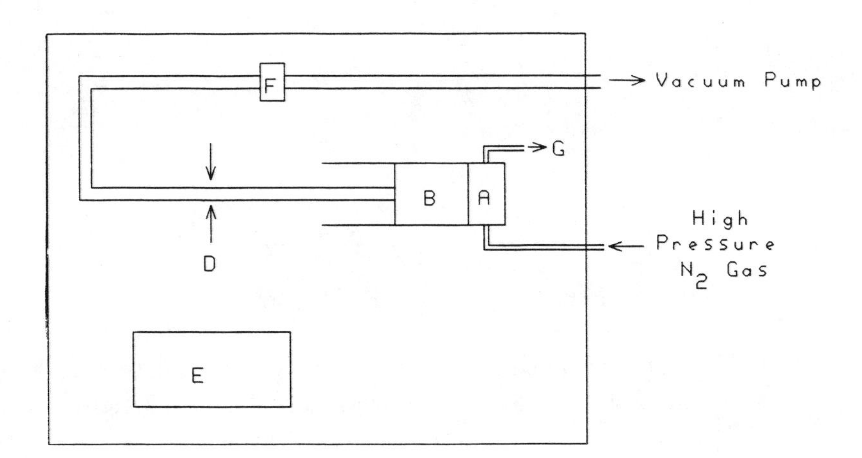

Figure 3. Apparatus Schematic.

A schematic layout of the apparatus is shown in Figure 3. The refrigerator (A) with its vacuum jacket (B) is mounted in the guide-holder (C) and connected to the vacuum line by a rubber hose connection at point (D). A vacuum gauge (F) is mounted in the vacuum line and the line is connected to a rotary vane vacuum pump located on the floor by vacuum rubber tubing. The high pressure N_2 gas is provided by a cylinder. The gas from the cylinder is passed through a regulator and then a line filter before being fed to the refrigerator. The line filter removes the last traces of impurities, hydrocarbons and water, from the gas. The low pressure exhaust from the refrigerator (G) is connected to a flow meter to monitor the flow rate. The refrigerator is electrically connected to the K-77 controller (E), which provides a read-out of the temperature of the sensor mounted within the cold stage of the refrigerator and also can supply current to the heater mounted within the cold stage, and connections to the computer data acquisition system are made to the cable connecting the controller and refrigerator.

NOTE: WHENEVER THE VACUUM JACKET IS REMOVED FROM THE REFRIGERATOR, THE REFRIGERATOR MUST REMAIN HELD IN THE GUIDE AND THE SPACER BOARD INSERTED IN THE GUIDE UNDER THE HEAT EXCHANGER. THIS PROVIDES SUPPORT FOR THE GLASS PORTION OF THE REFRIGERATOR TO PREVENT STRESS INDUCED BREAKAGE. (The lab instructor should demonstrate the sample changing procedure to the students.)

The front view of the K-77 Controller is shown in Figure 4.

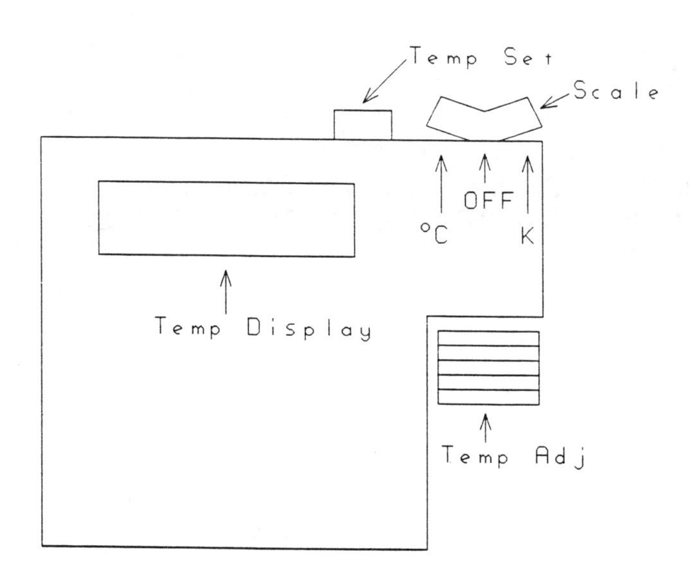

Figure 4. Front view of the K-77 Controller.

The experiment can be done manually by having one student record the time that a second student reads the temperature from the digital display on the K-77 Controller. However, the addition of the computer will allow the data to be taken in a manner suitable for the use of Fourier transform smoothing methods.

Data is to be acquired with the N_2 pressure maintained at 1800 psi throughout the cooldown by continuous adjustment of the regulator.

Ge 7040 varnish has long been used as a non electrically conducting thermal bonding material for use at low temperatures where the thermal expansion coefficients of the materials being bonded is not too different or clamping is also used. We have found that a slurry made by mixing varnish (thinned with a 50–50 mixture of ethanol–toluene) and Cu powder is an excellent bonding agent to mount samples on the refrigerator cold stage. The Cu powder is added to increase the thermal contact of the varnish. However, the presence of Cu can make this material electrically conducting, thus it must not contact the leads coming into the stage from the heat exchanger.

Mounting procedure: **Note: This is a hot cure procedure. Be careful that the student does not get burned by the temperature of the sample and stage.** Blank - Apply a drop of the slurry to the center of the stage at 30–35 °C, then apply two thin strips of plain varnish along the sides. Set controller to ≈ 90 °C, once temperature is achieved allow to air dry for 5 min. Seal the system and pump down. Sample - Apply a drop of slurry to the center of the stage at 30–35 °C and smooth out with sample. Set controller to ≈ 90 °C and use a wooden stick to apply pressure to sample while temperature increases. Once ≈ 90 °C has been reached, continue applying pressure for 5 min. Note: Be sure that the slurry does not flow across the leads from the stage to the heat exchanger. Remove pressure and apply plain varnish along sides of the sample where it meets the stage (note: it bubbles when applied at this temperature) and apply pressure with the wooden stick for 5 min. Remove pressure, seal the system and pump down.

To remove sample which have been bonded to the stage using GE varnish or slurry you must use a solvent made up of 50 – 50 % by volume mixture of ethanol and toluene. With the sample and stage at ≈ 25 °C, and using a small wooden stick, apply a small amount of the solvent along the sides of the sample where it meets the stage. Allow to set for a few minutes and lightly try to move the sample. Repeat this process until the sample can be slid off of the stage. Use a Chem-Wipe coated with solvent to clean both the stage and the sample. **Note: do not allow the solvent to contact the sides of the glass plates below the stage nor the region where the stage meets the heat exchanger. Excessive amounts of solvent can debond the laminations of the heat exchanger.**

Operation/Reduction Procedures

In order to determine the effect of different operation and reduction procedures on the results obtained by this method, different techniques will be used by student groups. The laboratory instructor should assign which operation/reduction procedure the student will use. The different procedures to be used are:

A. Operating/reduction procedure for multiple data read acquisition and no FFT (Fourier transform) reduction:

1. Obtain the T versus t data using multiple read averaging for the empty refrigerator, the standard Cu sample mounted and the sample (Cu or Gd) for which the heat capacity is to be determined mounted.

2. Compute the (dT/dt) versus T values at the same selected temperatures for each of the above.

3. Using the (dT/dt) data compute the C_p versus T for the sample (Cu or Gd).

B. Operating/reduction procedure for single data read acquisition and FFT reduction:

1. Obtain the T versus t data using single readings taken at evenly spaced times for the empty refrigerator, the standard Cu sample mounted and the sample (Cu or Gd) for which the heat capacity is to be determined mounted.

2. Do a Fourier transform smoothing of the T versus t data for each sample.

3. Compute the (dT/dt) versus T values at the same selected temperatures for each of the above.

4. Using the (dT/dt) data compute the C_p versus T for the sample (Cu or Gd).

The experiment is to be conducted in the following manner:

Prepare the system as described above for the blank run. The temperature should be at ≈ 90 °C and pumping down. Allow the system to pump for 10 min to reduce the pressure sufficiently to conduct the cooldown experiment (≈ 1-2 millitorr).

Turn off the heater on the refrigerator by holding the Set Temp button and turning the Temp Select knob counter clockwise until a reading of -198 °C is obtained. Do this quickly.

Open the nitrogen cylinder main valve and bring the regulator pressure to 1800 psi. **NOTE: DO NOT EXCEED 1800 psi AS DAMAGE TO THE REFRIGERATOR MAY OCCUR.** As soon as the regulator pressure is set to 1800 psi, begin the acquisition of the T vs time data. At this time the students should check the regulator pressure and adjust to read 1800 psi and continuously maintain it at 1800 psi during the cooldown to 80 K.

Immediately upon the completion of the blank run, the student should increase the set point for the controller to 25 °C by holding down the Temp Set button and turning the Temp Select knob clockwise. Then turn the N_2 gas regulator to zero and close the main cylinder valve. When the temperature reaches ≈ 25 °C, the students should wait for 5 minutes to insure that the refrigerator is sufficiently warm. This is to prevent water condensation on any cool portions of the refrigerator. Turn off the vacuum pump and disconnect the vacuum line from the refrigerator vacuum jacket at the position indicated (D) in the schematic of the apparatus.

It is now time to load the Standard Copper sample onto the cold stage using the assigned bonding material in the manner described above. THE LABORATORY INSTRUCTOR SHOULD SHOW THE STUDENTS HOW THIS IS DONE.

The remainder of the experiment consists of repeating the above for the Standard copper sample, the experimental copper sample and the experimental sample provided by the laboratory instructor (We have used Gd as it exhibits a λ type transition in the heat capacity near room temperature associated with the magnetic Curie point).

Safety

The refrigerator cold pad operates at temperatures ranging from 100 °C to - 199 °C and thus the students must be careful not to come into contact with the cold stage at temperatures other than when it has been brought to room temperature. Warnings as to this matter are indicated in the description using bold print.

Since high pressure gas and vacuum are used in this experiment, safety glasses must be worn at all times during this experiment.

Data Reduction Procedure

As discussed in the methodology section above, the raw data consists of either sensor emf or temperature versus time. The students must convert this raw data into (dT/dt) values versus T which are then combined to give the heat capacity of the sample versus T as explained in the theory section above. In addition, if the Fourier transform smoothing is to be done a fast Fourier transform calculation must be done on the T versus t data for each run. The students can also use the Debye theory the fit the copper sample heat capacity and determine both the electronic heat capacity coefficient, γ, and the Debye characteristic temperature, Θ_D.

Questions

What is the effect of adding a sample to the refrigerator on the cooling rate of the refrigerator? Explain your answer.

Compare the results obtained by averaging many readings as the data is obtained with those obtained by taking many single readings and then using Fourier Transform smoothing. What do you deduce about the merits of each data reduction method?

If the experiment were not limited by the four hour class time or available materials, what changes in the experimental procedure would you suggest to make the data:

more precise?

more accurate?

Explain in detail.

How could you perform this experiment so as to reduce the temperature shift (i.e. error) between the sample and the refrigerator?

Discuss the differences in the properties of copper and gadolinium which give rise to the different heat capacity results observed for these materials.

Acknowledgements

The author wishes to acknowledge the considerable effort expended by Ms. Cindy Wiesner in the development of this experiment.

Literature Cited

"Specific heats at low temperatures" by E. S. R. Gopal, Plenum Press (1966). [A very good discussion of low temperature heat capacities.]

Griffel, Skochdople and Spedding, *Phys. Rev.* **1954**, *93*, 657. [Heat capacity data for Gd].

Hardware List

MMR Technologies Model K-2205 Cryogenic Microminiature Refrigeration System IIB.
MMR Technologies Model K-77 Temperature Indicator/Controller.
PC-XT/AT Computer with printer and high resolution monitor.
Data Translation DT-2805 D/A - A/D Interface board.
High Pressure gas regulator.
Rotary vane vacuum pump capable of reducing the pressure to less than 10 microns.
Thermocouple vacuum gauge and controller.

RECEIVED October 14, 1992

Try a Different Approach

CHAPTER 30

A Monte Carlo Method for Chemical Kinetics

S. Bluestone

Kinetic studies of chemical reactions are undertaken primarily to elucidate or confirm a postulated reaction mechanism. For simple chemical reactions concordance of experimental concentration-time curves with those obtained from a supposed mechanism is tantamount to confirmation of the mechanism. For complex or difficult rate equations numerical integration techniques are employed to abstract the concentration–time profiles *(1)*. This treatment of chemical kinetics, called the deterministic approach, yields information concerning the average concentration of molecules, and is not applicable to cases where large fluctuations from the mean concentration occur *(2)*. On the other hand, the Monte Carlo (MC) method *(3)* can provide concentration–time curves for complex reactions, not by solving coupled differential equations, but by monitoring random picks of molecules represented by digits in a computer, and employing criteria that accepts or discards a potential conversion. The MC or stochastic approach and the deterministic approach yield comparable results when employed to solve a set of reaction rate equations *(4)*. However, the MC method should present less difficulty when used as a component of an undergraduate physical chemistry laboratory. The objective of this paper is to present kinetic simulations utilizing a MC procedure for a series of complex chemical reactions that are usually discussed in undergraduate physical chemistry courses. The method is inherently attractive as it is simple, yet it illustrates and reinforces the major concepts concerning chemical kinetics.

Monte Carlo Method

We begin the discussion of the MC method by considering the first order reaction $A \rightarrow B$. When n particles of reactant A are mixed with s particles of solvent S and labelled so that the A and S particles are represented in the computer by the digits 1 and 0, respectively. Schaad *(5)* has shown that if each A molecule in a first-order reaction has a constant probability P of conversion, then the number of A species reacting in time Δt is $nP\Delta t$. Each pick of a 1 or 0 digit in the computer has a constant probability P'. The number of 1's being converted to 0's in $\Delta t'$ is $nP'\Delta t'$. The expressions for the disappearance of A's and 1's have the same form, and except for a proportionality constant (which may be evaluated *(6)*) have an identical rate law. The rate law constant k can be associated with the quantity P, the probability that particle A reacts in unit time. We take the rate law constant k associated with P' to be weighted by a Boltzmann factor exp $(-E_a/RT)$, where E_a is an activation energy (a positive energy barrier), R the gas constant and T the temperature. The Boltzmann factor, exp $(-E_a/RT)$, is regarded as the fraction of molecules having the necessary energy for reaction. The rate law constant obtained by the MC method is therefore given by an Arrhenius expression

2503–6/93/0434$08.00/0

$$k = f \exp(-E_a/RT) \qquad (1)$$

where f, the pre-exponential factor, is called the frequency factor. This permits the investigation of how temperature or the activation energy parameter effect the kinetics.

The computer program requires the student to input the activation energy E_a and temperature T for the reaction. If an S particle is chosen, then no transition occurs and another cycle, random number, is selected. Suppose one of the A molecules is picked by a generated random number. The A particle may be converted to product B only if a computed Boltzmann factor is greater than a second random number between 0 and 1. Reaction occurs if

$$\exp(-E_a/RT) > \text{random number between 0 and 1} \qquad (2)$$

If an A particle is selected and overcomes the Boltzmann test of eq (2), it is converted to product B and its count decreases by one and the product B is increased by one unit. If $\exp(-E_a/RT)$ is less than a random number between 0 and 1, the original count of A is retained and another cycle (time unit) is started. Periodically the collection of particles is sampled every 200, 500, or 1000 cycles (picks) to give the number of A molecules as a function of the number N of cycles (picks). Comparable results are obtained from each sampling interval (within the expected MC fluctuations) and one sampling time frame is used consistently for given kinetic runs as a function of temperature. Each reaction in a mechanism is assigned a probability weighted by a Boltzmann factor. The goal is to obtain the rate law constants k_1, k_2, k_3, ... at several temperatures for activation energies E_{a1}, E_{a2}, E_{a3}, ... that are put into the computer program by the student. The velocity of the reaction is controlled at a selected temperature by making the activation energy choices. The individual rate constants obtained by the MC approach are proportional to experimental rate constants (the time parameter in the computer program is proportional to real time), however, the ratio of rate constants may represent experimental data.

The simulations were performed on a MacIntosh SE/30 Personal Computer. Approximately 10^4 cycles were generated per minute and the longest runs required some 2×10^5 cycles. Most importantly the periodic sampling results were stored in the computer's "Clipboard" and were conveniently transferred to a spreadsheet (EXCEL) for data analysis or to a graphics program (Cricket Graph). The number of reactant particles was chosen to be 10^4, a set sufficiently large to yield smooth data in the kinetic runs *(7)*. To reduce the rejection rate for A not being converted to product and to obtain computational efficiency the activation energies were no larger than 8000 J/mol for temperatures ranging from 100 to 1200 K. For E_a = 8000 J/mol and T = 300 K, $\exp(-E_a/RT)$ = 0.04. Utilizing the above calculation, 96% of attempts for conversion of A to product will be discarded and the same configuration retained. In the complex kinetic runs the largest value of the activation energy chosen was 5000 J/mol.

The MC simulations can serve as dry laboratory exercises to augment regular laboratory experiments. The data generated, concentrations as a function of a time parameter, can also be applied as classroom exercises or presented to students as homework projects. Several of the computer programs discussed in this paper are compiled in the appendix and are written in BASIC. In each case, the computer program pattern is similar and may be adapted to other chemical reactions. It is highly recommended that the student work the exercises in the Questions section.

Kinetics

The Monte Carlo method is applied to the following chemical reactions.

1. First order-first order reversible kinetics.

$$A \underset{k_2}{\overset{k_1}{\rightleftharpoons}} B \tag{3}$$

2. First order-first order consecutive kinetics.

$$A \xrightarrow{k_1} B \xrightarrow{k_2} C \tag{4}$$

3. First order-first order consecutive kinetics with a reversible step.

$$A \underset{k_2}{\overset{k_1}{\rightleftharpoons}} B \xrightarrow{k_3} C \tag{5}$$

4. Simple enzyme kinetics.

$$E + S \underset{k_2}{\overset{k_1}{\rightleftharpoons}} X \overset{k_3}{\rightleftharpoons} P + E \tag{6}$$

where E = enzyme, S = substrate, X = enzyme-substrate complex and P = product(s).

Reaction 1.

For first order-first order reversible kinetics (Equation 3) the student inputs, at a selected temperature, activation energies E_{a1} and E_{a2} for the forward and backward reaction, respectively. Each potential reaction transformation or conversion must meet the Boltzmann test criteria (Equation 2). The data, the number of A and B particles at cyclic intervals, are printed out and stored in the "Clipboard." A visual presentation of the data is produced by transferring the data to a graphics program. A typical run is shown in Figure 1(a).

At any time during a run, the number of A and B particles are equal to n,

$$A_0 + B_0 = A + B = A_{eq} + B_{eq} = n \tag{7}$$

where A_0 and B_0 are the initial number of A and B species and A_{eq} and B_{eq} are the equilibrium (average) number of A and B particles, respectively.

The rate equation for A is

$$d[A]/dt = -k_1 [A] + k_2 [B] \tag{8}$$

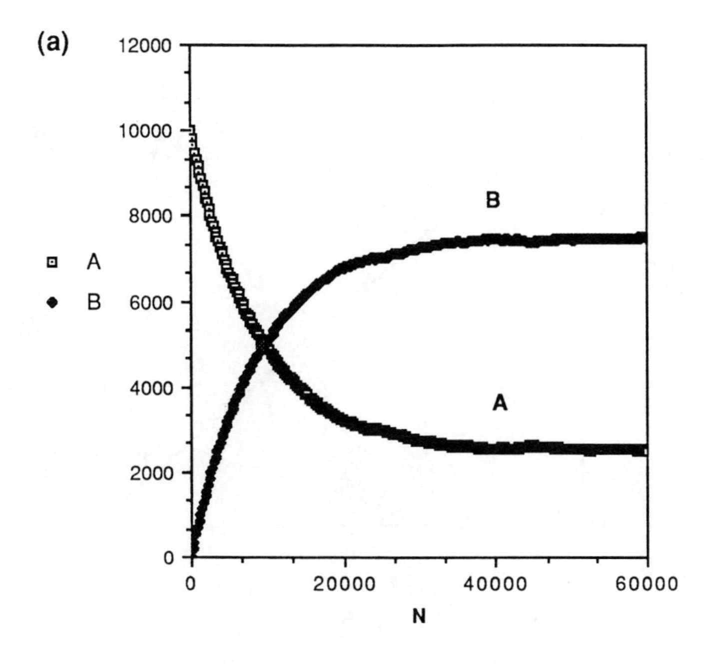

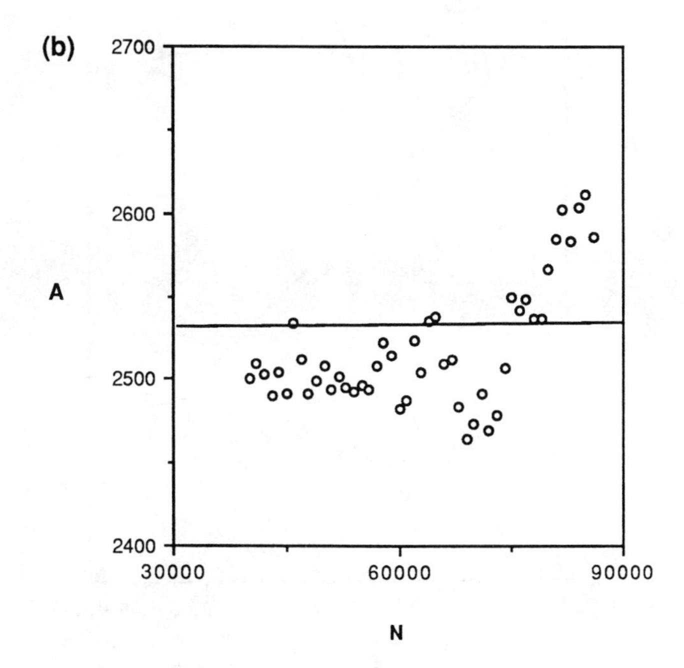

Figure 1. First order-first order reversible kinetics. E_A = 500 J/mol and E_B= 5000 J/mol. (a) T=500 K (b) The line indicates the value of A_{eq} at 500 K.

Continued on next page

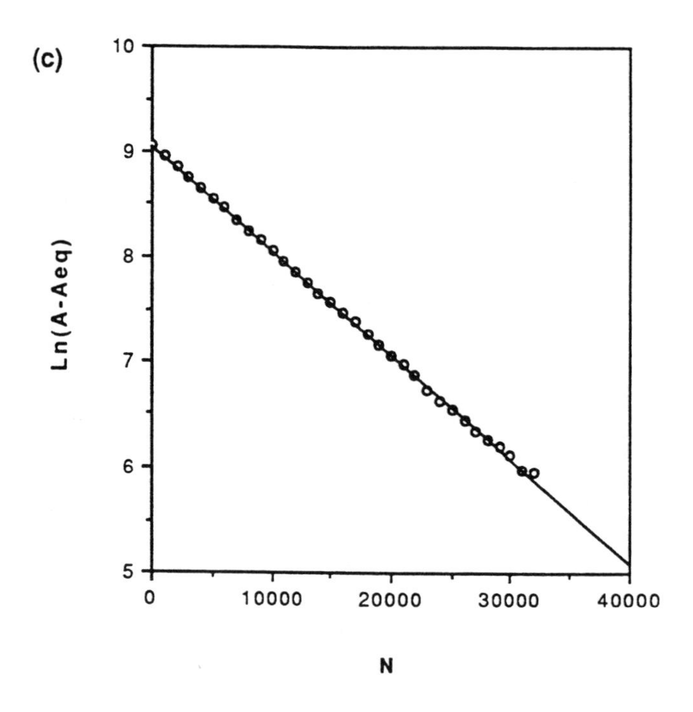

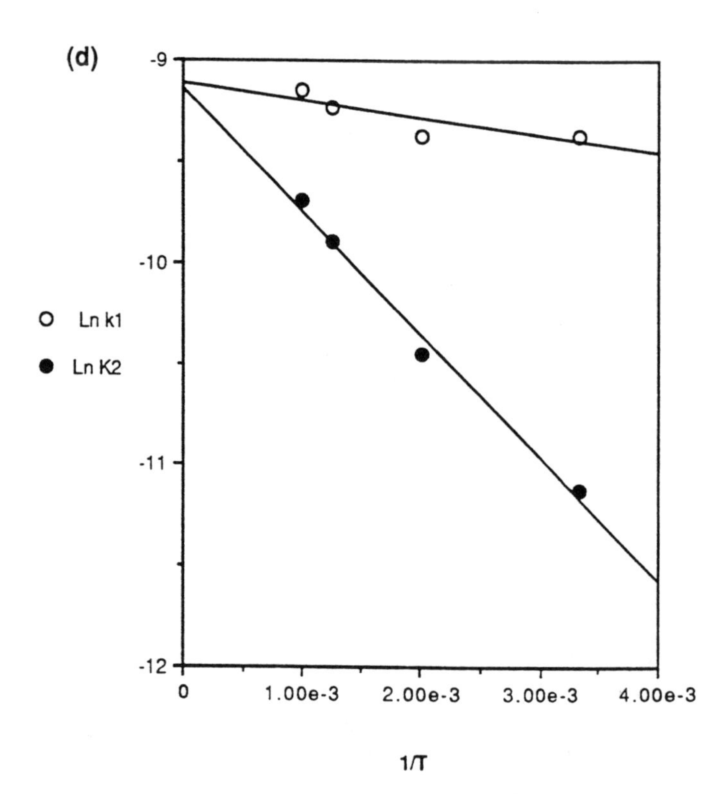

Figure 1 *continued.* First order-first order reversible kinetics. E_A = 500 J/mol and E_B = 5000 J/mol. (c) T = 500 K (d) Arrhenius plot.

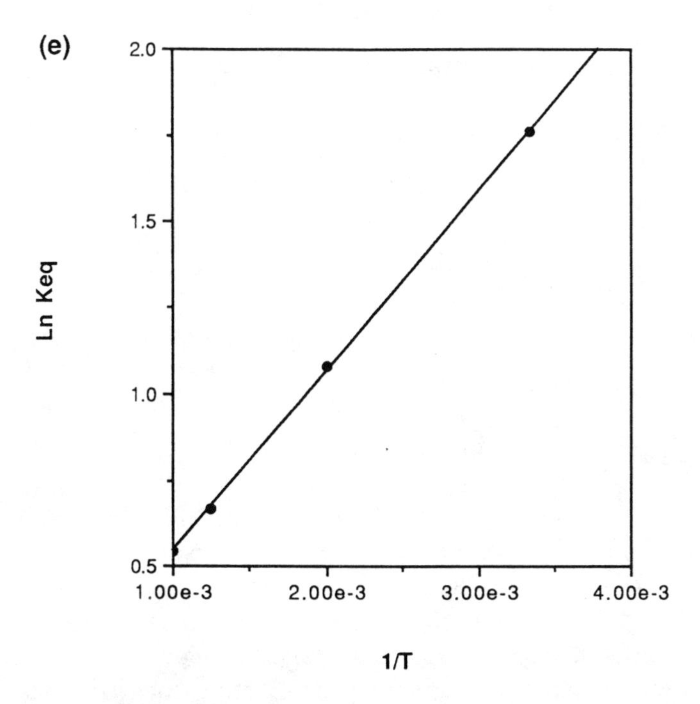

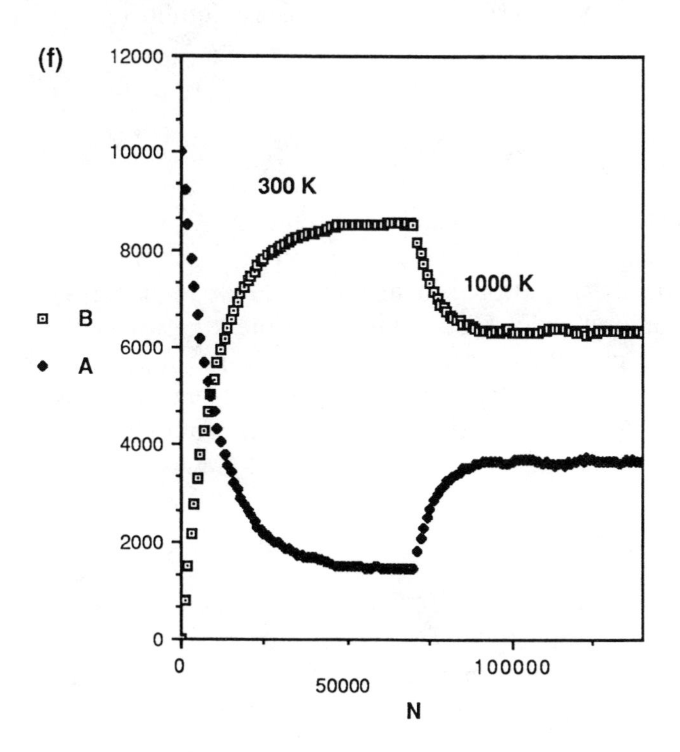

Figure 1 *continued.* First order-first order reversible kinetics. E_A = 500 J/mol and E_B = 5000 J/mol. (e) van't Hoff isochore (f) Illustration of temperature jump method in chemical kinetics.

At equilibrium

$$k_1\,[\mathrm{A}]_{\mathrm{eq}} = k_2\,[\mathrm{B}]_{\mathrm{eq}} \tag{9}$$

Equation (8) is transformed algebraically to

$$\mathrm{d}[\mathrm{A}]\,/\mathrm{d}t = -\,(k_1 + k_2)\,([\mathrm{A}] - [\mathrm{A}]_{\mathrm{eq}}) \tag{10}$$

and integration gives

$$\ln\,[([\mathrm{A}] - [\mathrm{A}]_{\mathrm{eq}})/\,([\mathrm{A}]_0\,-[\mathrm{A}]_{\mathrm{eq}})] = -\,(k_1 + k_2)\,t \tag{11}$$

Equation (11) may be written in terms of number of particles instead of concentrations as

$$\ln\,(A - A_{\mathrm{eq}}) = -\,(k_1 + k_2)\,t + \ln\,(A_0 - A_{\mathrm{eq}}) \tag{12}$$

The term $\ln\,(A - A_{\mathrm{eq}})$ is plotted against N, the number of cycles (proportional to t), and the slope of the line is a measure of the sum of the rate constants $(k_1 + k_2)$ in units of reciprocal cycles. To obtain A_{eq}, the data are examined to determine the number of cycles that elapse before equilibrium is achieved and subsequent values of A are averaged using a spreadsheet to yield A_{eq} and its standard deviation σ_{A}. Depending on starting conditions, usually $B_0 = 0$, approximately 4×10^4 cycles are necessary to insure equilibrium. Since

$$B_{\mathrm{eq}} = n - A_{\mathrm{eq}}$$

$$K_{\mathrm{eq}} = B_{\mathrm{eq}}/A_{\mathrm{eq}} = (n - A_{\mathrm{eq}})\,/\,A_{\mathrm{eq}} = k_1/k_2 \tag{13}$$

where K_{eq} is the equilibrium constant. Equation 13 is used to separate the individual rate constants from $(k_1 + k_2)$.

The data in a simulation run are graphed in Figure 1(a), and show the expected results for first order-first order reversible kinetics *(8-9)* . A plot of the equilibrium data for particle A, from 4×10^4 to 9×10^4 cycles, as shown in Figure 1(b), clearly depicts the fluctuations that accrue on account of the MC procedure of randomly selecting A and B particles. The relative standard deviation of A is calculated according to equilibrium theory *(10)* as follows

$$\sigma_{\mathrm{A}}/A_{\mathrm{eq}} = (B_{\mathrm{eq}}/A_{\mathrm{eq}}\,n)^{1/2} \tag{14}$$

Using the data of Figure 1(b), the MC value of $\sigma_{\mathrm{A}}/A_{\mathrm{eq}}$ at 500 K is 1.3%, while equilibrium theory yields 1.7%. If $B_{\mathrm{eq}} = A_{\mathrm{eq}}$ (for the case $E_{\mathrm{a1}} = E_{\mathrm{a2}}$), then the relative standard deviation of A at equilibrium varies as $n^{-1/2}$. Figure 1(c) depicts a plot of $\ln\,(A - A_{\mathrm{eq}})$ versus t and the slope of the least square line measures $(k_1 + k_2)$.

A plot of $\ln k_1$ and $\ln k_2$ versus $1/T$ (see Figure 1(d)) shows the expected Arrhenius behavior for the equilibrium reaction. From the slopes of the graphs the values of E_{a1} and E_{a2} are calculated (for assigning homework projects they may be regarded as unknowns) and compared with the inputted choices. Since $E = -R\, d(\ln k)/d(1/T)$, a large numerical value of E is associated with an extensive slope. At infinite temperature, as $1/T$ approaches zero, both lines in Figure 1(d) converge to the same pre-exponential factor f and serves as a check on the Arrhenius plots. Some numerical results are collected in Table 1.

Table 1. Kinetic parameters as a function of temperature.
E_A = 500 J/mol, E_B = 5000 J/mol and the data are for an interval of 1000 picks. The rate constants are in units of reciprocal cycles. (a) K_{eq} calculated from Equation (13). At an interval of 200 picks K_{eq} = 2.92 (.04). Standard deviations are in parenthesis and are obtained from Equation (15). (b) K_{eq} calculated from Equation (16).

T/K	$(k_1 + k_2)/10^{-4}$	$k_1/10^{-5}$	$k_2/10^{-5}$	K_{eq} (a)	K_{eq} (b)
300	0.988	8.51	1.47	5.80	6.07
500	1.14	8.51	2.88	2.95(.04)	2.95
800	1.48	9.79	5.02	1.95	1.97
1000	1.67	10.6	6.15	1.72	1.72

The logarithm of the equilibrium constant K_{eq} is plotted against $1/T$ and the graph is depicted in Figure 1(e). The slope of the van't Hoff isochore affords a value of $\Delta E^0 = E_B - E_A$. The standard deviation σ_k is calculated, according to propagation error theory, from eq (15) ($\sigma_A = \sigma_B$).

$$\sigma_k = (1/A_{eq}^2 + 1/B_{eq}^2)^{1/2}\, \sigma_A K_{eq} \tag{15}$$

The reaction A $\leftrightarrows$ B with E_{a2} = 5000 J/mol and E_{a1} = 500 J/mol is exothermic and as the temperature increases the amount of B decreases. k_1 and k_2 each increase with temperature, while the ratio k_1/k_2 (K_{eq}) decreases. The difference in activation energies $E_{a2} - E_{a1}$ and K_{eq} are about equal to the values for the cis to trans isomerization of dichloroethylene at 1200 K *(11)*.

Statistical-mechanical theory states that the probability p_i for a system in an energy state ϵ_i is

$$p_i = \exp(-\epsilon_i/RT)/\, \Sigma \exp(-\epsilon_j/RT)$$

The sum is over all energy states of the system. The most important quantity in statistical thermodynamics is the partition function q and is defined to be

$$q = \Sigma \exp(-\epsilon_j/RT)$$

An evaluation of the partition function allows the calculation of all thermodynamic properties of the system, including the equilibrium constant. For the case of an isomeric equilibrium, A $\leftrightarrows$ B reaction, the equilibrium constant is equal to *(12)*

$$K_{eq} = (q_A/q_B) \exp[-\Delta E_0/RT]$$

where ΔE_0 is the difference in zero point energies, $\epsilon_B - \epsilon_A$, and q_B and q_A are the partition functions of particles B and A, respectively. Since A and B are structureless molecules in the MC model, $q_A = q_B$.

Therefore

$$K_{eq} = \exp(-\Delta E_0/RT)$$

or

$$K_{eq} = \exp(\Delta E/RT) \tag{16}$$

where $\Delta E = E_{a2} - E_{a1}$, the difference in activation energies, is equal to $-(\epsilon_{a2} - \epsilon_{a1})$. K_{eq} is computed from the inputted activation energies, and the numerical results calculated from Equation (16) are given in the last column of Table 1. The equilibrium constants obtained by the MC method and the statistical mechanical calculation are in excellent agreement.

Figure 1(f) depicts the number of A and B molecules as a function of cycles at two temperatures. At 7×10^4 cycles the computer program abruptly changes the temperature from 300 K to 1000 K and the A $\rightleftharpoons$ B equilibrium is disrupted. The system undergoes a relaxation to new equilibrium concentrations. The approach of both A and B to their new equilibrium concentrations is first order with a rate constant given by $\mathfrak{T}^{-1}$. The relaxation time $\mathfrak{T}$ is $(k_1 + k_2)^{-1}$ *(13)* and Equation (12) is used to obtain $k_1 + k_2$. The kinetic reaction A $\rightleftharpoons$ B may thus be characterized in terms of relaxation kinetics, and affords an excellent illustration of the technique. The extension of the temperature jump method by the MC simulation technique to other chemical reactions is straight forward.

Reaction 2

For irreversible first order-first order consecutive kinetics (Equation 4) the student inputs at a selected temperature E_A (A → B) and E_B (B → C). The algorithm is similar to that of Reaction 1. At any time during a run the number of A, B and C particles are equal to A_0, the initial number of A molecules (B_0 and C_0 are zero).

$$A_0 = A + B + C \tag{17}$$

The rate equations

$$d[A]/dt = -k_1[A] \tag{18}$$

$$d[B]/dt = k_1[A] - k_2[B] \tag{19}$$

$$d[C]/dt = k_2[B] \tag{20}$$

have the well known solutions *(14)*

$$A/A_0 = \exp(-k_1 t) \tag{21}$$

$$B/A_0 = [k_1/(k_2 - k_1)] [\exp(-k_1 t) - \exp(-k_2 t)] \tag{22}$$

k_1 is obtained from the slope of the plot of ln (A/A_0) versus N, the number of cycles. To facilitate solving for k_2, Equation (22) is first parameterized. Using the definitions $\mathfrak{T} = k_1 t$ and $\kappa = k_2/k_1$, Equation (22) evolves to

$$B/A_0 = (\kappa - 1)^{-1}[\exp(-\mathfrak{T}) - \exp(-\kappa\mathfrak{T})] \quad (23)$$

$\mathfrak{T}$ is a measure of the number of half-lives for the first order decay of A ($\mathfrak{T} = 0.693\ t/t_{1/2}$). Equation (23) contains one unknown, κ, and it is necessary for the student to use a spreadsheet to find the best value for κ. One solution is to systematically vary κ until the variance (Var) in Equation (24) gives a minimum value.

$$\text{Var} = \Sigma\ [(B/A_0)_{MC} - (B/A_0)_{calc}]^2 \quad (24)$$

Another solution is to rearrange Equation (23) algebraically.

$$(B/A_0)\exp(\mathfrak{T}) = (1-\kappa)^{-1} - (1-\kappa)^{-1}\exp[(1-\kappa)\mathfrak{T}] \quad (25)$$

A plot of $(B/A_0)\exp(\mathfrak{T})$ versus $\exp[(1-\kappa)\mathfrak{T}]$ will give a straight line, assuming the best value of $1-\kappa$ has been selected, and the intercept and slope should be equal except for sign. After k_2 has been obtained from k and k_1, the value of B/A_0 is calculated from Equation (23) and compared to the generated MC B/A_0 values.

Figure 2(a) gives the familiar plot of A, B, and C as a function of time (cycles). Figure 2(b) depicts a plot of ln (A/A_0) against t to evaluate to k_1. Figure 2(c) shows a plot of $(B/A_0)\exp(\mathfrak{T})$ versus $\exp[(1-\kappa)\mathfrak{T}]$ and Figure 2(d) depicts a plot of $(B/A_0)_{MC}$ and $(B/A_0)_{calc}$ (Equation 23) against the number of cycles, and the concurrence is excellent. Arrhenius type plots are given in figure 2(e) for typical MC runs.

Reaction 3

The rate equations for a first order-first order consecutive kinetics with a reversible step are

$$d[A]/dt = -k_1[A] + k_2[B] \quad (26)$$

$$d[B]/dt = k_1[A] - (k_2 + k_3)[B] \quad (27)$$

$$d[C]/dt = k_3[B] \quad (28)$$

with $[A] = [A]_0$ and $[B]_0 = [C]_0 = 0$ at $t = 0$. It follows that at time t

$$[A]_0 = [A] + [B] + [C] \quad (29)$$

The set of the above equations can be solved explicitly *(15-16)* for [A], [B] and [C] as a function of time, however, the resultant expressions for [A], [B], and [C] are extremely complicated for any practical amenable treatment. The steady-state (ss) approximation is usually adopted to simplify the kinetic equations. The intermediate substance B is regarded to be so reactive that it cannot accumulate in any

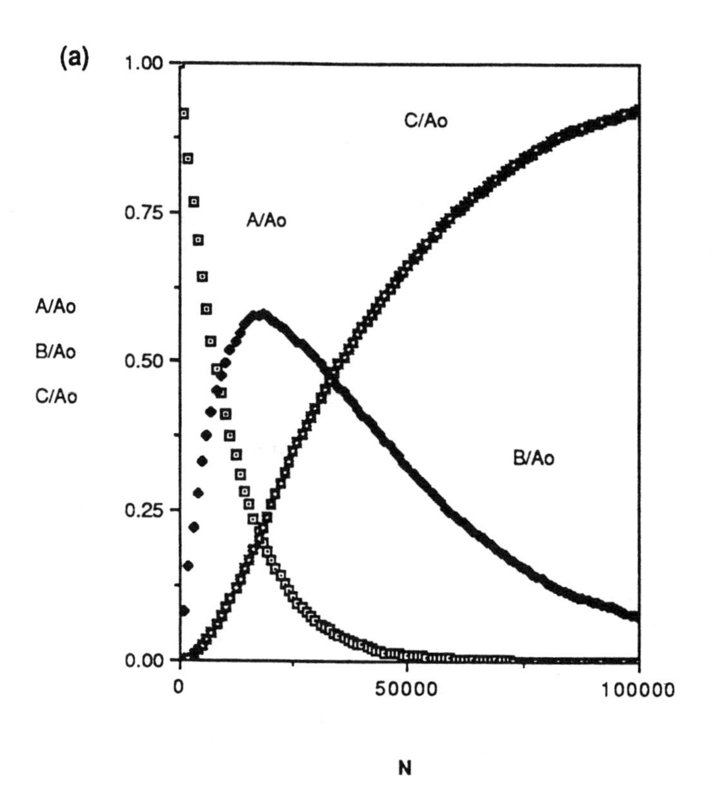

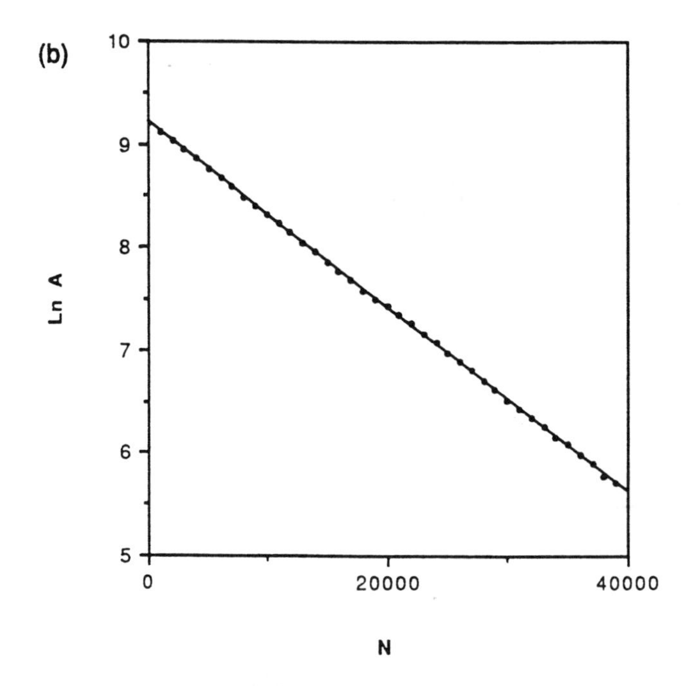

Figure 2. First order-first order consecutive kinetics. E_A = 500 J/mol and E_B = 5000 J/mol. (a) and (b) T = 500 K.

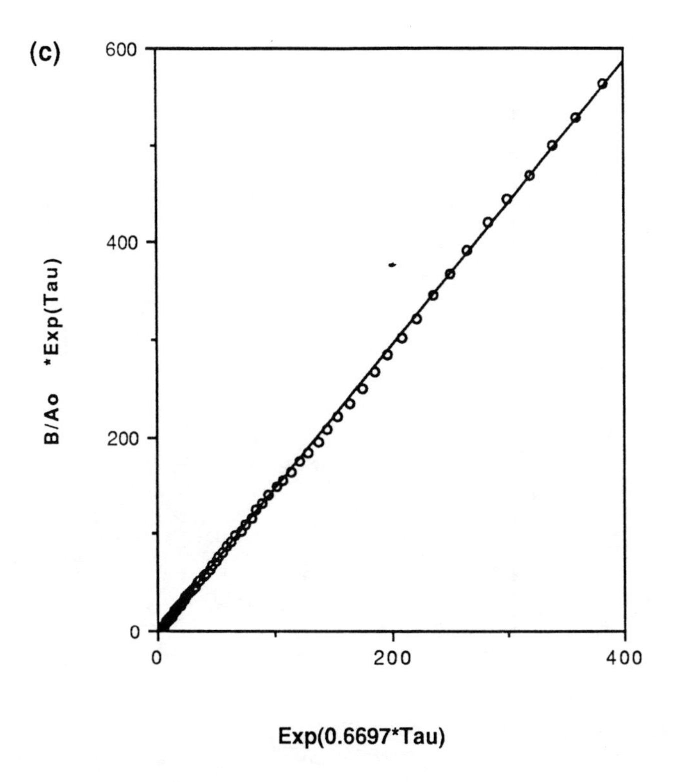

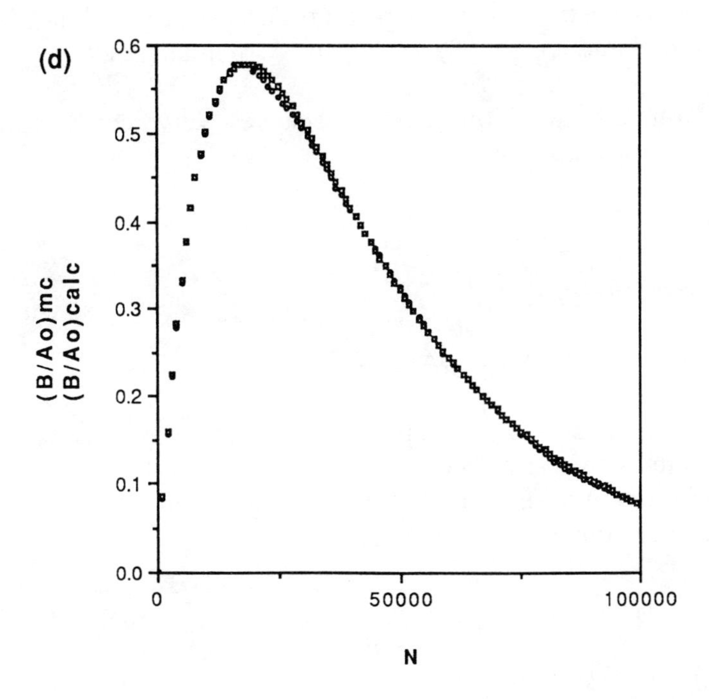

Figure 2. *continued* First order-first order consecutive kinetics. E_A = 500 J/mol and E_B = 5000 J/mol. (c) Test of Equation (25) (d) $(B/A_0)_{MC}$ obtained at 500K from the MC simulation and compared to $(B/A_0)_{calc}$ calculated from Equation (22).

Continued on next page

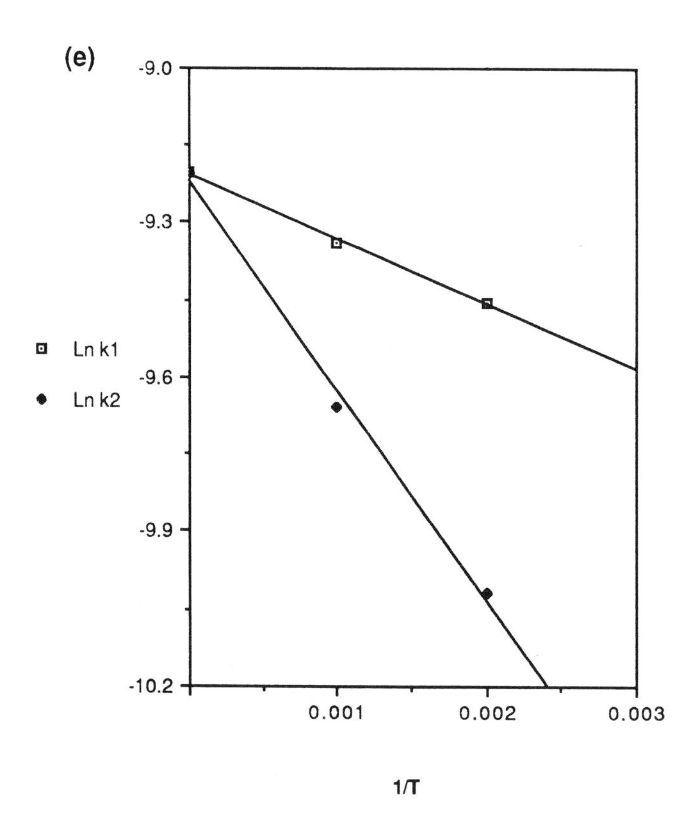

Figure 2. *continued* First order-first order consecutive kinetics. E_A = 500 J/mol and E_B = 5000 J/mol. (e) Arrhenius plot.

appreciable amount, and consequently changes very slowly with time. The steady-state approximation is

$$d[B]/dt = 0 \tag{30}$$

It can be shown *(15–16)* that

$$[B]_{ss} = [(k_1 /(k_2 + k_3)] [A] \tag{31}$$

The criterion for the steady-state approximation to be valid is $k_2 + k_3 >> k_1$. B should be formed slowly and disappear quickly. If the value of $[B]_{ss}$ from Equation (31) is substituted into Equation (26) and Equation (28), then the following equations are developed

$$d[A]/dt = - k[A] \tag{32}$$

$$d[C]/dt = [A]_0\, k \exp(-kt) \tag{33}$$

where

$$k = k_1k_3/(k_2 + k_3) \tag{34}$$

(k is an apparent rate constant).

Integrating Equation (32) and Equation (33) leads to

$$[A]/[A]_0 = \exp(-kt) \tag{35}$$

$$[C]/[A]_0 = 1 - \exp(-kt) \tag{36}$$

After substituting the value of [A] in Equation (35) into Equation (31), [B] is given by the expression

$$[B]/[A]_0 = [k_1/(k_2 + k_3)] \exp(-kt) \tag{37}$$

It follows that

$$\ln(A/A_0) = -kt \tag{38}$$

$$\ln(1 - C/A_0) = -kt \tag{39}$$

$$\ln(B/A_0) = \ln[k_1/(k_2 + k_3)] - kt \tag{40}$$

After the steady-state condition is assumed to prevail, that is, the cycle in which B reaches its maximum value, Equation (38) or Equation (39) is employed to evaluate the apparent rate constant k. If the logarithmic expressions on the left side of Equation (38) or Equation (39) are plotted against t (beyond the attainment of the ss condition), straight lines should result. The slope of either plot affords a value of k, the apparent rate constant. Since the number of B particles slowly decreases with time and fluctuates, Equation (40) gives the least precision.

The MC algorithm is programmed similar to Reaction 2 except for an additional step. The student inputs at a selected temperature E_{AB} (A → B), E_{BA} (B → A) and E_{BC} (B → C). When an A particle is chosen by a random number, it is tested for a potential reaction according to Equation (2). Given that $E_{BC} > E_{BA}$, and a B particle is picked by a random number, the Boltzmann test is first applied to the B → C reaction.

$$\exp(-E_{BC}/RT) > \text{random number between 0 and 1.}$$

If the conversion B → C fails, then the Boltzmann test for B → A is addressed. If the reaction B → A fails, the same configuration is retained and a new cycle begins. Given that $E_{BC} < E_{BA}$, the reverse sequence from that given above applies for B. The reaction involving B with the higher activation energy is tested first.

The other choice, testing the lowest energy barrier first, seems more natural in the sense that systems tend to flow in the direction of least resistance. However, a 'hot' molecule in the current kinetic model is programmed to spend its energy on the highest barrier, not 'wasting' its potential on a shallower barrier. The computer program simulates kinetics and the MC output is compared with experimental trends. Revisions are usually required to improve kinetic models.

Figures 3(a) and 3(b) depict the concentration changes A, B and C undergo with time at 200 K and 500 K, respectively. At 200 K, B attains its maximum number at about 4×10^4 cycles and then exponentially decreases to zero. Beyond 4×10^4 cycles the steady-state condition is assumed to exist. Notice in Figure 3(a)

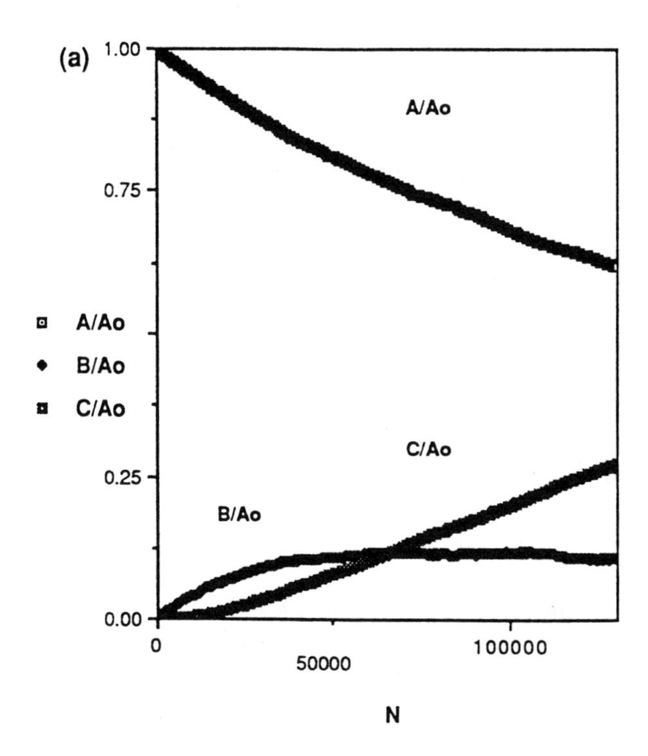

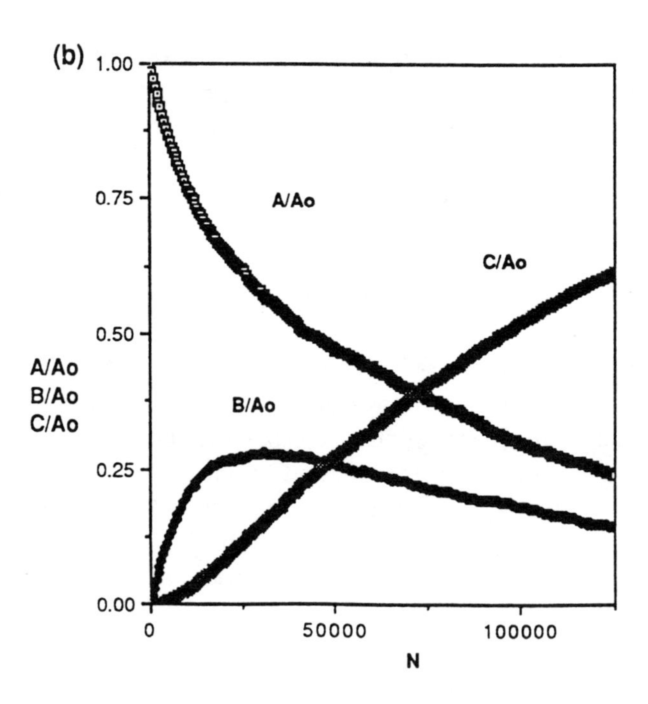

Figure 3. First order - first order consecutive kinetics with a reversible step. E_{AB} = 5000 J/mol, E_{BA} = 4000 J/mol and E_{BC} = 2000 J/mol. (a) 200 K (b) 500 K.

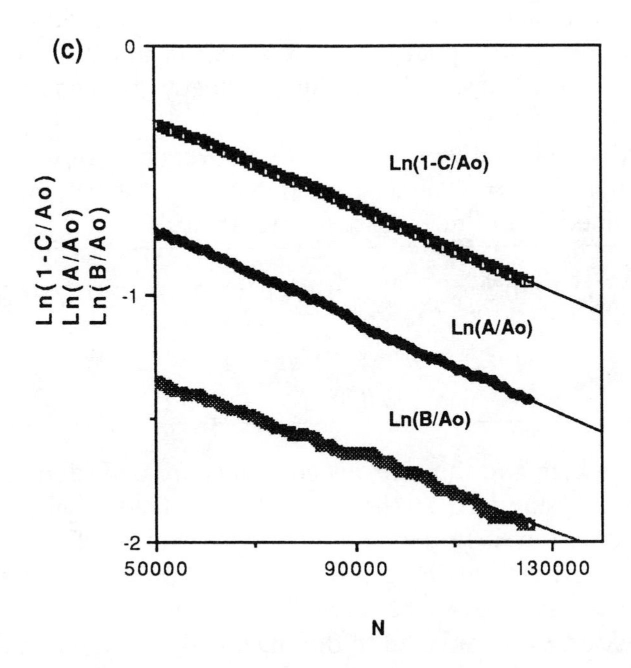

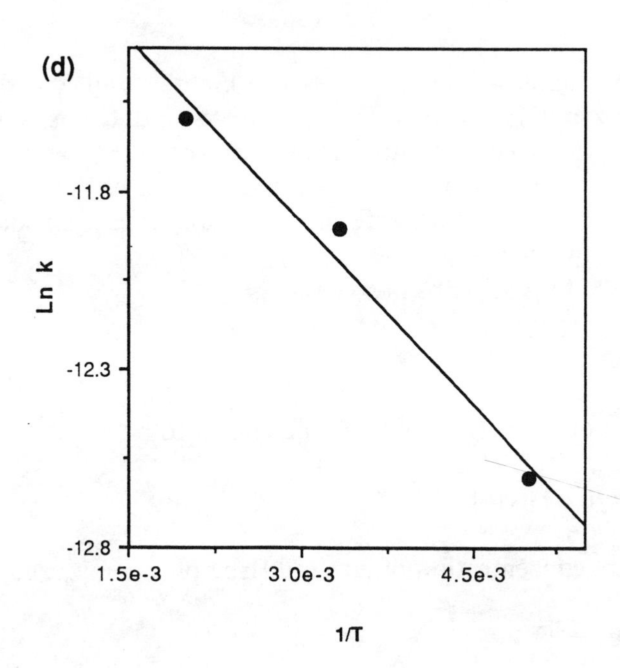

Figure 3 *continued.* First order-first order consecutive kinetics with a reversible step. E_{AB} = 5000 J/mol, E_{BA} = 4000 J/mol, and E_{BC} = 2000 J/mol. (c) 200 K (d) Arrhenius plot of ln k (apparent rate constant) versus reciprocal temperature. $k = k_1k_3/(k_2 + k_3)$.

the fraction of B is substantially less than the fraction of A, due primarily to the slow formation of B, and particles of B take a longer time to reach their maximum number when compared to the time required at 500 K (see Figure 3(b).

The log plots, as shown in Figure 3(c), provide [see Equation (38), Equation (39) and Equation (40)] the apparent rate constant k and $k_1/(k_2+k_3)$. The time parameter starts at 5×10^4 cycles (ensuring the steady-state condition). k_3 is the

Table 2. First order consecutive kinetics with a reversible step. E_{AB} = 5000 J/mol, E_{BA} = 4000 J /mol and E_{BC} = 2000 J/mol. The rate law constants are obtained from linear least square analysis.

T/K	$k/10^{-6}$	$k_1/(k_2+k_3)$	$k_3/10^{-5}$
200	3.36	0.12	2.8
300	6.74	0.26	2.6
600	9.19	0.38	2.4

ratio of $k/[k_1/(k_2+k_3)]$, and the rate law constants are collected in Table 2. k and $k_1/(k_2+k_3)$ show the standard Arrhenius behavior (see Figure 3(d)), while k_3 decreases with temperature. In the computer algorithm the Boltzmann test is first applied to the B → A reaction ($E_{BA} > E_{BC}$). At infinite temperature all B → A conversions are successful and the B → C reactions will not occur. As the temperature increases, k_3 is expected to diminish and the system approach the A ⇆ B equilibrium system. The student is asked to demonstrate this trend in an exercise.

Reaction 4.

Reaction 4 is the simplest model for enzyme kinetics and is a slight extension of Reaction 3. Enzyme E reacts with substrate S and forms a substrate-enzyme complex X, which can decompose to free enzyme and substrate or to product(s) and free enzyme. The enzyme E is regenerated at the end of the reaction, and consequently is a catalyst, yet the rate of conversion of X to product(s) depends on the concentration of enzyme E.

The rate of formation of product, v, is

$$v = d[P]/dt = k_3[X] \quad (41)$$

The steady-state concentration of X can be shown to be

$$[X] = [k_1/(k_2+k_3)][E][S] \quad (42)$$

and $[E]_0$, the total concentration of enzyme (free plus complexed) is

$$[E]_0 = [E] + [X] \quad (43)$$

After a few algebraic steps, involving solving for [X] in terms of $[E]_0$ and substituting the result in Equation (41), we obtain

$$v = k_1k_3[E]_0[S]/(k_2+k_3+k_1[S]) \quad (44)$$

$$v = k_3[E]_0[S]/(K_m + [S]) \tag{45}$$

where the Michaelis constant K_m is

$$K_m = (k_2+k_3)/k_1$$

Equation 45 is rearranged to a form of the Lineweaver–Burk equation

$$[E]_0/v = 1/k_3 + (K_m/k_3)(1/[S]) \tag{46}$$

From a plot of $[E]_0/v$ against $1/[S]$, k_3 and K_m can be obtained. No differential equations have to be solved, however, the velocity of reaction v needs to be calculated.

The MC algorithm is quite similar to Reaction 3, yet differs in the following ways. First, the free enzyme count increases by one unit when product P is formed. At all times during a run Equation (43) is obeyed. Second, two random numbers are generated to see whether a double 'hit' occurs. To form the enzyme-substrate complex one random number must pick an enzyme particle, the other a substrate species (there are two ways of accomplishing the double 'hit'), and then the Boltzmann test is applied to the potential formation of complex X.

Figure 4(a) and Figure 4(b) illustrate the behavior of E, S, X and P as a function of time and are in general agreement with computer solutions *(17)* of experimental data. The pictorial displays are strikingly similar. After obtaining the velocity of the reaction v in the initial stages of the reaction at a given number of cycles and after the time for the induction period of the product to pass, a Lineweaver–Burk plot can be drawn by plotting $[E]_0/v$ versus $[S]^{-1}$. The substrate concentration is taken at the same number of cycles that v is calculated. The greatest source of error in this procedure accrues in calculating v, due to the fluctuations of P in the MC simulation. The reaction velocity can be determined by a least square fitting of P to a polynomial of degree two in time and subsequently differentiating. The least square data fitting tends to smooth out fluctuations. However, it is necessary to repeat several of the MC runs to obtain reliable velocities and Figure 4(c) illustrates typical results. On the other hand, since the reaction velocity is measured at the beginning of the reaction, the MC simulation needs to be run only to about 20,000 cycles. Values of k_3 and K_m are presented in Table 3. An Arrhenius plot is shown in Figure 4(d) and E_3 ($X \rightarrow P+E$) obtained from the slope (2020 J/mol) is in excellent agreement with the input activation energy (2000 J/mol).

The K_m values of Table 3 show a decrease as temperature increases. In the MC simulation model K_m approaches k_3/k_1 (and is expected to be unity) as temperature becomes infinite (all $X \rightarrow P+E$ conversions succeed, $k_2=0$ and E_2 ($X \rightarrow S+E$) is less than E_3 ($X \rightarrow P+E$)).

Supplemental Questions

Reaction 1.

(a) Run program 1 to demonstrate that the amounts of A and B are the same at equilibrium regardless of the direction of approach. One MC simulation

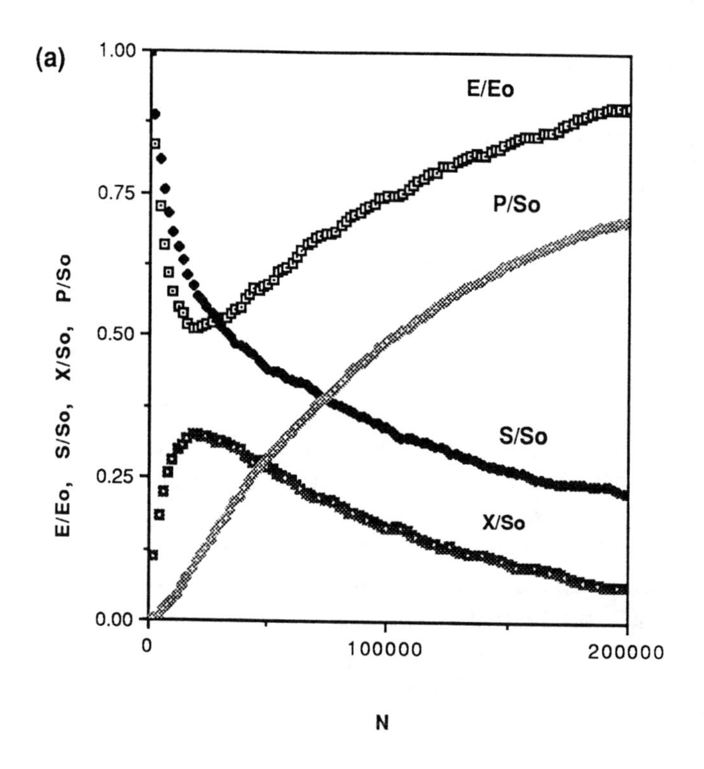

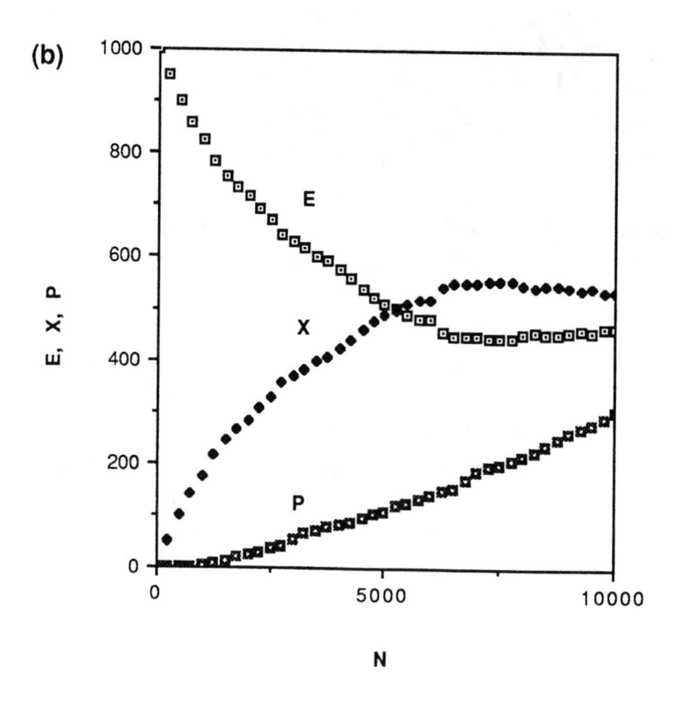

Figure 4. Enzyme Kinetics. E_1 = 500 J/mol, E_2 = 1000 J/mol and E_3 = 2000 J/mol. (a)T = 298K (b) Plot of the data from the initial stages of Figure 4(a).

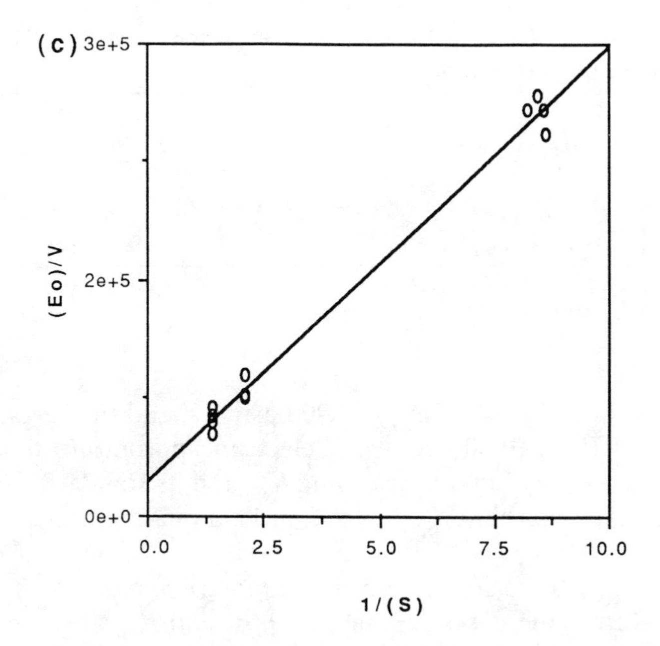

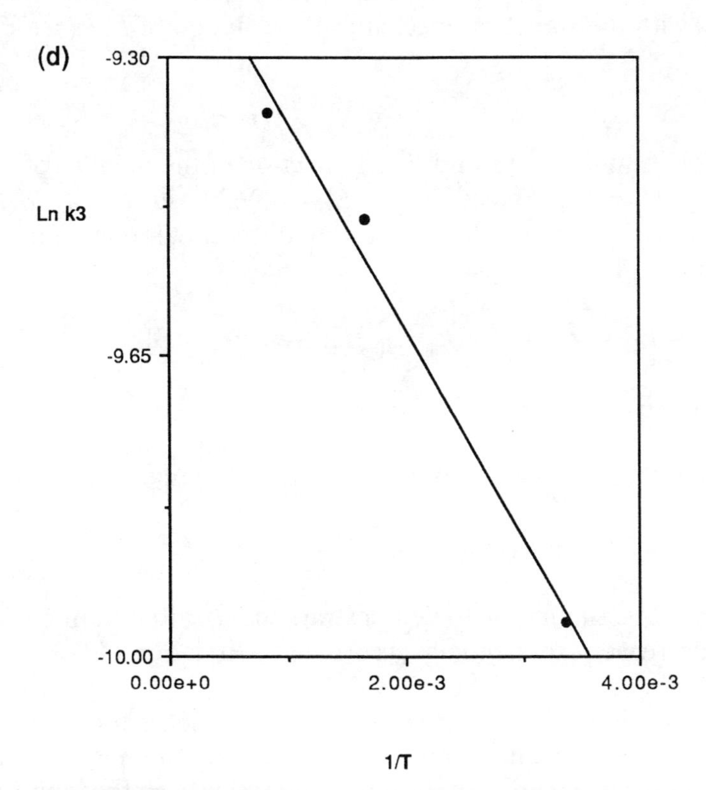

Figure 4 *continued*. Enzyme Kinetics. E_1 = 500 J/mol, E_2 = 1000 J/mol and E_3 = 2000 J/mol. (c) Lineweaver–Burk type plot (d) Arrhenius plot.

Table 3. Enzyme Kinetics. The parameters in the MC simulation are: E_1 (S+E→X) = 500 J/mol, E_2(X→S+E) = 1000 J/mol, E_3 (X→P+E) = 2000 J/mol and E_0+S_0 = 1 × 10^4 particles. The initial velocity of the reaction, dP/dN, is measured at 4500 cycles by a least square polynomial fitting procedure. The standard deviations given in parenthesis are obtained from a least square fitting of the data to the Lineweaver–Burk equation.

T/K	K_m	$k_3/10^{-5}$
298	1.3(.2)	4.7(.2)
600	1.3(.1)	7.6(.7)
1200	1.14(.04)	8.6(.9)

is run with A_0 = 2000 and B_0 = 8000 particles, and the second run with A_0 = 9000 and B_0 = 1000 species. Select an appropriate temperature and activation energies. In both cases find K_{eq} and its standard deviation. Show that if $A_0+B_0 = n$, then $A_{eq} = n/(K_{eq}+1)$, a quantity independent of A_0.

(b) Run program 1 to verify that for an endothermic reaction K_{eq} increases as the temperature increases. Evaluate k_1, k_2 and K_{eq} at three temperatures and find E_A(A→B), E_B(B→A) and ΔE. Compare K_{eq} found by the MC method with the statistical mechanical calculation of K_{eq} (see Equation 16)

Reaction 2.

(a) Run a MC simulation to find B as a function of the number of cycles for the condition $k_1 = k_2$. Select $E_{AB} = E_{BC}$. Demonstrate that $B/A_0 = \mathfrak{T} \exp(-\mathfrak{T})$, $\mathfrak{T} = k_1 t$, is a solution of the differential equation for the rate formation of B.

$$d[B]/dt = k_1[A] - k_2[B] = k_1[A]_0 \exp(-k_1 t) - k_2[B]$$

Compare $(B/A_0)_{MC}$ and $(B/A_0)_{calc}$.

(b) Assuming that $k_2 >> k_1$, show that the steady-state solution for B is

$$B_{ss}/A_0 = (k_1/k_2) \exp(-k_1 t) \quad (47)$$

Perform a MC simulation to demonstrate that B, after attaining its maximum value, decreases exponentially according to Equation (47).

(c) Run a MC simulation at three temperatures, solve for k_1 and k_2 at each temperature and evaluate the Arrhenius activation energies. Calculate the relative percent deviation of the latter quantities from the inputted activation energies.

Reaction 3.

(a) Run the MC simulation for Reaction 3 to demonstrate that A and B will establish a pre-equilibrium system (except for leakage of B to C), if $k_3 << k_1, k_2$. In the MC model at high temperatures the B→A conversion will become almost exclusive in comparison to the B→C reaction. Select a temperature of 10,000 K, E_{BA} = 3000 J/mol, E_{BC} = 2000 J/mol, E_{AB} = 5000 J/mol and $A_0 = 1 \times 10^4$ particles.

(b) For comparison, run Reaction 3 with the same conditions as in (a), except let E_{BA} = 2,000 J/mol and E_{BC} = 4,000 J/mol.

(c) Investigate what occurs if $E_{BC} > E_{BA}$ and the B → A reaction is tested first (a few steps in the BASIC program need to be rearranged).

Reaction 4.

(a) Run program 4 to establish a Lineweaver–Burk plot. Select E_1 = 1000 J/mol, E_2 = 2000 J/mol and E_3 = 4000 J/mol, and vary the initial concentration of substrate particles, $[S]_0$, such that $[S]_0 > [E]_0$

$([S]_0 = S_0/(S_0+E_0)$ and $[E]_0 = E_0/(S_0+E_0))$.

What method are you employing to measure the rate of reaction?

Concluding remarks

A simple Monte Carlo technique has been introduced to illustrate the method and to apply it to a few complex chemical reactions. The MC computer simulations yield information richer in detail than that provided by kinetic experiments. The concentration-time data obtained can be used as student exercises to enhance laboratory kinetic experiments or as homework projects. The extension to other chemical reacting systems is easily handled, and can serve to complement the method of solving differential equations numerically *(18)* in complex kinetic reactions. Other Monte Carlo computer programs that simulate processes occurring randomly in nature, handle problems that are mathematically intractable or study problems in which fluctuations are important may be investigated. Some of the best known systems studied with the Monte Carlo (MC) method are the structure of the liquid state *(19)*, the Ising lattice problem *(20)* and polymer configurations *(21-22)*.

Acknowledgment

The students at CSUF in the Spring 91 Physical Chemistry Laboratory class are gratefully acknowledged for their patience and helpful criticism.

Literature Cited

1. Edelson, D. *J. Chem. Educ.* **1975**, *52*, 642.
2. McQuarrie, D. A. *J. Appl. Prob.* **1967**, *4*, 413.
3. Boucher, E. A. *J. Chem. Educ.* **1974**, *51*, 580.
4. Steinfeld, J.I.; Francisco, J.S.; Hase, W.L. *Chemical Kinetics and Dynamics*; Prentice Hall: Englewood Cliffs, NJ, 1989; Ch. 2.
5. Schaad, L. J. *J. Am. Chem. Soc.* **1963**, *85*, 3588.
6. Rabinovitch, B. *J. Chem. Educ.* **1969**, *46*, 262.
7. Schaad, L. J. *J. Am. Chem. Soc.* **1963**, *85*, 3588.
8. Levine, I.N. *Physical Chemistry, 3rd. Ed.*; McGraw-Hill Book Company: New York, 1988; Ch. 17.
9. Atkins, P.W. *Physical Chemistry, 4th Ed.*; W.H. Freeman and Company: New York, 1990; Ch. 26.
10. Hill, T.H. *An Introduction to Statistical Thermodynamics*; Addison-Wesley Publishing Company: Reading, MA, 1960; pp. 181-182.
11. Jeffers, P.M. *J. Phys. Chem.* **1972**, *76*, 2829.
12. Hill, T.H. op. cit., Ch. 10.
13. Atkins, P.W. op. cit., pp. 794-795.
14. Atkins, P.W. op. cit., pp. 798-799.
15. Pyun, C.W. *J. Chem. Educ.* **1971**, *48*, 194.
16. Volk, L.; Richardson, W.; Lau, K.H.; Hall, M.; Lin, S.H. *J. Chem. Educ.* **1977**, *54*, 95.
17. Lehninger, A.L. *Biochemistry*; Worth Publishers: New York, 1975; 2nd Ed., pp. 191.
18. Steinfeld, J.I.; Francisco, J.S.; Hase, W.L. op. cit., pp 56-67.
19. Allen, M.P.; Tildesley, D.J. *Computer Simulation of Liquids*; Clarendon Press: Oxford, 1987.
20. Gordon, R. in *Cooperative Phenomena in Biology*; Kasseman, G., Ed.; Pergamon Press, New York, 1990; Ch. 5.
21. Engström, S.; Lindberg, M. *J. Chem. Educ.* **1988**, *65*, 973.
22. Allan, D.S. *J. Chem. Educ.* **1991**, *68*, 623.

Appendix

BASIC Programs

1. PROGRAM 1. REACTION 1.

```
REM Reversible First order Kinetics and  Relaxation Kinetics
REM Input forward and reverse activation energies.
OPEN "clip:" FOR OUTPUT AS#1:          REM to put results in memory
INPUT" any number"; U:                 REM to stop and change program
LPRINT "First order kinetics,  A <----> B   ; reversible reaction with forward"
LPRINT "and backward reactions having different activation energies"
PRINT "First order kinetics,  A <----> B    ; reversible reaction with forward"
PRINT "and backward reactions having different activation energies"
PRINT"Input the forward activation energy (between 500 to 5000 joules )"
```

```
PRINT" E(forward) = ": INPUT EF:          REM EF= forward activation energy
PRINT"Input the backward activation energy (between 500 to5000 joules)"
PRINT" E(backward = ":INPUT EB:           REM EB = backward activation energy
PRINT" input temperature T/K":INPUT T
C=EXP(-EF/(8.314*T)):                     REM Boltzmann factors
D=EXP(-EB/(8.314*T))
PRINT"Temperature  T = "; T
PRINT"A  <----> B  : reversible : program randomly chooses A and B molecules."
DIM x(10001),y(10001)
P=1:q=1:PRINT
10 INPUT"INPUT the sampling parameter";k:REM count parameter
PRINT" input number of A molecules, set parameter l between 0 to 10,000"
:INPUT g:                                 REM g=number of A particles
INPUT"INPUT the No. of A+ B molecules to be 10000"; r
h = r-g:                                  REM h=number of B particles
LPRINT"initial A and B molecules, respectively, = ";g,h
FOR i=1 TO g:                             REM assigning A=1 and B=2
x(i)=1
NEXT i
FOR i= g+1 TO r
y(i)= 2: NEXT i
RANDOMIZE TIMER:PRINT:                    REM start random number generator
PRINT" INPUT number of cycles n  =  ":INPUT n:REM n=total number of picks
in  program
LPRINT"No. of cycles n= ";n: LPRINT "Temperature =";T :LPRINT
LPRINT"E(forward) = ";EF;"  E(backward) = ";EB:LPRINT
LPRINT"     Molecules"
LPRINT"No. A         No. B        No. Cycles"
LPRINT g,h,0
WRITE #1,0,g,h:                 REM save in memory to transfer to other programs
100 FOR i=q TO n:                         REM q = 1 initally
A=INT(RND*r+1):                           REM choose a random number
IF x(A)=1 THEN GOTO 200
GOTO 300
200 IF C>RND*1 THEN GOTO 250:             REM Boltzmann test A-->B
GOTO 340
250 g=g-1:                                REM A converted to B
x(A)=0
y(A)=2
h=h+1
GOTO 340
300 IF D>RND*1 THEN GOTO 310:             REM Boltzmann test B-->A
GOTO 340
310 x(A)=1:                               REM B converted to A
y(A)=0
g=g+1
h=h-1
340 IF i=P*k THEN GOTO 350:REM test for sampling cycle, P=1and k=inputted
sampling time
GOTO 400
```

```
350
PRINT g,h,i:          REM may be omitted and replaced with a screen graph
LPRINT g,h,i:         REM record of number A's, B's and cycles (for homework
problems,etc.)
P=P+1
WRITE #1, i,g,h:                        REM Save in memory
400 NEXT i
REM  Relaxation Kinetics: Temperature change- input a new temperatue and
number of cycles to end at the new  selected temperature
INPUT"Do you wish to select a new temperature and continue?(yes or no)";A$
IF A$="no" THEN GOTO 500
PRINT: INPUT" The new Temp. run will stop at n cycles; INPUT n .";n:
IF i=n THEN GOTO 500
INPUT"INPUT New TEMPERATURE T";T:PRINT
PRINT"TEMP T = ";T
LPRINT"TEMP T= ";T
C=EXP(-EF/(8.314*T)):D=EXP(-EB/(8.314*T))
 q=i+1:  GOTO 100:          REM back to random number generation, etc.
500 CLOSE #1:               REM close writing to memory
PRINT"End of kinetic run"
LPRINT"End of kinetic run"
END
```

2. Program 2. Reaction 2.

```
REM Consecutive First order Kinetics: Input A--->B and B--->C activation
energies.
REM same nomenclature used as in program 1.
OPEN "clip:" FOR OUTPUT AS#1
INPUT" any number"; U
LPRINT "First order kinetics,  A ----> B ----> C, having different activation
energies"
PRINT "First order kinetics,  A ----> B ----> C"
PRINT"Input the first activation energy (between 500 to 5000 joules )"
PRINT" E(A--->B) = ": INPUT EF
PRINT"Input the second activation energy (between 500 to 5000 joules)"
PRINT" E(B--->C) = ":INPUT EB
PRINT" input temperature T/K":INPUT  T
c=EXP(-EF/(8.314*T))
D=EXP(-EB/(8.314*T))
PRINT"Temperature  T = "; T
PRINT"A ----> B ---->C  : reversible : program randomly chooses A and B
molecules."
DIM x(10001),y(10001),Z(10001)
P=1:q=1:gb=0:gc=0:
INPUT"INPUT the sampling parameter";k : REM count parameter
PRINT" input number of A molecules, set parameter g between 5000 to 10000"
INPUT"INPUT the number of A Molucles";g
INPUT"input the No. of A +B molecules to be 10000"; r
```

```
h = r-g:                                    REM number of B particles
LPRINT"initial A and B molecules, respectively, = ";g,h
FOR i=1 TO g
x(i)=1
NEXT i
FOR i= g+1 TO r
y(i)=2:Z(i)=0
NEXT i
RANDOMIZE TIMER
PRINT" input number of cycles n  =  ": INPUT n
LPRINT"No. of cycles n= ";n: LPRINT "Temperature =";T :LPRINT
LPRINT"E(A--->B) = ";EF;"   E(B--->C) = ";EB:LPRINT
LPRINT"      Molecules"
LPRINT"No. A          No. B          No. C        No. Cycles"
LPRINT g,gb,gc,0:                           REM number of A's,B's, C's and cycles
WRITE #1,0,g,gb,gc
100 FOR i=q TO n
A=INT(RND*r+1):                             REM choose a random number
IF x(A)=1 THEN GOTO 200
GOTO 300
200 IF c>RND*1 THEN GOTO 250
GOTO 300
250 g=g-1
x(A)=0
y(A)=2
gb=gb+1:                     REM increase the count of particle B by 1
GOTO 340
300 IF y(A)=2 THEN GOTO 305:                REM test if B particle is chosen
GOTO 340
305 IF D>RND*1 THEN GOTO 310
GOTO 340
310 Z(A)=1                                  REM product C is formed
y(A)=0
gb=gb-1
gc=gc+1:                                    REM product C is formed
340 IF i=P*k THEN GOTO 350
GOTO 400
350 LPRINT g,gb,gc,i
PRINT g,gb,gc,i:             REM alternative is to have a screen graph
P=P+1
WRITE #1, i,g,gb,gc
400 NEXT i
500 CLOSE #1
PRINT"End of kinetic run"
LPRINT"End of kinetic run"
END
```

3. Program 3. Reaction 3.

```
REM First-order first-order kinetics with a reversible step
REM Same nomenclature as in program 1
INPUT" any number"; U
OPEN "clip:" FOR OUTPUT AS#1
LPRINT " A <----> B---->Pr "
LPRINT " Input activation energies Eab,Eba,Ebc": REM for A-->B, B-->A,
B-->C, respectively
PRINT"Input the activation energy (between 500 to 8000 joules )":PRINT
INPUT "Eab,Eba,Ebc"; Eab,Eba,Ebc
PRINT" input temperature T/K":INPUT T
Ca=EXP(-Eab/(8.314*T)):                 REM Boltzmann factors
Cb=EXP(-Eba/(8.314*T)): Cpr=EXP(-Ebc/(8.314*T))
PRINT"Temperature T = "; T
PRINT" A <----> B---->Pr : program randomly chooses A, and B molecules."
DIM X(10001),Y(10001),Z(10001)
P1=1:q=1:Pr=0
INPUT"INPUT the sampling parameter";k: REM sampling parameter- flexible you
may change.
INPUT" input number of A molecules, set parameter g between 5000 to
10000";g:     REM usually set A initially = 10,000 particles
INPUT"input the No. of A+ B molecules to be 10,000 particles"; r: REM h =
number of B particles
h = r-g:                                REM usually h = 0
LPRINT"initial A and B molecules, respectively, = ";g,h
FOR i=1 TO g
X(i)=1
NEXT i
FOR i= g+1 TO r
Y(i)= 2: NEXT i
FOR i=1 TO r: Z(i)=0:NEXT i
RANDOMIZE TIMER
PRINT" input number of cycles n = ": INPUT n
LPRINT"No. of cycles n= ";n: LPRINT "Temperature =";T :LPRINT
LPRINT"Eab, Eba, Ebc = ";Eab, Eba, Ebc :LPRINT
i=0
LPRINT"     Molecules"
LPRINT"No. Cycles    No. A     No. B      No. Pr"
LPRINT i,g,h,Pr
PRINT"No. Cycles          No. A           No. B             No. Pr"
PRINT i,g,h,Pr
WRITE #1,0,g,h,Pr
100 FOR i=q TO n
R1=INT(RND*r+1)
IF Y(R1)=2 THEN GOTO 250:               REM B particle selected
IF X(R1)=1 THEN GOTO 200:               REM A particle chosen
GOTO 500
200 IF Ca>RND*1 THEN GOTO 210
GOTO 500
```

```
210 g=g-1:h=h+1
X(R1)=0:Y(R1)=2
GOTO 500
250 XY=RND*1
if Eba>Ebc then GOTO 255 ELSE 300: REM Branch point, chose higher activation
energy first
 255 IF Cb>XY THEN GOTO 265:          REM  test B--> A reaction first
IF Cpr>XY THEN GOTO 275:              REM Test B--> C reaction
GOTO 500
265 h=h-1:Y(R1)=0:g=g+1:X(r1)=1
GOTO 500
275 Pr=Pr+1:h=h-1:Y(R1)=0:Z(r1)=3: GOTO 500
300 IF Cpr>XY THEN  GOTO 375:         REM Test B-->C reaction first
IFCb>XY THEN GOTO 365                 REM test B-->A reaction
GOTO 500
365 h=h-1:g=g+1:X(R1)=1: Y(R1)=0:GOTO 500
375 h=h-1: Pr=Pr+1:Y(R1)=0:Z(R1)=3
500 IF i=P1*k THEN GOTO 550
GOTO 800
550 LPRINT i,g,h,Pr
PRINT i,g,h,Pr
P1=P1+1
WRITE #1, i,g,h,Pr
800 NEXT i
900 CLOSE #1
PRINT"End of kinetic run"
END
```

RECEIVED October 14, 1992

CHAPTER 31

An Integrated Writing Program in the Physical Chemistry Laboratory

Thomas M. Ticich

"Oh, those lab reports!" This common response from students reflecting on their experience in physical chemistry points to the pervasiveness and notoriety of a writing component in these courses. Laboratory reports have certainly exacerbated students' anxiety and contempt for physical chemistry. Are they worth it? What do students learn about writing and what do instructors typically teach them about writing in the physical chemistry laboratory?

The motivation for revising the writing program in my physical chemistry courses arose both the drudgery with which students approached writing and the poor quality of their final product. The questions above helped to expose the primary culprit in my initial writing "program" - the absence of a concrete method and purpose. I explained very little to students about the process of writing beyond a cursive explanation of the structure of a journal article, saving my comments and insight for a ruthless post mortem in red ink.

Traditional Approach

Freshman Composition --writing skills--> STUDENT

STUDENT ----------> History, Philosophy, Chemistry

Writing across the curriculum

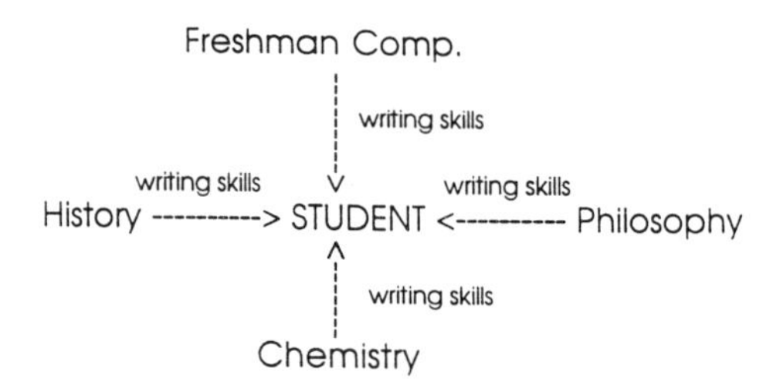

Figure 1. Two approaches to teaching writing in higher education.

Figure 1 puts my revised integrated program into perspective by examining two approaches to teaching writing that are prevalent in higher education. The traditional approach relies on a freshman composition course to equip students with all the writing skills they will need during their college careers. The burden is then on the student to apply those skills in a variety of contexts and disciplines. When they encounter poor writing skills, instructors in subsequent courses often scold their students and then berate the English department for not teaching students how to

2503–6/93/0462$06.00/0

write properly. "Writing Across the Curriculum" is a more recent movement that distributes the task of teaching writing among all disciplines in the academic community rather than placing the responsibility solely on a freshman composition course. This approach allows students to develop and adapt their writing skills in a variety of contexts throughout their college careers *(1,2)*.

Gustavus Adolphus College adopted a writing across the curriculum program in the fall of 1985 *(3)*. Currently, this program requires students to complete three "W" courses for graduation. These courses, which are offered in nearly every department in the college (81 courses at last count), must devote at least two class periods to discussing writing in the discipline and require at least 15 pages of writing. The instructor must evaluate the written assignments for both content and writing style. At least one of the three "W" courses must be at or above the intermediate level. The chemistry department offers three "W" courses: Physical Chemistry I & II and Instrumental Analysis.

Given the tremendous amount of writing done by professional chemists, few individuals would question the appropriateness of plugging chemistry courses into a writing program. But why does the task of teaching students writing in the discipline so often fall to physical chemists? Physical chemistry comes at an ideal time in the curriculum and in the cognitive development of the student to teach writing. Physical chemistry courses typically have small enrollments, making it practical for instructors to carefully evaluate the writing assignments. By the junior year, most students have mastered the art of keeping a laboratory notebook. For the most part, they are comfortable and confident in the laboratory, ready to tackle complex, open-ended laboratories. At the same time, many have made the transition in cognitive development from Perry's dualist *(1)*, who views every problem as having one solution, to the multiplist, who recognizes a multiplicity of solutions to the same problem. The results of many physical chemistry experiments allow for several interpretations, not all of which are equal in the eyes of the scientific community. An important goal of advanced chemistry courses is to help students recognize that experimental results may be interpreted differently, and then to teach them the criteria used by chemists to judge the merit of various interpretations. Writing helps engage students in these issues, bringing them to what Perry terms the relativist stage, at which point they can critically reflect on their own work and that of others.

The yeast for the program I will describe came from a workshop that I attended on "Teaching Writing and Higher Order Reasoning" which is sponsored by the University of Chicago as part of their Institute on Issues in Teaching and Learning. Although not geared specifically to scientific writing, the workshop did provide several important insights.

1. *Good writing is intimately related to clear thinking.* When faced with the task of teaching writing, it is easy to get caught up in the details of spelling, punctuation, grammar, and citation format. Students produce muddled, disorganized writing largely because they haven't clearly thought about what they want to say. The act of writing is in itself a learning process since it requires students to organize their thoughts.

2. *Good writing is discipline specific.* This statement neither endorses jargon nor suggests that there is no common ground for good writing across disciplines.

However, there are discipline-specific formats which an expert superimposes on an article he/she is reading. If this expert encounters information in unexpected places (experimental details in a discussion section, for example) he/she will likely perceive the article as being poorly written *(4)*.

3. *Instructors need to teach students explicitly what good writing is in the discipline and how to think critically.* The problem with the post mortem approach is that it assumes that students know the process and criteria for analyzing their results, and evaluates their work based on this tacit assumption. Instructors also need to socialize students into their disciplines by teaching them the formats used by professionals in the discipline. Note that the socialization process does not necessarily result from a transition in cognitive development. Consequently, students need both good and bad models of scientific writing.

The integrated writing program for the two physical chemistry courses at Gustavus described below is built upon these three principles.

PHYSICAL CHEMISTRY I

The goal of the writing component of Physical Chemistry I at Gustavus is a fairly traditional one - to teach students the journal article format for presenting their laboratory results. However, I've added two new twists - a writing workshop and a scientific review framework. They serve to address the issues listed above and to engage students in these issues throughout the semester.

1. Writing Workshop.

The writing workshop, held during the first or second laboratory period, challenges students to think about several important topics in scientific writing through small group exercises and class discussions. The writing workshop is not a crash course in technical writing. Rather it focuses on three general topics: writing a discussion, the attributes of scientific writing, and some practical guidelines for achieving the goals of good scientific writing.

A. Writing a discussion. My experience is that this is the section of the laboratory report that students find the most challenging. In her scheme for assessing the difficulty of writing tasks, Rosenthal provides some insight into this problem *(5)*. Her criterion for ranking a writing task as a low, medium, or high level cognitive task is the extent of generalization, abstraction, or analysis it requires. Chemistry courses taken prior to physical chemistry generally focus on low level tasks, such as keeping a laboratory notebook. Physical chemistry is often the first course in which students encounter a high level task, such as writing a discussion, that requires analysis and argument.

In this portion of the workshop, students read the first three sections (Introduction, Experimental Method, and Results) of a laboratory report that describes a density measurement *(6-7)*. I then give them three different versions of the discussion section and ask students first to rank the three versions individually, and then to form small groups and reflect on those attributes that influenced their rankings. We group these attributes as a class and compile the qualities of a good discussion. Usually there is a consensus that a good discussion addresses questions or puzzles suggested by the results. Therefore, a good starting point for writing a discussion is to extract meaningful questions and puzzles from the data and then to sketch out answers to or reflections on these issues. I encourage students to concern themselves with the actual mechanics of presentation only after completing these two steps. My experience is that when students try to circumvent these two steps, their attempts at "discussion" degenerate into a vague summary tacked onto a litany of error sources.

The three discussions I distribute are those in Shoemaker, 3rd and 5th editions, respectively *(6-7)*, and a version I generated in the style of a poorly written student discussion (see Appendix A). Students consistently ranked the latter as the least well-written. A surprising result was their preference for the 3rd edition version over the 5th edition version. The 5th edition version is the longer of the two, largely because it uses atomic mass and X-ray crytallographic data to compute a more accurate literature value for the density than that given in the CRC Handbook of Chemistry and Physics. Some students apparently equate questioning the data in the CRC Handbook with chemical apostasy. This provides an ideal opportunity to stress the importance of seeking out the most current and reliable literature data available.

B. Attributes of scientific writing. Now that they have examined a piece of scientific writing I ask students what makes scientific writing different from other types of writing, and, in particular, what makes it difficult. After compiling and organizing a list of attributes from the class, I introduce the seven goals of language in scientific writing given by Michael Alley in The Craft of Scientific Writing *(8)* (Figure 2). In this diagram, the proximity of the goal to the top of the figure indicates its importance, and the lines connect goals that are strongly coupled. I briefly discuss

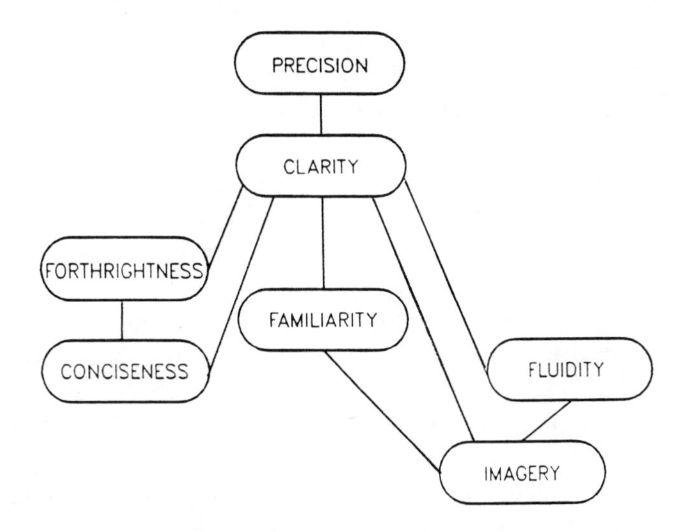

Figure 2. The seven goals of language in scientific writing. Reproduced with permission from reference 8. Copyright 1987 Prentice Hall.

each goal, and show how the students' list of attributes of scientific writing map onto Alley's scheme. Alley's diagram is compact enough to provide students with the broad picture of what their writing should accomplish without overwelming them with numerous rules. These goals also provide a cohesive framework for more specific guidelines of technical writing.

C. Two guidelines for producing good writing. I present two concrete guidelines to assist students in composing or editing their drafts.

i. Express crucial actions as verbs. Avoid (or at least question) nominalizations. *(9)* A nominalization is a noun derived from a verb. For example, "conclusion" is a nominalization of the verb "conclude". One can often trace verbosity and complex sentence structure (such as the proliferation of prepositional phrases) to the excessive use of nominalizations. I present the following examples.

"Monitoring of the laser power was performed by a volume-absorbing calorimeter."

The crucial action is "monitor"; yet, it is expressed as a nominalization. Note that the following revision is clearer and shorter.

"A volume-absorbing calorimeter monitors the laser power."

I encourage students to seek out nominalizations (which often end in -ing, -ion, or -ment) and to attempt revision whenever possible. This process can efficiently nudge their first drafts toward a clearly written final product.

ii. Avoid (or at least question) the use of the passive voice. Not unrelated to the problems caused by nominalizations is the excessive use of the passive voice. There is a school of thought among scientists that equates use of the passive voice with objectivity, debasing active writing as "unscientific". This is nonsense. In fact, passive constructions often produce awkward, muddled, and lengthy prose. I present students with the following example.

"The laser power was monitored by a volume-absorbing calorimeter."

They note that although this version avoids a nominalization, the active revision given earlier is shorter and eliminates the prepositional phrase. Having introduced these guidelines, I divide the students into groups to revise additional examples.

2. Scientific Review Process Framework.

I use the scientific journal review process as a framework for evaluating the writing assignments. The rationale for this is twofold. Students need good and bad models of scientific writing. What better source is there of good and bad models than their own reports? It is easier to recognize poor writing than to produce good writing, and for some students an important first step in composition is recognizing how not to write. The framework also serves to introduce students to the mechanics of the scientific review process and to another important format - the review itself. I find that most students profess ignorance of how articles are selected for publication in journals. They are genuinely surprised to learn that the fate of a manuscript submitted for publication is largely determined through a peer review process.

I distribute the masthead of a fictitious journal, for which the students themselves serve as the editorial board (Figure 3). The mechanics of report submission and review in my class loosely parallels that of a journal. Students submit two copies of each report, one of which goes to an anonymous reviewer that I select. The reviewer turns in two copies of a review. I grade the student's report,taking into account the reviewer's comments, and return to the student the graded report and a copy of the review. The report author has the option of resubmitting the manuscript. I also return a graded copy of the review to the reviewer. The practice of grading the reviews discourages students from being either overzealous or unrealistically benevolent in their critiques. Each member of the editorial board reviews three manuscripts during his/her term, which accounts for 5% of the total grade (lecture and laboratory). Note that this does deviate from the usual practice of providing no remuneration to reviewers for their services.

PHYSICAL CHEMISTRY II

The second semester course exposes students to the variety of forms that professional chemists use to communicate their work. Learning to write in different forms is closely connected to learning to write for different audiences. As in the first semester course, I use a laboratory period for a writing workshop, which includes the following topics.

1. Survey of formats. Students read and compare articles on the cold fusion announcement from four different sources: The New York Times *(10)*, Time magazine *(11)*, Science News *(12)*, and the Journal of Electroanalytical Chemistry *(13)*. The focus in their small group discussion is on audience and the type of information presented. They note, for example, that the first two sources focus mostly on the controversy, a topic which is never raised in the journal article.

2. Comparison of writing styles. Students read and compare two descriptions of coherent light: one from Usborne's New Technology Series *(14)*, which is aimed at the junior high level, and one from Andrew's text Lasers in Chemistry *(15)*, which is geared toward senior undergraduates or graduate students. The focus in this exercise is on how the information is presented as opposed to what is presented. Students examine the vocabulary, images, sentence length, and sentence structure in the two descriptions.

3. Written exercise. After reading an abstract of an experiment they performed in Physical Chemistry I, students write an abstract appropriate for a non-scientist roommate.

Olmstead *(16)* describes many possible assignments. I use the following in my course.

1. Student Procedure. Students work through an experiment following a fairly sketchy procedure. They then write a detailed procedure that would require minimal assistance from the instructor for a student with their background.

The Journal of PHYSICAL CHEMISTRY LABORATORY

Thomas M. Ticich, Editor

The *Journal of Physical Chemistry Laboratory* is published semiannually by the Chemistry Department of Gustavus Adolphus College. Its purpose is to motivate chemistry students towards excellence in communicating their laboratory work and to introduce them to the reviewing process used in scientific journals.

Information for Contributors

Submit manuscripts in duplicate to Thomas M. Ticich, *Journal of Physical Chemistry Laboratory*, Department of Chemistry, Gustavus Adolphus College, St. Peter, MN 56082. Manuscripts may also be submitted in person or placed in the editor's mailbox in the chemistry office. The manuscript should be accompanied by a statement transferring copyright from the author to the Chemistry Department. A suitable form for copyright transfer is available from the editor's office. This written transfer of copyright is necessary under the 1978 U.S. copyright law in order for Gustavus Adolphus College to continue disseminating chemical research results as widely as possible.

Publication Charge: To support the cost of wide dissemination of laboratory results, the author's institution is requested to pay a *page charge* of $45 per page (with a one-page minimum) and an *article charge* of $20 per article. The publication charge will be waived for manuscripts submitted as a result of a course requirement for Chemistry 71.

All mathematical expressions and formulas, insofar as possible, should be typed, with unavailable Greek letters or symbols inserted neatly in ink. Use fractional exponents instead of root signs.

References should appear as consecutively numbered footnotes and should be collected on a separate sheet at the end of the manuscript. Consult *The Art of Scientific Writing* (Ebel *et al.*) for the correct format.

Figure 3. Masthead for the Journal of Physical Chemistry Laboratory (with apologies to the Journal of Chemical Physics).

2. Popular Science Article. Students present their laboratory results as if it were a new discovery being described in Discover or New Scientist.

3. Research proposal. Students use the ACS Petroleum Research Fund grant application to request funding to continue work on the experiment at another institution. They present their results as a preliminary investigation in the body of the proposal.

4. Technical report. Students present a summary of their results to a fictitious laboratory supervisor in industry, making recommendations to continue or terminate the project.

ASSESSMENT

I have several observations from implementing this writing program for the past three years. First, students are very enthusiastic about writing reviews. They are extremely interested in the comments of their peers (more so than those of their instructor). Students also find the different formats in the second semester course a welcome change of pace from the journal article format.

On the other hand, I've found that it is very easy to overload students with writing assignments in a course that is already bulging with content. Therefore, it is extrememly important to adjust students' workload so that they have sufficient time to tackle writing assignments. Currently, I require a full journal report for three experiments in Physical Chemistry I, and only a results and discussion section for the remaining four. I also allow students to collaborate on all results sections. If the assigned format does not allow for detailed presentation of data, calculations, and analysis (in Physical Chemistry II), students attach additional sheets with the information (no formal presentation required).

Computers can facilitate data analysis and editing, allowing students to devote more time to writing. A powerful program, such as Mathematica, can eliminate the tedium of grinding through partial derivatives for error propagation. However, computers can also be a burden if students need to acquire most of their computer skills during their physical chemistry course. Introducing students to computer techniques (wordprocessing, spreadsheets, graphical and numerical analysis) early in the chemistry curriculum ensures that they will have these skills in hand upon entering advanced courses.

There is an excellent annotated bibliography on scientific writing in reference 17. The references listed therein, together with Alley's text *(8)* and The Little Red Schoolhouse *(9)* are excellent resources for building a writing program.

Literature Cited

1. Perry, W. G. *Forms of Intellectual and Ethical Development in the College Years: A Scheme*; Holt, Rinehart, and Wilson: New York, NY, 1967.
2. Zinsser, W. K. *Writing To Learn*; Harper & Row: New York, NY, 1989.
3. Zinsser, W. K. *The New York Times* **1986**, *135*, 4/13, Section 12, 58.
4. Gopen, G. D.; Swan, J. A. *Am. Scientist* **1990**, *78*, 550.
5. Rosenthal, L. C. *J. Chem. Educ.* **1987**, *64*, 996.

6. Shoemaker, D. P.; Garland, C. W.; Steinfeld, J. I. *Experiments in Physical Chemistry*; 3rd edition, McGraw-Hill: New York, NY, 1974; pp 15-21.
7. Shoemaker, D. P.; Garland, C. W.; Steinfeld, J. I. *Experiments in Physical Chemistry*; 5th edition, McGraw-Hill: New York, NY, 1989; pp 16-22.
8. Alley, M. *The Craft of Scientific Writing*; Prentice-Hall: New York, NY, 1987.
9. Williams, J. M.; Colomb, G. G.; Kinahan, F. X.; *The Little Red Schoolhouse Syllabus*; University of Chicago, 1992. (Available on request from Joseph M. Williams, Department of English, University of Chicago, Chicago, IL 60637)
10. Browne, M. B. *The New York Times* **1989**, *138*, 3/28, C1.
11. Elmer-DeWitt, P. *Time* **1989**, *133*, 72.
12. Amato, I. *Science News* **1989**, *135*, 196.
13. Fleischmann, M.; Pons, S.; Hawkins, M. J.; *J. Electroanalyt. Chem.* **1989**, *261*, 301.
14. Myring, L. M.; Kimmitt, M. *Lasers*; Usbourne New Technology Series; Usbourne Publishing Ltd.: London, UK, 1984.
15. Andrews, D. L. *Lasers in Chemistry*; Springer-Verlag: New York, NY, 1986.
16. Olmsted, J. *J. Chem. Educ.* **1984**, *61*, 799.
17. Shires, N. P. *J. Chem. Educ.* **1991**, *68*, 494.

Appendix A

This experiment shows how one can obtain the density of a substance using a gravimetric technique. Weighing a pycnometer while empty and again after filling with a sample of germanium gives the mass of the sample. The volume of the sample is determined by then filling the pycnometer containing the sample with water. Since the density of water is known and the volume of the pycnometer is given, the density of the germanium sample is readily computed.

The values and limits of error obtained for the density of germanium at 25 ^{o}C are

Sample I: 5.310 ± 0.003 g cm^{-3}
Sample II: 5.332 ± 0.003 g cm^{-3}

The value given by the Handbook of Chemistry and Physics is 5.35 g cm^{-3}. Possible reasons for deviation of our results from the literature value include variation in room temperature during the weighing, the purity of the water used, human error in weighing, purity of the germanium samples, error in the value for the density of water used, and error in the volume of the pycnometer. Since the two measurements differ from each other by more than the stated error limit, either the error analysis is wrong or the two samples may differ enough in purity to give different densities.

This experiment was fairly easy to carry out. Accurate weights can be obtained by using the analytical balance and the procedure is simple and gives good values for densities of samples.

RECEIVED October 14, 1992

Indexes

Author Index

Affiliation Index

Subject Index

D

Q

R

S

Text produced by Richard W. Schwenz
Production: Betsy Kulamer and Margaret J. Brown
Indexing: Deborah H. Steiner
Acquisition: Cheryl Shanks
Cover design: Alan Kahan

Printed and bound by Kirby Litho, Arlington, VA